AF342279

LIBERTY MUTUAL
RESEARCH CENTER
LIBRARY

UPPER EXTREMITIES ORTHOTICS

UPPER EXTREMITIES

ORTHOTICS

Fourth Printing

Written and Illustrated By

MILES H. ANDERSON, Ed.D.

Director, Prosthetics-Orthotics Program

Department of Surgery
School of Medicine
University of California
Los Angeles, California

Technical Consultants

John J. Bray, C.P. and O.
Associate Director, Prosthetics-Orthotics Program

Department of Surgery
School of Medicine
University of California
Los Angeles, California

Lee Roy Snelson, C.O.
Chief, Orthotics Research

Rancho Los Amigos Hospital
Downey, California

Leonard Marmor, M.D.
Associate Professor, Department of Surgery (Orthopedics)

School of Medicine
University of California
Los Angeles, California

Edited By

RAYMOND E. SOLLARS

**Associate Director,
Prosthetics-Orthotics Program**

Department of Surgery
School of Medicine
University of California
Los Angeles, California

CHARLES C THOMAS • PUBLISHER

Springfield • Illinois • U.S.A.

Published and Distributed Throughout the World by
CHARLES C THOMAS • PUBLISHER
BANNERSTONE HOUSE
301-327 East Lawrence Avenue, Springfield, Illinois, U.S.A.

This book is protected by copyright. No part of it may be reproduced in any manner without written permission from the publisher.

© 1965, by CHARLES C THOMAS • PUBLISHER
ISBN 0-398-00044-1
Library of Congress Catalog Card Number: 65-14155

First Printing, 1965
Second Printing, 1970
Third Printing, 1974
Fourth Printing, 1979

With THOMAS BOOKS careful attention is given to all details of manufacturing and design. It is the Publisher's desire to present books that are satisfactory as to their physical qualities and artistic possibilities and appropriate for their particular use. THOMAS BOOKS will be true to those laws of quality that assure a good name and good will.

Printed in the United States of America
00-2

INTRODUCTION

Upper Extremities Orthotics supersedes Functional Bracing of the Upper Extremities which was published in 1958. The 1958 textbook was used in 11 three-week classes for orthotists, and 14 classes for physicians and therapists given at U.C.L.A. during the seven year period from 1958 through 1964. A total of 136 orthotists, 293 physicians, and 223 therapists completed their respective training classes in those years. They came from all parts of the United States and many foreign countries, and comprise approximately 60 clinic teams actively engaged in making upper extremities orthotics service available to handicapped people in their areas. Prior to 1958 this service could only be obtained at a few of the major polio respiratory centers. There can be no doubt that as a result of this expansion and improvement of service in upper extremities orthotics, many hundreds of individuals have been rehabilitated by having the usefulness of their hands and arms improved. Many of these patients would otherwise have gone through life suffering from serious handicaps, unable in many cases to care for their most elemental activities of daily living. Certainly this is an accomplishment in which great pride can be taken by the U. S. Vocational Rehabilitation Administration that sponsored much of the orthotic research, and all of the educational activities by Los Angeles County's Rancho Los Amigos Hospital where most of the research and development was done, and by the University of California at Los Angeles, Prosthetics-Orthotics Program, which organized and administered the educational activities.

In addition to its use with orthotists, physicians, and therapists, Functional Bracing of the Upper Extremities was used as a text in 23 five-day classes for vocational rehabilitation counselors from 1958 through 1964. In these classes, vocational rehabilitation counselors and other rehabilitation personnel learned how upper extremities bracing could be applied in the rehabilitation of their handicapped clients, how the service could be obtained, and what might be expected in the way of results from it. A total of 477 rehabilitation personnel were trained in these classes.

Since the start of the upper extremities orthotics education program at U.C.L.A. in 1958, approximately 350 patients have been fitted with splints and braces as part of the laboratory instruction in the classes for orthotists. Many more have been fitted as part of the normal flow of upper extremities orthotics research and service at Rancho Los Amigos Hospital.

The splints and braces included in the 1958 textbook <u>Functional Bracing of the Upper Extremities</u> had one very serious shortcoming. They were dependent upon the patient's remaining muscle power for their operation, and in those numerous instances where the patient had no residual power, they were of little functional value to him. This problem has been very frustrating in working with patients with flail hands and arms, particularly those affected bilaterally. The development by the orthotics research group at Rancho Los Amigos Hospital of the artificial muscle, powered with compressed carbon dioxide gas, and the small electric motor powered by rechargeable batteries have made a source of outside power available to help these patients. The artificial muscle or the electric motor can be applied to the flexor hinge splint to restore grasp to a flail hand, and to the locking joint functional arm brace to enable the patient to move the hand from one position to another to accomplish useful activities. Material on the principles of the artificial muscle and the electric motor and their application are included in this text, <u>Upper Extremities Orthotics</u>.

The approach to functional arm bracing in the 1958 textbook proved to be unnecessarily complex and cumbersome, and as a result the application and patient acceptance of these devices fell short of expectations. The development of the custom-fitted plastic shoulder cap as a stable foundation for a functional arm brace, and of the locking joint brace applied to the shoulder cap, made available an appliance that in our experience has proven more acceptable to the patient with serious arm involvement. While these improvements have increased patient acceptance of functional arm braces, the resistance to their use is still much greater than that to hand splints. The more complex arm braces apparently exceed the "gadget tolerance" of many patients who find the value of the function gained outweighed by the inconvenience of the devices. It has been found that the degree of "gadget tolerance" in each case is directly related to the amount of functional loss, and to the amount of training given to the patient in the control and use of his brace. The unilateral patient with one unimpaired arm is easier to fit with a brace but does not adopt it for regular use as readily as the bilateral patient who is more difficult to fit but accepts his braces more readily as he is so much more helpless without them. In both cases, training plays a key role. If the unilateral can be taught to gain enough function from his brace, he will use it. Only with patient and persistent training can the bilateral develop enough skill to realize all the potentials of his equipment. Experience has shown that programs of upper extremities orthotics are seldom successful without the services of skilled physical and occupational therapists to train the patients in the use of their splints and braces.

An interesting relationship between orthotics and surgery began to develop soon after the start of the educational program in 1958, as a result of the lectures on reconstructive hand and arm surgery given by the orthopedic surgical staff at Rancho Los Amigos Hospital during the field trip to that institution which was a part of each class for physicians given in upper extremities orthotics at U.C.L.A. The feasibility of a number of surgical procedures could be tested by first using hand splints, arm braces, or both, and if satisfactory function could be obtained by their use, they could be duplicated by various combinations of tendon transplants and joint fusions, and the splints and braces discarded. This approach to the problem of reconstructive surgery of the hand and arm enabled both the surgeon and the patient to study the function the surgery would provide prior to making an often irreversible commitment to the actual surgical procedure itself. A significant increase and improvement in upper extremities reconstructive surgery appears to have resulted from this development, and there is evidence that many patients are now enjoying improved hand function as a result of surgery, who otherwise would not have been able to obtain such help.

In the past seven years we have learned much more about upper extremities orthotics than we knew when we started. However, as is true in most professional fields, the more we learn about upper extremities orthotics the more we realize how little we know. Present equipment and techniques barely scratch the surface, and many difficult problems remain unsolved. Research on these problems is being pressed forward by the engineers and orthotists at Los Angeles County's Rancho Los Amigos Hospital under the sponsorship of the U. S. Vocational Rehabilitation Administration. Just as the past seven years have seen much progress as a result of their efforts, no doubt the next seven years will be productive of even more impressive advances. Improvement in current equipment and techniques will no doubt be accompanied by success in present efforts to apply upper extremities orthotics to dysfunctions caused by arthritis, cerebral palsy, cerebral vascular accidents, and other handicapping impairments.

While admittedly far from perfect, the 1958 volume <u>Functional Bracing of the Upper Extremities</u> was the first book published in this field. Coupled with the educational program at U.C.L.A., it accomplished its goal of introducing and disseminating the then known devices and techniques to the physicians, therapists, orthotists, and rehabilitation personnel who might use them in the rehabilitation of handicapped people. <u>Upper Extremities Orthotics</u> is not perfect, but it is sincerely hoped that it will help advance this professional field to greater effectiveness in the work of assisting those unfortunates who have lost part or all of the use of the hands and arms.

The title of this book, <u>Upper Extremities Orthotics</u>, was adopted to give recognition to the orthotics profession and to the orthotists who are chiefly responsible for the great improvements in orthotics service that have taken place during the past seven years.

Miles H. Anderson, Ed.D.,
Director, Prosthetics-Orthotics Program,
University of California, Los Angeles

January 1, 1965

ACKNOWLEDGEMENT

No book covering a field as broad and highly technical as upper extremities orthotics can be prepared without the cooperation of many individuals and institutions, and certainly the authors of this volume are very grateful to many for the willing assistance they received.

We are deeply indebted to the Vocational Rehabilitation Administration of the Department of Health, Education, and Welfare, sponsor of the research and education programs in orthotics at both Los Angeles County's Rancho Los Amigos Hospital and the University of California at Los Angeles Prosthetics-Orthotics Education Program, for unfailing support and encouragement from the start of the program seven years ago to the present time. Particularly helpful have been Miss Mary E. Switzer, Commissioner; Dr. James F. Garrett, Assistant Commissioner; Dr. J. Warren Perry, Deputy Assistant Commissioner, and Miss Cecile Hillyer, Chief, Division of Training, all of the Vocational Rehabilitation Administration.

Increasing cooperation over the past seven years has resulted in a closer working relationship between the staff members of Los Angeles County's Rancho Los Amigos Hospital and the University of California at Los Angeles. Particularly helpful have been John E. Affeldt, M.D., Medical Director, Vernon L. Nickel, M.D., Chief of Surgical Services and Head Orthopedist, Jacquelin Perry, M.D., Orthopedic Surgeon, Betty Yerxa, O.T.R., Occupational Therapy Instructor, and Patrick Marer, Administrative Assistant, Orthotics Laboratory. We are grateful to Vice-Chancellor Stafford L. Warren of U.C.L.A. (now Special Assistant to the President of the United States in charge of the Program for the Mentally Retarded), for without his timely assistance in 1958, Functional Bracing of the Upper Extremities and the educational program would have expired before getting started. Substantial contributions have been made by Charles O. Bechtol, M.D., Chief, Division of Orthopedics, Ralph E. Worden, M.D., Chairman, Department of Physical Medicine, Sam Colachis, M.D., Assistant Professor of Physical Medicine, Bernard R. Strohm, R.P.T., Lecturer in Physical Therapy, Jeannine F. Dennis, O.T.R., Research Occupational Therapist, and Fred Sanders, Laboratory Mechanician, all of the U.C.L.A. Medical School.

Special gratitude is expressed to the three men who served as technical consultants during the preparation of this manuscript. Leonard Marmor, M.D., Associate Professor of

Surgery (Orthopedics), Medical School, University of California, Los Angeles, made himself available for repeated day to day consultation as the manuscript progressed. Mr. Lee Roy Snelson, C.O., Chief of Orthotics Research, Rancho Los Amigos Hospital, who is the originator of many of the devices treated herein, supplied current information and design details in order to make certain that this text would be truly up to date. Mr. John J. Bray, C.P. and O., Associate Director, Prosthetics-Orthotics Program, Medical School, University of California, Los Angeles, assisted greatly in the design of this text for efficient instructional use, and also deserves thanks for his assistance in the preparation of the section on hand anatomy.

Without a reliable source of supply for parts and materials for fabricating splints and arm braces the program in upper extremities orthotics could not operate in the field. Those suppliers who have cooperated by designing, production engineering, and producing the necessary components are Lloyd Brown and Jerry Leavy of H. J. Hosmer Corp., George W. Robinson, C.O., Robin-Aids Manufacturing Co., Jack Conry, C.O., Jaeco Co., and Lee Roy Snelson, C.O., Orthopedic Supplies Company. We express our gratitude to these manufacturers, who have gone to great lengths to make upper extremities orthotics components readily available.

The preparation of reproduction manuscript, line drawings, page layout, final editing and proofreading and all the tasks related to getting a book ready for publication must be done perfectly or the resulting publication will not be effective as a teaching aid. The excellent work in this department was done by Frances Schaaff and Shirley Fink, Technical Illustrators, June Hepner, Reproduction Typist, and Florence Conot, Editor. Our thanks to them for a job well done.

Last, but not least, we thank the many patients who so graciously consented to be photographed and allowed the photographs to be used for illustrations.

M. H. A.

UPPER EXTREMITIES ORTHOTICS

CHAPTER I. FUNCTIONAL ANATOMY OF THE HAND

Introduction

A thorough understanding of the anatomy of the hand and arm is essential for the orthotist who is studying upper extremities orthotics. Before he can fit and modify functional splints so they conform to the hand without causing discomfort while they assist the weakened or paralyzed upper extremity to regain function, the orthotist must have a reasonably clear understanding of the characteristics of each bone, muscle, and joint in that very complex part of the body. Only then can he really understand the relationship between the splints he makes and fits, and the dynamics of the musculo-skeletal system of the hand and arm. An understanding of anatomy will enable the orthotist to work out solutions to the most complex splinting and bracing problems in a logical manner, based on anatomical facts.

Anatomical terminology is complex and strange to the uninitiated. Derived largely from Latin and Greek, the vocabulary of anatomy is like a new language to the student when he first encounters it. The effort to master this new language is essential in all branches of orthotics, but it is particularly so in upper extremities bracing as the orthotist must work very closely with the physician and therapist if the patient is to be rehabilitated successfully. Since the physician and therapist speak the language of anatomy routinely, the orthotist must also become adept in this language or he will not be able to function efficiently as a member of the clinic team.

This chapter on functional anatomy of the hand is intended only as a brief introduction to the subject. It is hoped that the orthotist will be stimulated by it to undertake additional study, including enrollment in laboratory and lecture courses if at all possible.

<u>Hand Prehension Patterns</u> [1]

It is evident equally from a study of the muscle-bone-joint anatomy and from observation of the postures and motions of the hand that an infinite variety of prehension patterns is possible. For purposes of analysis, however, it suffices to describe the principal types. Seeking a logical basis for defining the major prehension patterns, Keller, et al.[2] found that the object-contact pattern furnishes a satisfactory basis for classification. From photographic observation of the prehension patterns naturally assumed by individuals when picking up and holding for use common objects used in everyday life, three types were selected from among those originally classified by Schlesinger[3]. These, appearing in Figure 1, are palmar, tip, and lateral prehension, respectively. The frequency with which each of these types occurred in the investigation cited is given

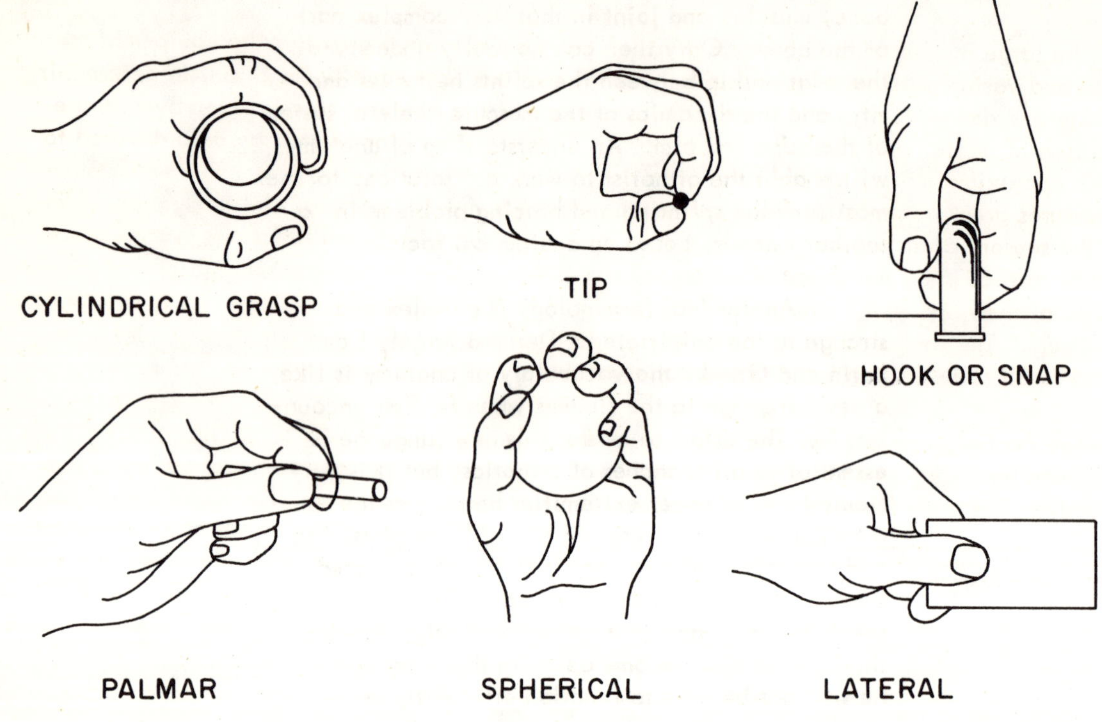

Figure 1.

---

[1] Taylor, Craig L. and Schwarz, Robert J., "The Anatomy and Mechanics of the Human Hand," <u>Artificial Limbs</u>, Vol. 2, No. 2, May, 1955, pp. 29-32.

[2] Keller, A. D., Taylor, C. L., and Zahm, V., <u>Studies to Determine the Functional Requirements for Hand and Arm Prostheses</u>, Department of Engineering, University of California at Los Angeles, 1947.

[3] Schlesinger, F., <u>Der mechanische Aufbau der kunstlichen Glieder in Ersatzglieder und Arbeitshilfen</u>, Springer, Berlin, 1919.

in Figure 2. While the relative percentages differ in the two types of action, the order of frequency with which the prehension patterns occurred is the same.

FIGURE 2. FREQUENCY OF PREHENSION PATTERNS

Series	Occurrence of Prehension Type (%)		
	Palmar	Tip	Lateral
Pick up	50	17	33
Hold for use	88	2	10

The large number of muscles and joints of the hand obviously provides the equipment for numerous and varied patterns of movement.   Not so evident, but equally important in determining complexity and dexterity of motion, are the large areas of the cerebral cortex given over to the coordination of motion and sensation in the hand. Thus, in the motor cortex the area devoted to the hands approximately equals the total area devoted to arms, trunk, and legs.[1] This circum-stance ensures great potentiality for coordinated movement and for learning new activities. Sim-ilarly, the sensory areas are large, so that they determine such advanced functions as stereognosis, the ability to recognize the shape of an object simply by holding it in the hand. The great tactile sensitivity of the hand is, of course, in large part due to the rich supply of sense organs in the hand surface itself. The threshold for touch in the finger tip, for example, is 2 gm. per sq. mm., as compared to 33 to 26 for the forearm and abdomen respectively.[1]

The Penfield and Boldrey homunculus[2] (Figure 3) graphically illustrates the relative areas of the human brain required to adequately serve the sensory and motor needs of the various parts of the body. The body parts are drawn in size proportionate to the amount of brain matter connected to them. The large hands indicate that they require a very large amount of cerebral accommodation to properly control them.

---

[1] Best, C. H. and Taylor, N.D., *Physiological Basis of Medical Practice*, Williams and Wilkins, Baltimore, 1937, p. 1256.

[2] After Fulton, *Physiology of the Nervous System*, p. 405, Figure 84 (1938 Edition).

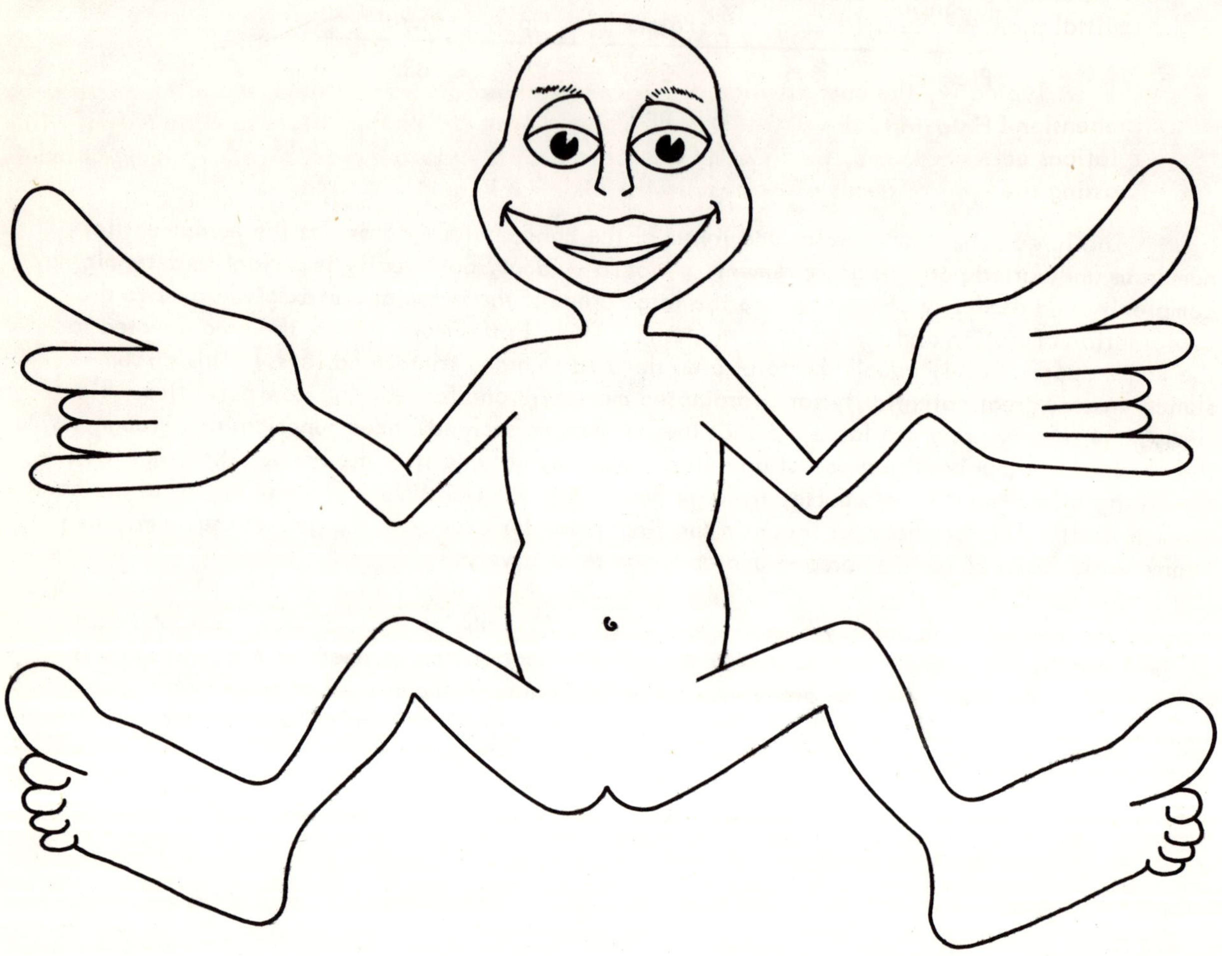

Figure 3.

The Position of Function, the Resting Hand Position

The resting hand assumes a characteristic posture, a feature easily seen when the hand hangs loosely at the side. The resting wrist takes a mid-position in which, with respect to the extended forearm axis, it is dorsiflexed 35 degrees (Figure 4). It is worth noting that this is the position of greatest prehensile force. The mid-position for radial or ulnar flexion appears to be such that the metacarpophalangeal joint center of the middle finger lies in the extended sagittal plane of the wrist (Figure 5).

Typically, the conformation of fingers and thumb is similar to that shown for palmar prehension (Figure 1), the fingers being more and more flexed from index to little finger. The relations between thumb, palm, and fingers are such as to permit grasp of a 1.75 inch cylinder crossing the palm at about 45 degrees to the radio-ulnar axis.

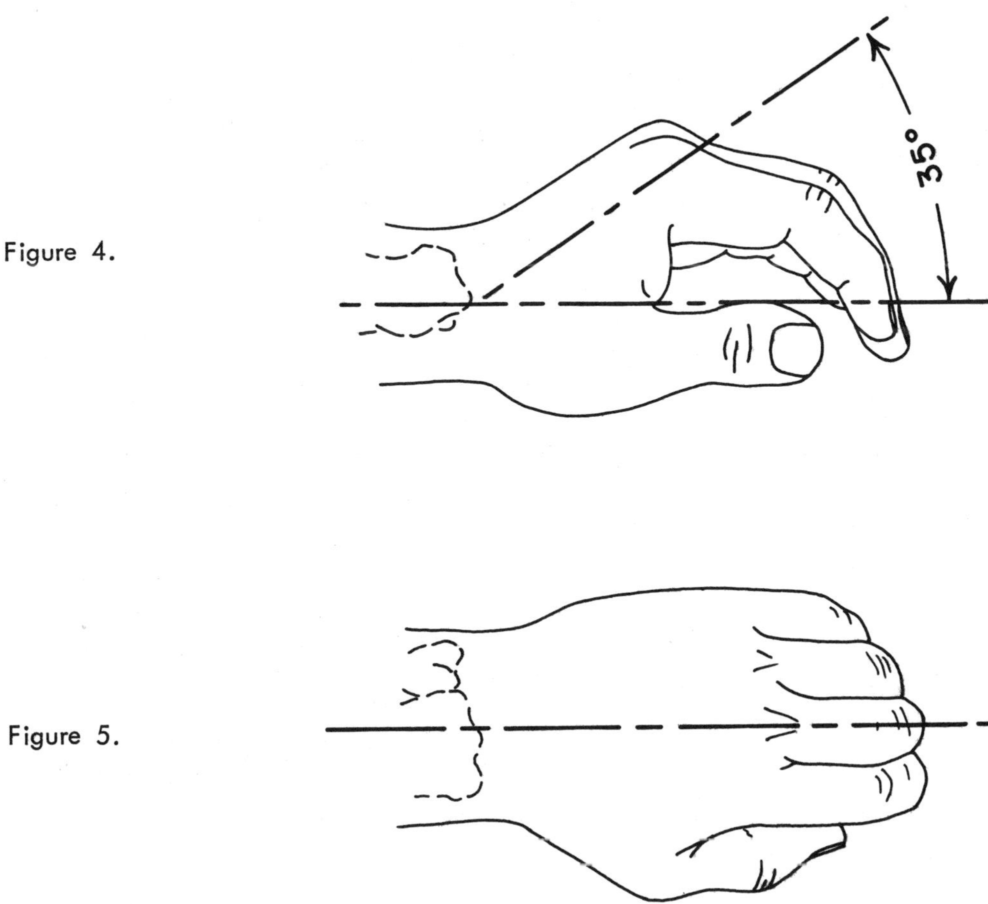

Figure 4.

Figure 5.

<u>Bones of the Hand and Wrist</u>

In the illustration below all of the bones of the hand and wrist are shown and named, except the names of the individual carpals.

The name <u>carpals</u> is from the Greek <u>karpos</u>, meaning wrist, hence the carpals are the wrist bones.

<u>Meta</u> is Greek for beyond; the metacarpals are the bones beyond the carpals.

<u>Phalanx</u> is Greek for a row of soldiers in line of battle; perhaps the bones of the fingers reminded some early anatomist of rows of soldiers, leading him to use the word phalanx in referring to any of the individual bones of the fingers. The plural of phalanx is <u>phalanges</u>.

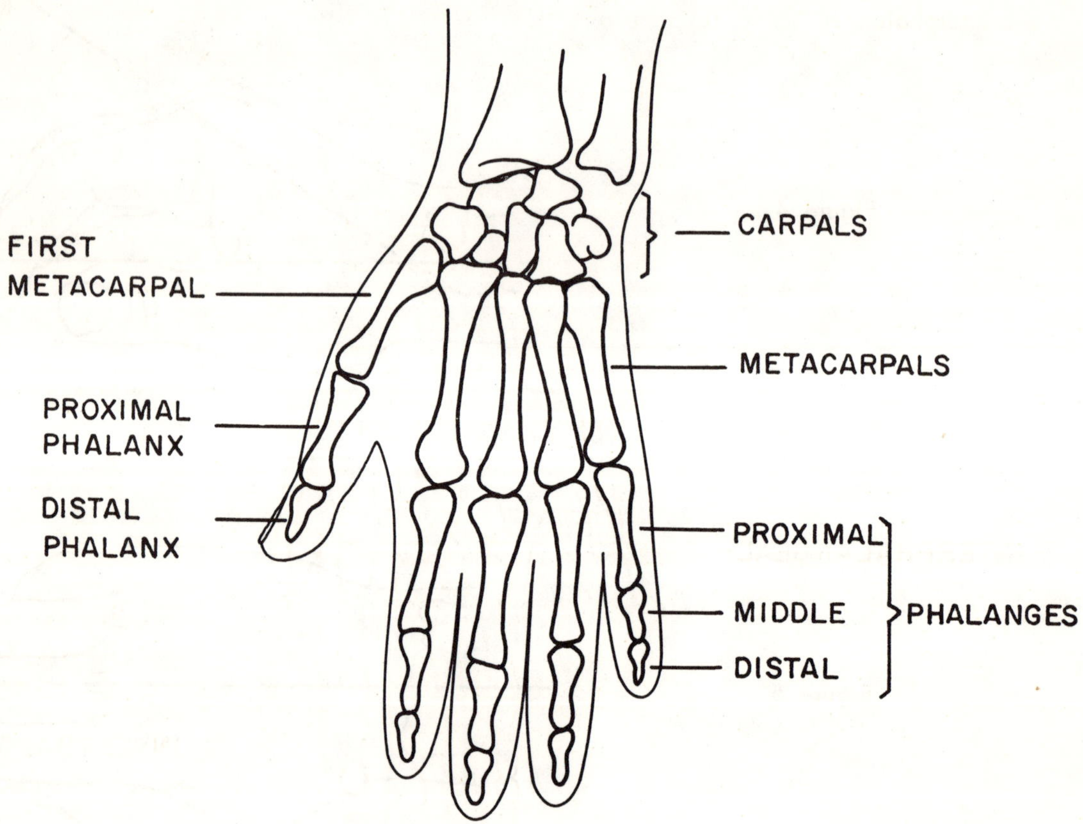

Figure 6. Bones of the Hand and Wrist

<u>Joints of the Hand and Wrist</u>

The joints of the hand and wrist are shown in the illustration below with labels indicating the name of each joint.

The radiocarpal joint is so named because it provides the articulation between the radius bone in the forearm and several of the carpal bones in the wrist.

The intercarpal joint provides articulation between the two rows of carpal bones. The carpometacarpal joint does the same for the distal row of carpal bones and the metacarpal bones in the hand.

The metacarpophalangeal joints (frequently abbreviated "M. P." joints), provide articulation between the metacarpals and the proximal phalanges of the fingers and thumb. The proximal interphalangeal joints (abbreviated "P.I.P." joints), are between the proximal and middle phalanges of the fingers, the distal interphalangeal joints (abbreviated "D.I.P." joints), between the middle and distal phalanges. Since the thumb has only one phalanx, it has only one interphalangeal joint, (abbreviated "I.P." joint).

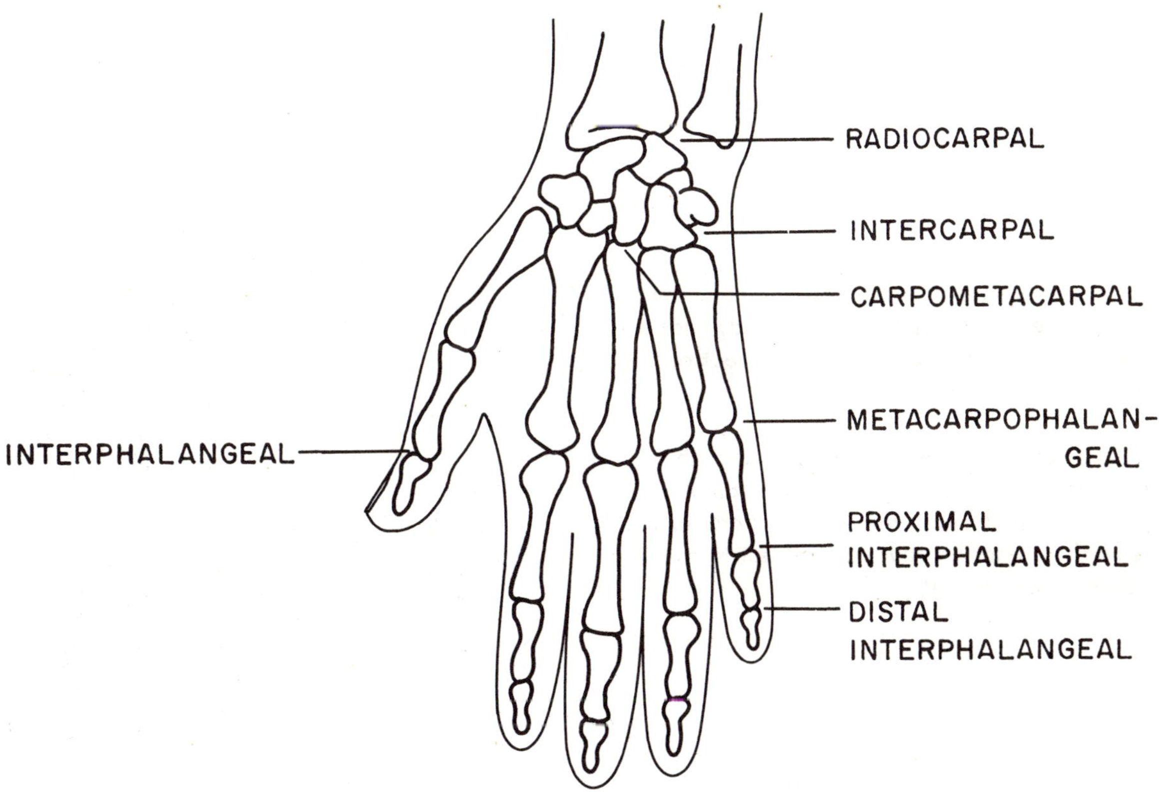

Figure 7. Joints of the Hand and Wrist

Terms of Anatomical Reference

The back of the hand is called the dorsum, a Latin word meaning back, and may also be referred to as the dorsal aspect of the hand.

The palm of the hand is referred to as the volar aspect of the hand, and comes from the Latin _vola_ --palm or sole. The word "palmar" is also used quite frequently instead of volar. Either is acceptable.

The fingers are identified as the index finger, next to the thumb, then the middle finger, the ring finger, and the little finger (Figure 8).

The border of the hand on the thumb side is the radial aspect of the hand, that on the little finger side, the ulnar aspect of the hand   (Figure 9).

The term distal is used to describe that part of an extremity farthest from the point of attachment, and the term proximal is used to describe that part closest to the point of attachment.

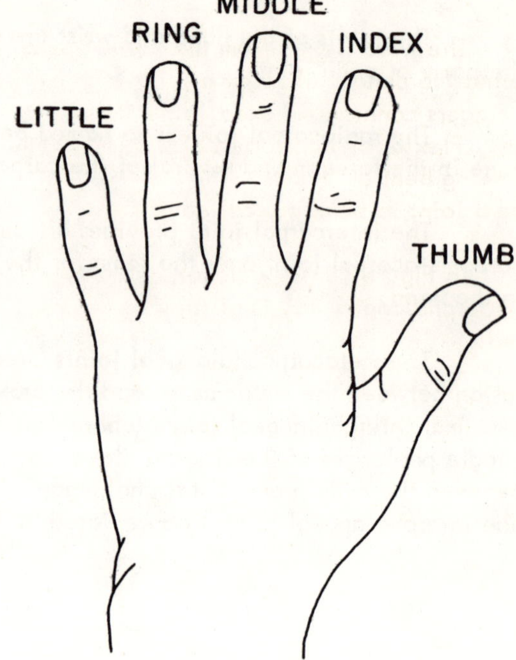

Figure 8. Dorsal Aspect of Hand

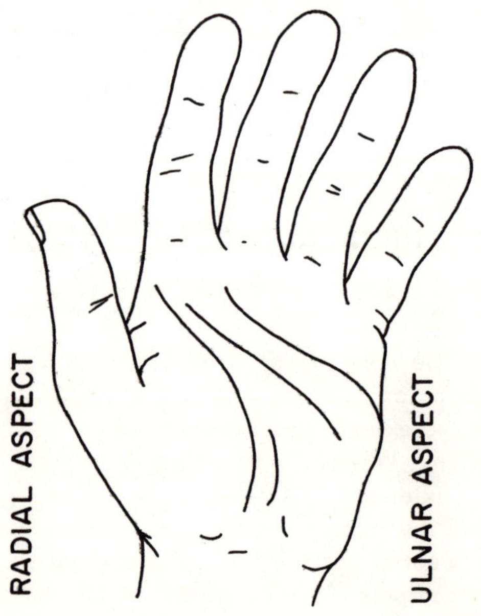

Figure 9. Volar or Palmar Aspect of Hand

Movements of the Fingers

The word <u>flex</u> is from the Latin flexus, meaning "to bend". To flex a joint is to bend it. Fingers cannot be flexed, only finger joints. Flexion is the motion of a joint when it is being bent or flexed. The muscles that cause a joint to bend are called "flexors".

In Figure 10 the metacarpophalangeal and interphalangeal joints of the fingers are flexed.

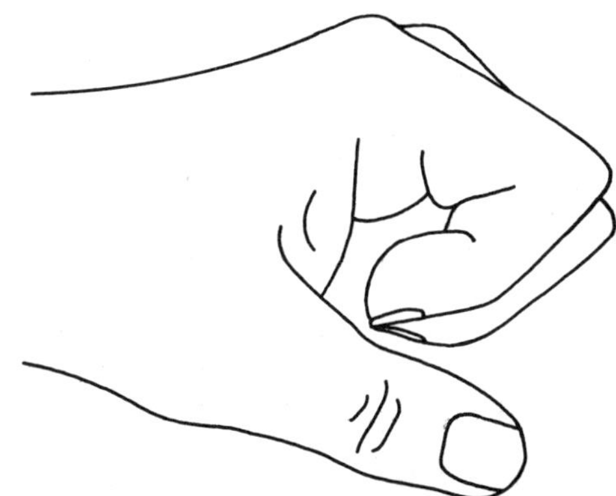

Figure 10. M.P. and I.P. Joints of Fingers Flexed

In Figure 11 the metacarpophalangeal joint of the index finger is flexed.

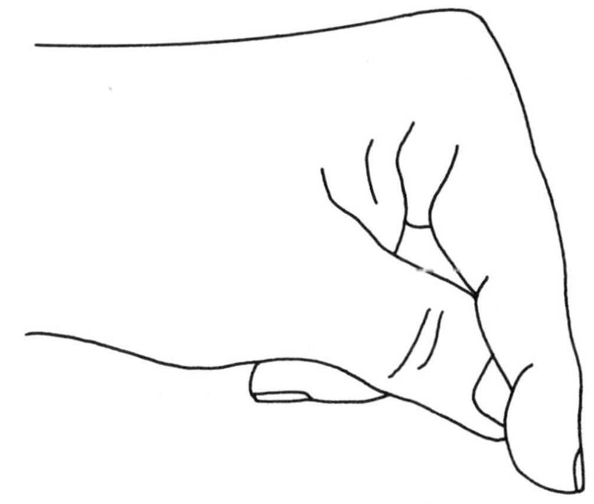

Figure 11. M.P. Joint of Index Finger Flexed

The word <u>extend</u> is from the Latin extendere, <u>ex</u> meaning out, <u>tendere</u> meaning stretch, so extend means stretch out or straighten. When a joint is flexed, it may then be extended or straightened. In Figure 12 all of the joints of the fingers are extended to the "neutral position", in alignment with the metacarpals. If they are extended past the neutral position, this is called "hyper-extension". "Hyper" is Greek for "more than the normal".

Figure 12. All Joints of Fingers Extended

Both the metacarpophalangeal and interphalangeal joints of the fingers can be flexed and extended. In addition, the fingers can be moved medially and laterally at the metacarpophalangeal joints. To describe these movements the center line down the middle finger is used as a reference line. Movements of the other fingers are described as M.P. joint abduction or adduction depending on whether the movement is away from or toward the reference line, (Figure 13).

Many students become confused when attempting to learn the meaning of the words "abduct" and "adduct". In Latin, "ab" means away from, "ducere" means lead or take, so abduct is from the word abducere, to take away from. In English, abduct may be used to mean "kidnap", or take a person away. Therefore, when a finger is moved <u>away</u> from the reference line, it is being abducted.

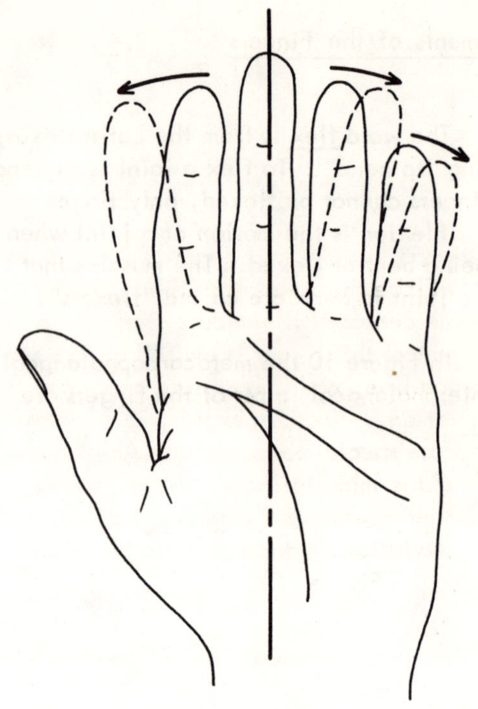

Figure 13. Abduction of Finger Metacarpophalangeal Joints

In Latin the word "ad" means "to" or "toward", so in this case adduct means moving toward the reference line.

In Figure 14, the index, ring, and little fingers are shown moving toward the reference line, therefore the M.P. joints are being adducted.

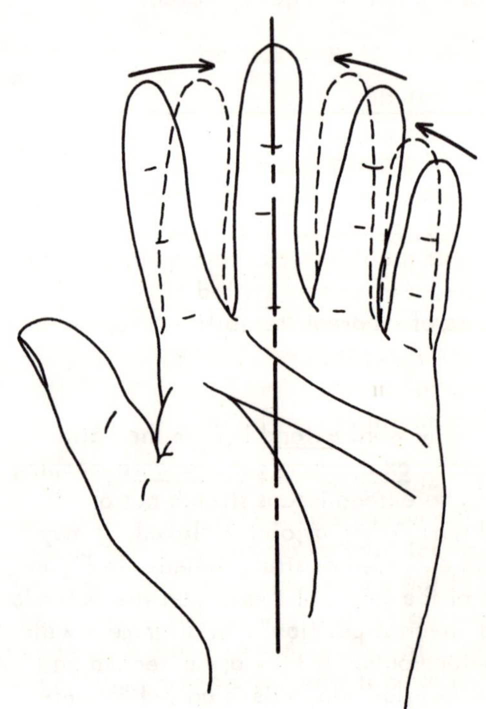

Figure 14. Adduction of Finger Metacarpophalangeal Joints

Since the reference line is placed down the center of the middle finger, any movement of this finger is away from the reference line, and is therefore abduction, or deviation. Because of this, it is necessary to use special terms to describe movements of the middle finger. Movement of this finger toward the thumb is called radial deviation, toward the finger, ulnar deviation (Figure 15).

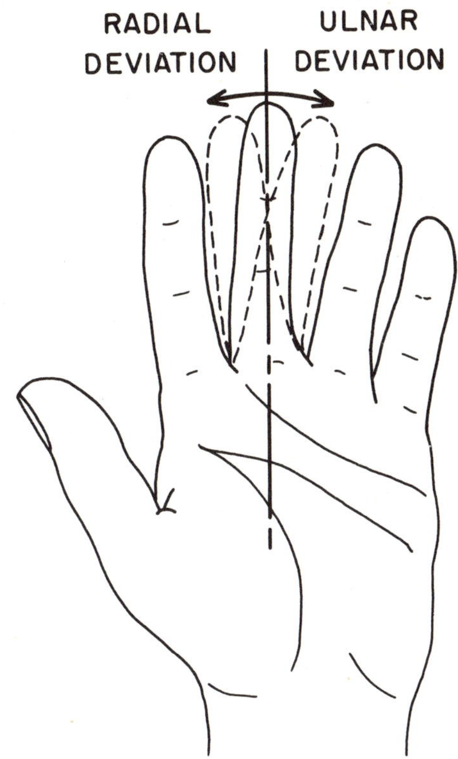

Figure 15. Middle Finger Metacarpophalangeal Joint Radial and Ulnar Deviation

Movements of the Thumb

The basic movements of the thumb are flexion, extension, adduction, abduction, and opposition. By combining these, the tip of the thumb can be made to describe a circle, a movement called circumduction. Because of its great flexibility, the thumb can, for example, combine flexion and abduction in varying degrees to obtain movements containing elements of both.

Abduction is illustrated in Figure 16. The movement takes place at the carpometacarpal joint of the thumb. The metacarpal bone of the thumb serves somewhat the same function as the proximal phalanges of the fingers. In fact, for many years some authorities believed it was a phalanx rather than a metacarpal.

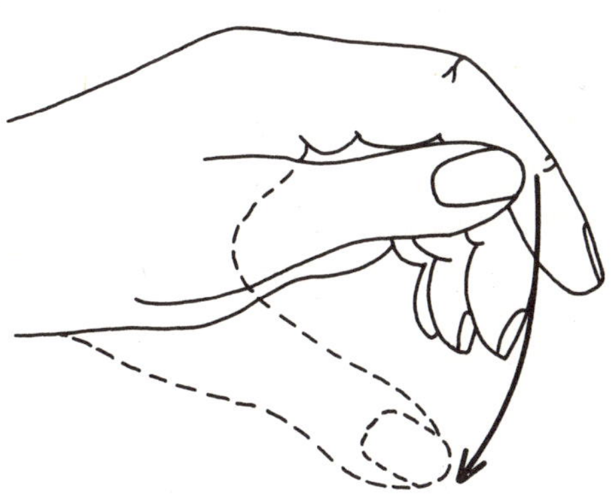

Figure 16. Abduction of Thumb Carpometacarpal Joint

Adduction, Figure 17, is the movement whereby the thumb is brought from abduction back to a position in which the thumb is on the palmar aspect of the hand.

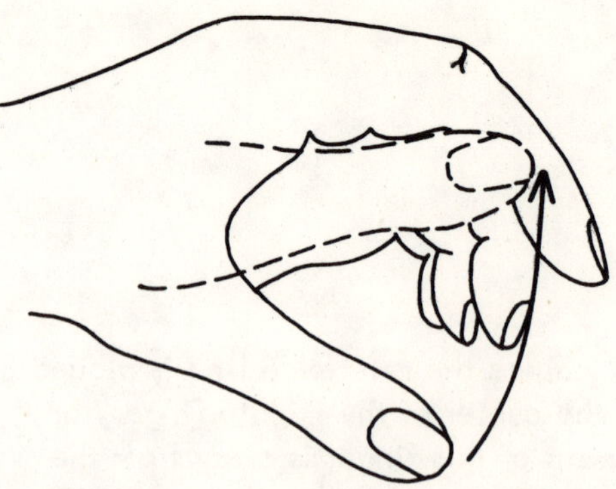

Figure 17. Adduction of Thumb Carpometacarpal Joint

Extension, Figure 18, is a movement of the thumb away from the center line of the middle finger, in which the thumb M.P. and I.P. joints are extended.

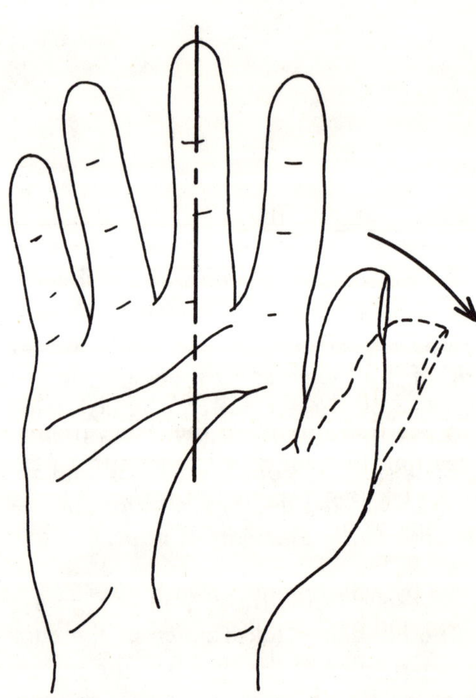

Figure 18. Extension of Thumb at the M.P. and I.P. Joints

Flexion, Figure 19, is movement of the thumb toward the center line of the middle finger, in which the thumb M.P. and I.P. joints are flexed.

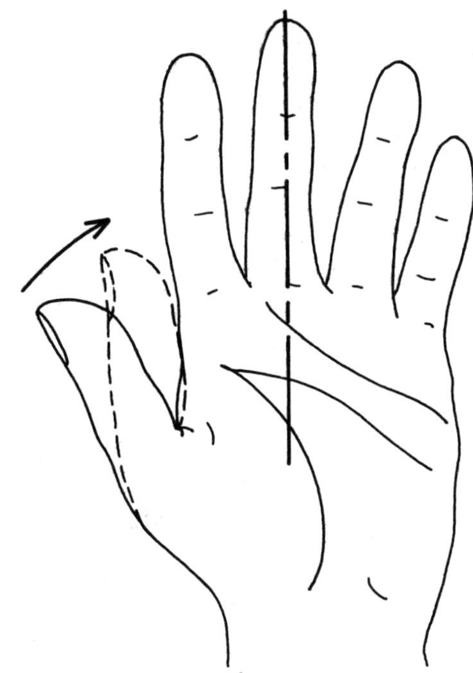

Figure 19. Flexion of Thumb M.P. and I.P. Joints

In opposition, the thumb moves from its position lateral to the index finger to a position in front of all the fingers. It can then be used to pinch or grasp in opposition to any or all of the fingers. The movement is a rotary one, which can be readily observed by moving one's own thumb into opposition and back several times   (Figure 20).

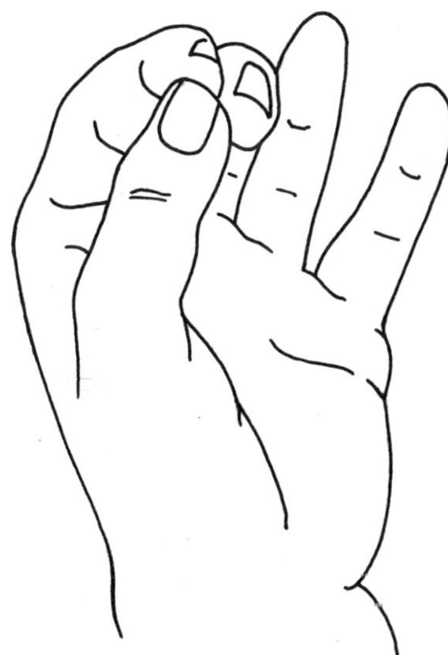

Figure 20. Thumb Opposition

<u>Movements of the Wrist Joints</u> (<u>Radiocarpal</u>, <u>Intercarpal</u>, <u>Carpometacarpal</u>)

With the hand and arm horizontal, palm down, flexion of the joints of the wrist moves the hand down, as illustrated in Figure 21.

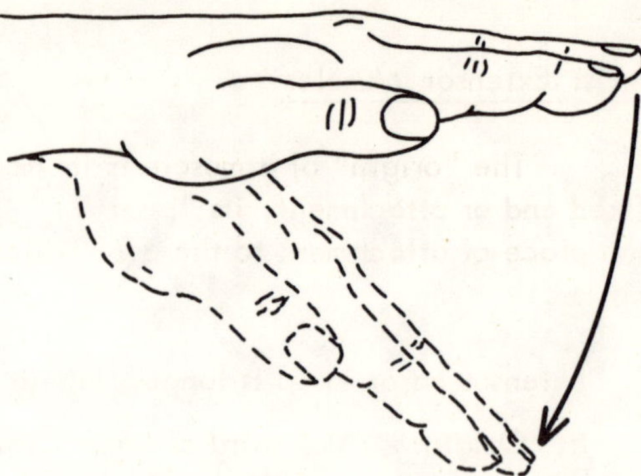

Figure 21. Flexion of Wrist Joint

With the hand and arm horizontal, palm down, extension of the joints of the wrist moves the hand up, as illustrated in Figure 22.

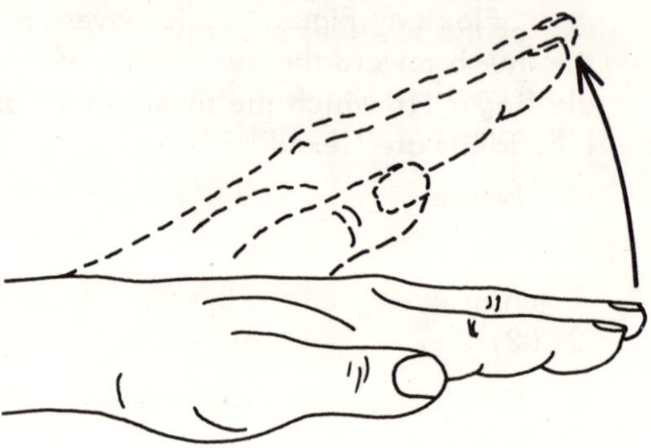

Figure 22. Extension of Wrist Joint

With the hand and arm horizontal, thumb pointing up, radial deviation of the joints of the wrist moves the hand up, as illustrated in Figure 23.

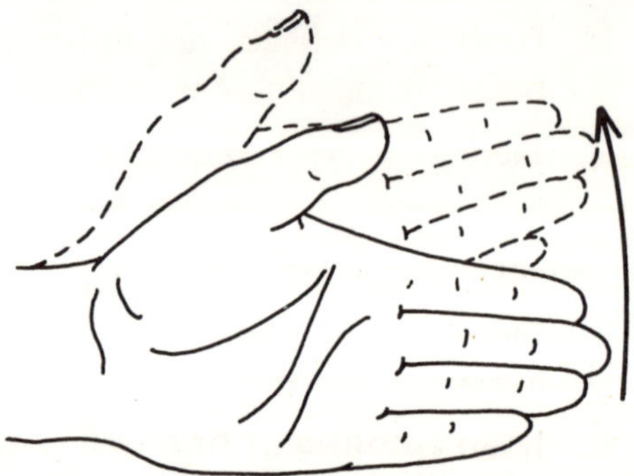

Figure 23. Radial Deviation of Wrist Joint

With the hand and arm horizontal, thumb pointing up, ulnar deviation of the joints of the wrist moves the hand down, as illustrated in Figure 24.

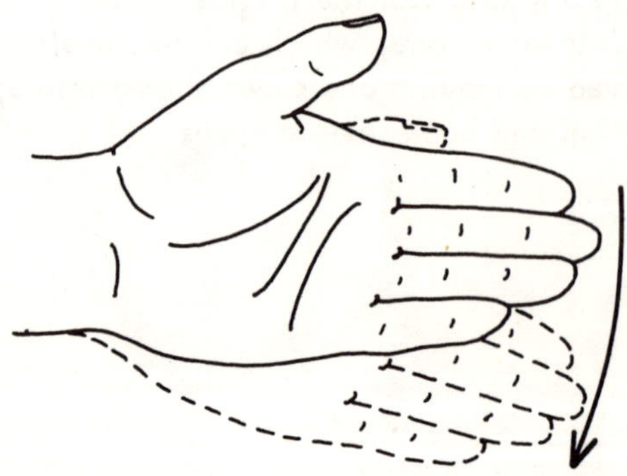

Figure 24. Ulnar Deviation of Wrist Joint

Wrist Extensor Muscles

The "origin" of a muscle is its more fixed end or attachment, its "insertion" is the place of attachment to the bone which it moves.

1. Extensor carpi radialis longus, Figure 25 (1)

    a. Origin: Distal third of lateral condyloid ridge of the humerus.

    b. Insertion: Dorsal surface of the base of the second metacarpal.

    c. Function: Extends wrist joint, with some radial deviation.

    d. Nerve supply: Radial nerve.

2. Extensor carpi radialis brevis, Figure 25 (2)

    a. Origin: Lateral epicondyle of humerus.

    b. Insertion: Base of third metacarpal.

    c. Function: Extends, and assists in radial deviation, of wrist joint.

    d. Nerve supply: Radial nerve.

3. Extensor carpi ulnaris, Figure 25 (3)

    a. Origin: Lateral epicondyle of humerus.

    b. Insertion: Base of fifth metacarpal.

    c. Function: Extends wrist joint, with some ulnar deviation.

    d. Nerve supply: Radial nerve.

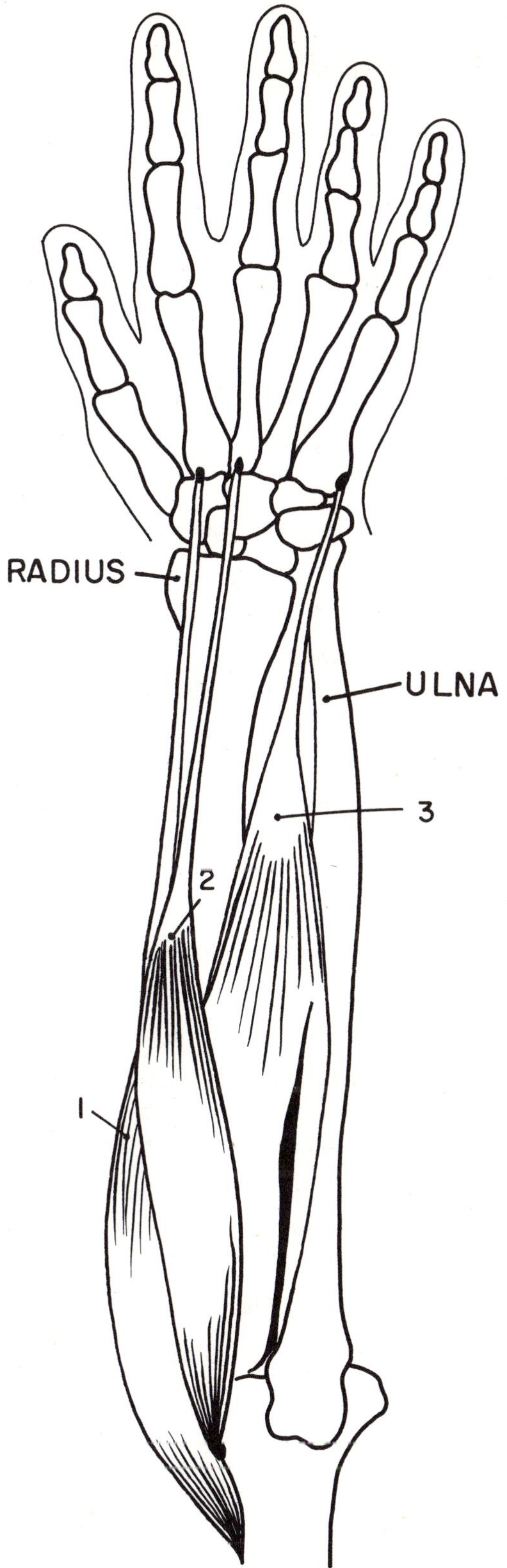

Figure 25. Wrist Extensor Muscles

Wrist Flexor Muscles

1. Flexor carpi radialis, Figure 26 (1)

   a. Origin: Medial epicondyle of the humerus.

   b. Insertion: Volar surface of the base of the second metacarpal.

   c. Function: Flexion and some radial deviation of the wrist.

   d. Nerve supply: Median nerve.

2. Palmaris longus, Figure 26 (2)

   a. Origin: Medial epicondyle of the humerus.

   b. Insertion: Annular ligament of the wrist and the fascia of the palm.

   c. Function: Contributes some assistance in flexion of the wrist and abduction of the thumb. Helps to cup the palm, and often lends dynamic origin to the intrinsic muscles of the thumb.

   d. Nerve supply: Median nerve.

3. Flexor carpi ulnaris, Figure 26 (3)

   a. Origin: Medial epicondyle of the humerus, proximal two-thirds of ridge on dorsum of the ulna.

   b. Insertion: Palmar surfaces of two of the carpals (pisiform, hamate), and the fifth metacarpal.

   c. Function: Flexes the wrist.

   d. Nerve supply: Ulnar nerve.

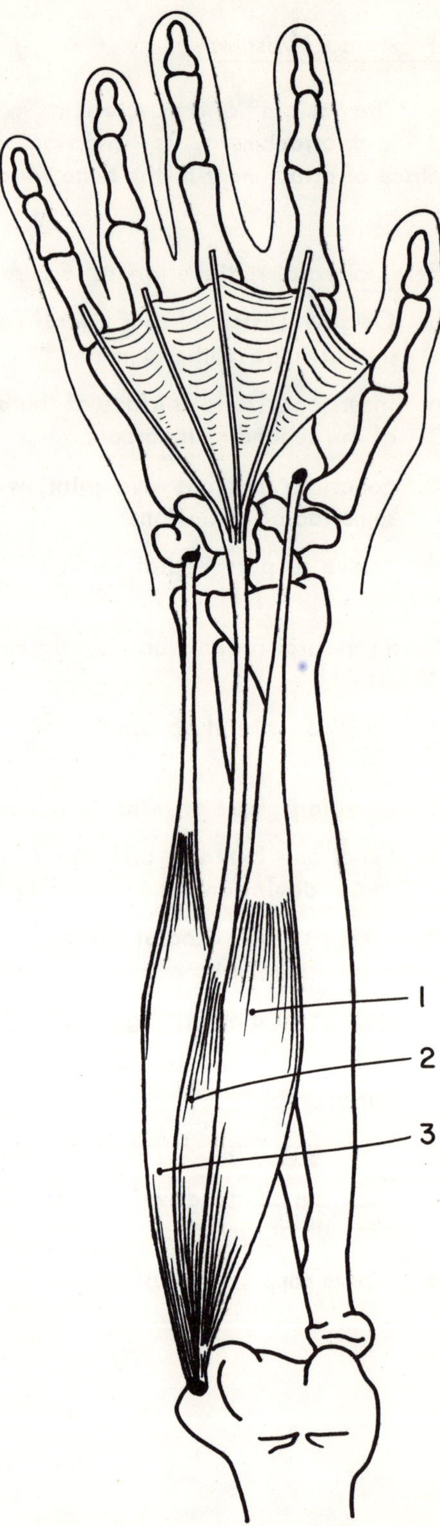

Figure 26. Wrist Flexor Muscles

Finger Extensor Muscles

1. Extensor digitorum communis, Figure 27

   "Extensor digitorum" means extensor of
   the fingers, "communis" means common,
   or belonging to several. In this case,
   the muscle acts to extend all four metacarpophalangeal joints.

   (Note that in Figure 27 you are looking
   at the <u>back</u> of your right hand.)

   a. Origin: Lateral epicondyle of the
       humerus.

   b. Insertion: By four tendons which
       pass through the wrist and then sep-
       arate to go to the four fingers. The
       tendons are attached by slips to the
       bases of the proximal phalanges;
       they then divide into three parts,
       the middle part attaching to the base
       of the middle phalanx and the two
       outer parts attaching to the base of
       the distal phalanx.

   c. Function: Extends the metacarpo-
       phalangeal joints. In some individ-
       uals it may extend the proximal and
       distal interphalangeal joints to some
       extent, but in most cases this action
       is weak. It tends to separate the
       fingers as it extends, after this it
       acts to extend the wrist. It stabi-
       lizes the hand for flexion.

   d. Nerve supply: Radial nerve.

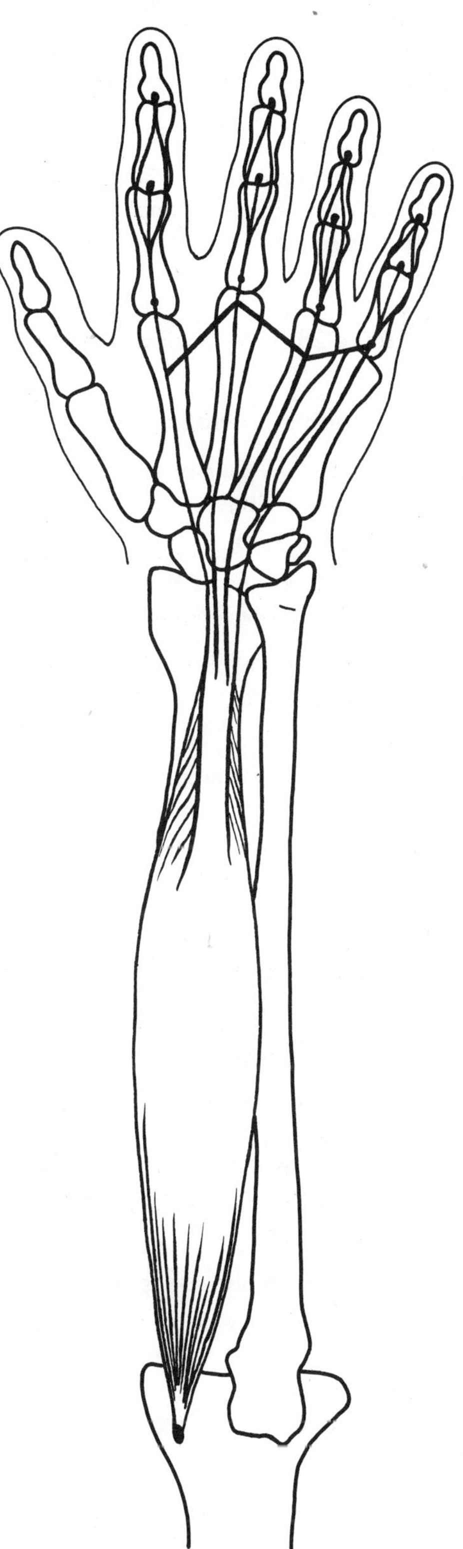

Figure 27. Extensor Digitorum Communis

2. Extensor digiti quinti proprius, Figure 28 (1)

   "Extensor digiti quinti" means the muscle
   is an extensor of the fifth or little finger.
   "Proprius" means belonging to one only,
   so this muscle acts on the little finger
   only.

   (Figure 28 is the back of the right hand.)

   a. Origin: Lateral condyle of humerus.

   b. Insertion: Dorsum of proximal phalanx
      of little finger.

   c. Function: Assists the extensor digi-
      torum communis to extend the joints
      of the little finger.

   d. Nerve supply: Radial (dorsal
      interosseous).

3. Extensor indicus proprius, Figure 28 (2)

   This term means the muscle is an extensor
   of the index finger only.

   a. Origin: Dorsal surface of distal
      half of the ulna.

   b. Insertion: Middle and distal
      phalanges of index finger.

   c. Function: Extends the proximal
      and distal interphalangeal joints,
      assists in extension of the meta-
      carpophalangeal joint of the index
      finger.

   d. Nerve supply: Deep radial.

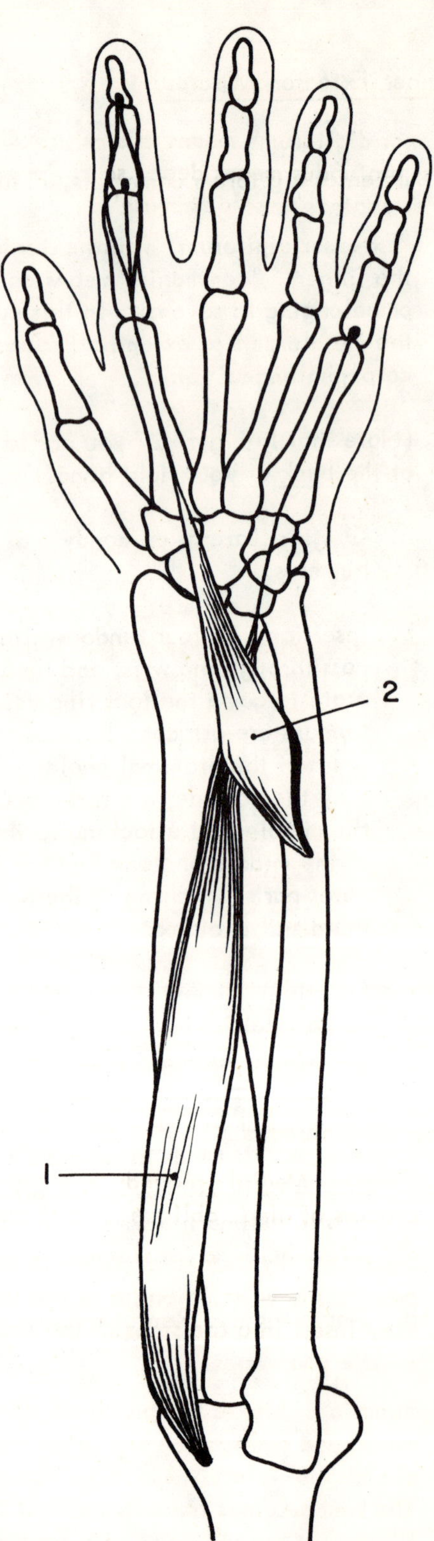

Figure 28. Extensor Digiti Quinti Proprius (1)
Extensor Indicus Proprius (2)

<u>Finger Flexor Muscles</u>

1. Flexor digitorum profundus, Figure 29 (1)

   "Flexor digitorum" means this muscle flexes the fingers; profundus means deep, so the muscle is located under another muscle or muscles.

   a. Origin: Middle half of volar and inner surfaces of the ulna.

   b. Insertion: By the four tendons which separate after passing the wrist, one going to each of the four fingers, where they are inserted into the dorsal surfaces of the bases of the distal phalanges.

   c. Function: Primarily flexes the distal interphalangeal joints of the fingers; closely associated with the flexor digitorum sublimis in flexing the proximal interphalangeal joints. The muscle fibers to which the profundus tendon of the index finger is attached are provided with nerve supply independent of the fibers that operate the other three fingers. Therefore the index profundus is separate, anatomically and functionally, enabling the index finger distal interphalangeal joint to flex independently from those of the other three fingers.

   d. Nerve supply: Median nerve to radial half of muscle, ulnar nerve to ulnar half.

2. Flexor digitorum sublimis, Figure 29 (2)

   "Sublimis" means high or superficial so this finger flexor is located on top of other muscles, in this case the profundus.

   a. Origin: Medial epicondyle of humerus, coronoid process of ulna, oblique line of radius.

   b. Insertion: By four tendons which separate after passing the wrist, then go to the four fingers where they insert into the sides of the bases of the four middle phalanges.

   c. Function: Flexes the proximal interphalangeal joints; continued contraction will then flex the metacarpophalangeal joints. Assists in flexing wrist and elbow. The four sublimis tendons connect to groups of muscle fibers that are provided with separate nerve supplies, making the action on each finger independent, both anatomically and functionally.

   d. Nerve supply: Median nerve.

Figure 29(1). Flexor Digitorum Profundus, Palmar View of Right Hand

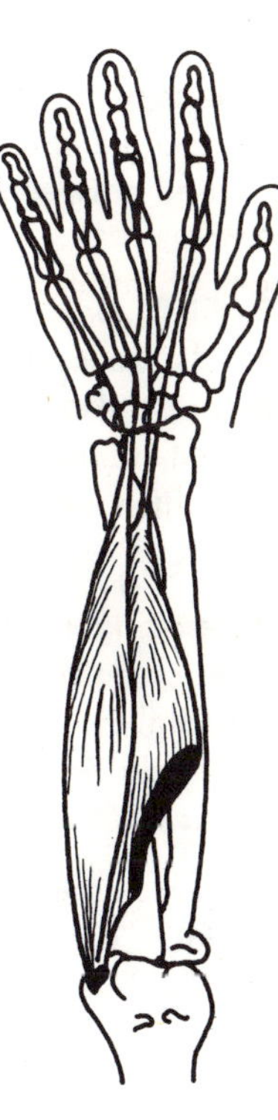

Figure 29(2). Flexor Digitorum Sublimis, Palmar View of Right Hand

Finger Abductor and Adductor Muscles

1. Palmar interosseous muscles, Figure 30

   There are three palmar interosseous muscles, the First (1), Second (2), and Third (3). The word "interosseous" means "between bones", as these muscles are located between the palmar sides of the metacarpal bones.

   a. Origin: First, ulnar side of second metacarpal; Second, radial side of fourth metacarpal; Third, radial side of fifth metacarpal.

   b. Insertion: First, base of ulnar side, proximal phalanx of index finger; Second, base of radial side, proximal phalanx of ring finger; Third, base of radial side, proximal phalanx of little finger.

   c. Function: All the interossei flex the metacarpophalangeal joints and extend the proximal and distal interphalangeal joints of the fingers. In addition the palmar interossei adduct the index, ring, and little fingers.

   d. Nerve supply: Deep branch of the ulnar nerve.

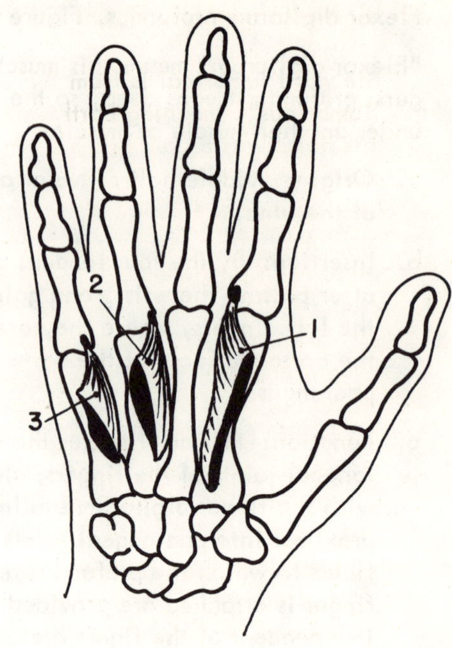

Figure 30. Palmar Interosseous Muscles

2. Dorsal interosseous muscles, Figure 31

   There are four dorsal interosseous muscles, the First (1), Second (2), Third (3), and Fourth (4). They are located between the dorsal sides of the metacarpal bones.

   a. Origin: First, adjacent sides of first and second metacarpals; Second, adjacent sides of second and third metacarpals; Third, adjacent sides of third and fourth metacarpals; Fourth, adjacent sides of fourth and fifth metacarpals.

   b. Insertion: First, radial side of base of first phalanx, index finger; Second and Third, radial and ulnar sides of base of first phalanx, middle finger; Fourth, radial side of base of first phalanx, ring finger. In addition, there is an insertion into the aponeurosis or "hood" of the long extensor tendons.

   c. Function: Flex metacarpophalangeal joints of index, middle and ring fingers; and distal interphalangeal joints of middle and ring fingers; abduct index and ring fingers; radial and ulnar deviation of middle finger.

   d. Nerve supply: Ulnar nerve, palmar branch.

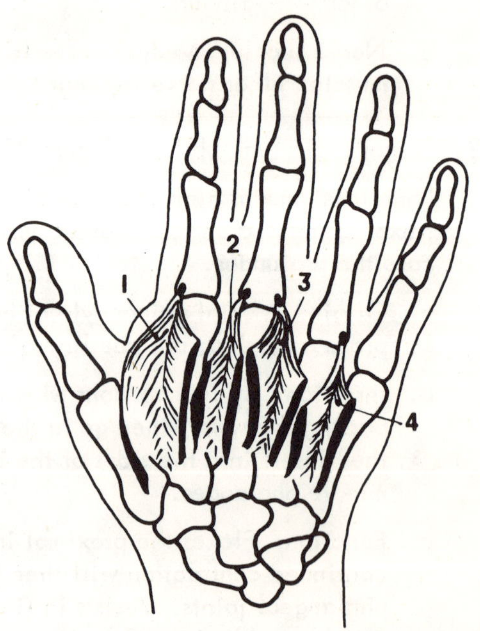

Figure 31. Dorsal Interosseous Muscles

3. Lumbrical muscles, Figure 32

The word lumbrical is from the Latin "lumbricus", meaning earthworm, and this name was probably given to these muscles because of their slender, worm-like appearance. One lumbrical muscle is a lumbricalis, the plural is lumbricales. There are four lumbricales in the hand.

a. Origin: First, tendon of flexor digitorum profundus of index finger; Second, tendon of same muscle of middle finger; Third, tendons of same muscle of middle and ring fingers; Fourth, tendons of same muscle of ring and little fingers.

b. Insertion: First, tendon of extensor digitorum communis of index finger; Second, tendon of same muscle of middle finger; Third, tendon of same muscle of ring finger; Fourth, tendon of same muscle of little finger.

c. Function: The lumbricales assist the interossei in flexing the metacarpophalangeal joints and extending the proximal and distal interphalangeal joints of the fingers.

d. Nerve supply: First and Second, median nerve; Third and Fourth, ulnar nerve.

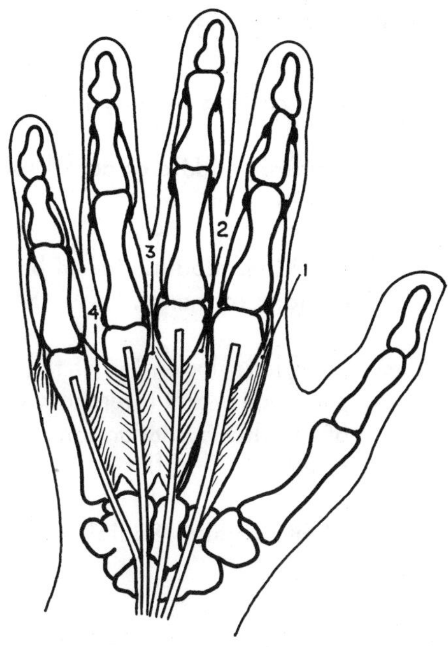

Figure 32. Lumbrical Muscles

Thumb Extensor Muscles

1. Extensor pollicis longus, Figure 33 (1)

   a. Origin: Middle of dorsal surface of ulna.

   b. Insertion: Radial side of base of distal phalanx of thumb.

   c. Function: Extension of the metacarpophalangeal and interphalangeal joints of the thumb, adduction of the entire thumb.

   d. Nerve supply: Radial nerve.

2. Extensor pollicis brevis, Figure 33 (2)

   a. Origin: Dorsal surface of radius, and interosseus membrane.

   b. Insertion: Base of proximal phalanx of thumb.

   c. Function: Extends metacarpophalangeal joint, flexes interphalangeal joint of thumb; abducts thumb. Assists in radial flexion of wrist after full abduction of the thumb.

   d. Nerve supply: Radial nerve.

3. Abductor pollicis longus, Figure 33 (3)

   a. Origin: Dorsal surfaces of radius and ulna.

   b. Insertion: Radial side of base of first metacarpal.

   c. Function: Abducts and assists in extending the thumb.

   d. Nerve supply: Radial nerve.

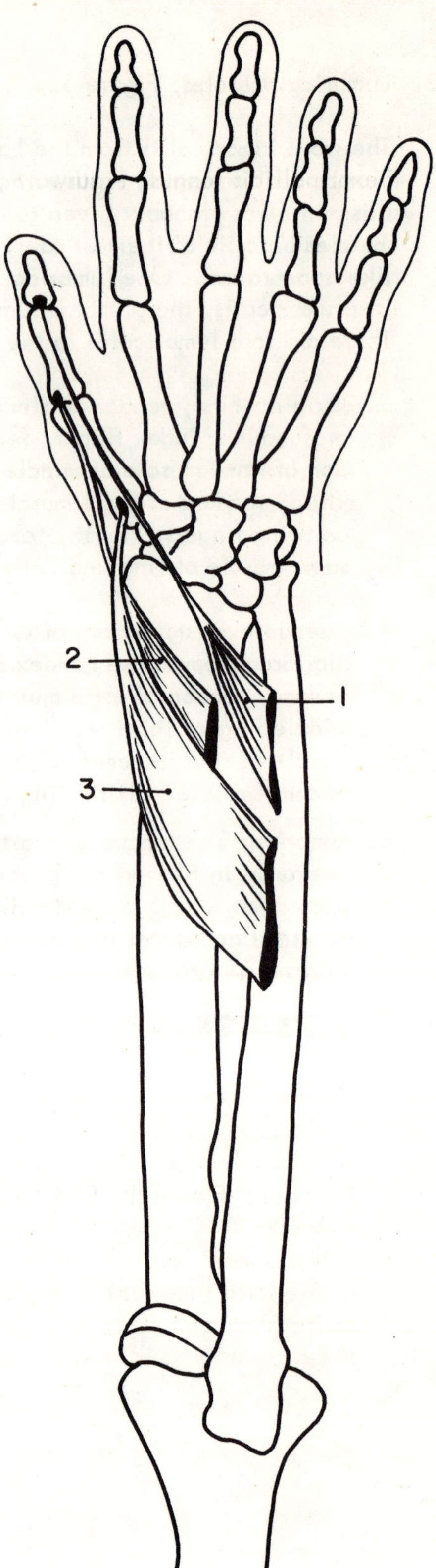

Figure 33. Thumb Extensor Muscles

Thumb Flexor Muscles

1. Flexor pollicis longus, Figure 34 (1)

    The word "pollicis" is Latin for "thumb",
    so the name of this muscle means "long
    flexor of the thumb".

    (The illustration in Figure 34 is of the
    palm of the right hand.)

    a. Origin: Volar surface of radius,
        coronoid process of ulna.

    b. Insertion: Volar base of the distal
        phalanx of the thumb.

    c. Function: Flexes the metacarpo-
        phalangeal and interphalangeal
        joints of the thumb.

    d. Nerve supply: Median nerve.

2. Flexor pollicis brevis, Figure 34 (2)

    a. Origin: Lower border of transverse
        carpal ligament, greater multangular
        bone (one of the carpal bones).

    b. Insertion: Radial side of proximal
        phalanx of thumb.

    c. Function: Flexes the first carpo-
        metacarpal joint and the metacar-
        pophalangeal joint of the thumb,
        at the same time rotates the thumb
        so its grasping surface or pulp faces
        the comparable surfaces of the
        fingers.

    d. Nerve supply: Median nerve.

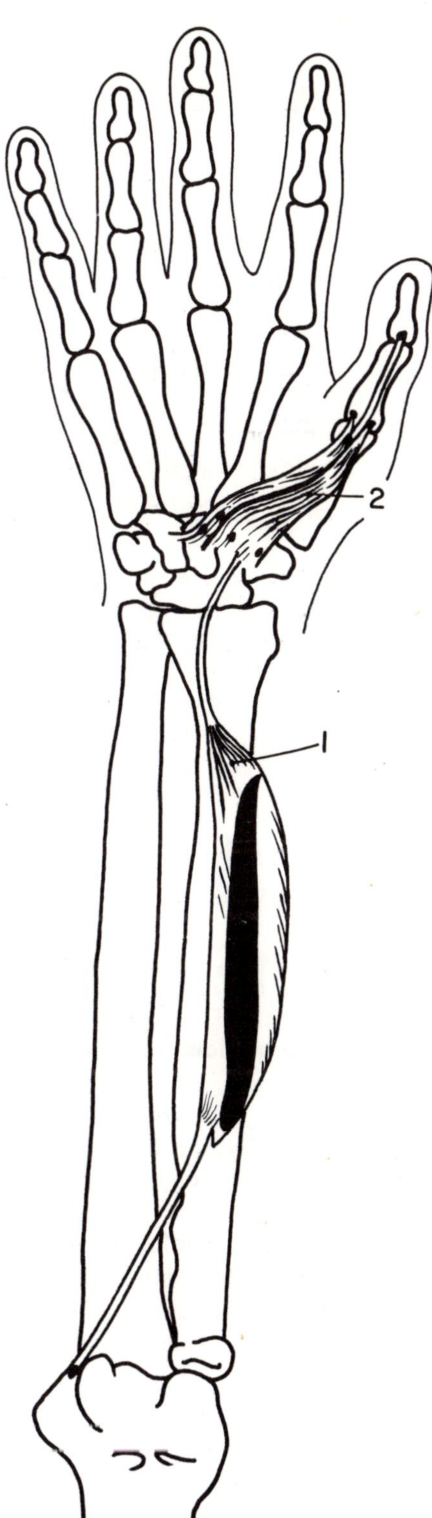

Figure 34. Thumb Flexor Muscles

<u>Thumb Opponens Muscles</u>

1. Opponens pollicis, Figure 35

   An opponens muscle is one that tends to
   draw a digit across the palm of the hand
   toward the other fingers.

   a. Origin: Transverse carpal ligament,
      greater multangular bone.

   b. Insertion: Radial side of first meta-
      carpal bone (thumb).

   c. Function: Flexes and inwardly ro-
      tates the first metacarpal, thus
      assisting in attaining opposition be-
      tween the thumb and fingers.

   d. Nerve supply: Median nerve.

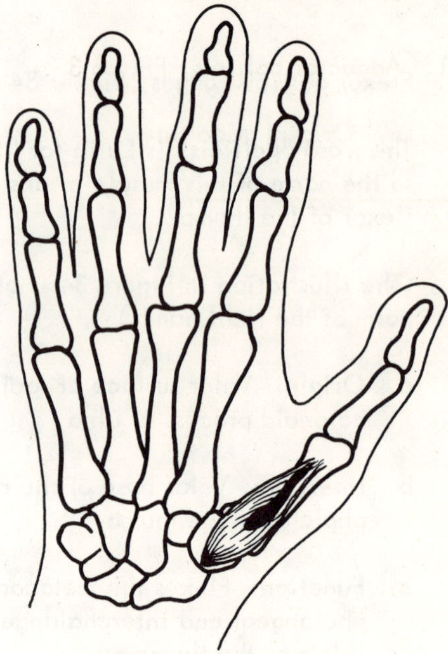

Figure 35. Opponens Pollicis

2. Abductor pollicis brevis, Figure 36

   a. Origin: Transverse carpal ligament,
      tuberosity of navicular, ridge of
      greater multangular (carpal bones).

   b. Insertion: Radial side of base of
      proximal phalanx of thumb.

   c. Function: Abducts the thumb.

   d. Nerve supply: Median nerve.

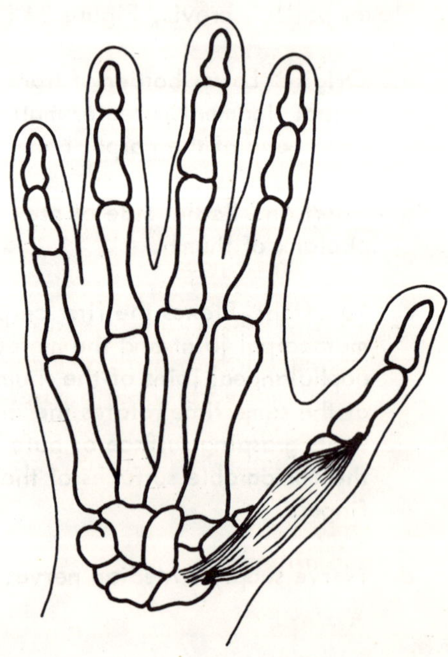

Figure 36. Abductor Pollicis Brevis

Thumb Adductor Muscle

1. Adductor pollicis, Figure 37

   a. Origin: Capitate bone, annular liga-
      ment, lower two-thirds of third meta-
      carpal.

   b. Insertion: Inner surface of base of
      first phalanx of thumb, tendon of
      extensor pollicis longus.

   c. Function: Adducts the thumb toward
      the index finger and on to the middle
      finger, close to the palm of the hand.

   d. Nerve supply: Ulnar nerve.

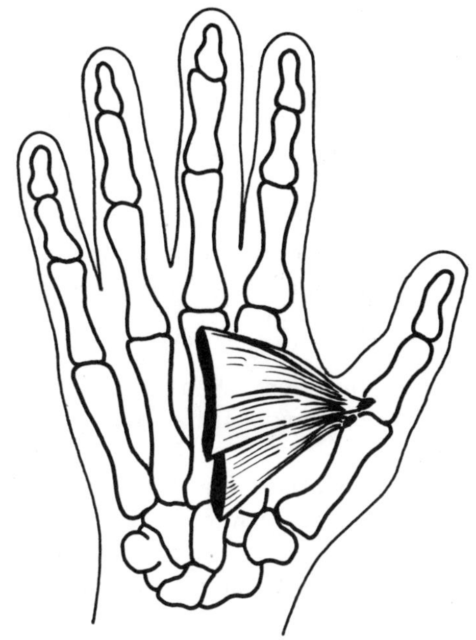

Figure 37. Adductor Pollicis

Hypothenar Muscles

   The hypothenar eminence is the ridge
on the palm at the base of the little finger.

1. Abductor digiti quinti, Figure 38

   These terms mean "the abductor of the
   fifth or little finger".

   a. Origin: Pisiform bone (one of the
      carpals in the wrist), and tendon of
      the flexor carpi ulnaris muscle.

   b. Insertion: Ulnar side of base of
      proximal phalanx of little finger;
      aponeurosis of extensor digiti
      quinti proprius.

   c. Function: Abducts little finger,
      helps flex its metacarpophalangeal
      joint.

   d. Nerve supply: Ulnar nerve.

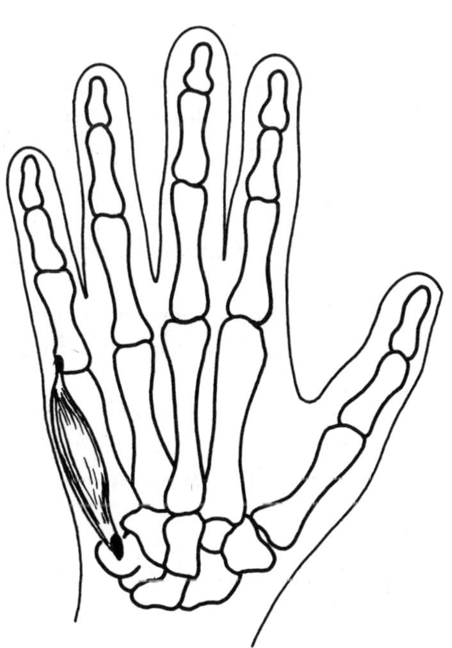

Figure 38. Abductor Digiti Quinti

2. Flexor digiti quinti brevis, Figure 39

These terms mean "the short flexor of the fifth or little finger".

a. Origin: Volar carpal ligament, hook of the hamate bone (one of the carpals), the pisiform-hamate ligament.

b. Insertion: Ulnar side of proximal phalanx of little finger.

c. Function: Flexes metacarpophalangeal joint of little finger.

d. Nerve supply: Ulnar nerve.

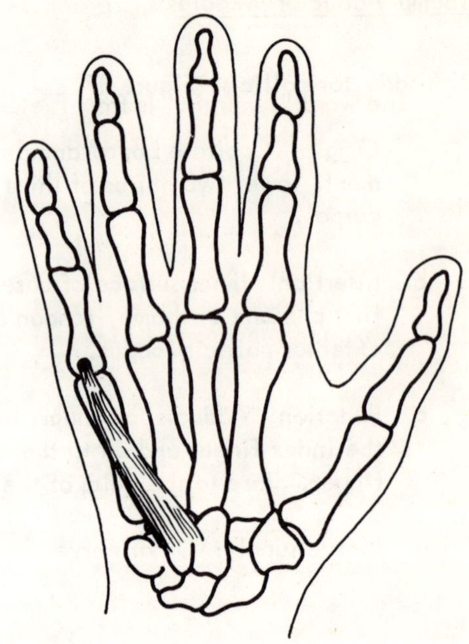

Figure 39. Flexor Digiti Quinti Brevis

3. Opponens digiti quinti, Figure 40

An opponens muscle is one which tends to draw a digit across the palm toward the other fingers. In this case, the muscle tends to draw the little finger across the palm toward the other fingers.

a. Origin: Hook of the hamate bone, transverse carpal ligament.

b. Insertion: Ulnar margin of the metacarpal of the little finger.

c. Function: Flexes and rotates the fifth metacarpal, causing the little finger to be drawn toward the center of the hand.

d. Nerve supply: Ulnar nerve.

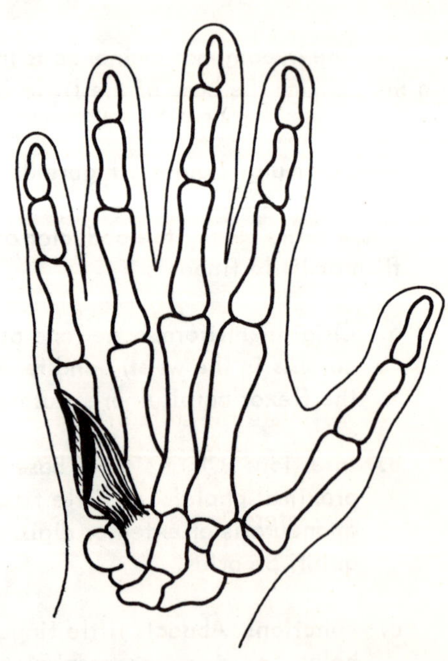

Figure 40. Opponens Digiti Quinti

Synergic Action of the Extrinsic and Intrinsic Muscles of the Hand

The word "synergic" is from the Greek "synergein", to work together. Synergic action of the muscles is their function of working together to accomplish more than the sum of their individual efforts. The extrinsic (outside) muscles are those that have their muscle bellies and origins in the forearm, and are connected to their insertions in the hand by long tendons. Because of this length, they are often referred to as the "long extensors" and the "long flexors". The intrinsic (inside) muscles are those that have their origins and insertions entirely within the hand itself. In the normal hand there is a delicate mechanical balance between the long flexors and extensors and the intrinsics, and as long as this balance is maintained, the hand will function efficiently. When any of these muscles are weakened or destroyed, the synergic balance is upset, the hand loses function and may become deformed.

In the descriptions in this chapter of muscles of the hand and their functions, the material on functions has been kept as simple as possible. However, in reality there is nothing simple about the functional anatomy of the hand. In this section we will discuss some of the more complex relationships between the intrinsic and extrinsic muscles of the hand that are of importance to the orthotist in fitting hand splints. To save time and space the metacarpophalangeal joints will be referred to as the M.P.'s, the proximal interphalangeal joints as the P.I.P.'s, and the distal interphalangeal joints as the D.I.P.'s.

The mechanism that coordinates the actions of the intrinsic and extrinsic muscles of the hand is the aponeurosis, Figure 41, sometimes called the extensor hood. It is a thin membranous sleeve made of the expanded fibers of the lumbrical and interossei tendons and the long extensor tendon. When the long extensor tendon pulls it proximally it slides well back over the M.P. joint as the latter extends   (Figure 41-A). On the other hand, when the lumbricales and interossei contract, they pull the M.P. joint into flexion, and as flexion is initiated, the aponeurosis slides distally, giving muscles a greater mechanical advantage   (Figure 41-B).

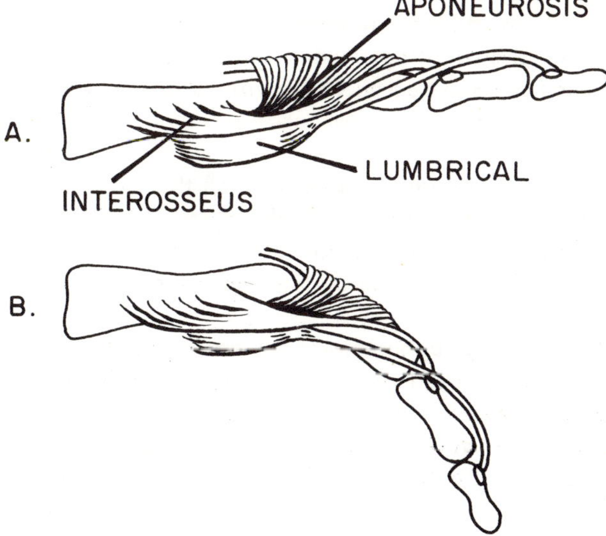

Figure 41. Synergic Action of the Extrinsic and Intrinsic Muscles of the Hand. Note the shift in direction of force exerted on the M.P. joint by the interosseous and lumbrical muscles as the aponeurosis slips distally as in "B".

The aponeurosis makes it possible for the lumbricales and interossei to perform three separate functions in coordination with the long extensors and flexors, which none of them could do as well alone. First, the lumbricales and interossei flex the M.P.'s when the long extensor is relaxed, by pulling on the aponeurosis. They can initiate M.P. flexion even when those joints are in hyperextension. The M.P.'s can flex to 90 degrees, and through antagonistic action between the long extensor and the interossei-lumbricales, they can be stabilized at any point in that range. The P.I.P.'s and D.I.P.'s are flexed by the flexor digitorum sublimis and profundus respectively. The latter muscles can flex these joints without action by the interossei-lumbricales, but the tips of the fingers end up at their bases as the muscles do not have enough excursion to flex the M.P.'s. When the three are coordinated through synergic action, the finger tips can be placed anywhere on the palm or thumb.

Second, the long extensor extends the M.P.'s, and in doing so slides the aponeurosis back over the M.P. joints. The long extensor uses almost all of its excursion in extending the M.P.'s, and has little left to extend the I.P.'s. However, with the M.P.'s stabilized by the long extensor, the interossei-lumbricales are now in position, because of the proximal shift of the aponeurosis, to pull on the extensor tendon and complete the job by extending the P.I.P.'s and D.I.P.'s. In this remarkable manner the same muscles serve as the flexors of one joint and the extensors of two others.

Some interesting variations in the phasic action of the muscles are worth noting. When the M.P.'s, P.I.P.'s, and D.I.P.'s are in the fully flexed position, the extensor digitorum communis tendon can extend the I.P.'s effectively until the M.P.'s reach 45 degrees. The long extensor is at its maximum efficiency at about 67 degrees of M.P. flexion. At 45 degrees of M.P. flexion the interossei-lumbricales take over and continue until the I.P.'s are fully extended. The long extensor completes the extension of the M.P.'s and is responsible for stabilizing the M.P. joints, without which the interossei-lumbricales could not function. The lumbricales also aid by drawing the flexor profundus distally to allow easier I.P. extension. It is interesting to observe that when both of the I.P.'s are fully flexed, the long extensor is stretched so much it cannot extend the D.I.P. alone.

Third, the interossei-lumbricales adduct the index, ring, and little fingers, abduct the index, and ring fingers, and cause radial and ulnar deviation of the middle finger. These movements can only be made when the extensors stabilize the M.P.'s in extension, and seldom exceed 15 degrees.

The interossei perform much the same functions as the lumbricales, but are stronger in all motions, particularly M.P. abduction-adduction. The lumbricales have a good angle of approach into the aponeurosis for initiating M.P. flexion, and they have a relatively long excursion if you count that contributed by the profundus.

An exception to the general description of the function of the aponeurosis should be mentioned. The first dorsal interosseous flexes the M.P. joint of the index finger when the long extensor is slack, and it abducts it when the long extensor stabilizes the M.P. in extension, but it does not extend the P.I.P. and D.I.P. joints. In the case of the index finger, this is done by the palmar interosseous and the lumbricale.

References

Bunnell, Sterling, Surgery of the Hand, Philadelphia, J. B. Lippincott Co., 1949.

Flatt, Adrian E., The Care of Minor Hand Injuries, St. Louis, The C. V. Mosby Co., 1959.

Marble, Henry C., The Hand, A Manual and Atlas for the General Surgeon, Philadelphia, W. B. Saunders Co., 1960.

CHAPTER II. FUNCTIONAL HAND SPLINTS

PATIENT EVALUATION

Introduction

The term "splint" is usually used to mean "any device used to hold a broken bone in place or to keep a part of the body in a fixed position, as a thin, rigid strip of wood or metal"; it is so defined in Webster's dictionary. Such a splint might be described as "static", as its purpose is to prevent movement, and static splints are used extensively for that purpose, although plaster casts are more frequently used than rigid strips of wood or metal. A functional hand splint is quite different, as it is designed to encourage movement rather than prevent it. The functional hand splint is a dynamic device that can be used to accomplish any or all of four objectives in any given case:

1. Prevent or help to correct deformities of the hand and wrist.

2. Prevent joint stiffness by maintaining mobility.

3. Increase the strength of weak muscles by encouraging their use through proper positioning and assistance, and by discouraging or preventing the substitution of stronger but less appropriate muscles for those weakened by paralysis.

4. Increase the functional capacity of the hand by stabilizing it in better positions for function, and by providing mechanical assists to aid the weakened muscles.

Weakness, paralysis, or deformity of the hand and wrist may result from any of a number of causes, such as poliomyelitis, cerebral vascular accident or stroke, injuries to the spinal cord or other nerve, polyneuritis, infectious multiple neuritis, syringomyelia, leprosy, arthrogryposis, cerebral palsy, muscular dystrophy, and arthritis. One or more of the four objectives mentioned above may be achieved to some degree in these dysfunctions through the use of functional hand splints.

There are three basic types of functional hand splints, the short opponens, long opponens, and flexor hinge splint. There are fourteen attachments for use with the short and long opponens splints, each designed to ameliorate a certain muscular loss or combination of losses, to correct certain deformities, or to place the wrist, hand, and fingers in functional position. The flexor hinge splint is a highly specialized version of the long opponens splint, in which provision is made for an active "three jaw chuck" grasp with the thumb, index, and middle fingers.

The combinations of muscle loss and actual or threatened deformity that may be encountered are quite numerous, and for each such combination it is necessary to select the splint and attachments that will give the best results. To do this with a minimum of lost time and error, the first step is to take a thorough inventory of the patient's hand and arm functions. This is called "evaluation", and is done by carefully and systematically testing the strength and range of motion in all possible directions at each joint. For example, the wrist is tested for extension, flexion, and ulnar and radial deviation. An "Evaluation Check List" form is used to note all findings. This information is used in combination with medical, social, and vocational facts known about the patient to decide which splint components will be of greatest benefit to him.

The evaluation procedure is commonly carried out by the members of the clinic team, the physician, therapist, and orthotist, in the clinical conference. Such clinics confer at regular intervals, the length of time between meetings depending on the size of the case load. At any given meeting new patients may be evaluated and rehabilitation programs worked out for them, and old patients may be re-evaluated to determine progress made and to decide what changes in the rehabilitation program are needed, if any. Functional bracing of the upper extremities service is never successful unless each patient is systematically followed up at regular intervals until his rehabilitation is completed and he is discharged. Improvement is usually progressive as the effects of the splint or brace, in combination with physical and occupational therapy, bring about alterations in the patient's condition. These changes require that the splints and braces be adjusted, revised, or replaced to accommodate to the new conditions. Normal wear and tear, particularly with children, requires occasional repair or replacement.

Evaluation is not a mysterious ritual to be performed only under certain prescribed clinical conditions, however. It is very important that each member of the clinic team be able to do an evaluation. The orthotist must learn to do it so he can check on his own work when he is fitting, revising, or repairing a splint or brace. The therapist is usually responsible for evaluation in the clinic, and the physician has occasion to check on the patient's condition as part of his over-all supervision of the patient's progress. The standard procedure followed in making an evaluation of a patient's muscles is adequately described in the book by Daniels, Williams, and Worthingham, <u>Muscle Testing</u>, (Second Edition), published by W. B. Saunders Co. of Philadelphia in 1956, on pages 94 through 161. It is suggested that this procedure be studied very carefully, and used in evaluating patients who are candidates for functional arm braces or hand splints.

An evaluation check list form for recording medical, social, and vocational information, the muscle strength evaluation and range of motion data, and the prescription is shown on the following three pages. A form of this type should be completed for each patient every time he is evaluated.

UPPER EXTREMITIES ORTHOTICS

ORTHOTIC EVALUATION

Name _______________________________ Sex _____ Age _____ Height _________ Weight _______

Diagnosis: _________________________________ Onset: ________________

Paralysis
(Loss of Function)

Spastic
(Function – No Control)

Total   ( )   Shoulder   ( )   Reflex   ( )   Shoulder   ( )

Elbow   ( )   Elbow   ( )

Partial   ( )   Hand   ( )   Other   ( )   Hand   ( )

Sensory ____

Comments ____

		PASSIVE RANGE		MUSCLE GRADE		ASSISTANCE NEEDED	
		Right	Left	Right	Left	Right	Left
Scapular	Abduction						
	Adduction						
	Elevation						
	Depression						
Gleno-Humeral Joint	Flexion						
	Extension						
	Abduction						
	Adduction						
	Subluxation	Present in Left ☐ Right ☐					
Rotation	Internal						
	External						
Elbow	Flexion						
	Extension						
Forearm	Pronation						
	Supination						
Wrist	Extension						
	Flexion						
	Ulnar Deviation						
	Radial Deviation						

THUMB	RIGHT				LEFT			
	PASSIVE RANGE		MUSCLE GRADE		PASSIVE RANGE		MUSCLE GRADE	
Opposition								
Adduction								
Abduction								
	MP	I P	MP	I P	MP	I P	MP	I P
Flexion								
Extension								

FINGER			PASSIVE RANGE			MUSCLE GRADE		
			MP	I P	DIP	MP	I P	DIP
RIGHT	Index	Flexion						
		Extension						
	Middle	Flexion						
		Extension						
	Ring	Flexion						
		Extension						
	Little	Flexion						
		Extension						
LEFT	Index	Flexion						
		Extension						
	Middle	Flexion						
		Extension						
	Ring	Flexion						
		Extension						
	Little	Flexion						
		Extension						

5 - N	Normal	Complete range of motion against gravity with full resistance.
4 - G	Good	Complete range of motion against gravity with some resistance.
3 - F	Fair	Complete range of motion against gravity.
2 - P	Poor	Complete range of motion with gravity eliminated.
1 - T	Trace	Evidence of slight contractility. No joint motion.
0 - O	Zero	No evidence of contractility.

PERSONAL INFORMATION

INTERVIEW CHECK LIST

1. Education and field of specialization:

2. Occupation prior to onset:

3. Present occupation (if any):

4. Future occupational plans:

5. Hobbies, recreational interests:

6. Activities patient is able to do now:

   Personal   Occupational   Recreational

7. Activities patient cannot do now, and would like to be able to do:

   Personal   Occupational   Recreational

8. Medical and surgical considerations related to bracing or splinting:

9. Degree of patient interest in proposed bracing and splinting program, and willingness to devote necessary time and effort:

10. Prescription:

Clinic Chief ________________________________

TYPES OF FUNCTIONAL HAND SPLINTS AND ATTACHMENTS

Introduction

After completing the patient evaluation the clinic team decides which splint and attachments are needed to start the patient on the road to rehabilitation. To do this, each member of the team must be familiar with the different types of splints and attachments and their uses. In the following section each device is illustrated and its functions and applications are described. This information can then be used in conjunction with the patient evaluation data to select the best equipment for the patient.

It is important for the physician, therapist, and orthotist to be in complete agreement in their understanding of the characteristics and applications of the splints and attachments, and their indications and contra-indications.

Short Opponens Hand Splint

The short opponens hand splint without attachments is shown in Figure 42. This splint is made of two pieces of .064 inch 2024-T4 sheet aluminum, the palmar piece and the wrist piece. Both pieces are bent to fit the hand with an allowance for 1/4 inch padding. In the volar view it can be seen that the radial extension (1) fits snugly against the radial aspect of the hand proximal to the index finger M.P. joint. The opponens bar (2) fits proximal to the metacarpophalangeal joint of the thumb, and assists in holding the thumb in opposition to the fingers. In the palmar view the curve of the palmar arch (3) can be seen. The palmar arch supports the metacarpals in their normal position, preventing the collapse which may occur in a weakened hand. A strap attached to the wrist piece holds the splint in place on the hand. Two holding studs are riveted to the dorsal aspect of the palmar piece for mounting the various attachments that may be used. The attachment bar and attachment studs are so designed that the attachments may be quickly removed and replaced without the use of tools.

The short opponens hand splint is used in cases where there is no significant involvement of the wrist, and the problem is confined to the muscles of the hand and fingers only. It is seldom used without one or more of the various attachments that may be applied to it.

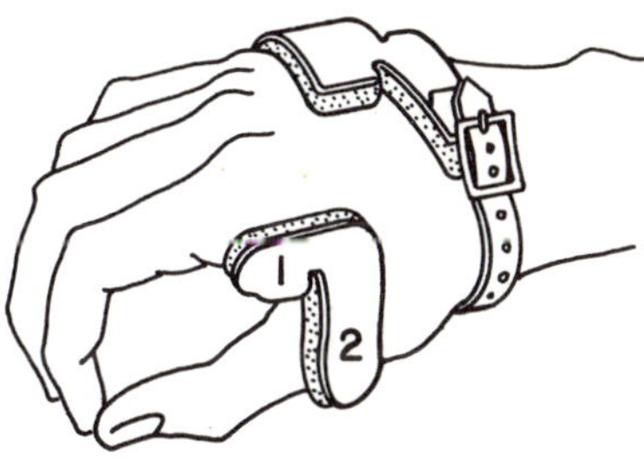

Dorsal view

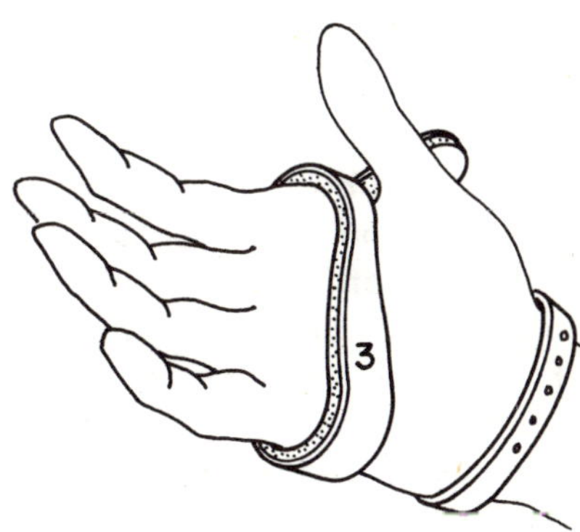

Palmar view

Figure 42. Short Opponens Hand Splint

While sheet aluminum is the material most used, functional hand splints may be made of stainless steel, plastic, or any of a number of suitable materials. However, kits of pre-formed aluminum parts are available at reasonable cost, and the use of these kits saves much time and work.

Long Opponens Hand Splint

The long opponens hand splint without attachments is illustrated in Figure 43. Like the short opponens, it is made of two pieces of .064 inch 2024-T4 sheet aluminum, but a long forearm piece (1) is used instead of a wrist piece. The palmar piece is the same on both types of hand splint, and the functions of the long opponens splint are the same as the short except that it immobilizes the wrist, thus preventing any significant amount of wrist flexion, extension, or deviation, either radial or ulnar.

The long opponens hand splint is useful where a stable wrist is desired, and it has been used successfully on cases with spastic wrist flexors as well.

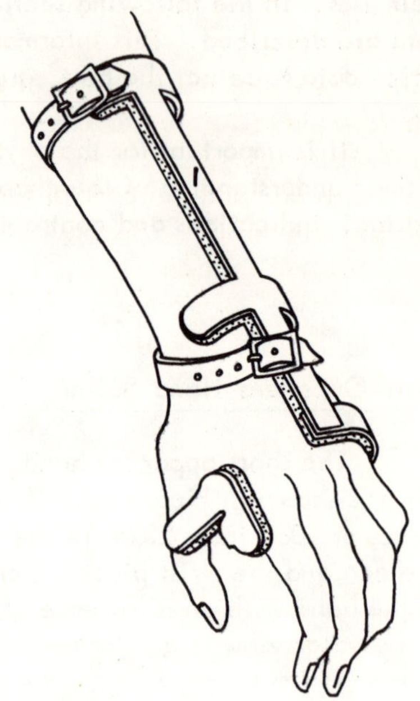

Figure 43. Long Opponens Hand Splint

"C"-Bar

The "C"-bar (Figure 44) is a curved piece of .037 inch stainless steel riveted to the palmar arch of either the short or long opponens hand splint in such a position that it tends to hold the M.P. joint of the thumb in abduction, but without extending beyond the I.P. joint, so the latter can be flexed. The "C"-bar is used when the thumb abductors are zero to poor and the adductor is fair to normal. In this situation the thumb is drawn close to the radial aspect of the hand, and the thumb web soon becomes tight, making it impossible to abduct the thumb. The "C"-bar can be used to correct this tightness by bending it out a little at a time as the web is stretched to regain the full range of motion, and to permit the thumb to oppose the fingers.

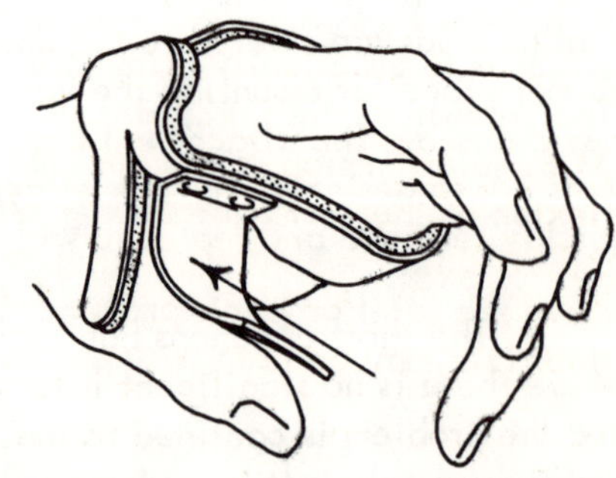

Figure 44. "C" Bar

Thumb Post

The thumb post (Figure 45) is a stainless steel extension riveted to the opponens extension on either the long or short opponens hand splint for the purpose of statically holding the thumb in a position of prehension. The distal end of the post is shaped to partially encircle the distal phalanx of the thumb enough to stabilize it without interfering with contact between the thumb pad and the pads of the fingers. It is used in cases where active opposition and thumb flexion are not sufficient for prehension.

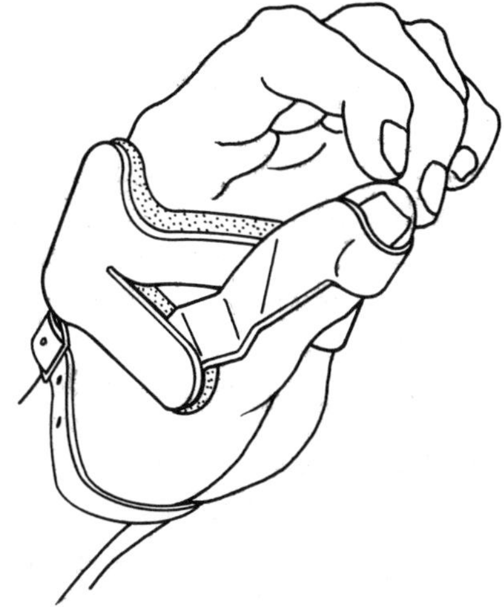

Figure 45. Thumb Post

Interphalangeal Joint Stabilizer

The interphalangeal joint stabilizer (Figure 46) is made up of two (1) or three (2) stainless steel rings, depending on whether one or both I.P. joints of a finger are to be stabilized. The rings are held together with a stainless steel rod, and when worn on the finger the effect is to practically eliminate I.P. joint motion. The purpose of the I.P. joint stabilizer is to obtain M.P. joint flexion in the absence of independent flexion of the M.P. joints. By blocking the motion of a distal joint the muscle acting on that joint will then move the joint or joints proximal to it. The I.P. joint stabilizer may be used with extension assists.

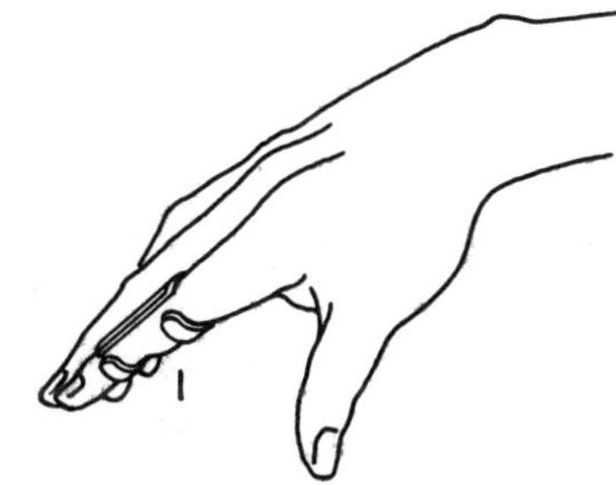

Figure 46. I. P. Joint Stabilizer

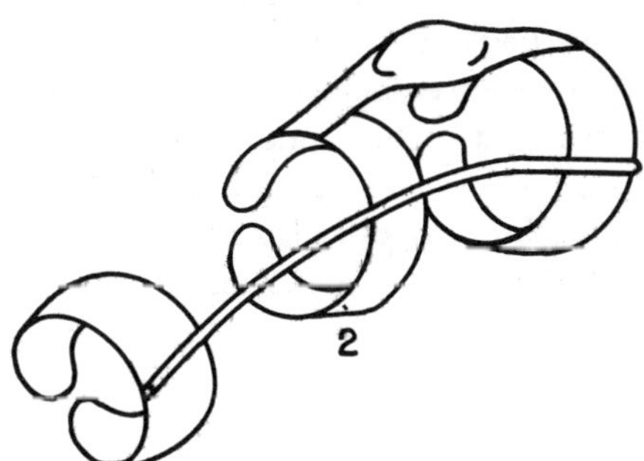

Spring Swivel Thumb

The spring swivel thumb (Figure 47) is made of
.056 inch stainless steel piano wire into which a coil
spring has been wound. One end swivels in a support
bracket that holds it in a position over the index finger
M.P. joint. The other end has a horse-shoe shaped
stainless steel finger piece soldered to it, shaped to
fit the proximal phalanx of the thumb. The spring
action tends to hold the thumb in abduction against the
opponens bar. The spring swivel thumb is used on either
the long or short opponens hand splint to abduct the
thumb when the thumb abductors are zero to poor but
the adductors and long flexor and extensor are fair to
normal. The pull of the flexor moves the thumb into
opposition and the spring permits grasp between the
thumb and fingers. The hand must be flexible and free
from deformity, and the thumb web must be loose for
the spring swivel thumb to function, as the spring is
not strong enough to stretch the web, or to maintain
the gains made by manual stretching. This is done
with the "C"-bar, as explained earlier. The spring
swivel thumb is a more dynamic attachment than the
"C"-bar, as the spring action allows more freedom of
movement. The two are never used together. In some
cases an overactive long thumb flexor and I.P. joint
instability will cause flexion at the I.P. joint and none
at the M.P. joint. To correct this situation an I.P.
joint stabilizer (Figure 46) is substituted for the finger
piece to stop the I.P. motion and transfer it to the M.P.

A number of splint attachments use spring or
elastic assists, and it is important to understand the
basic principle of this device. It is a method of bor-
rowing energy from an active muscle to help one that
is too weak to function. If a joint has good flexors
but zero extensors some of the flexor energy can be
stored in a rubber band or spring to be released when
the patient wants to extend the joint. These elastic
and spring devices also serve a therapeutic use by
providing resistive exercise.

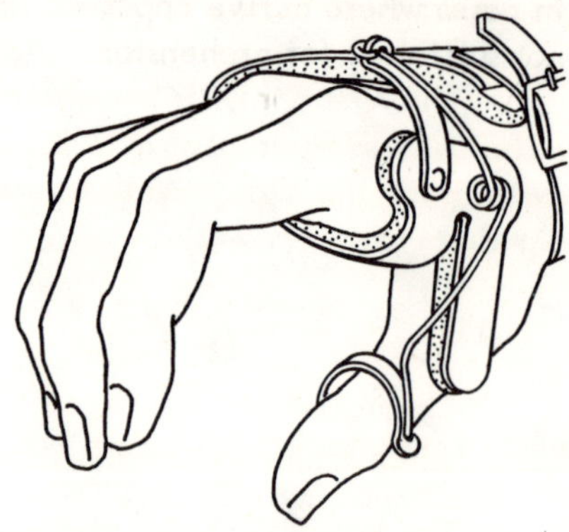

Figure 47. Spring Swivel Thumb

First Dorsal Interosseous Assist

The first dorsal interosseous assist (Figure 48) is a spring device similar to the swivel thumb, except it applies force to the index finger to abduct it. It is made of .047 inch stainless steel piano wire with a coil spring wound into one end, which is riveted to the radial extension of either of the two types of hand splints. A finger piece made of plastic or stainless steel is fitted to the proximal phalanx of the index finger and is attached to the other end of the spring. When the first dorsal interosseous is zero to poor the index finger tends to adduct out of functional position. The first dorsal interosseous assist abducts it so it can oppose the thumb.

The first dorsal interosseous assist can be adjusted to also assist flexion of the M.P. joint. It is helpful in cases where the first dorsal interosseous is in the range from zero to poor.

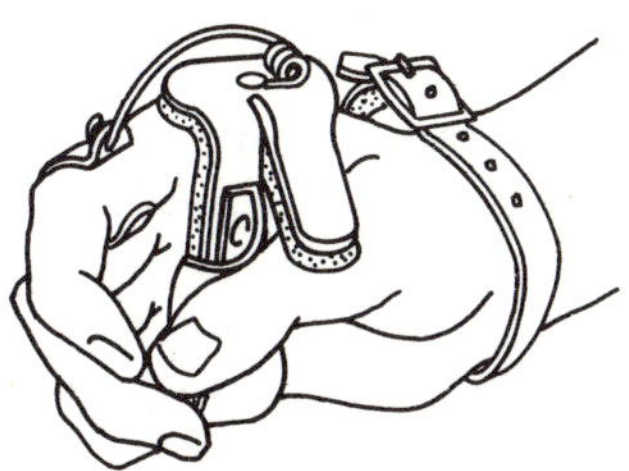

Figure 48. First Dorsal Interosseous Assist

Thumb Interphalangeal Extension Assist

The thumb interphalangeal extension assist (Figure 49) is a spring wire device which is fastened to the opponens bar of either the short or long opponens type of hand splints. It is shaped much like the first dorsal interosseous assist, except the plastic finger piece is fitted to the distal phalanx of the thumb and the spring action tends to hold the thumb I.P. joint in extension. An aluminum stop (1) is riveted to the opponens bar so it extends along the thumb almost to the I.P. joint. It confines the extension force to the I.P. joint, thus preventing hyperextension of the M.P. joint.

Zero to poor extensor pollicis longus in combination with fair to normal flexor pollicis causes the thumb I.P. joint to remain in excessive flexion, in which position it cannot oppose the fingers and provide grasp. The spring tension on the assist is adjusted so it is strong enough to extend the I.P. joint, but not strong enough to prevent the flexor pollicis from flexing the joint. The thumb I.P. extension assist is used only with a "C"-bar.

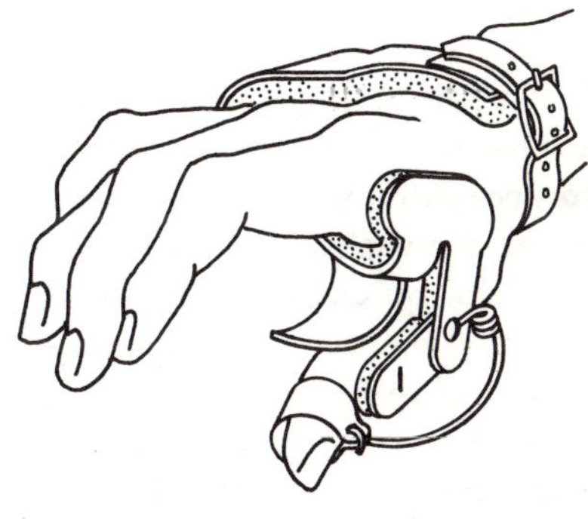

Figure 49. Thumb I.P. Extension Assist

Metacarpophalangeal (M.P.) Stop

The M.P. stop (Figure 50) is an attachment for use on either the short or long opponens hand splint. The curved, padded aluminum lumbrical bar (1) is positioned slightly proximal to the proximal I.P. joints of the fingers so the M.P. joints are flexed about 15 degrees. It is riveted to an aluminum attachment bar (2) which fastens to the dorsal extension of the hand splint with quickly detachable slip locks. The M.P. stop is used when the intrinsic extensors (interossei and lumbricales) of the hand are zero to poor, but the extensor digitorum communis and the flexors are fair to normal. The action of the extensor digitorum communis under these conditions tends to cause the M.P. joints to hyperextend, leaving the I.P. joints flexed, forming a "claw hand", (Figure 51). The normal action is for the long extensor to start extension of the M.P. and I.P. joints and for the intrinsics to complete it. However, if the M.P. joints are not allowed to extend completely, the long extensor will have enough contraction left to extend the I.P. joints. It has been found that if the M.P. stop is set to hold the M.P. joints in about 15 degrees of flexion, the I.P. joints can be extended enough to provide good function.

I.P. Extension Assist with M.P. Stop

The I.P. extension assist with M.P. stop (Figure 52) is primarily a therapeutic attachment for providing extension assist, for exercising the finger flexors and for maintaining joint mobility. It is made up of an aluminum attachment bar (1) provided with slip lock holes for quick removal and installation, to which is fastened the curved and padded M.P. stop (2). It is located just proximal to the proximal I.P. joints, and to prevent hyperextension of the M.P. joints that might be caused by the pull of the rubber bands. The attachment bar is bent up and a curved aluminum extension assist support (3) is riveted to the end of it so it is located well above and slightly beyond the distal phalanges of the fingers. Rubber bands are fastened to holes in the extension assist support and to plastic finger pieces fitted to the distal phalanges. The rubber band tension is adjusted so it is not too great for the finger flexors to overcome, yet strong enough to extend the joints. Sometimes it is necessary to use strips of thin rubber sheeting at the start of the exercise program if the patient cannot stretch regular rubber bands. This attachment can be used on both short and long opponens hand splints.

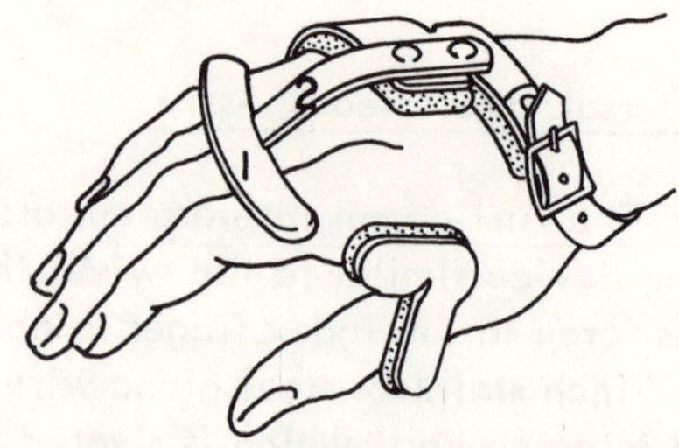

Figure 50. Metacarpophalangeal (M.P.) Stop

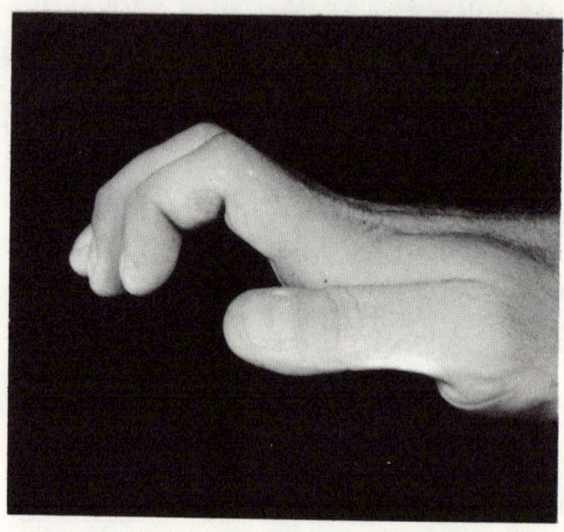

Figure 51. Claw Hand

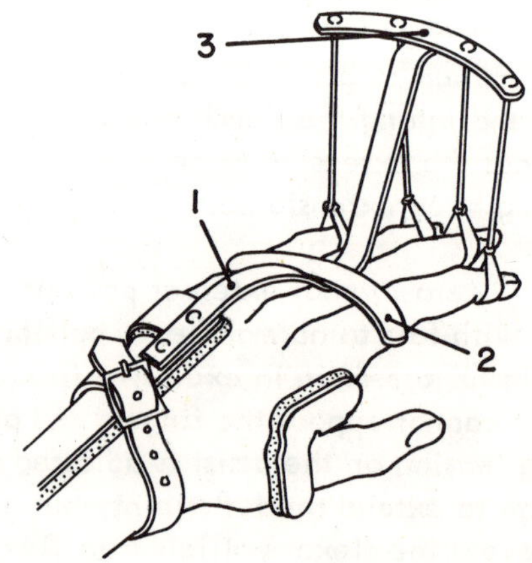

Figure 52. I.P. Extension Assist with M.P. Stop

Adjustable M.P. Flexion Control with I.P. Flexion Assist

The adjustable M.P. flexion control with I.P. extension assist (Figure 53) is mechanically the same as the I.P. extension assist with M.P. stop, except that the M.P. stop (1) is adjustable and of heavy-duty construction. It is used with the long opponens splint only, because of the leverage needed to maintain the pressure of the M.P. stop. In case of contracture of the extensor digitorum communis the M.P. stop maintains the gains made by manual stretching by the therapist.

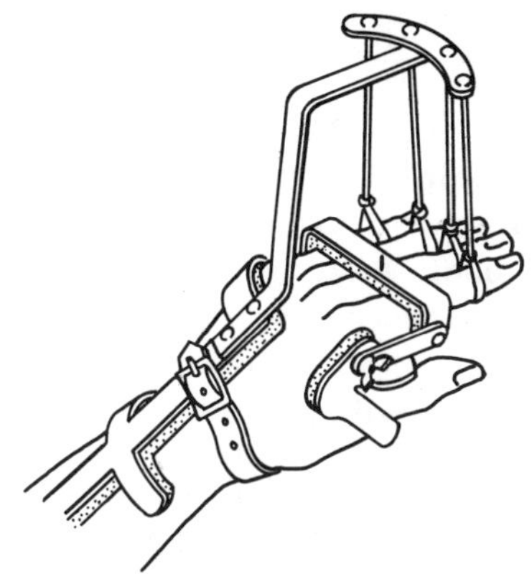

Figure 53. Adjustable M.P. Flexion Control with I.P. Extension Assist

Metacarpophalangeal Extension Assist

The metacarpophalangeal extension assist (Figure 54) is the same mechanically as the I.P. extension assist with M.P. stop, except the M.P. stop is omitted and the plastic finger pieces are fitted to the proximal rather than the distal phalanges of the fingers, so that extension assist is provided for the M.P. joints only. It can be fitted to provide assistance to one or more fingers. It is used when the extensor digitorum communis is zero to poor and the long flexors and the intrinsic M.P. flexors are fair to good, and the intrinsic I.P. extensors are fair to good.

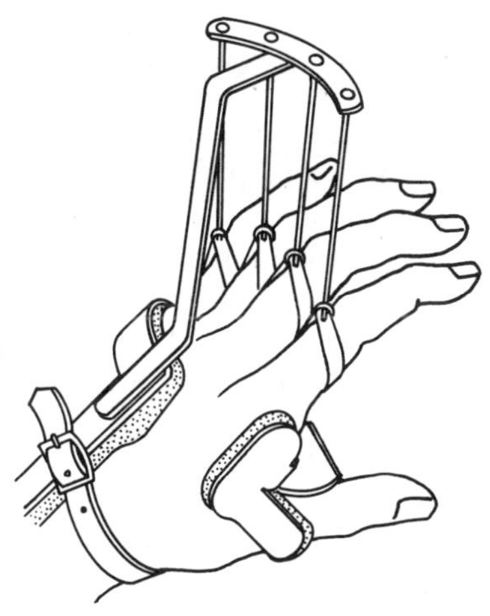

Figure 54. M.P. Extension Assist

Adjustable Metacarpophalangeal Control

The adjustable metacarpophalangeal control (Figure 55) is the same mechanically as the adjustable M.P. flexion control with I.P. extension assist, except the I.P. extension assist is omitted. It is used only with the long opponens hand splint because of the leverage required to maintain the necessary pressure on the M.P. control. It is used in cases of contracture of the extensor digitorum communis to maintain the gains made by manual stretching by the therapist.

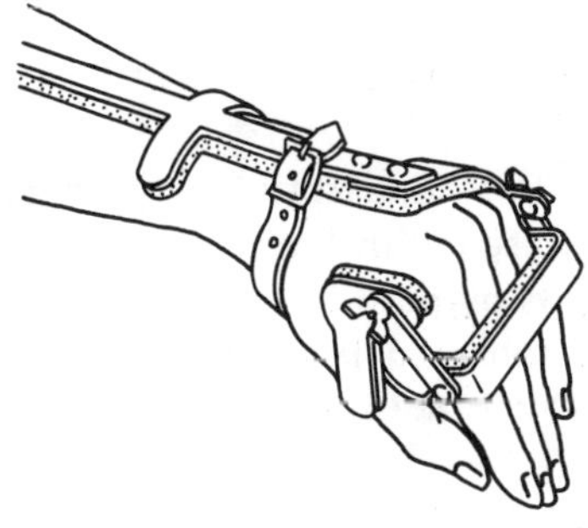

Figure 55. Adjustable M.P. Flexion Control

Action Wrist with Dorsiflexion Assist

The action wrist with dorsiflexion assist (Figure 56) is a modification of the long opponens hand splint designed to assist wrist extensors when the latter are in the poor to fair range. A hinge (1) is inserted into the forearm piece directly over the wrist joint which allows the wrist to flex. A lever (2) on the part of the splint distal to the hinge is connected with a rubber band (3) to a hook (4) on the proximal end of the forearm piece in such a way that the rubber band is stretched when the wrist is flexed. The tension of the rubber band then assists the weakened muscles to dorsiflex the wrist.

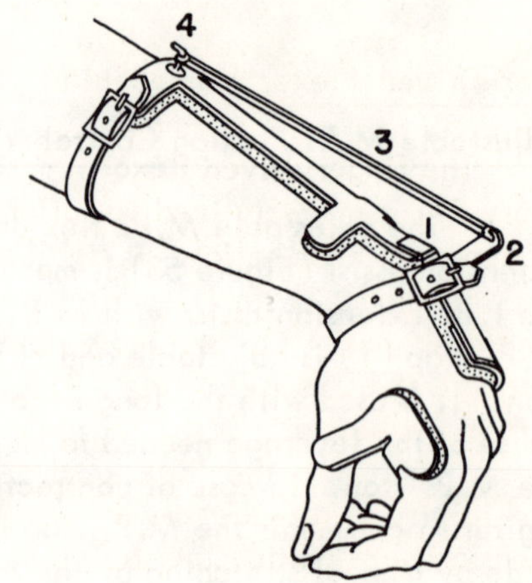

Figure 56. Action Wrist with Dorsiflexion Assist

In some cases it is necessary to add a volar flexion stop (Figure 57) to the action wrist to prevent strong wrist and finger flexors from pulling the wrist into flexion when it is desired to flex the fingers only. The stop is a simple bracket (1) against which the lever rests at the point of maximum flexion desired.

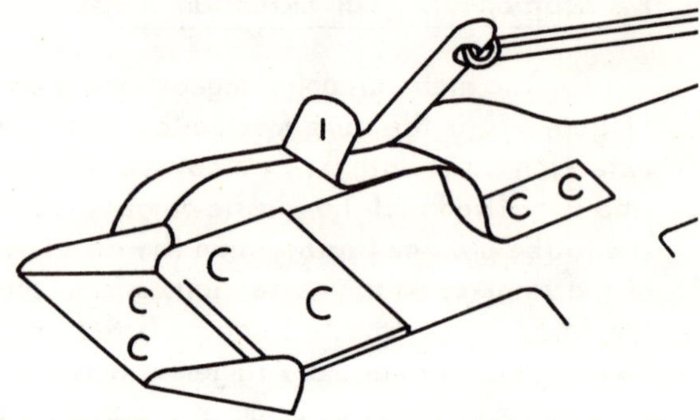

Figure 57. Volar Flexion Stop for Action Wrist

Hand Splint Prop

The hand splint prop (Figure 58) is a bracket used with the long opponens hand splint only. Its purpose is to hold the hand above the bed or table so the weight of the hand, arm, and splint is not borne on the thumb.

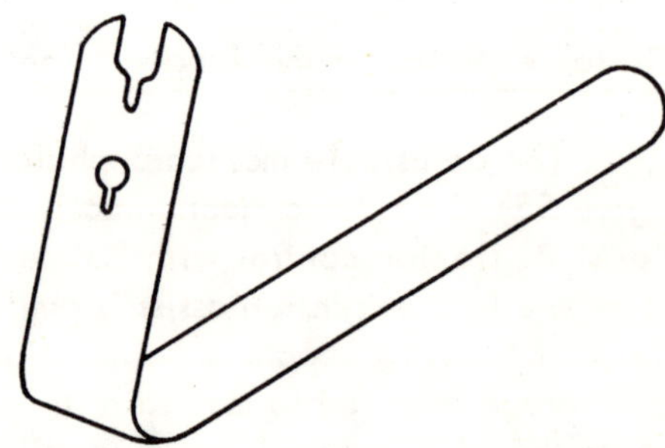

Figure 58. Hand Splint Prop

Finger-Driven Flexor Hinge Splint

The finger driven flexor hinge splint
(Figure 59) is a specialized variation of the
long opponens hand splint. It stabilizes the
wrist, the thumb is stabilized in opposition
by a thumb post (1), the index and long
fingers are stabilized in fitted finger pieces
(2) that are hinged (3) to the palmar
piece in line with the index and middle
M.P. joints, to provide "three jaw chuck"
prehension between the thumb, index, and
long fingers.   The finger section is hinged
and the ulnar side of the splint is open to
make it easier for the patient to remove and
put on the splint without help. To make it
easier to fasten the straps, loops are provided into which a finger of the opposite
hand may be inserted, then simple pressure of the hole in the strap over the head
of the attaching stud completes the job.
(There are five variations of the flexor hinge
splint: Finger driven, flexion or extension
assist; shoulder-driven tenodesis; wrist driven;
and artificial muscle powered.) The finger
driven flexor hinge splint transfers the force of a muscle acting on one joint to another having
no muscle able to deliver adequate force. For example, in the splint illustrated in Figure 59,
the index and middle fingers are joined as one by the finger piece. If the index finger had
fair to good flexion and zero to poor extension, and the middle finger had zero to poor flexion
and fair to good extension, the index finger could flex both fingers and the middle finger could
extend both fingers, thus providing a functional "three jaw chuck" grasp. Sometimes the ring
and/or little fingers have power that can be used by extending a bar from the finger piece over
or under them. The finger driven flexor hinge splint may be used when the wrist extensors and
flexors and thumb muscles are zero to poor, but some active finger flexors and extensors in the
range of fair to good remain. This splint is not indicated when the hand or wrist are spastic.

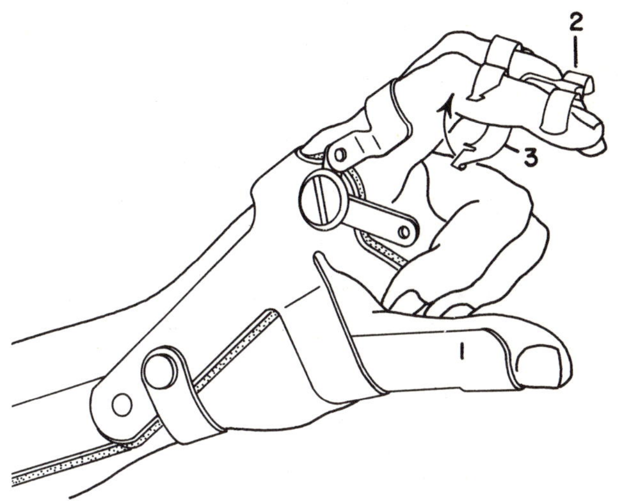

Figure 59. Finger Driven Flexor
Hinge Splint

Flexor Hinge Splint with Flexion or Extension Assist

The flexor hinge splint with flexion assist (Figure 60) or extension assist (Figure 61) is the same in all respects to the finger driven flexor hinge splint, except for the addition of the spring assist. The flexor hinge splint with flexion assist may be used with wrist flexion and extension zero to poor, M.P. flexion zero to poor, and M.P. extensors fair to good. The active extensors extend the index and middle finger M.P. joints, and at the same time stretch the assist spring (1). When the extensors are relaxed, the spring flexes the M.P. joints using the energy stored in it by the extensors. The spring pulls on a lever (2) which is part of the finger piece. The chief disadvantage of this arrangement is that the amount of pinch is limited to the amount of energy the extensors can store in the spring during extension, rarely more than 8 to 16 ounces. However, there are instances where this amount of prehension will serve most of the patient's needs, in which case the comparatively simple flexion assist is quite satisfactory.

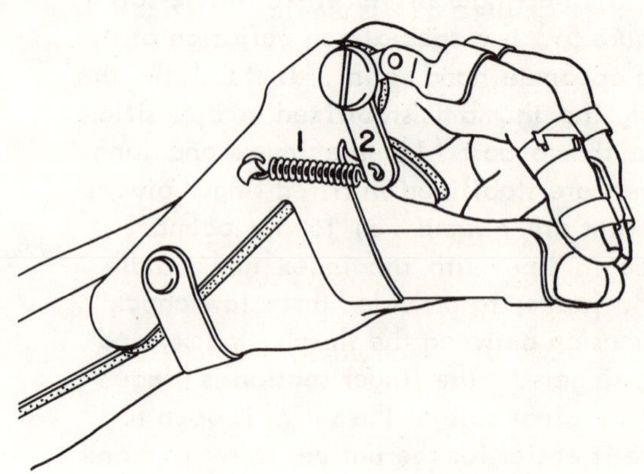

Figure 60. Flexor Hinge Splint
with Flexion Assist

The flexor hinge splint with extension assist (Figure 61) may be used when wrist flexion and extension are zero to poor, M.P. extension zero to poor, and the strength of one or more flexors is fair to good and free of spasticity. The extension assist spring (1) is stretched when the M.P.'s are flexed to enable the fingers to grasp an object. Some of the flexion power is absorbed by the spring, and is then released to extend the M.P.'s when the flexors are relaxed. The spring is connected to the finger piece by a special wire pivoted lever (2).

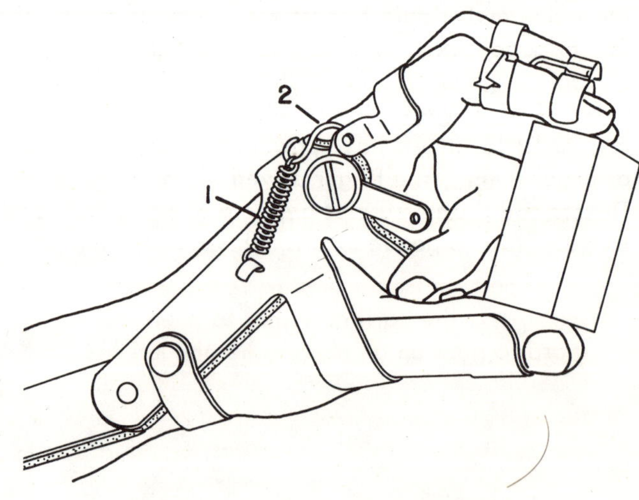

Figure 61. Flexor Hinge Splint
with Extension Assist

Shoulder-Driven Flexor Hinge Splint

The shoulder-driven flexor hinge splint (Figure 62) is similar to the one with spring flexion assist for prehension, except it is designed for patients who have neither flexors nor extensors stronger than zero to poor. This makes it necessary to extend the M.P.'s by some other means. If the patient has good back and shoulder muscles and can abduct his scapulae, a simple "butterfly" or shoulder-to-shoulder double loop harness (1) made of 1 inch dacron tape will provide enough force and excursion through a Bowden cable (2) to extend the M.P.'s and stretch the flexor spring (3). The Bowden cable housing, a stainless steel flexible tube, is anchored to one of the shoulder loops by a threaded crossbar held in place by a leather retainer (4) sewed to the dacron of the loop. The stainless steel cable slides through this housing, and is anchored to the other shoulder loop by a stainless steel hanger (5) which is buckled to the dacron tape. The buckle provides a means for adjusting the tension on the cable. The cable passes down through the flexible housing to the pressure relief control (6), which in turn is connected to the finger piece operating lever (7) on the splint. The flexor spring (3) is attached to the other end of this lever, which is pivoted in the middle.

When the system is relaxed the flexor spring keeps the M.P.'s flexed so the thumb, index, and middle fingers can maintain grasp in proportion to the tension of the flexor spring. If the patient wishes to open this grasp to pick up an object, he abducts his scapulae which in effect increases the distance between the housing crossbar (4) and the cable hanger (5) which causes the cable (9) to be withdrawn from the housing an amount approximately equal to the amount of scapular abduction. The distal end of the housing is secured to the forearm piece of the splint by another crossbar and leather (8). With all the slack adjusted out of the

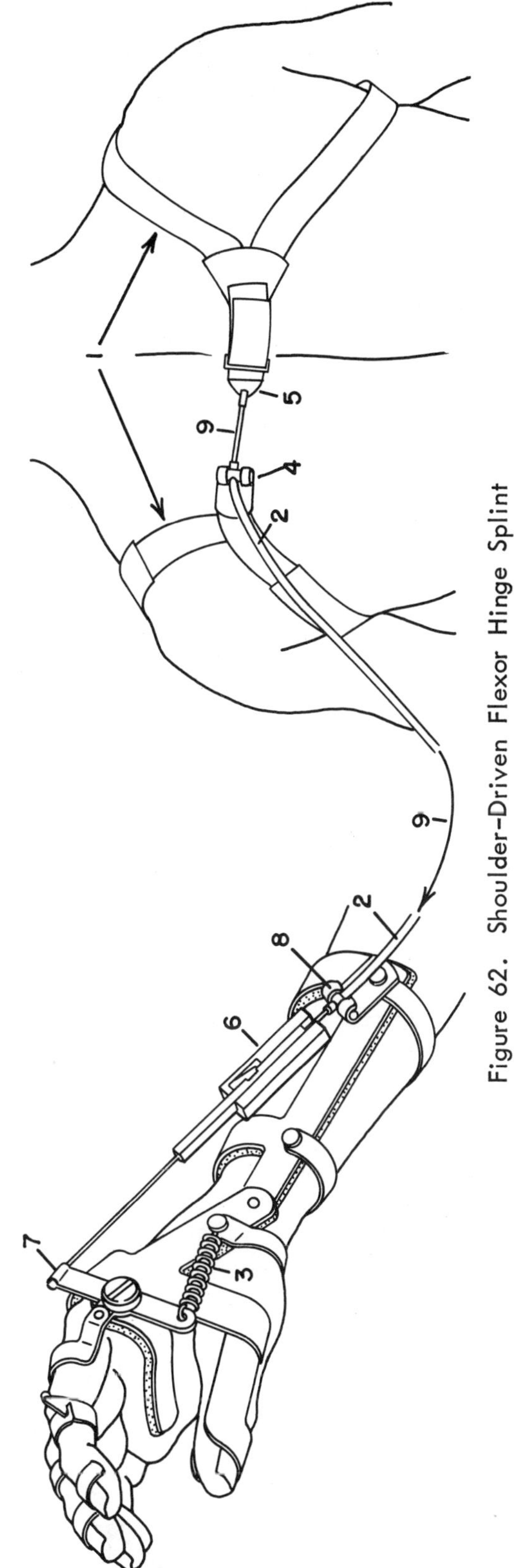

Figure 62. Shoulder-Driven Flexor Hinge Splint

cable and with both ends of the housing held fast, all of the excursion of the cable inside will be transmitted to the pressure relief control (6) regardless of the positions into which the patient may put his hand, which is the advantage of the Bowden type cable control. When the pull of the cable is transmitted to the lever (7) the M.P.'s are extended and the spring (3) is stretched.

At first shoulder-driven flexor hinge splints were fitted without the pressure relief control, but it was quickly discovered that the patients spent much of their time in the relaxed position. The flexor spring forced the pads of the thumb, index, and middle finger together with considerable pressure, and over a period of time these pads became too tender and sore to use the splint. The pressure relief control was developed to overcome this problem. It is a simple alternator locking mechanism that enables the patient to lock the thumb and fingers slightly apart. A slight pull on the cable by abducting the scapulae releases the lock. A detailed drawing illustrating the pressure relief control mechanism is shown in Figure 63.

The shoulder-driven flexor hinge splint has been used with success by slightly spastic hemiplegics with motor control problems as well as by other types of paralytics.

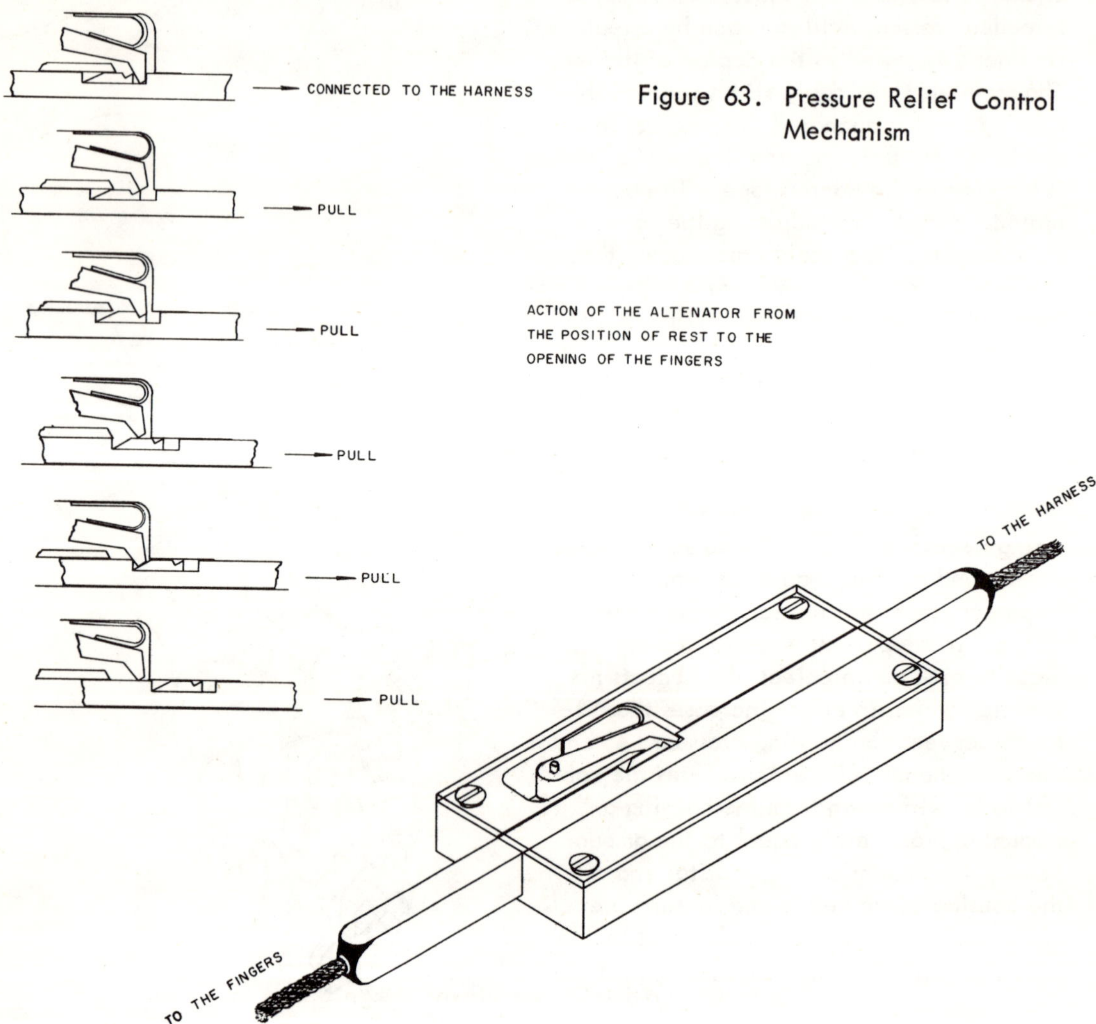

Figure 63. Pressure Relief Control Mechanism

Wrist-Driven Flexor Hinge Splint

The wrist-driven flexor hinge splint (Figure 64) was developed primarily to help quadri-plegics, as it is not unusual to find many of them with flail hands but fair to good wrist flexors and/or extensors. The purpose of the wrist-driven flexor hinge splint is to translate this wrist motion into active prehension. The thumb post and finger pieces are fitted in the same way as with the other flexor hinge splints, but the wrist pivot (1) is set up so the splint can flex and extend with the patient's wrist joint. A fixed lever (2) that is part of the forearm piece (3) is connected by a link (4) to the finger piece operating lever (5). The finger piece and its operating lever are free to rotate on the finger piece pivot screw (6). In drawing "A" the wrist is flexing, and as it does so the finger piece operating lever (5) is forced to rotate counter-clockwise on its pivot (6) since it is connected to the fixed lever (2) by the link (4). In drawing "B" the wrist is extending, and as the hand moves up the operating lever is rotated clockwise, flexing the finger piece, and bringing the thumb and fingers into prehension.

In cases where either wrist flexion or extension but not both are in the fair to good range, a spring assist can be added to make up the deficiency.

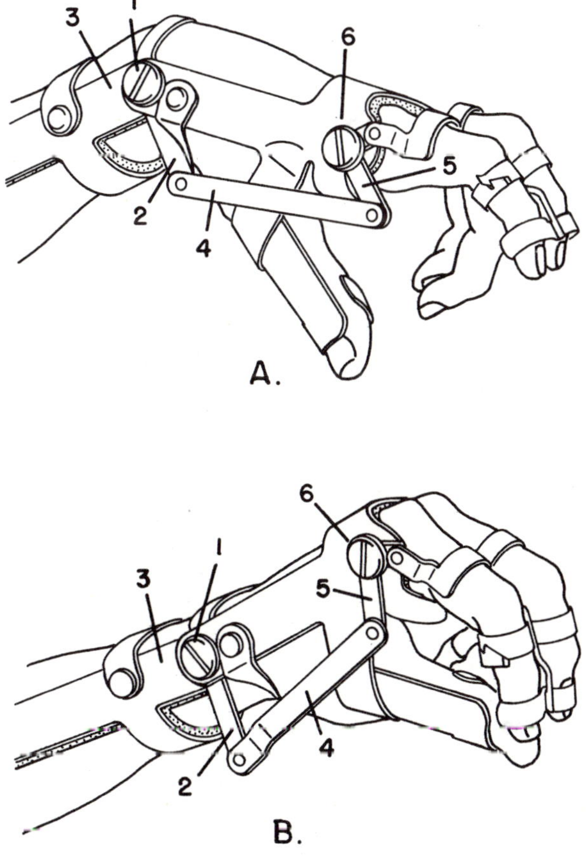

Figure 64. Wrist-Driven Flexor Hinge Splint

Artificial Muscle Power Flexor Hinge Splint

All of the flexor hinge splints discussed up to now were designed to be operated by the patient's own muscle power. Some patients are found, however, who have no source of power on or in their bodies that can be used to operate the splint. Some high spinal cord injury quadriplegics are so nearly completely paralyzed that it is often difficult to locate a muscle capable of exerting an ounce or two of force to operate a microswitch.

The artificial muscle power flexor hinge splint (Figure 65) was designed to help these seriously involved patients by utilizing a source of outside power, compressed carbon dioxide gas.

The artificial muscle (1) is a rubber tube covered by a nylon mesh and sealed at one end with a closed ferrule, and at the other end with a similar ferrule with an opening to take a fitting for a plastic tube. When the artificial muscle is at rest as in "A" it is elongated and the fingers of the splint are open. When carbon dioxide gas under pressure is admitted to the artificial muscle it expands and shortens, and in doing so pulls on the finger piece operating lever (2), flexing the M.P.'s and closing the grasp. At the same time the extension spring (3) is extended. When the gas is exhausted from the artificial muscle, the extension spring extends the M.P.'s and opens the grasp.

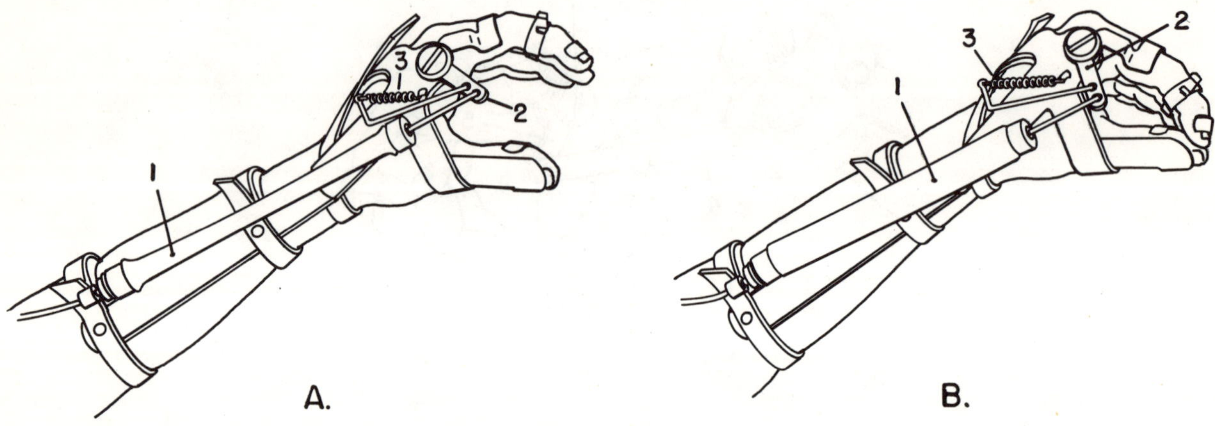

Figure 65. Artificial Muscle Power Flexor Hinge Splint

The compressed carbon dioxide gas is stored in a small steel cylinder (Figure 66) that can be easily carried in a pocket or small pouch. The gas in the cylinder is under 850 pounds per square inch pressure at 70 degrees Fahrenheit, so a regulator is provided to reduce the pressure to a maximum of 90 psi.

Figure 66. CO_2 Cylinder and Regulator

Several types of control valves are available to enable the patient to operate the artificial muscle. Some patients may have enough scapular ab-duction force to operate a harness slide valve (Figure 67) but not enough to operate the splint through a cable. The valve is mounted between the two loops of the harness so scapular abduc-tion opens it, relaxing closes it, open-ing it half way exhausts the gas from the muscle. The lever valve (Figure 68) can be operated by the foot, toe, knee, or chin. Several lever valves can be set up in tandem (Figure 69) where several artificial muscles are needed. A microswitch finger valve (Figure 70) can be mounted on the finger piece of the splint and oper-ated by the ring or little finger.

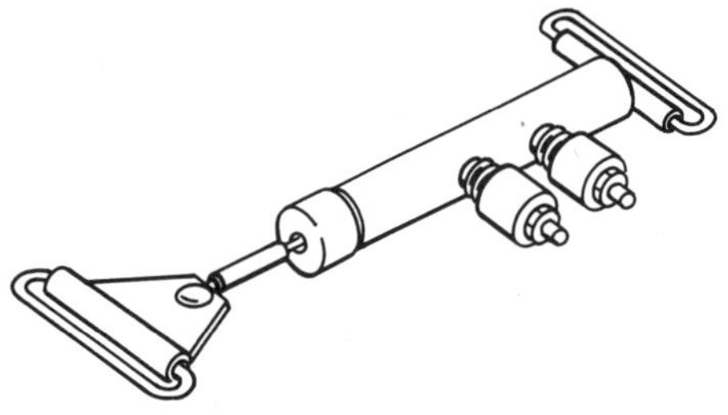

Figure 67. Harness Slide Valve

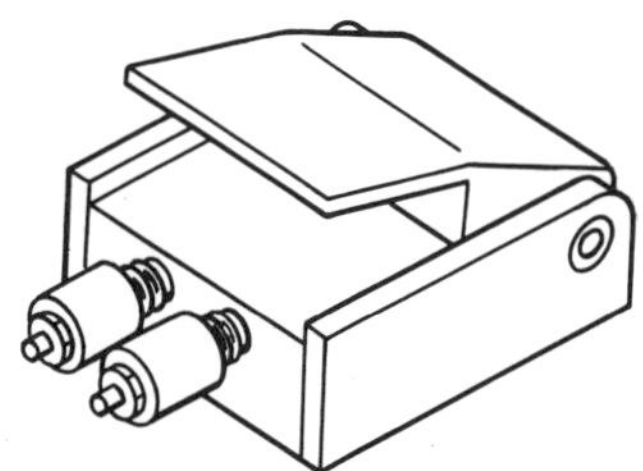

Figure 68. Lever Valve

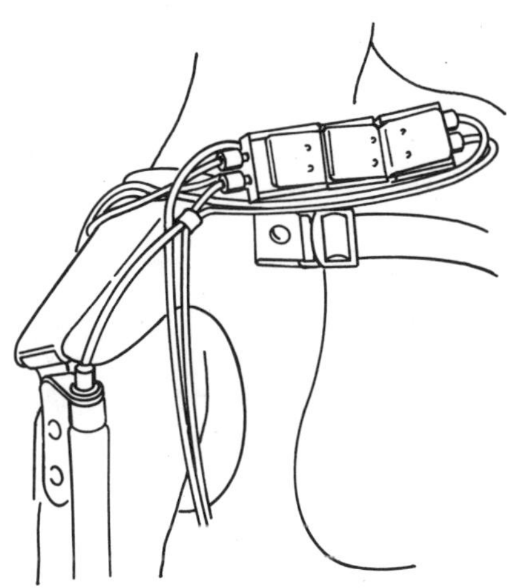

Figure 69. Chin Control Assembly

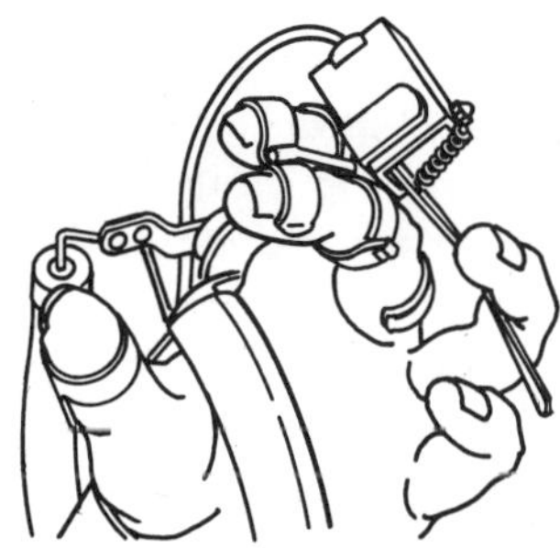

Figure 70. Finger Valve

A cylinder to provide piston power wrist flexion (Figure 71) can be added. The problem in broadening the application of outside power is that of control sites.

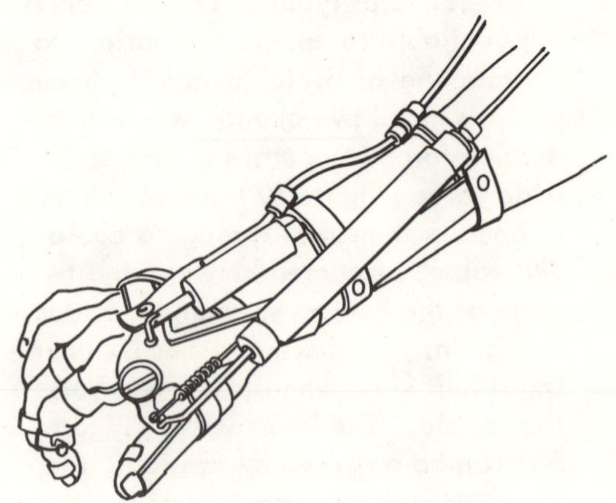

Figure 71. Artificial Muscle Power Flexor Hinge Splint with Piston Power Wrist Flexion

Electric Power Flexor Hinge Splint

With rechargeable light weight nickel cadmium batteries now available, it is practical to drive a flexor hinge splint with a small electric motor. The major components of the system are shown in the illustration (Figure 72). The 12 volt direct current reversible electric motor drives a pulley through reduction gears. The cable is wound on the pulley, and the cable housing is anchored to the gear cover. Running the motor one direction winds in the cable, reversing it unwinds the cable.

The motor is controlled through a double micro-switch (Figure 73). When the paddle is pressed part way down the ball stud attached to it closes the first switch. Pressing it all the way down opens the first switch and closes the second.

The cable runs through a sheave or pulley attached to the hinge joint that drives the finger piece on the splint, and is then anchored to the spring hook on the splint center bar. The action of the sheave gives the motor a 2 to 1 mechanical advantage, and slows down the speed of opening and closing the fingers, giving the patient better control.

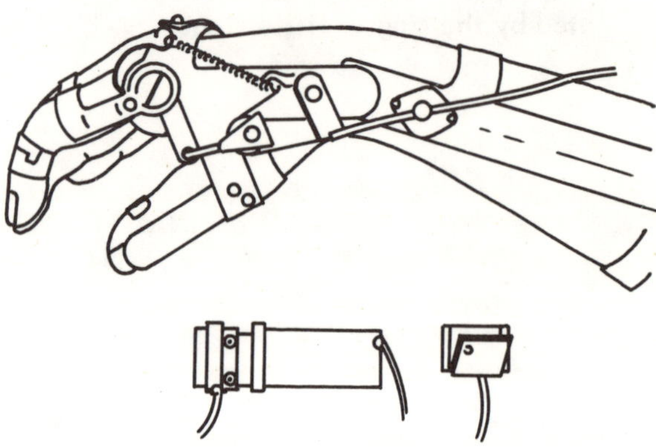

Figure 72. Flexor Hinge Splint with Electric Motor Flexion Control. The small electric motor and control switch are also illustrated.

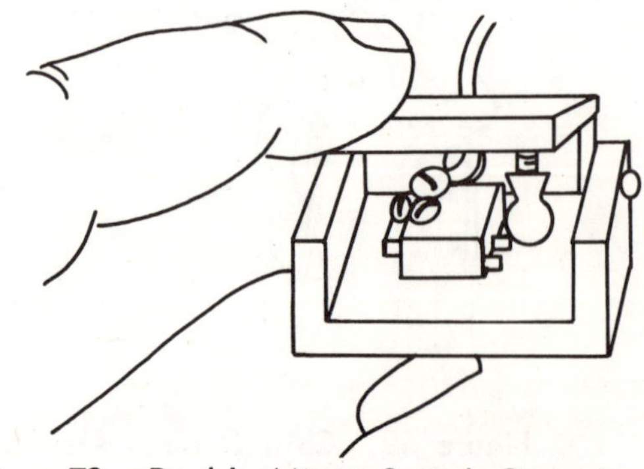

Figure 73. Double Micro-Switch Control. As the lever is depressed it closes one micro-switch, further depression opens it and closes a second switch.

BASIC SKILLS NEEDED IN UPPER EXTREMITIES ORTHOTICS FABRICATION

Introduction

There are a number of frequently used operations in fabricating and fitting hand splints and arm braces that are not explained in the instructions in this text-book because it has been assumed that they are already known to experienced orthotists. This book was primarily designed for use in extension classes for experienced orthotists because upper extremities orthotics has usually been considered a specialized branch of the field, and not one commonly learned by the orthotist during his training period. However, as more beginners are now taking training courses in upper extremities orthotics, there is increasing need for some guidance in basic skills. In this section a few of the more important and useful operations will be explained for the benefit of the beginner.

1. Cutting metal with a hacksaw, (Figure 74). In splint and brace work the hacksaw is often used to cut sheet aluminum and stainless steel, as well as steel rods. Do not use a blade with too few teeth per inch when cutting thin material or the blade will hang up on the material. Use a blade with 32 teeth per inch so that at least two teeth contact the work at all times. Be sure the teeth point away from you when sawing, and use long, even strokes, at a rate of about 40 strokes per minute. Do not press the saw blade to the material on the return stroke. Clamp material to be cut in a smooth jaw vise as close to the area of cut as possible.

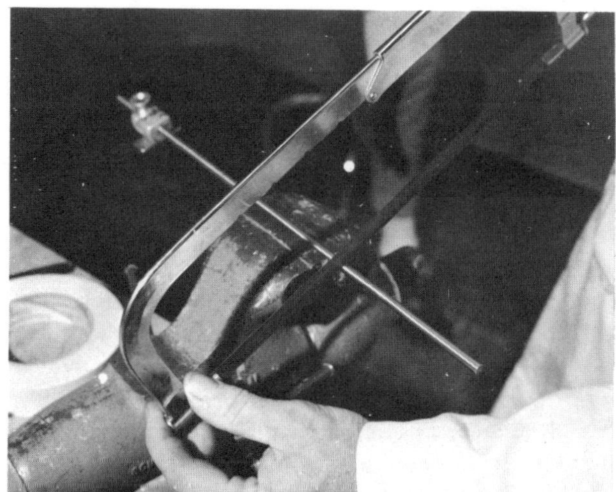

Figure 74. Cutting Stainless Steel Rod with a Hacksaw

2. Drilling holes in metal, (Figure 75).
It is often necessary to drill holes in
aluminum and stainless steel for rivets
and screws. Always center punch the
exact location of the hole center so the
drill will not "wander". Be sure the
drill is the right size, is sharp, and is
clamped tightly in the drill press
chuck. Drill aluminum with a drill
speed of around 1200 R.P.M. Hold
the material with a clamp, and press
lightly and evenly on the feed lever,
particularly as the drill goes through.
There is a tendency for the drill to
"grab" and spin the work or for the
work to climb the drill. Drill stainless steel at around 800 R.P.M.,
and lubricate the drill with "Petrochem Microfinish" (Rutland Tool
Co., 949 S. 5th St., Montebello,

Figure 75. Drilling a Hole in a Splint
Part with a Drill Press

Calif.). When the drill starts cutting keep an even, steady pressure and never let up until
the hole is completed. If pressure on the drill is released and then reapplied, it will dull
the cutting edge of the drill causing it to overheat and be ruined. Always de-burr the hole
with a hand-held countersink or a larger size drill.

3. Tapping, (Figure 76). Taps are used for
cutting threads inside of holes, whereas dies
are used for cutting threads on the outside
of rods. The tap is clamped in the chuck of
a tap wrench and is carefully screwed into
the hole at right angles to the surface. The
tap should be lubricated with "Petrochem
Microfinish", and screwed in until the hole
is threaded. Care must be used to never
force a tap, as they break off easily and the
broken piece is often difficult to remove.
This is particularly true with the small size
taps, such as 8-32.

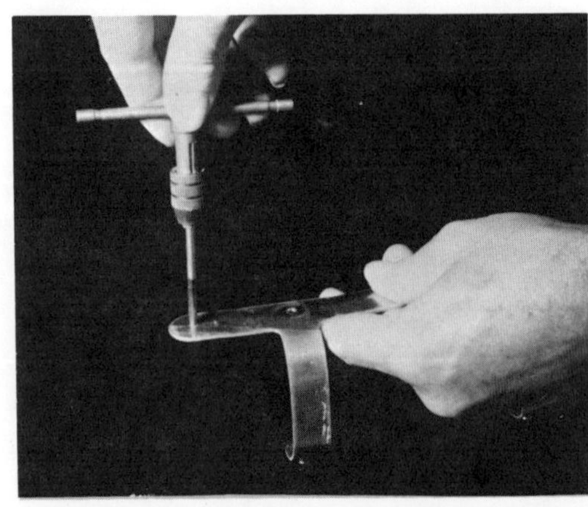

Figure 76. Tapping Threads in a Hole
in a Splint Part

It is very important to use the correct size drill for the size tap to be used. The tap and drill sizes used in splints and braces are:

Tap	Drill
6-32	35
8-32	29
10-32	21

4. Riveting, (Figure 77). Stainless steel 3/16 x 3/32 inch rivets are used exclusively in assembling hand splints. No. 40 holes are drilled for this type rivet, the rivet is inserted, and the excess length is cut off with diagonals. About 1/16 inch of rivet is left for heading with the hammer. A "T" riveting bar is essential for doing this and may be obtained from Jaeco, 7939 Chatfield Ave., Whittier, Calif. The base of the bar is held in a vise, with the "T" at right angles to the vise opening. The head of the rivet is centered over one of the arms, and the end is flattened with a hammer.

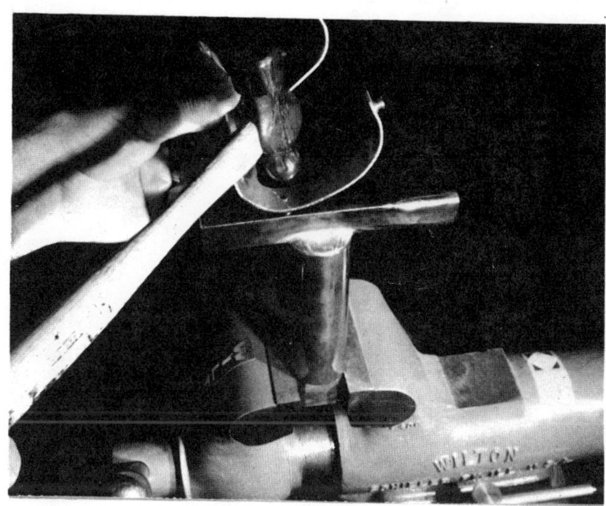

Figure 77. Using the "T" Riveting Bar

5. Bending and shaping metal. The special bending fork (Figure 78), available from Jaeco, is used for shaping sheet metal extensions into various contours.

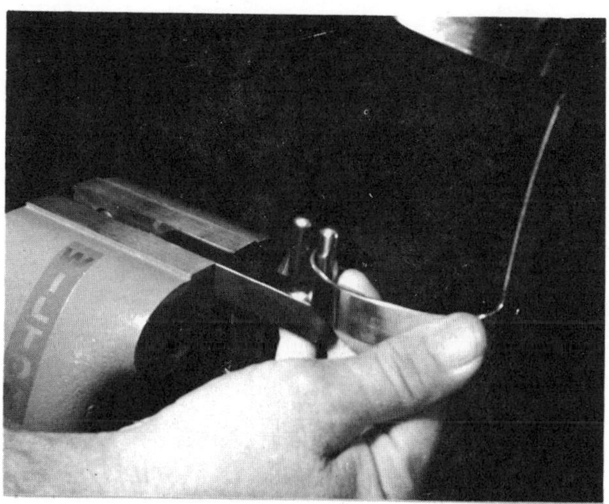

Figure 78. Using the Special Bending Fork

The lead forming block and mandrel (Figure 79) are used for forming a concave surface in a sheet metal part. Radial extensions on hand splints should be made concave in this manner to fit the contour of the hand.

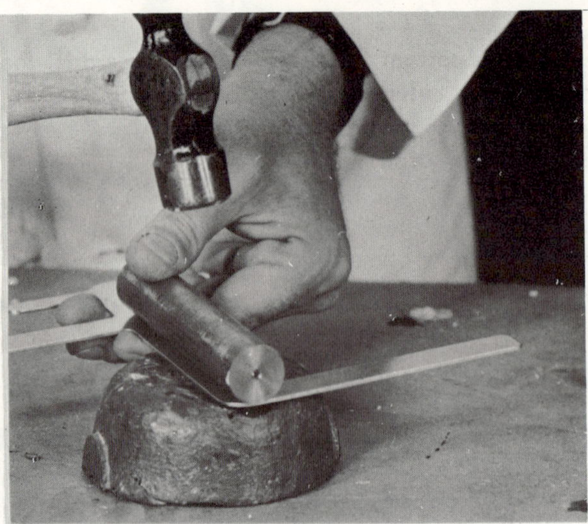

Figure 79. Lead Forming Block and Mandrel Used to Form Concave Surface in Splint Part

Round nose pliers (Figure 80) are very useful in shaping rings, loops in wire, and the like. The rounded jaws of the pliers shape the metal without galling the surface, and are much better for this purpose than ordinary long nose pliers.

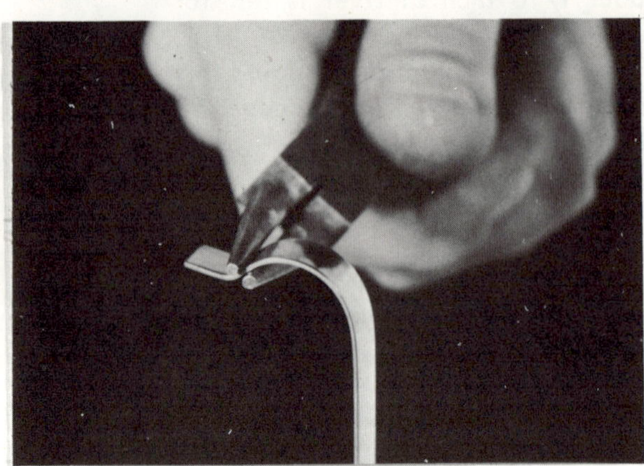

Figure 80. Round Nose Pliers are Useful for Shaping Rings

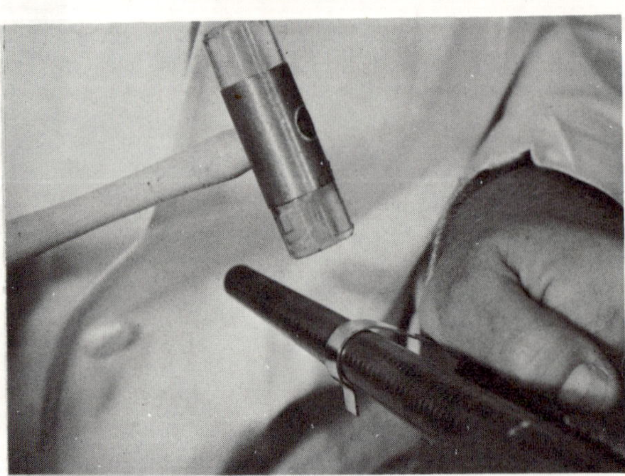

Figure 81. The Ring Mandrel Used to Form Rings to Correct Size

The ring mandrel and plastic tip hammer (Figure 81) are useful for finishing shaping rings and finger pieces to exact size. The part is shaped over the portion of the ring mandrel that corresponds to the size wanted.

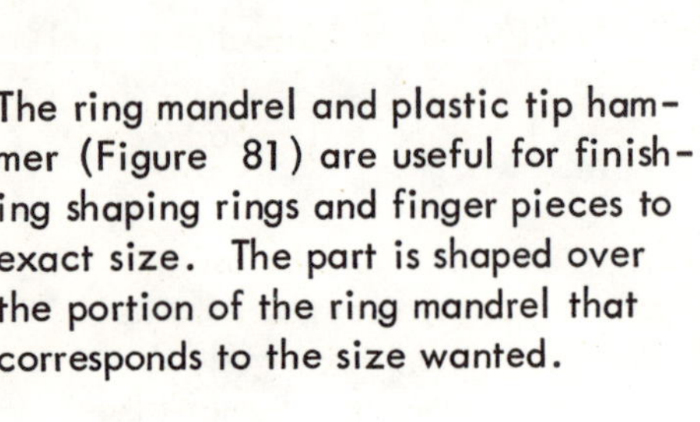

6. Silver soldering, or low temperature brazing, is used in assembling stainless steel parts for hand splints, and is better than ordinary solder because it is much stronger. The parts must be thoroughly cleaned, then clamped or held in position. They are then given a thorough coating with "Handy Flux" (Handy and Harmon, 850 3rd Ave., New York) over all surfaces, not just the area to be brazed. This will protect the stainless steel from being stained and will save much buffing time later. A welding or brazing torch is used to heat the parts. When the flux becomes liquid the silver alloy (Easy-Flo 45 .062, Handy and Harmon) is applied, (Figure 82). When the proper temperature is reached the silver alloy will flow smoothly and join the parts, making a strong joint. The parts should then be thoroughly washed in hot water to remove all the flux.

Figure 82 . Silver Soldering Finger Pieces

7. Alumaweld soldering. In some situations it is necessary to solder steel screws to an aluminum splint. A special alloy, "Alumaweld All-Metal Solder" and "Alumaweld Flux", manufactured by Johnson Manufacturing Company, Mt. Vernon, Iowa, are used for this purpose. The parts to be soldered must be clean, and free of all grease and moisture. The aluminum must be sanded until bright to remove all oxidation. Do not touch with your fingers after sanding, as the oil from the skin will adhere to the metal and prevent a good bond. After cleaning and sanding both parts to be bonded, place in position with pliers to avoid touching with the fingers. Stir the Alumaweld Flux until thoroughly mixed and apply enough to bridge all around the joint to be bonded. Sand all four sides and the end of the Alumaweld metal solder to remove the oxidation. Do not touch with your fingers. With clean diagonal cutters, cut a piece of the solder large enough so that when melted it will flow over and bond the parts to be welded.   Place the solder in the flux previously applied to the joint. Wherever possible position the parts to be joined in such a way that the flux and solder are placed on the upper side of the parts to be bonded. Apply the brazing torch flame to the under side of the parts, holding the tip about three inches below the parts. When the proper heat is reached, the flux will give off a dense white smoke. As soon as dense smoking starts remove the torch.   The Alumaweld will flow out on the material and make a smooth sound bond.

When using a welding torch to heat Alumaweld, use a smaller tip than would be used for welding metal of the size being bonded. In addition to the smaller tip, use a soft flame achieved by reducing the oxygen supply until the cone of the flame is approximately 1 inch long. Hold the torch about 3 inches away from the joint to be bonded. Remove the flame as soon as the dense white smoke appears.

8. Rounding corners, smoothing edges, removing burrs. Splint and brace parts come in contact with the patient's skin, and if they are even slightly sharp or rough, the skin will be irritated and painful sores may result. Damaging a patient's skin in this way is inexcusable. In addition to the unnecessary additional treatment that will be required to clear up the damage, the patient's mental attitude is likely to take a strong turn against any kind of splints and braces, making it very difficult to get him to wear them in the future. The 1 inch de-burring machine (Figure 83), a small belt sander, is the most efficient device for smoothing splint and brace parts. It will

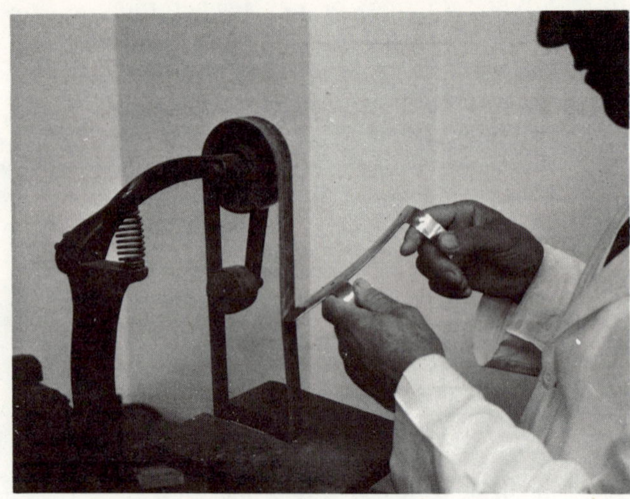

Figure 83. De-burring Machine

quickly round corners, smooth rough edges, and even remove undesirable surface scratches from both aluminum and stainless steel parts. It is strongly recommended that all parts be de-burred before starting fabrication and fitting, as the sharp edges may cut the orthotist's hands as well as the patient's hands during fitting. When the parts are shaped, a final smoothing and rounding should be done prior to buffing.

9. Buffing: Applying the composition to the buffing wheel.

On a new wheel: Be sure the wheel is revolving toward you and has attained full speed. Then pass tube of composition lightly across wheel face in front of, and a little below spindle, until face is slightly coated. Now hold a clean piece of old metal lightly against wheel face for a few seconds to spread coating. Do this several times until face is uniformly coated. After face has been adequately charged with composition, you are ready to buff. It will be necessary to apply more composition from time to time as you buff, and the best way to do this is with a wiping motion, being careful not to overload the wheel.

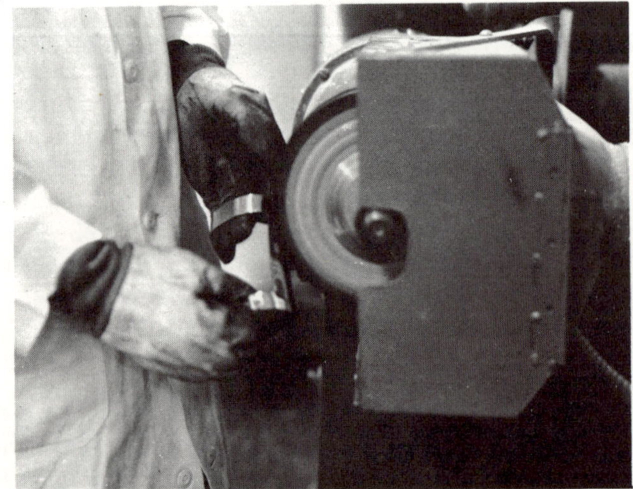

On an old wheel: First clean the face of your wheel as well as possible. Thus you will avoid marring work with coarse cutting compositions previously used and also will be assured of a more uniform coating. The old face may be removed by holding the edge of a file or other tool against the wheel as it revolves. When clean, proceed as with a new wheel. Never use a wheel for coloring which has previously been used for coarse buffing or cutting down. Particles of grit remaining in the face may scratch your final finish. You will save time and avoid costly mistakes by using a separate wheel for each type of composition.

Be sure your wheel is revolving toward you and at a high speed. Coat your wheel with the right composition and let it run for a few seconds so the composition will settle on the buff. (Never put the composition on the article to be buffed.) Now grasp your work firmly and apply it lightly against face of wheel, being sure to keep it below the center of the wheel, otherwise the buff might catch in your work and pull it out of your hands. Keep your work constantly in motion, always removing it from the wheel with a slanting downward stroke. This will blend buffing marks and help avoid spotty or streaked results. Never allow wheel to strike upper edge of work as work may be torn from your hands. Wipe additional composition across wheel face as needed, but do not overload. Do not use too much pressure against the wheel--let the composition and the wheel do the work. Clean residue of composition from work when changing from cutting to coloring composition to prevent coarse grits from becoming imbedded in coloring wheel, thus scratching your final finish. When finished, wipe the surface buffed with a soft flannel cloth dipped in powdered whiting to remove all traces of composition. A little practice will enable you to obtain the results you want.

A few suggestions for safe and efficient buffing:

1. Hold your work firmly at all times. Cotton work gloves will protect your hands and keep them clean.

2. Use minimum pressure against wheel. Let wheel and composition do the work.

3. Always hold work below center of wheel.

4. Never take your eyes OFF work for an instant.

5. Use caution when buffing an article with sharp projections or openwork, such as jewelry. It might catch in wheel and be torn out of your hands.

6. If wheel slows down, you are pressing too hard or your motor is too light for your wheel.

7. Inspect plated work frequently, as thin plating is easily buffed off.

8. Use great care buffing plastics. Never permit them to overheat; it causes scoring and discoloration.

9. Do not wear a necktie. It may catch in machinery and cause an accident.

10. Do not lean over wheel. Composition may be thrown into your face. Wear goggles if possible.

11. Never allow your body or clothing to touch a revolving buff.

12. Work in a well-ventilated place to avoid dust accumulation.

10. Laboratory safety and housekeeping: The orthotics laboratory can be a safe and pleasant place to work if each student will be personally responsible for keeping his own work station clean and orderly and for cooperating with his fellow students in keeping the equipment used by all in good condition, ready for use, and in its proper storage place when not being used. The one sure sign of the good technician is a clean and orderly work place. Dirt and disorder are the greatest causes of work slow-downs. Time spent in hunting for a tool or part lost in a tangle of debris on the workbench is time that can never be recovered. Look at the way in which the efficient surgeon arranges his work environment and instruments so he can perform surgery with a minimum of delay and waste motion. Imitate his techniques for they are the mark of the true professional. On the other hand, an orthotist with a soiled smock, disorderly work station, dirty floor, tools in poor condition, and equipment battered and spattered with plaster is judged by the observer as being an amateur who not only takes little pride in his work but probably knows very little about it.

Some general safety precautions that should be learned and observed:

a. The chief hazards connected with hand tools are:

--Being struck by the tool you are using because you used the wrong tool for the job, using the right tool in the wrong way, or using a defective tool. Dull cutting tools are apt to slip instead of cutting.

--Being hit by chips or dust from material or tool. The eyes are the most vulnerable to injury from these hazards, and protective goggles should be worn when doing work when flying material is unavoidable, as in grinding. Defective tools, such as a mushroomed head on a chisel, are also a source of flying particles, and such tools should always be ground smooth to avoid this difficulty.

--Being struck by a tool slipping from its handle. Hammers and similar tools must be kept tight in their handles.

--Being struck by a tool being used by another student. Stand away from others who are swinging hammers or other tools that might hit you. When you use such a tool, see that no one is close enough to be struck.

--Stumbling over, being cut or hit by a tool left on the floor, a bench, or a stool. Keep tools where they belong. Hang them up or place them in a proper storage facility when not in use.

b. Precautions in using plastics and other toxic materials:

-- Avoid getting plastics and other chemicals on the skin, as they are toxic and may cause a dermatitis or skin rash. If some does get on the skin, wash it off with alcohol as soon as possible, then thoroughly cleanse with Boraxo or other good hand cleaner.

-- Be sure the suction from the overhead fume collecting hood is functioning so that all toxic fumes will be evacuated.

-- Never mix the catalyst and promoter before adding them to a plastic mix, as there is a possibility of explosion if this is done.

-- When mixing plastics, do not put the container on the bench above your face when you add ingredients or stir. This puts your mouth, nose and eyes in the direct path of the rising fumes. Avoid this by holding the container well out in front of you, at eye level or slightly above.

-- Keep all containers capped or closed when not in actual use and return them to a place of safe keeping to prevent breaking or misplacement.

-- Use particular care in handling epoxy resins and foaming plastics as they are unusually toxic, both the materials themselves and the fumes they create.

-- Handle all chemicals, including plaster, with great care to avoid spilling them on the floor, bench, and clothing. If any is spilled, clean it up immediately.

-- Plaster must never be flushed down sink drains as it will plug them up. Allow plaster to harden in the plastic containers, then discard the debris into the waste can.

-- Measure quantities of chemicals, such as resins, promoters, and catalysts with care and accuracy. This is no place for guesswork.

-- If any chemicals of any kind are splashed into the eyes, get <u>immediate</u> medical treatment. In the case of acids, such as soldering flux, emergency flushing with water is recommended, followed by medical treatment as soon as possible.

c. Precautions in using power tools:

-- Be sure hand power tools have a ground terminal on the plug before using.

-- Grip work firmly when grinding, polishing, de-burring, or cutting on the bandsaw.

-- Never talk to or bump into anyone working at a power tool.

-- Clear away all debris from the power tool before leaving it.

-- Always turn off a power tool before leaving it. In the case of a bandsaw, use the brake to stop the blade.

-- Report any power tool that appears to be dangerous, defective, or making unusual noises.

TOOLS AND EQUIPMENT FOR UPPER EXTREMITIES ORTHOTICS

1 set Allen Wrenches

1 Bending Fork

1 Brush, Bench

1 Center Punch

5 Yates Clamps

1 Combination Square, 12"

1 Diagonal Cutters, 7"

1 Die and Holder, 6-32, 8-32, 10-32

1 Dividers, Spring 10"

1 Hammer, Ball Peen, 8 oz.

1 Hammer, Riveting, Tinners, 6 oz.

1 Hammer, Plastic Tip, 3/4 inch Tips

1 Hacksaw

1 Rotary Leather Punch

1 Pliers, Round Nose

1 Pliers, Combination

1 Pliers, Long Nose

1 Pliers, Vise Grip

1 set Pin Punches

1 Rivet "T" Bar

1 Rule, 6', 1/2" Flexible

1 6' Tape Measure

1 Screwdriver, 4"

1 Screwdriver, 6"

1 Shears, Wiss #28

1 Shears, Wiss, Metalmaster

1 P.V.A. Bag Jig

1 P.V.A. Bag Sealing Iron

1 Scale, 1000 gram capacity

1 Cake Blue Carpenter's Chalk

1 Skin Marking Pencil

1 Sharpening Stone

1 Tee Tap Wrench

3 Taps, 6-32, 8-32, 10-32

1 Torch Liter

1 Wrench, Crescent 6"

1 Countersink, 80°, high speed

1 Round Sureform

1 Flat Sureform

1 Staple Gun

1 Solder Gun

1 Fractional Drill Set, high speed

1 Number Drill Set, high speed

1 Heat Gun

1 Plumber's Torch Outfit

1 Bottle Lloyd's Stainless Steel Soldering Flux

1 Jar "Handy Flux" for Silver Solder

1 Jar Rubber Cement

1 Set Ring Gauges

1 Ring Mandrel

1 Knife

1 Scratch Awl

1 File, 8" Flat Mill

1 File, 4" Round

1 Lead Forming Block

1 Mandrel, 3/4" x 8" Cold Rolled Steel

1 Machinist Bench Vise, 4"

1 Devil Level Goniometer

1 Bandsaw

1 Bench Grinder

1 Buffer

1 Belt Sander, 1"

1 Drill Press

Sources of tools, equipment, and materials are:

1. General hardware and tool supply firms for most commonly used tools and equipment, such as files, drill press, pliers, saws, etc.

2. A. J. Hosmer Corp., P.O. Box 152, Santa Clara, Calif.

   Special functional arm brace parts and assemblies.

   Upper extremities prosthetics parts used on functional arm braces, such as cables, ball terminals, and the like.

   P.V.A. bag fabricating equipment. All plastics and related supplies.

3. Orthopedic Supplies Co., Inc., 9126 E. Firestone Blvd., Downey, Calif.

   Parts and kits of parts for hand splints, artificial muscles, electric power systems, ball bearing and suspension feeders.

4. Jaeco Orthopedics, 7939 Chatfield Ave., Whittier, Calif.

   Hand splint and functional arm brace parts, special "T" Bar Anvil and bending fork.

HOW TO TAKE ORTHOTIC MEASUREMENTS FOR FUNCTIONAL HAND SPLINTS

1. Obtain a copy of the form "Orthotics Measurements for Functional Hand Splints" (see sample form on next page) and fill in the patient's name and the date in the appropriate blanks. With a rule or a tape measure, measure the patient's hand width across the M.P. joints. If both hands are to be fitted, measure both, and enter the dimensions in the blanks provided opposite item one on the form.

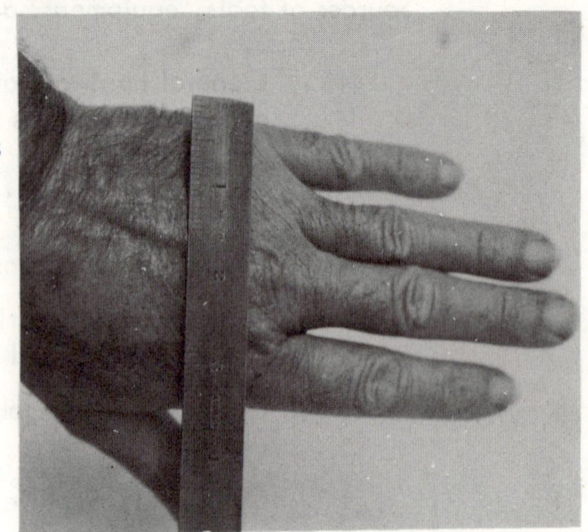

2. Measure the anterior-posterior dimension of the patient's hand at the M.P. joints. The measurement is the "thickness" of the hand at the M.P. joints, and can be taken by placing a scale on the radial aspect of the hand and sighting across it, as in the illustration. Enter the measurement on item two on the form.

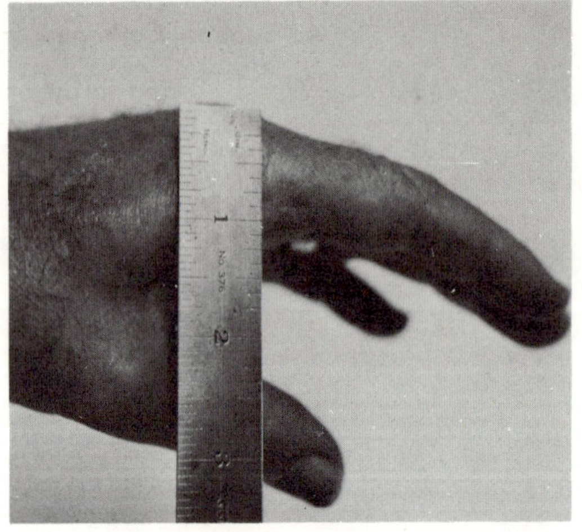

3. Measure the distance between the radial aspect of the hand at the M.P. joint of the index finger and the M.P. joint of the thumb, and record the measurement on item three on the form.

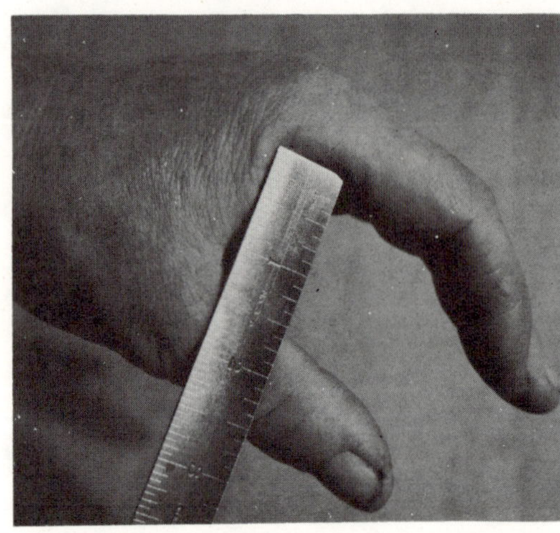

ORTHOTICS MEASUREMENTS FOR FUNCTIONAL HAND SPLINTS

PATIENT'S NAME _______________________________________ DATE _____________

Measurement	Left Hand	Right Hand
1. Width of hand across metacarpophalangeal joints at widest point	_________	_________
2. Anterior-posterior distance of hand at the M.P. joints	_________	_________
3. Distance between palmar aspect of index finger metacarpophalangeal joint and M.P. joint of thumb	_________	_________
4. Thumb circumference proximal to distal interphalangeal joint (use ring gauge)	_________	_________
5. Finger circumference at middle phalanx (use ring gauge)		
Index	_________	_________
Middle	_________	_________
Ring	_________	_________
Little	_________	_________
6. Finger circumference at proximal phalanx (use ring gauge)		
Index	_________	_________
Middle	_________	_________
Ring	_________	_________
Little	_________	_________
7. Distance from ulnar styloid to metacarpophalangeal joint of little finger	_________	_________
8. Circumference of wrist proximal to ulnar styloid	_________	_________
9. Circumference of forearm at midpoint	_________	_________
10. Distance between M.P. joint of the middle finger and the wrist joint.	_________	_________

TABLE OF RING GAUGE DATA

RING SIZE	CIRCUMFERENCE	METAL LENGTH	
		0.036" STOCK	0.125" STOCK
0	1 – 15/32"	1 – 19/32"	1 – 27/32"
0 – 1/2	1 – 17/32"	1 – 5/8 "	1 – 29/32"
1	1 – 9/16"	1 – 11/16"	1 – 15/16"
1 – 1/2	1 – 5/8 "	1 – 23/32"	2 "
2	1 – 21/32"	1 – 25/32"	2 – 1/32"
2 – 1/2	1 – 23/32"	1 – 13/16"	2 – 1/8 "
3	1 – 3/4 "	1 – 7/8 "	2 – 5/32"
3 – 1/2	1 – 13/16"	1 – 29/32"	2 – 3/16"
4	1 – 27/32"	1 – 31/32"	2 – 7/32"
4 – 1/2	1 – 29/32"	2 "	2 – 9/32"
5	1 – 15/16"	2 – 1/16"	2 – 5/16"
5 – 1/2	2 "	2 – 3/32"	2 – 3/8 "
6	2 – 1/32"	2 – 5/32"	2 – 13/32"
6 – 1/2	2 – 3/32"	2 – 3/16"	2 – 15/32"
7	2 – 1/8 "	2 – 1/4 "	2 – 1/2 "
7 – 1/2	2 – 3/16"	2 – 9/32"	2 – 9/16"
8	2 – 7/32"	2 – 11/32"	2 – 19/32"
8 – 1/2	2 – 9/32"	2 – 3/8 "	2 – 21/32"
9	2 – 5/16"	2 – 7/16"	2 – 11/16"
9 – 1/2	2 – 3/8 "	2 – 15/32"	2 – 3/4 "
10	2 – 13/32"	2 – 17/32"	2 – 25/32"
10 – 1/2	2 – 15/32"	2 – 9/16"	2 – 27/32"
11	2 – 1/2 "	2 – 5/8 "	2 – 7/8 "
11 – 1/2	2 – 9/16"	2 – 21/32"	2 – 15/16"
12	2 – 19/32"	2 – 23/32"	2 – 31/32"
12 – 1/2	2 – 21/32"	2 – 3/4 "	3 – 1/32"
13	2 – 11/16"	2 – 13/16"	3 – 1/16"
13 – 1/2	2 – 3/4 "	2 – 27/32"	3 – 1/8 "
14	2 – 25/32"	2 – 29/32"	3 – 5/32"

4. Measure the circumference of the thumb slightly proximal to the I.P. joint. Use a set of ring gauges, trying different sizes until one is found that fits snugly. The circumferences in inches for each ring size will be found in a table on back of the form. Select the correct one, and enter it in item four on the form.

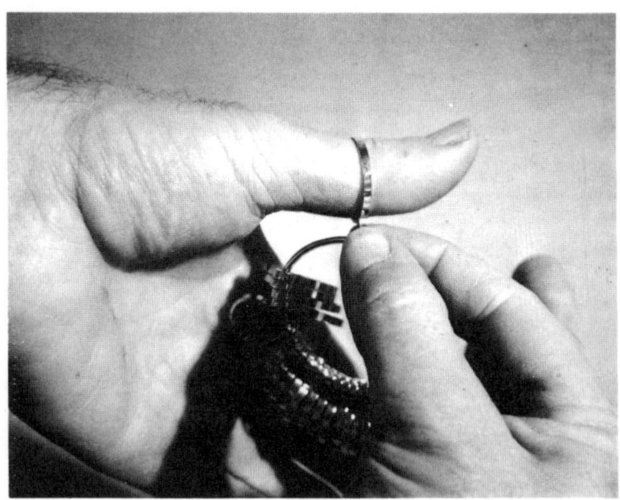

5. Measure the circumference of the proximal and middle phalanges of the four fingers, using the ring gauges, and enter them under items five and six on the form.

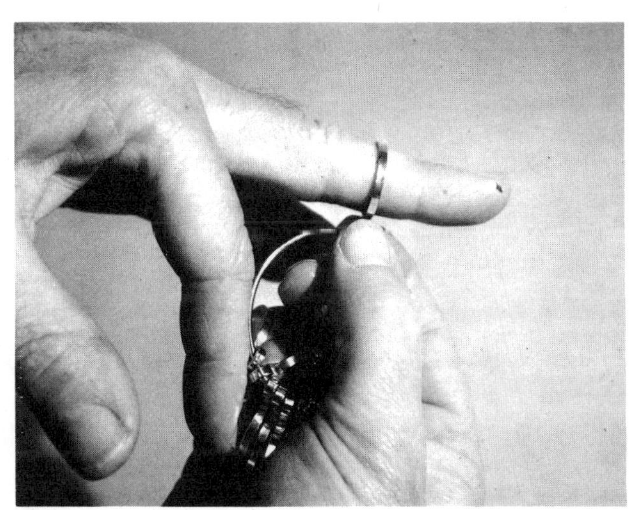

6. Measure the distance from the ulnar styloid to the M.P. joint of the little finger and enter the measurement under item seven on the form.

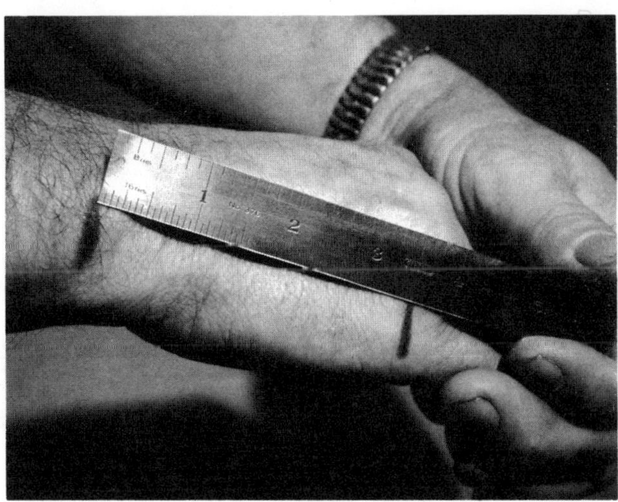

7. Measure the circumference of the wrist slightly proximal to the ulnar styloid, using the tape measure, and enter the measurement as item eight on the form.

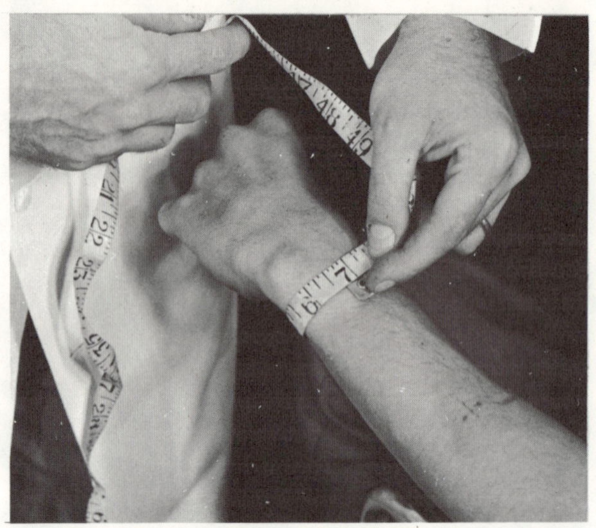

8. Measure the circumference of the forearm half way between the elbow and wrist, using the tape measure, and enter the measurement as item nine on the form.

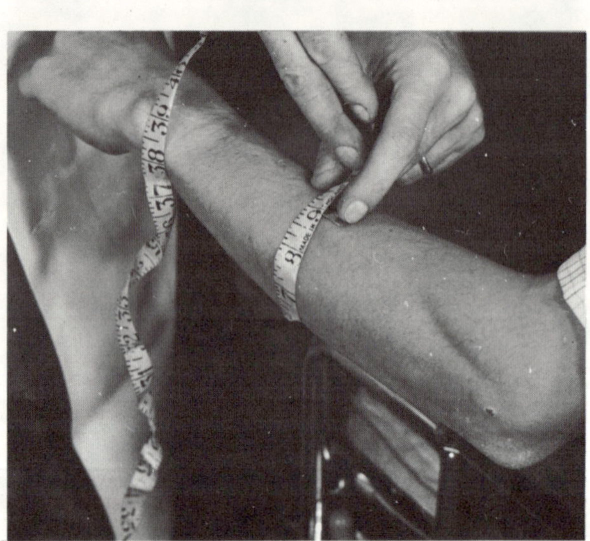

9. Measure the distance between the M.P. joint of the middle finger and the wrist joint, and enter the measurement as item ten on the form.

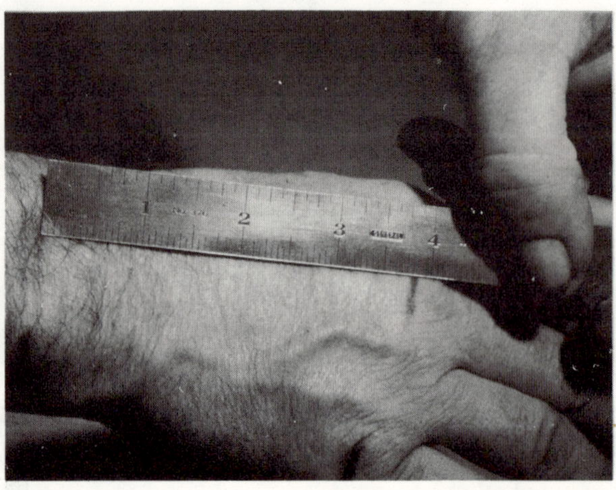

HOW TO MAKE AND FIT A SHORT OPPONENS HAND SPLINT

1. Select the correct size palmar and wrist pieces. If you are using a kit of pre-shaped parts, select the kit according to the following table:

Hand Width at M.P. Joints	Kit No.
1-3/4" — 2-1/4"	1
2-1/4" — 3-1/4"	2
3-1/4" — 4-1/4"	3

If you prefer to make your own parts, patterns for the palmar and wrist pieces are on the following pages, Figures 85 and 86. The pieces are numbered 1 through 3 the same as the kits in the table above. These patterns are full size, and can be traced onto .064" 2024-T4 sheet aluminum. Cut the pieces to shape on a metal cutting band saw, hand file the rough edges to the pattern outlines, then finish the edges and corners by sanding with a belt sander or by hand sanding. Pattern shapes are the same for both left and right hands.

2. The various parts of the palmar piece are identified in Figure 84 to make it easier to follow instructions for shaping it.

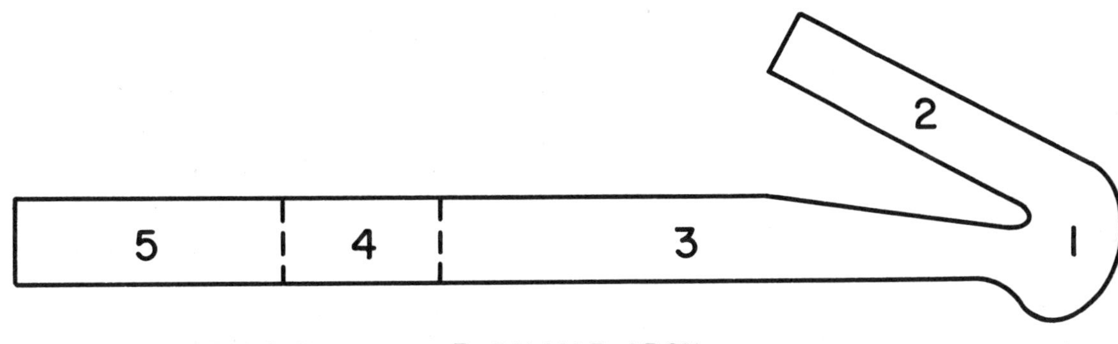

1. RADIAL EXTENSION
2. OPPONENS BAR
3. PALMAR ARCH
4. ULNAR EXTENSION
5. DORSAL EXTENSION

Figure 84. Palmar Piece for Short and Long Opponens Hand Splints, Showing Major Parts

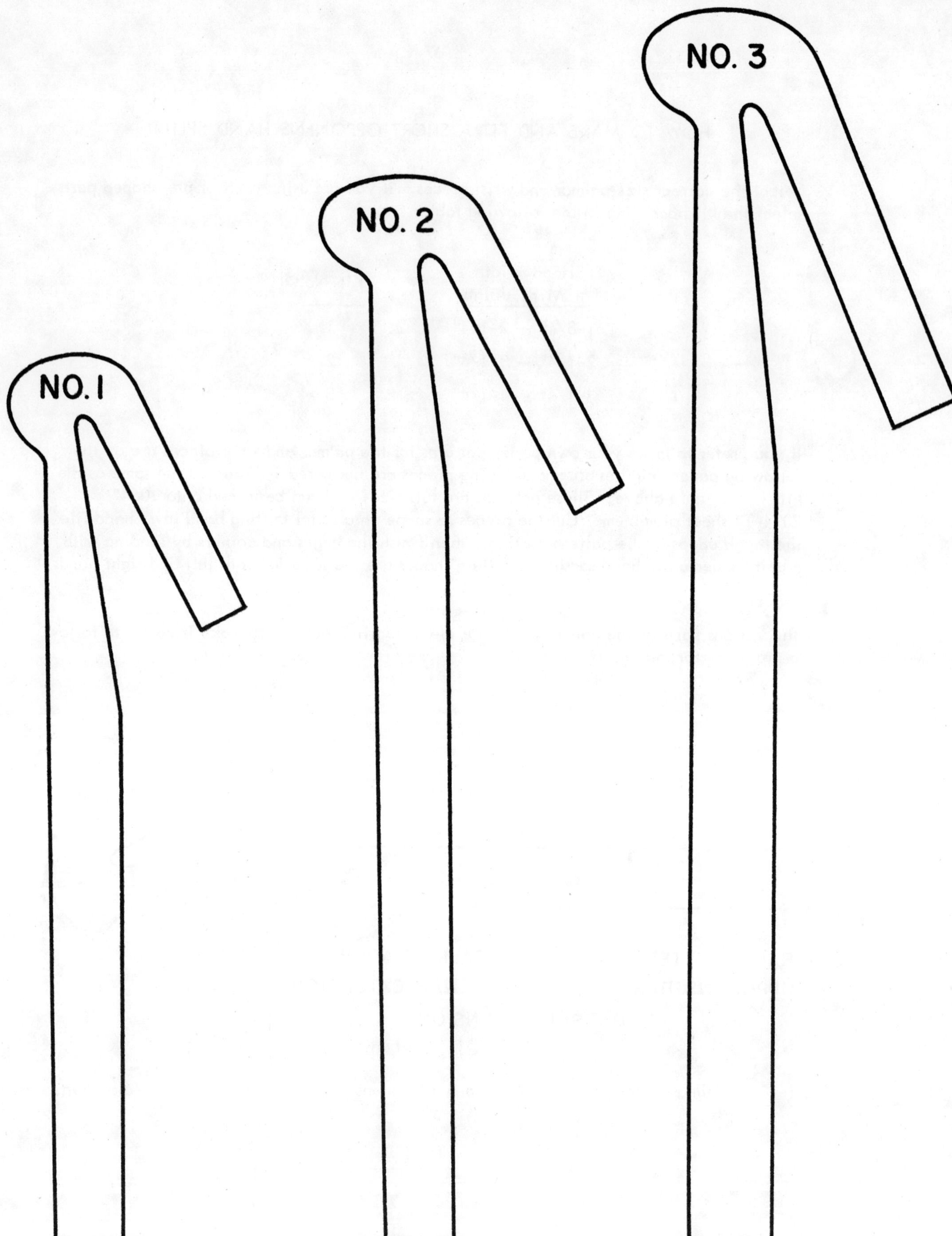

Figure 85. Patterns for the Three Sizes of Palmar Pieces. Palmar, wrist, and forearm pieces are made in sizes 1, 2, and 3, and only pieces of one size can be used in combination.

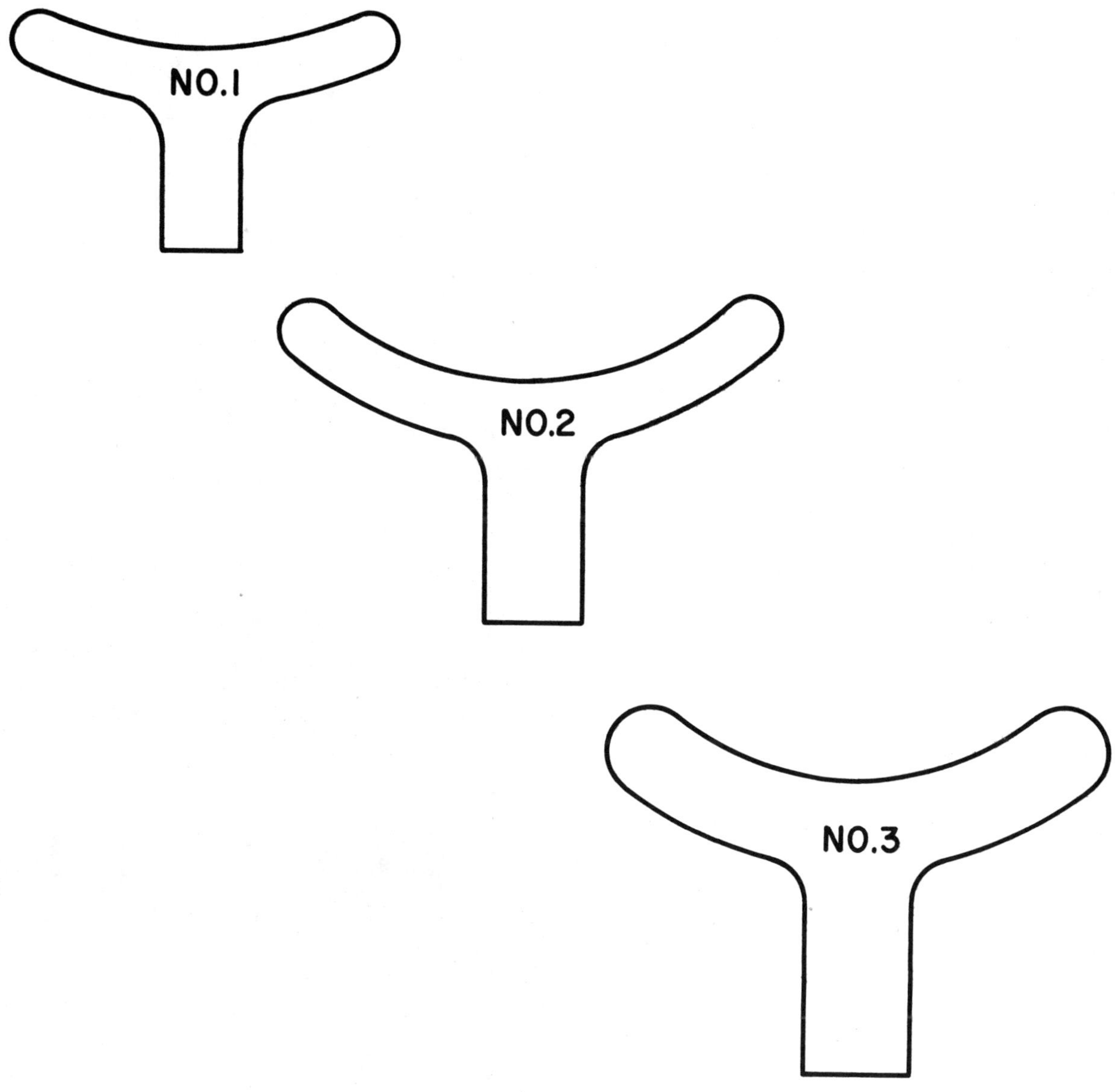

Figure 86. Patterns for the Three Sizes of Wrist Pieces

3. Bend the radial extension at right angles
   to the palmar arch. To determine how to
   bend for a right hand or a left hand, hold
   the palmar piece horizontally with the
   opponens bar to the left and pointing up.
   If the splint is for a right hand, bend the
   radial extension away from you, if for a
   left, toward you. In the illustration the
   splint is being shaped for a left hand.
   Clamp a small (1/8 inch opening) bending fork in the vise, insert the radial
   extension into the opening, and bend it
   at right angles to the palmar piece, as
   illustrated. Bend the radial extension
   concave to conform to the hand, either
   on the bending fork or by forming it with
   a 3/4 inch mandrel struck with a hammer
   with a lead block for a backing.

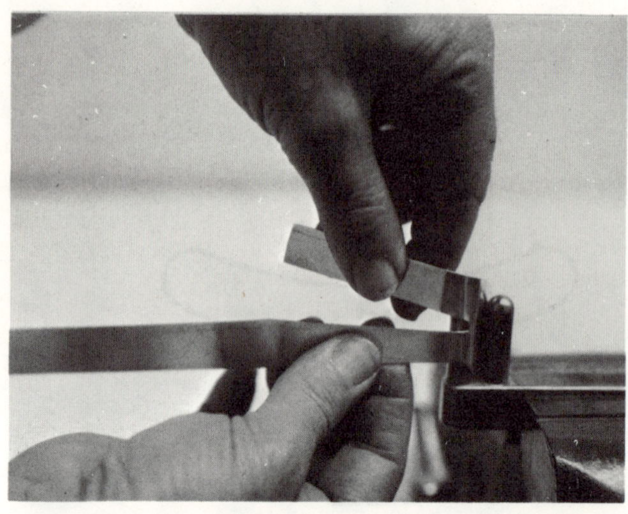

4. Bend the opponens bar until it and the
   radial extension are in a straight line
   and at right angles to the palmar arch.

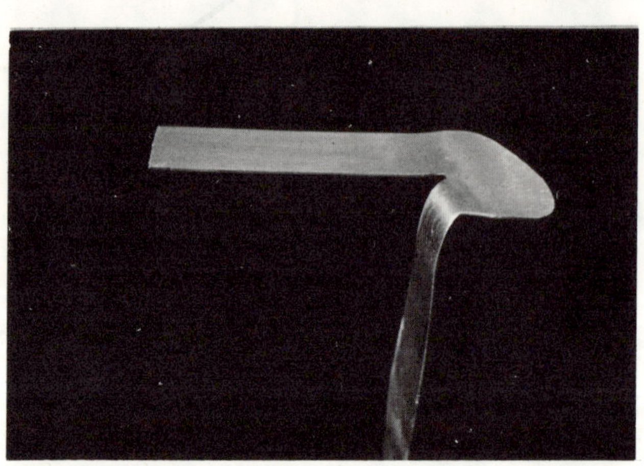

5. Bend the palmar piece about ten degrees
   proximally where it joins the radial
   extension so the palmar arch will clear
   the M.P. joints of the ring and little
   fingers.

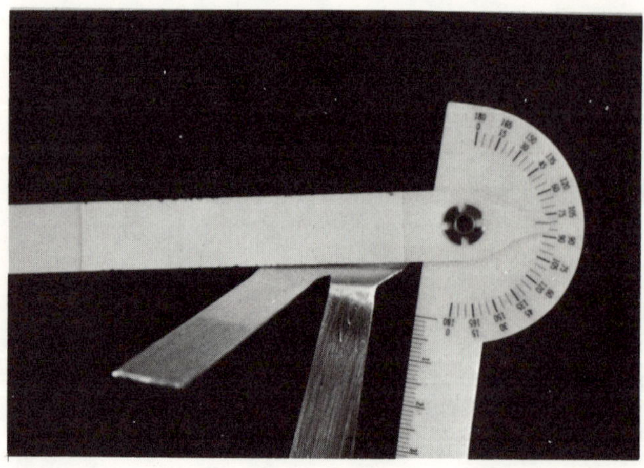

6. Shape the palmar arch with enough
   curve to support the center of the
   palm of the hand in normal position.
   The arch formed by the M.P. joints
   tends to collapse in paralyzed hands,
   so this support is important. If the
   patient is available, try the splint
   on his hand at this point.   Be sure
   the palmar arch has enough curve,
   the radial extension fits the contour
   of the radial aspect of the hand, and
   the opponens bar fits the thumb.
   When doing this, remember to allow
   1/4 inch for the felt padding that is
   put in place later.

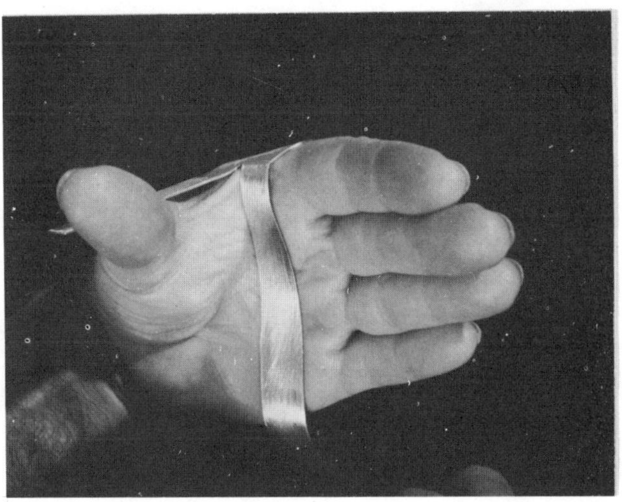

7. Measure from the inner surface of
   the radial extension a distance
   equal to the patient's hand width
   across the M.P. joints as recorded
   on the orthotic information form plus
   1/2 inch allowance for padding.
   Make a mark at this point with a
   pencil.

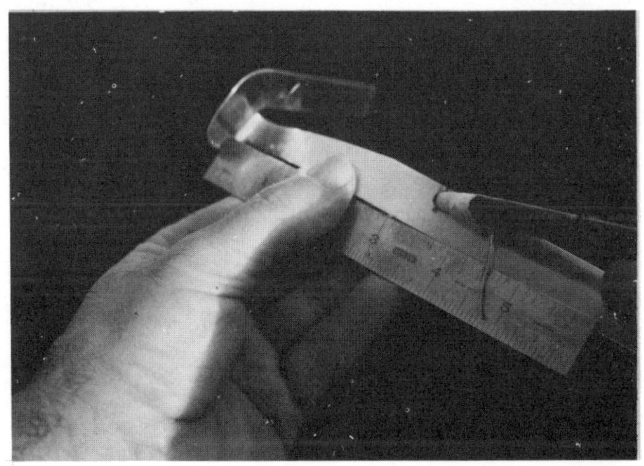

8. Start to bend the ulnar extension at a
   point about 1/2 inch short of the mark
   made in the previous step. Make the
   bend gradually, without sharp bends or
   kinks. The dorsal extension must extend across the dorsum of the hand
   almost to the radial extension.

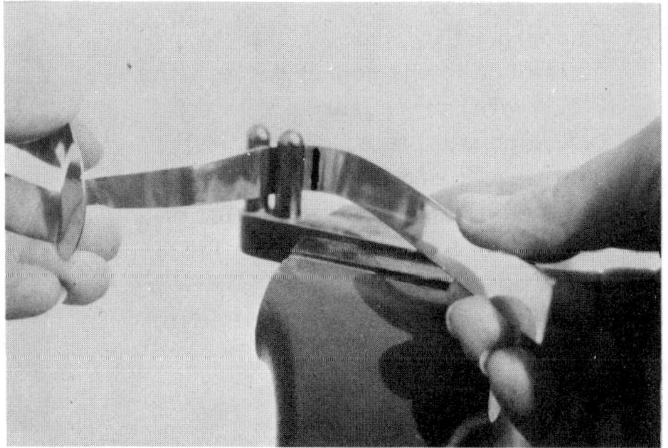

9. Continue bending the ulnar extension into a "U" shape. The dorsal extension should be approximately parallel to the palmar arch and about 1-1/2 inches from it if the splint is for a man. If for a woman, this distance should be about 1-3/8 inches, and in the case of a child, the dimension is proportionately smaller according to the size of his hand.

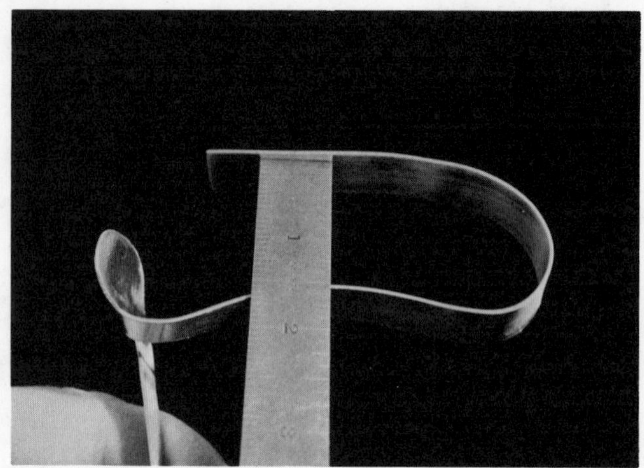

10. Bend the dorsal extension proximally at the ulnar extension so it will be at an angle of 15 degrees to the palmar arch. This is necessary so it will clear the M.P. joints on the dorsal aspect of the hand.

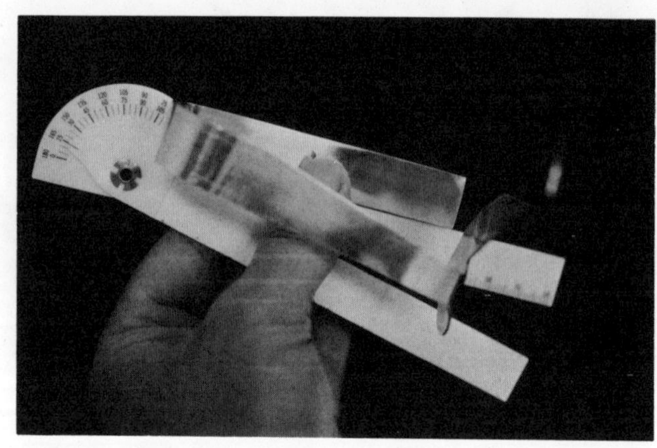

11. Select a wrist piece of size to correspond to the palmar piece. The No. 1 wrist piece is used with the No. 1 palmar piece, and similarly with sizes 2 and 3. Shape the wrist piece to conform to the contours of the patient's wrist by bending the wings on the bending fork. The wings fit distally from the ulnar styloid as illustrated.

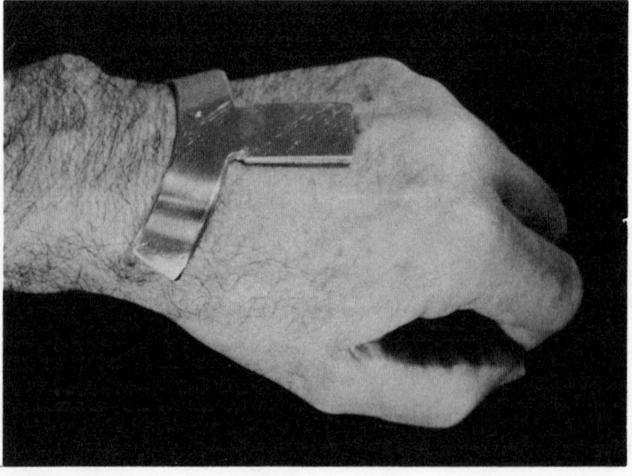

12. Clamp the wrist piece into position on the dorsal extension of the palmar piece. The center line of the wrist piece must be parallel to the inner surfaces of the radial extension and opponens bar, and so located on the dorsal extension of the palmar piece that it lies on the midline of the dorsum of the hand with the wings located comfortably on the wrist.

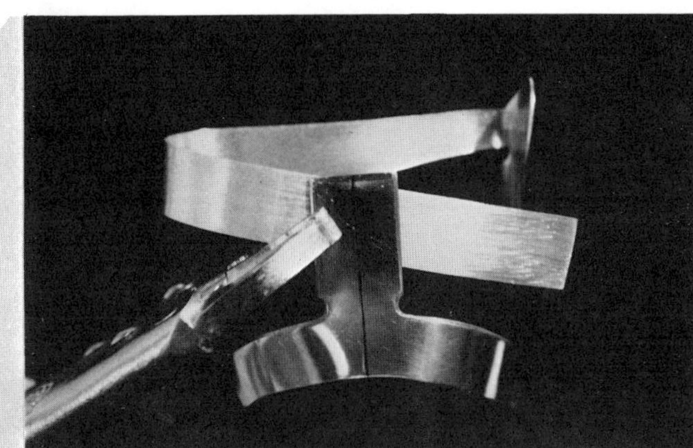

13. Drill No. 40 holes through the wrist piece and dorsal extension of the palmar piece at opposite corners of the wrist piece extension, then rivet the parts together with 3/32 x 3/16 inch stainless steel rivets. It will be necessary to cut off the excess length of rivet with diagonal cutters. Drill a No. 21 hole in the ulnar wing of the wrist piece for the strap, and a No. 29 hole in the radial wing for the truss stud used to attach the free end of the strap. Trim excess material where the parts join together, smooth all edges, round corners, and polish all outside surfaces on the buffer.

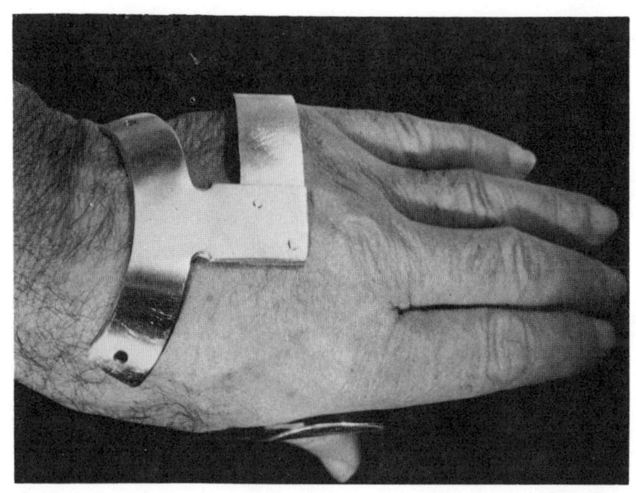

14. Fasten the strap (1/2 x 10 inch smoked elk) to the ulnar wing of the wrist piece with a No. 4829 (small) Speedy Rivet. A buckle may be used, but a 6–32 x 3/8 inch truss stud is easier for the patient to fasten. Cut off the excess length of the stud, then rivet it to the radial wing. Double back 2 inches of strap and rivet it to form a finger loop to make it easier for the patient to fasten the strap onto the truss stud, then punch and slit a No. 5 hole into the strap at a point where it will hold the splint snugly in place without being uncomfortably tight.

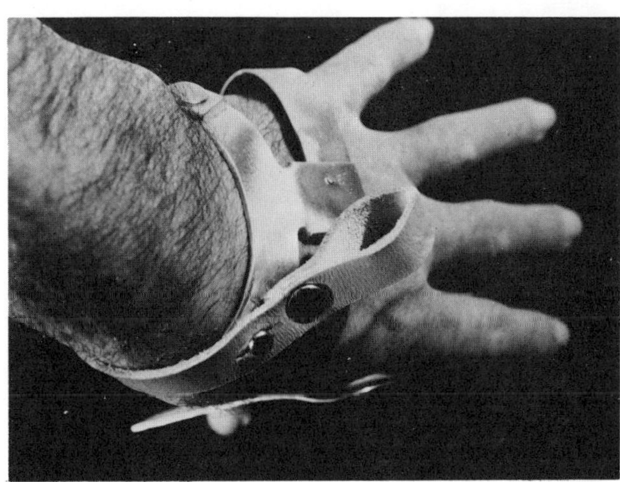

15. Cement the felt padding into place
    using rubber cement. Use either the
    pre-cut pads furnished in the kit, or
    make some from 1/4 inch white sur-
    gical felt. Try the splint on the
    patient again, and make adjustments
    necessary to achieve a snug, com-
    fortable fit. The patient must be
    able to flex his M.P. joints to 90 de-
    grees without interference by the
    splint, and the opponens bar must
    support the thumb proximal to the
    M.P. joint of the thumb. The splint
    is now ready for the installation of
    any attachments that may have been
    prescribed.

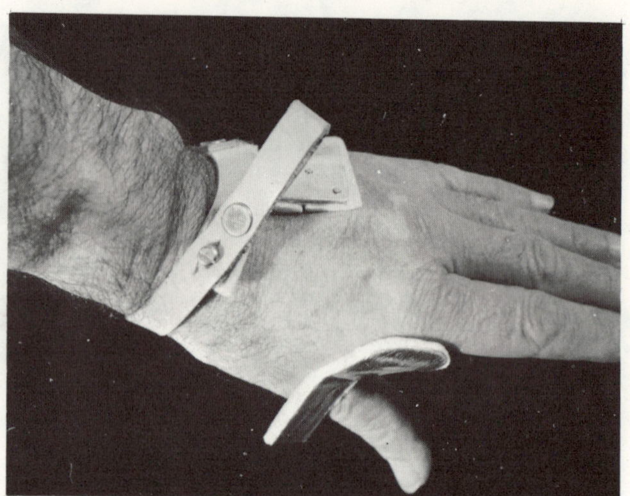

HOW TO MAKE A SPRING SWIVEL THUMB

1. Make the swivel support from a piece of .064 inch stainless steel 1/4 inch wide and 1-1/2 inches long, with a No. 29 hole drilled 3/16 inch from one end. Round the corners and smooth the edges.

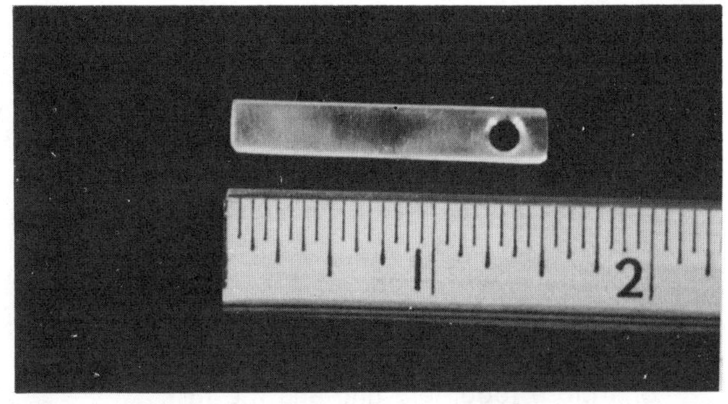

2. Bend the drilled end at right angles 1/4 inch from the end, then shape the remaining portion into approximately a 1-1/2 by 1/2 inch curve with a pair of round nose pliers.

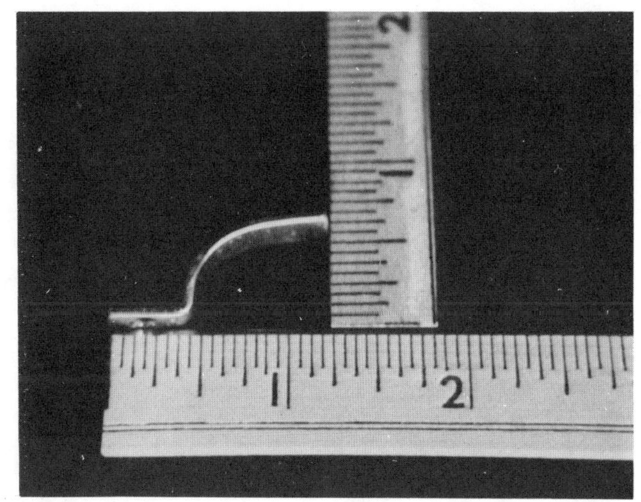

3. Attach the swivel support to the radial aspect of the splint with two 3/32 x 3/16 inch stainless steel rivets, then try the splint on the patient and shape the support so the hole is centered over the M.P. joint of the index finger.

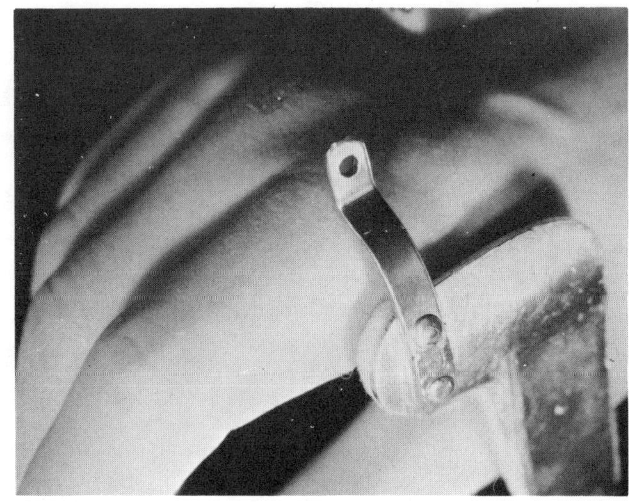

4. Make the spring of an 8 inch length of 21 gage (.047") stainless steel music wire for cases with average thumb flexors, for stronger ones use 24 gage (.055"), for weaker ones use 20 gage (.035"). Clamp the wire at midpoint between a 3/16 inch steel mandrel and the jaw of a vise, grip one end of the wire with a pair of Vise-grips, then wind three turns of wire around the mandrel. Draw the turns tight and close together, and end the last one so the entire length of the spring is in a straight line.

5. Bend a loop in one end of the spring with the round nose pliers, then use the loop to mount the spring on the swivel support, with the coils of the spring toward the hand. With the splint on the patient's hand, shape the spring into a curve, ending at the thumb.

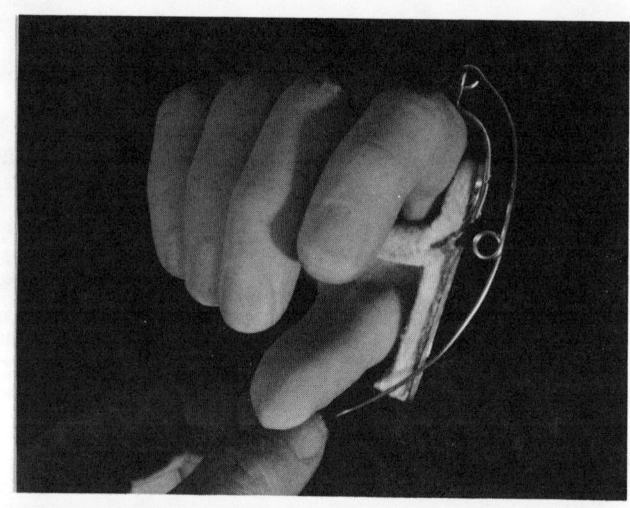

6. The length of the thumb piece is determined by dividing the circumference of the first phalanx of the thumb by 3/4. Cut a piece of .037 x 3/8 inch stainless steel to this length, then smooth the edges and round the corners.

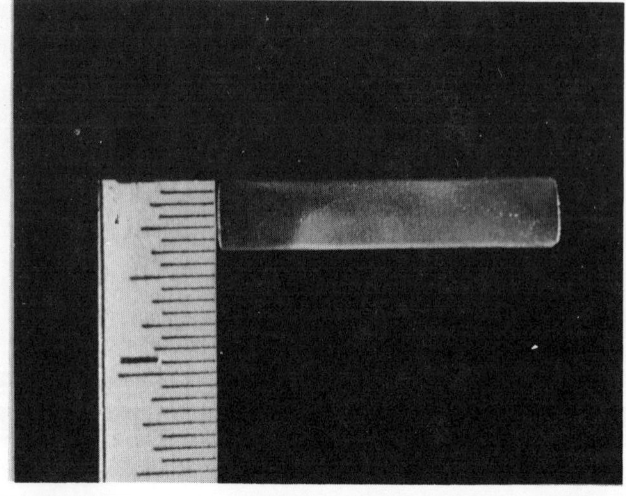

7. Bend the thumb piece into a "U" shape, curved to fit the contour of the proximal phalanx of the patient's thumb. Flare the ends of the "U" out slightly so the thumb piece will slip onto the thumb more easily.

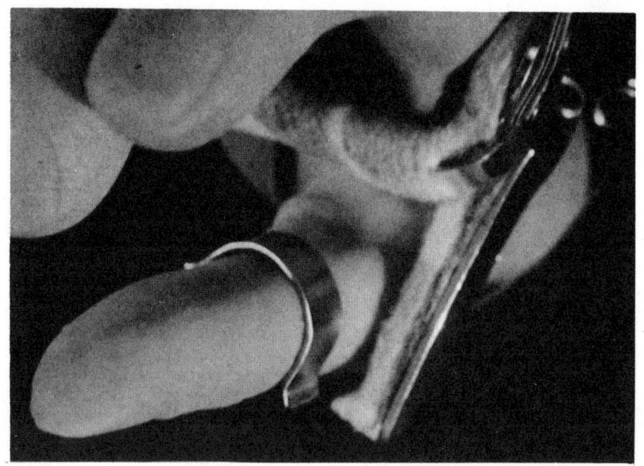

8. Locate the thumb piece on the spring a distance from the inside edge of the felt to the thumb piece equal to the distance from the palmar aspect of the index finger M.P. joint to the thumb M.P. joint as noted on the orthotic information form. Mark the location of the thumb piece on the spring.

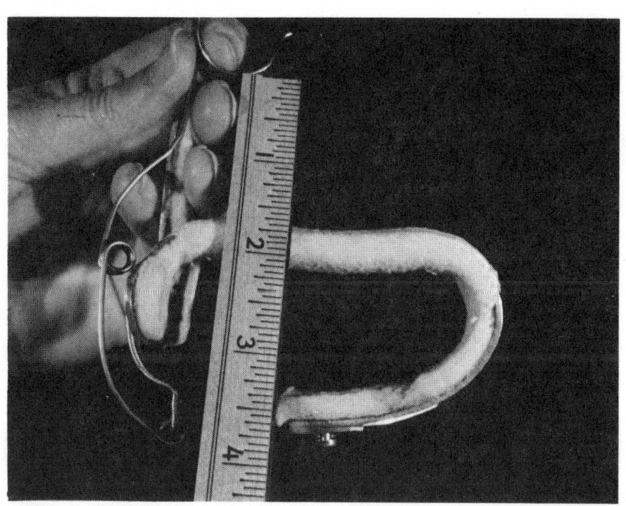

9. Solder the thumb piece to the spring, using soft solder, stainless steel soldering flux, and a soldering iron. Place the splint on the patient's hand and adjust the swivel support, spring, and thumb piece so the patient can flex his thumb against the spring tension. The spring should then return his thumb to the open position. Be certain the thumb piece fits comfortably on the thumb.

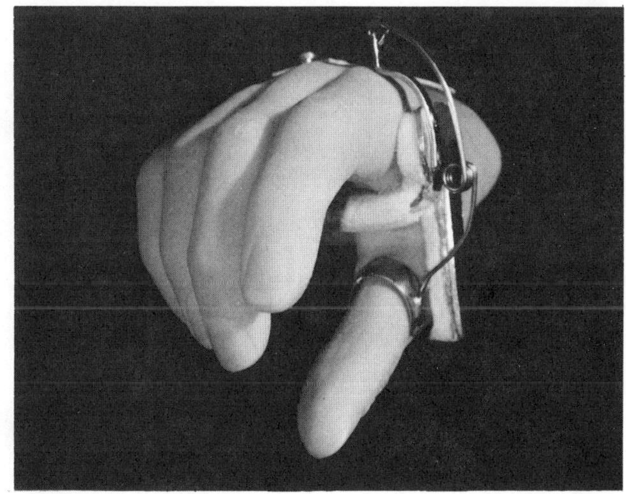

HOW TO MAKE AND FIT A FIRST DORSAL INTEROSSEOUS ASSIST

1. To make the assist spring, grip a 6 inch piece of .047 inch stainless steel piano wire against a piece of 3/16 inch drill rod in a vise. Grip the wire about an inch from one end, and make sure it is horizontal and the rod vertical. Grip the long end with a pair of vise-grips and wind three turns of wire around the rod to form the spring, ending with the straight parts at an angle of 90 degrees to each other.

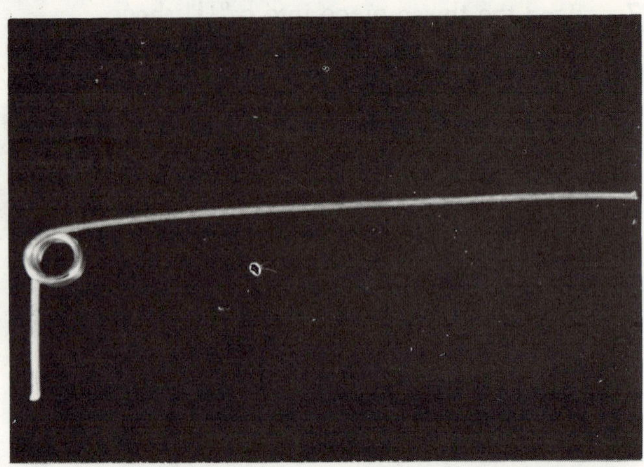

2. With round nose pliers, shape a 3/32 inch loop in the short end. It must be quite close to the coil spring and at right angles to it.

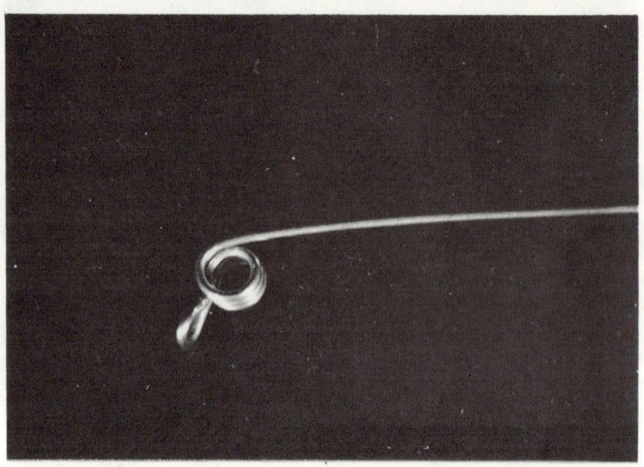

3. Drill a #40 hole in the center of the radial extension, then rivet the assist spring onto the extension using a 3/32 x 3/16 stainless steel rivet. Position it so the long arm of the spring will extend parallel to the index finger when it is bent to shape.

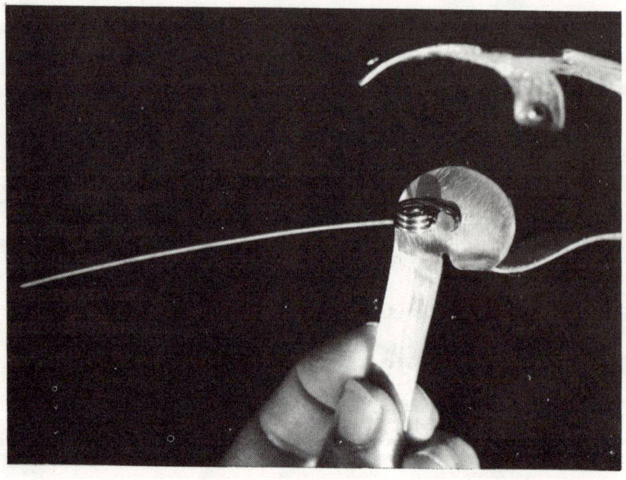

4. To make the finger piece, cut off a piece of .037 x 5/16 inch stainless steel equal in length to 3/4 of the circumference of the proximal phalanx of the index finger. Round the corners and smooth the edges, then drill a #40 hole about 1/8 inch from one end, and in the center of the piece.

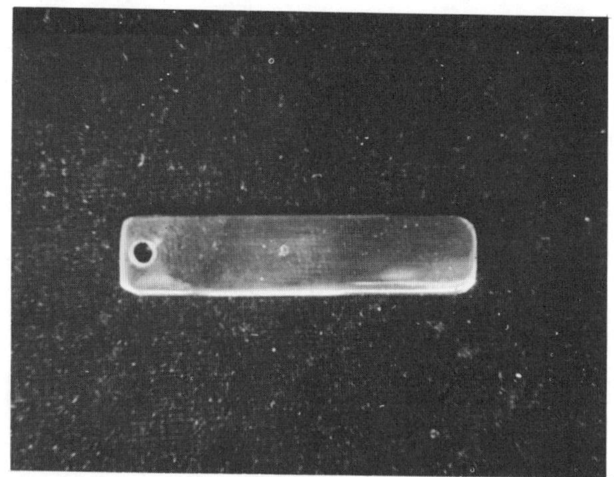

5. With the end of the piece with the hole gripped in the vise at a point about 3/8 inch from the end, bend the remainder to an angle of 60 degrees.

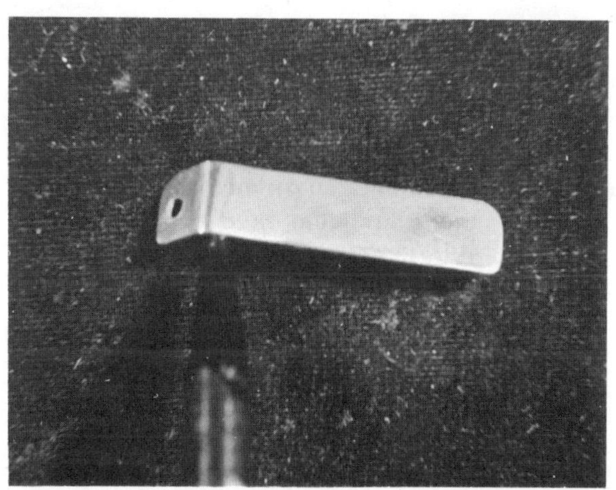

6. With round nose pliers, bend the piece into a half-circle.

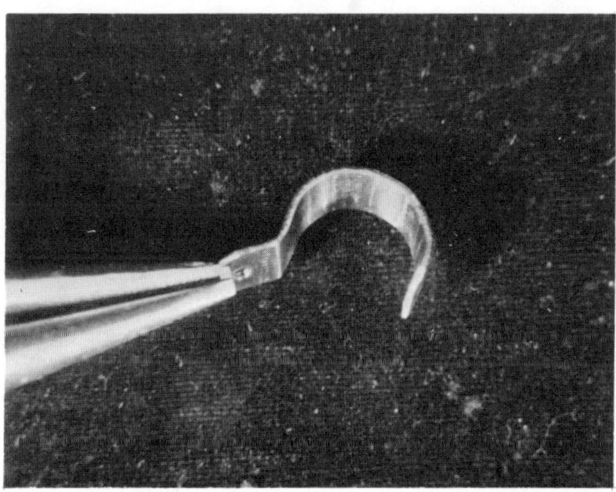

7. Use the round nose pliers to shape the spring arm into an arch that curves down to a point slightly proximal to the P.I.P. of the index fingers. Bend a loop in the end of the spring, then attach the finger piece to the loop.

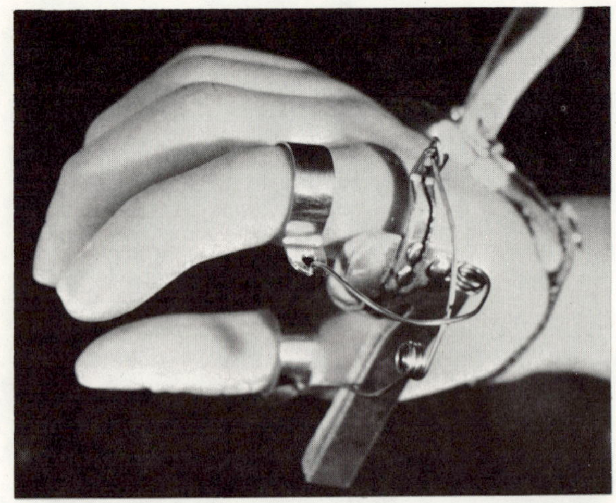

8. With the splint on the patient's hand, the spring should be aligned with the index finger and should exert enough force to hold the index M.P. enough in abduction to provide opposition with the thumb. The spring should not be too strong for the patient to adduct his finger.

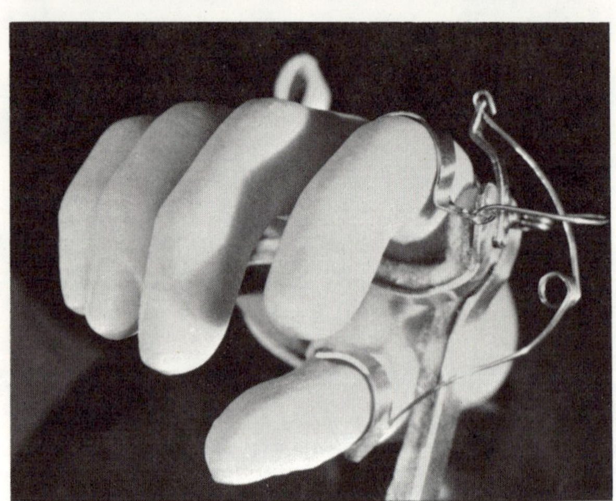

9. Check the fit of the finger piece. It should come down well over the ulnar side of the index finger proximal phalanx to provide stability and distribute the pressure over a wide area to prevent irritation.

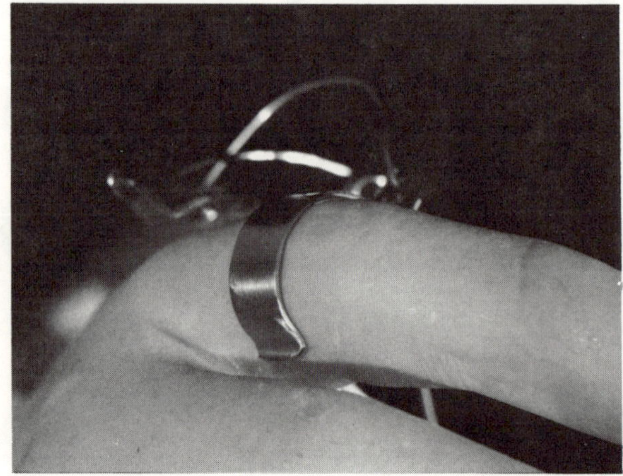

10. Some patients prefer a plastic finger piece to the stainless steel half-ring. A piece of .020 x 3/4 x 4 inch plastic tubing formed into a loop fastened with a grommet is satisfactory. A plastic finger piece is shown in the illustration at right. The disadvantage of this type of finger piece is that it is more difficult to get on and off.

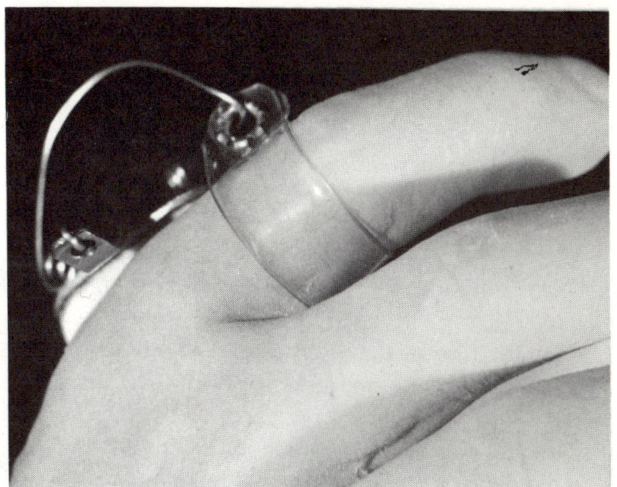

HOW TO MAKE AND FIT AN M.P. STOP

1. Lay out and drill the holes for the M.P. stop attachment studs on the center line of the wrist piece. Mark the center line and center punch on it 3/16 inch back from the distal edge. Measure 3/4 inch back from the first punch mark and center punch again at that point. Drill No. 40 holes at the center punch marks and rivet the attachment studs into the holes.

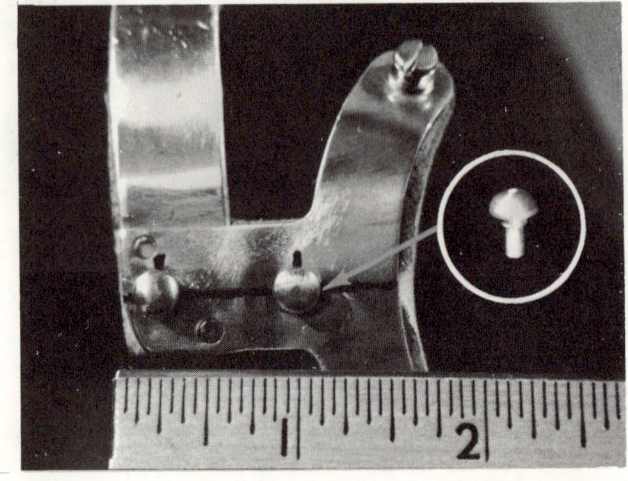

2. Try the attachment bar on the attachment studs. It should slip into place easily, but must be tight enough so it will not fall off when in use. If the bar is too loose, peen the studs slightly to make it tighter.

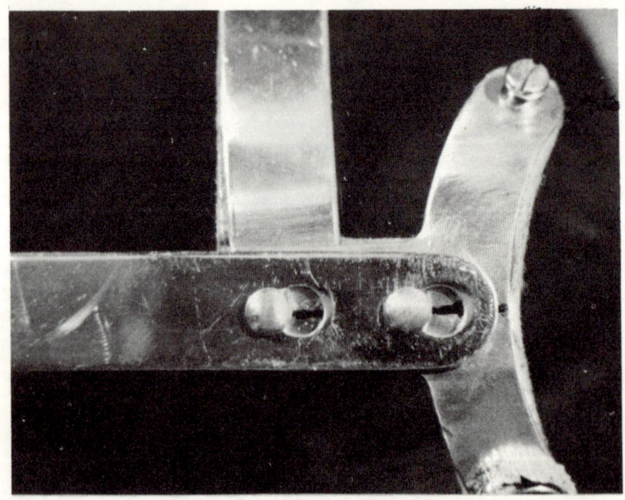

3. Make the M.P. stop of .064 inch sheet aluminum 3/8 inch wide, curved on a radius approximately 1-1/2 times the width of the patient's hand at the M.P. joints. Place the curved piece of aluminum on the patient's fingers slightly proximal to the first I.P. joints and mark it to length flush with the sides of the fingers.

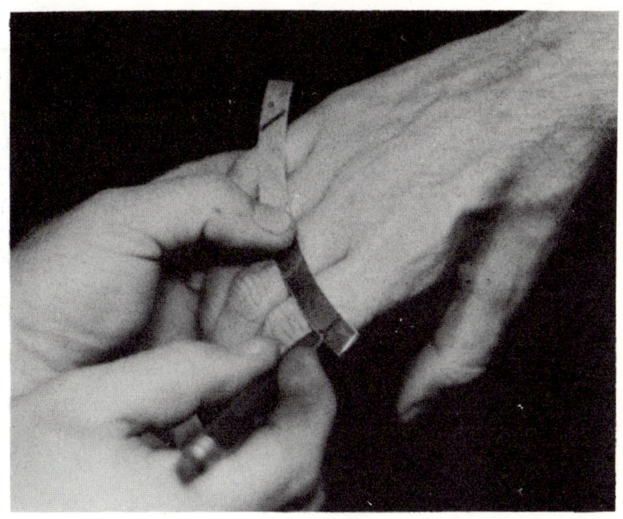

4. Trim the M.P. stop at the marks, round the ends and smooth the edges. Place the splint on the patient's hand with the attachment bar in place. Bend the bar down slightly at the M.P. joints, then place the M.P. stop under it, so it is just proximal to the first I.P. joints. Mark the position on both parts.

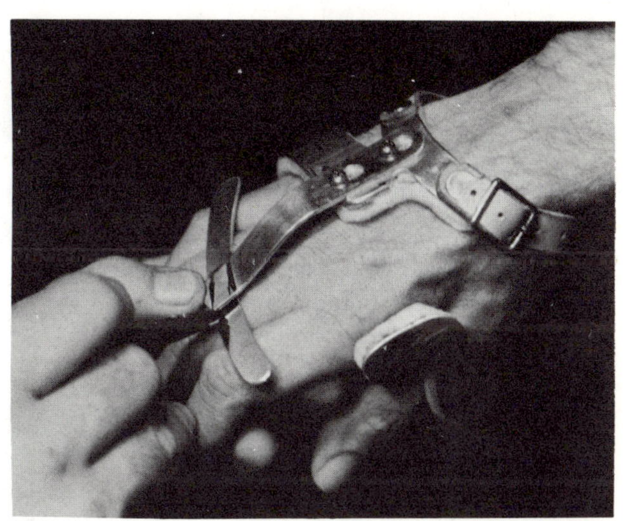

5. With the attachment bar and M.P. stop held in position with clamp or vise-grips, drill two No. 40 holes and rivet the parts together with 3/32 x 3/16 inch stainless steel rivets. Bend the ends of the M.P. stop down slightly to conform to the arch of the hand, then polish the entire assembly.

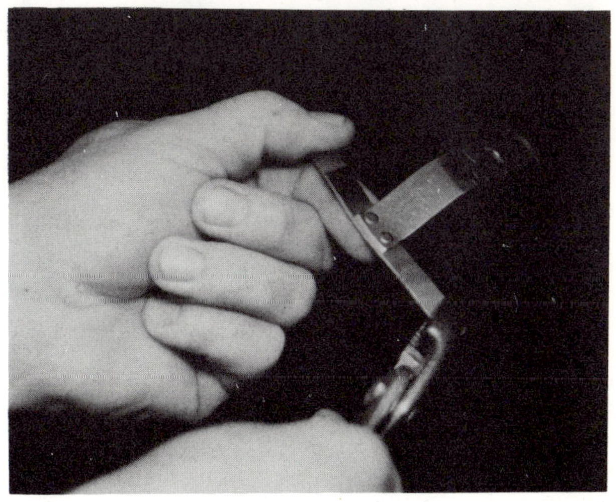

6. Cover the M.P. stop with moleskin.

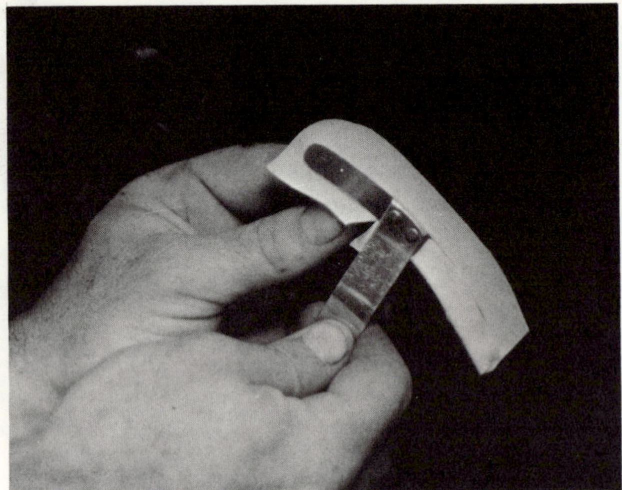

7. Fit the splint and M.P. stop on the patient's hand by bending the attachment bar and M.P. stop as necessary so the latter contacts the fingers proximal to the proximal I.P. joints and holds them so the M.P. joints are in about 15 degrees flexion, as shown in the illustration. The stop must lie flat on the fingers.

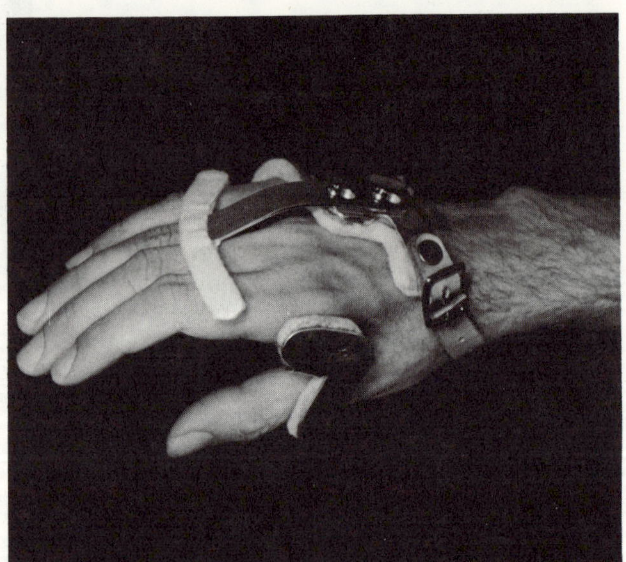

HOW TO MAKE AND FIT A LONG OPPONENS HAND SPLINT

Introduction

The Long Opponens Hand Splint is made of two parts, the palmar piece and the forearm piece. The palmar piece is identical to the one used on the short opponens hand splint, and the procedure for shaping and fitting it is the same as was described in the earlier section on the short splint. The forearm piece is included in the parts kit, or may be made by using the patterns on the following pages. There are three sizes of forearm pieces, identified as No. 1, the smallest, No. 2, and No. 3, the largest. These numbers correspond to the identifying numbers on the palmar pieces, and corresponding numbers must always be used together. In making and fitting a long opponens hand splint, the palmar piece is shaped and fitted first. For instructions on how to do this, refer to "How to Make and Fit a Short Opponens Hand Splint".

1. Bend the wings on the forearm piece to the contour of the patient's forearm, then locate the palmar piece on the forearm piece. Center the forearm piece on the dorsal extension of the palmar piece so the center line on the forearm piece is parallel to the radial extension, as illustrated. Maintain this relationship while using a ruler to adjust the distance from the distal edge of the ulnar wrist wing of the forearm piece to the proximal edge of the ulnar extension on the palmar piece so it is equal to the distance from the patient's ulnar styloid to his fifth M. P. joint. (See Orthotics Measurement Form.) Clamp the two parts together in this position.

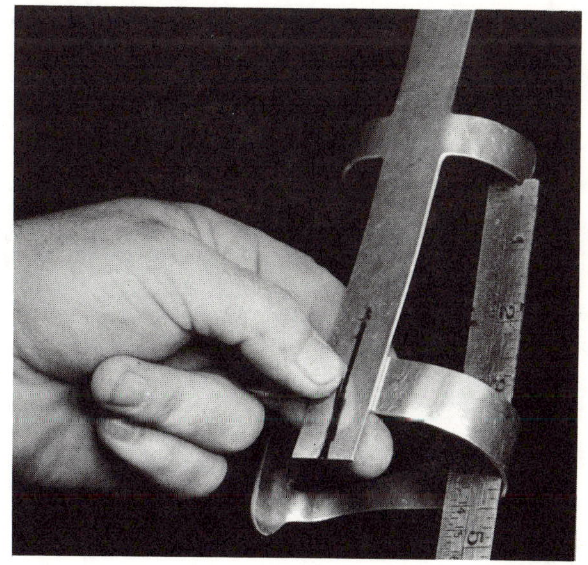

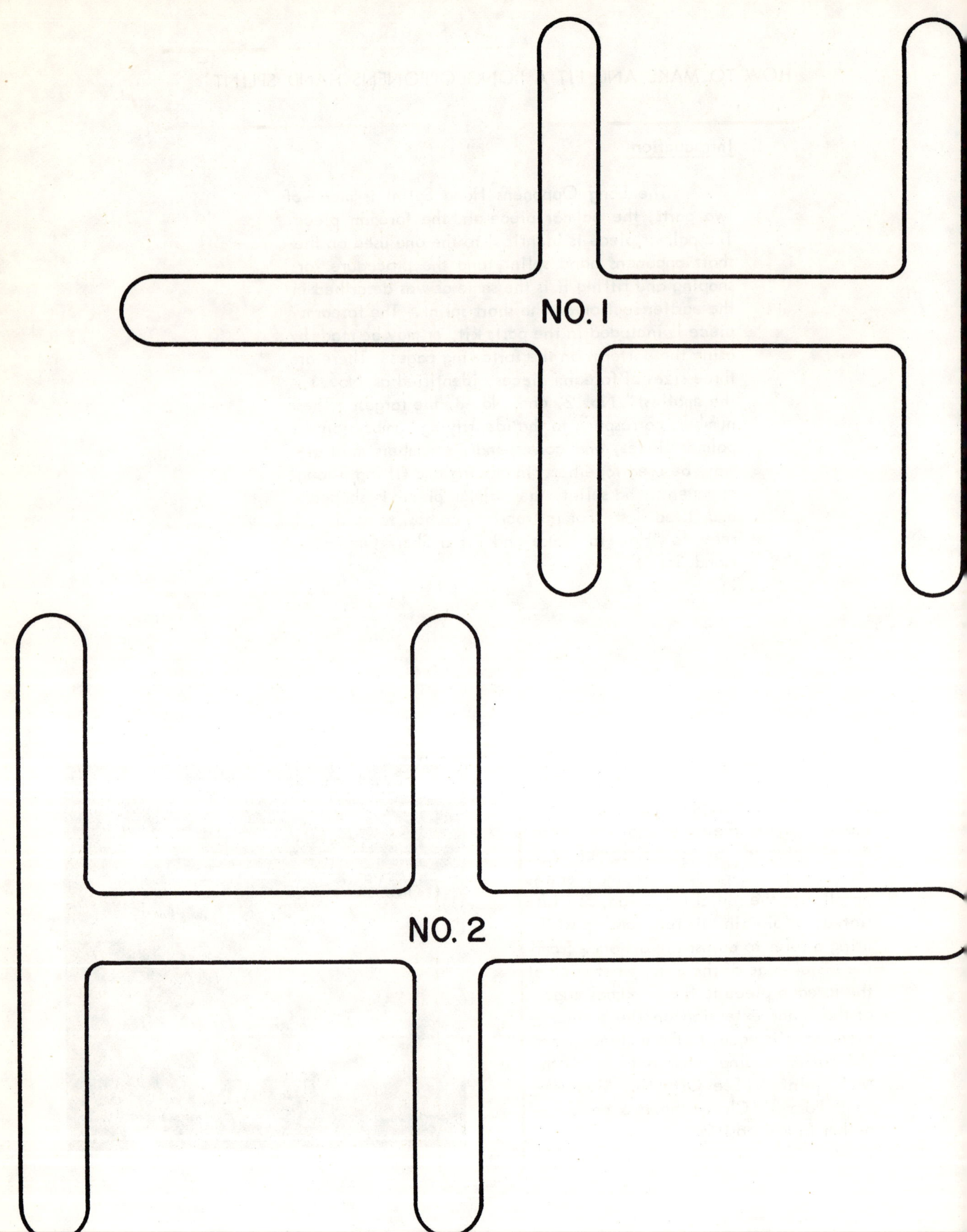

Figure 87. Patterns for No.'s 1 and 2 Sizes of Forearm Pieces

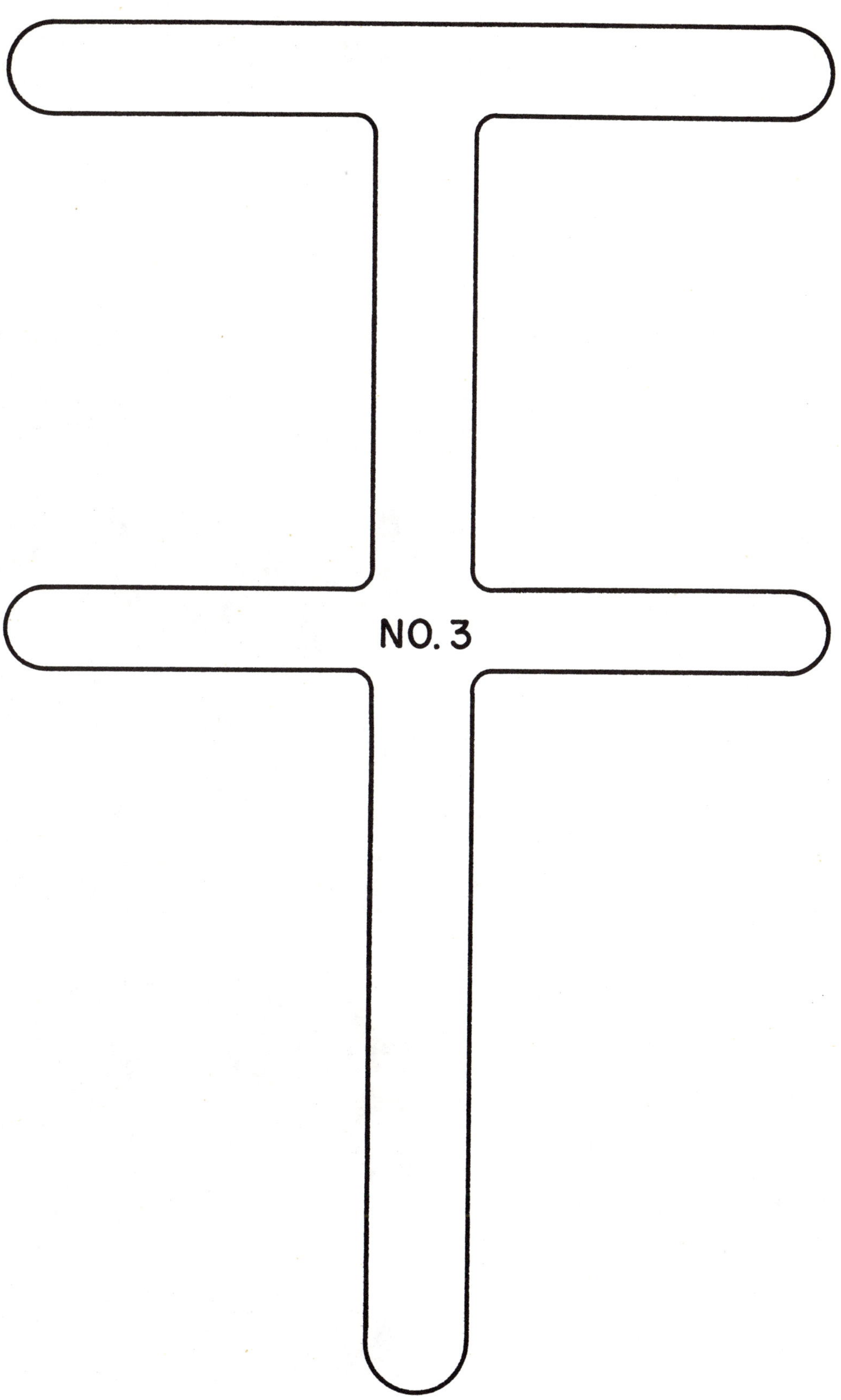

Figure 88. Pattern for No. 3 Size Forearm Piece

2. Drill two No. 40 holes through the parts while they are clamped in position, then rivet them together with stainless steel rivets, size 3/32 x 3/16 inch. Be sure to place the heads of the rivets inside the palmar piece, toward the hand, otherwise they will obstruct the attachment bar when the time comes to install it. Cut off the excess length of the forearm piece.

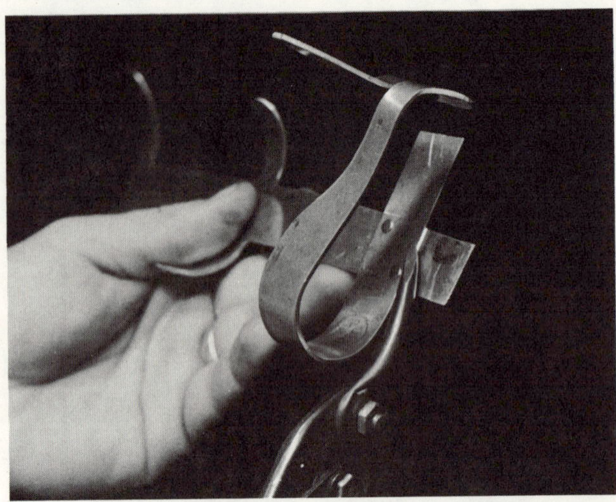

3. Cut off the excess portion of the dorsal extension, then round all corners and smooth all edges.

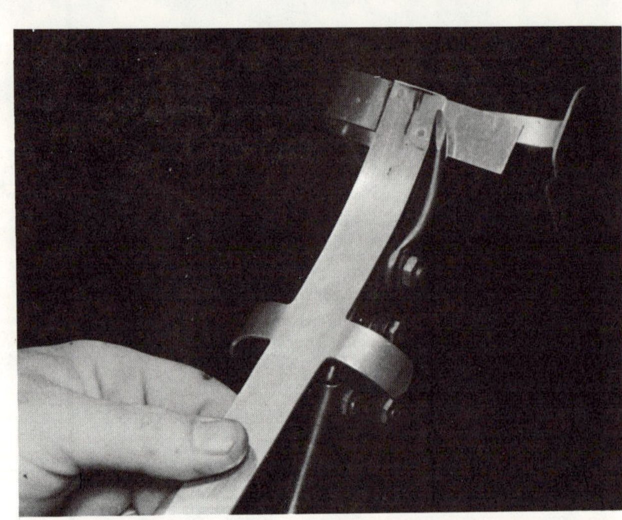

4. Bend the forearm piece approximately 30 degrees at a point about 1 inch distal to the wrist wings. This amount of bend will extend the wrist to the position of function. Polish the splint.

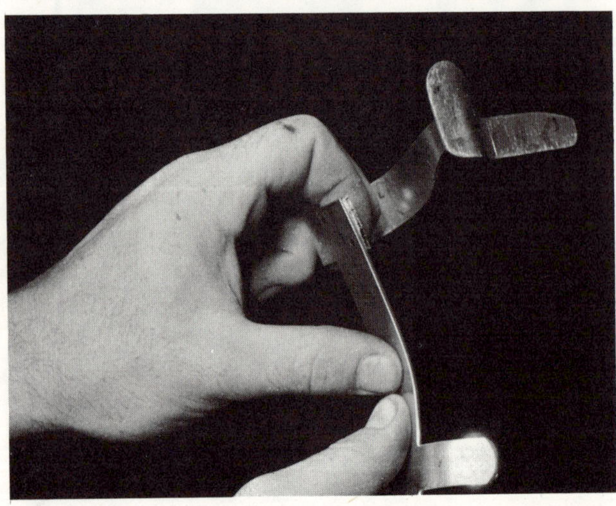

HOW TO MAKE AND FIT A C-BAR

1. Cut a piece of .037 inch stainless steel 1-1/8 x 1-3/4 inches. Square the ends, then mark a line on one end at a 5 degree angle with the edge. Make a second line 1/4 inch from and parallel to the first one.

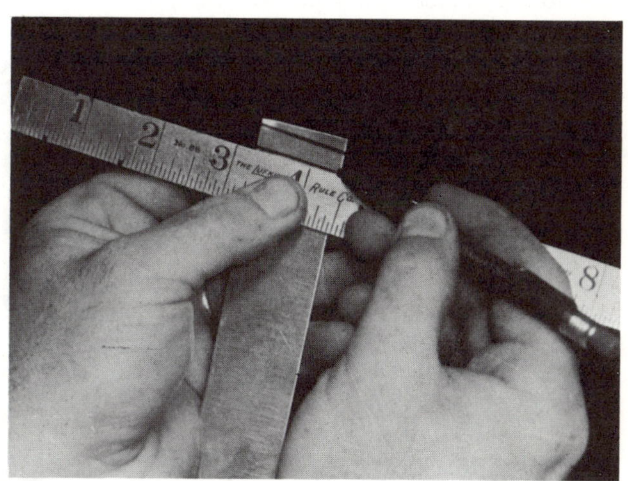

2. Trim the metal on the piece even with the first line, using shears.

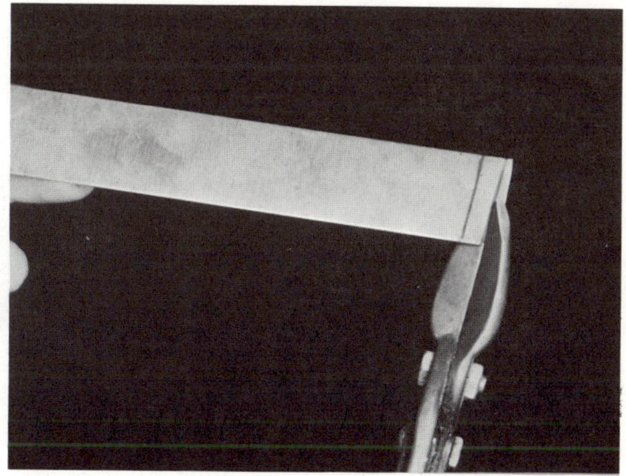

3. The reason for shaping the end of the bar at a 5 degree angle is to adapt it to the curve of the palmar arch, to which it will be riveted. The C-bar must be mounted on the arch so its sides are at right angles to the forearm piece and parallel to the opponens bar. It may be necessary to make slight changes in the angle of the end of the bar to accomplish this.

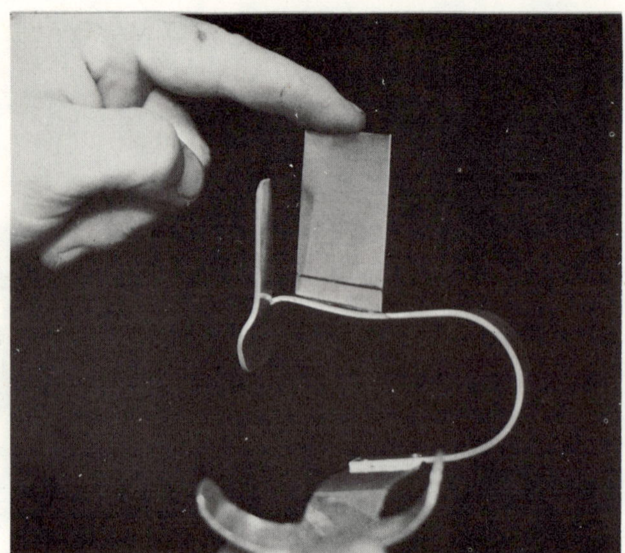

4. Make sure the bending line on the bar is 1/4 inch from and parallel to the end of the bar, then bend a flange at right angles, exactly on the line. Drill two No. 40 rivet holes in the ends of the flange, then remove the burrs and smooth the edges of the flange only.

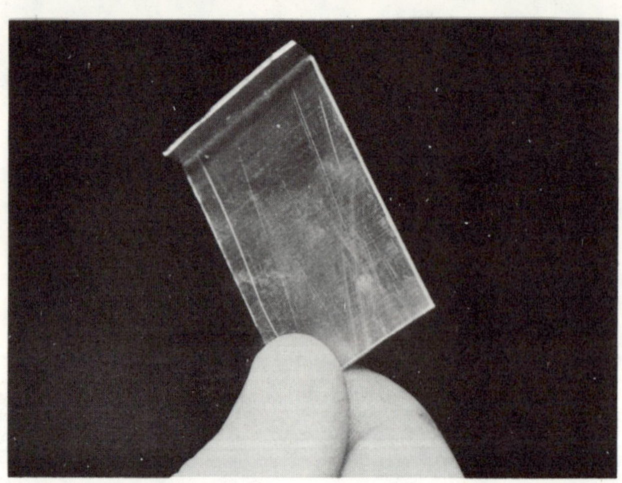

5. Place the C-bar on the palmar arch with the angle of the flange flush with the proximal edge of the arch, and the radial edge of the bar parallel to and about 1/4 inch from the opponens bar. While holding the bar in this position, mark the arch through the rivet holes with an awl, then drill a No. 40 hole in the center of each mark.

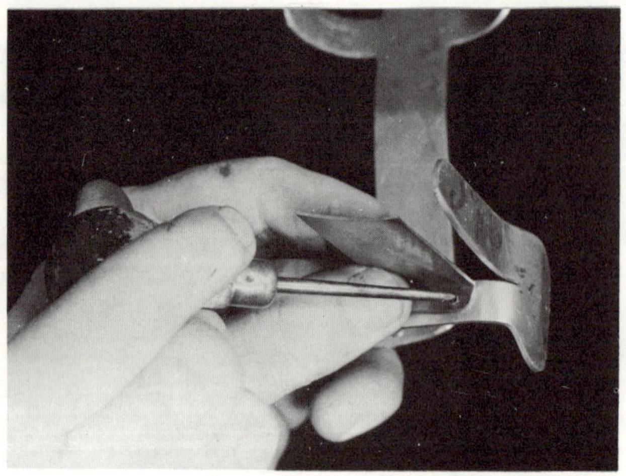

6. Rivet the C-bar to the palmar arch with 3/32 x 3/16 inch stainless steel rivets with the rivet heads on the C-bar side of the splint. With the splint on the patient's hand, shape the C-bar to conform to the curvature of the patient's thumb.

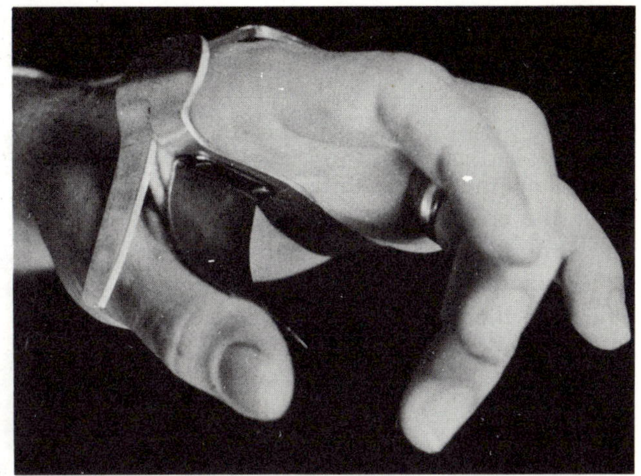

7. Attach the wrist strap to the splint with a 6-32 truss stud, (use No. 29 drill), locating the stud far enough distally so the strap will angle back and under the base of the hand. This will prevent the splint from working loose. Make a slit in the strap to slip over the truss stud at a point that will hold the strap snugly but not too tight for comfort.

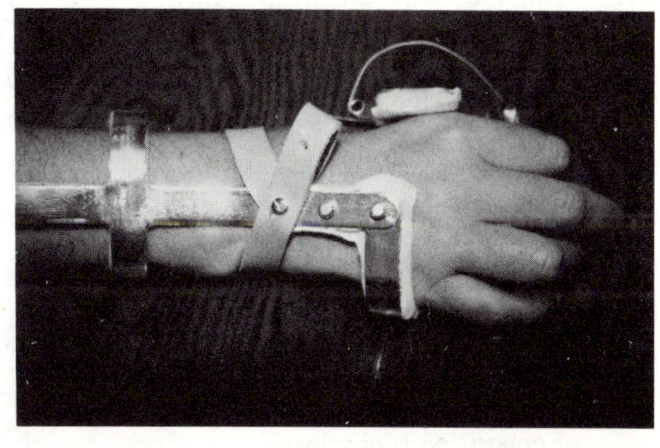

8. Install the felt padding provided in the kit (or cut out padding from 1/4 inch white surgical felt), using rubber cement to hold it in place temporarily. Try the splint on the patient, and mark the C-bar at a point slightly proximal to the thumb I.P. joint. Remove the splint and extend the mark into a curve, as illustrated.

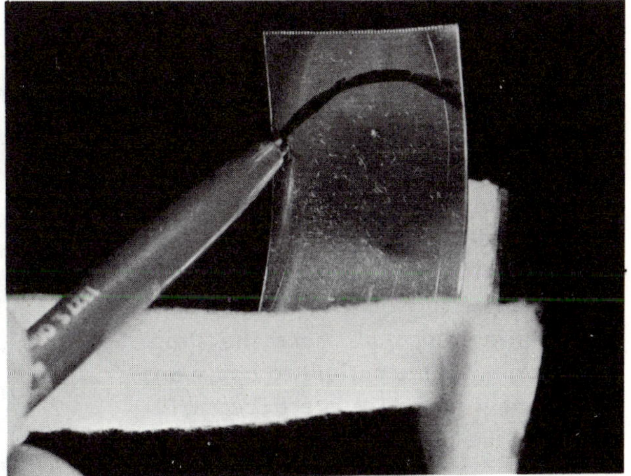

9. Trim the C-bar to the mark made in the previous step, then peen the distal edge of the bar forming a rolled edge to prevent it from cutting into the thumb. Round the corners and smooth all edges.

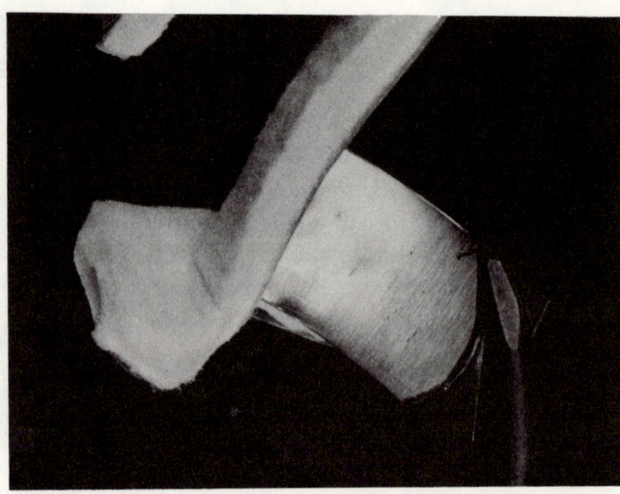

10. With the splint on the patient, check the C-bar and adjust if necessary. The curve of the bar should hold the thumb so the web between it and the hand is stretched enough to permit the thumb and fingers to come into opposition for useful prehension. If the web is tight, it should not be stretched painfully. In such cases the C-bar is adjusted over a period of time to slowly stretch the web until the desired amount of opposition is obtained.

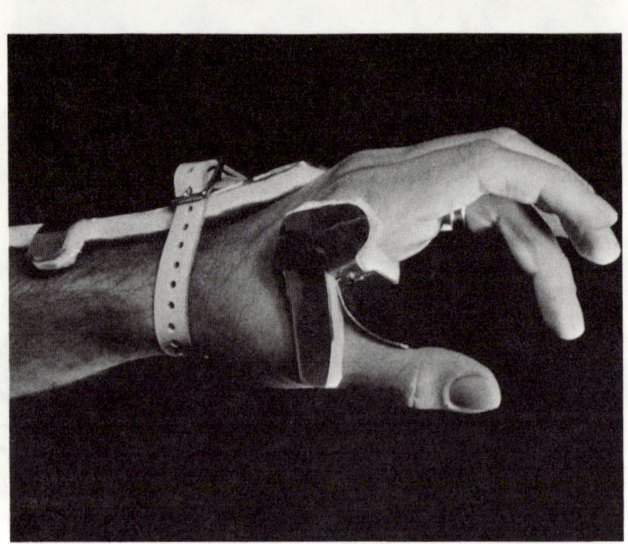

11. Check the straps for proper adjustment. The distal strap must be angled proximally so the pull of the strap is against the heel of the palm to prevent the splint from slipping off. The proximal strap should be just tight enough to hold the splint securely without excessive pressure on the patient's arm. While truss studs make the straps easier for the patient to apply and remove, buckles are acceptable and are often used.

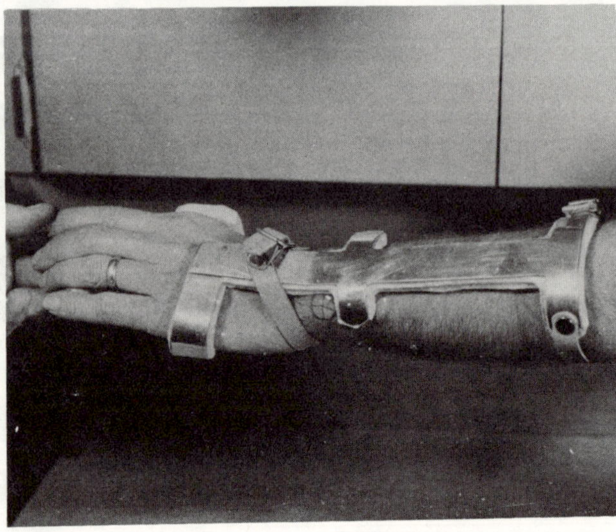

HOW TO MAKE AND FIT A THUMB I.P. EXTENSION ASSIST

1. Make the extension stop of .037 inch stainless steel, 1-1/2 x 3/4 inches. With the splint on the patient's hand, hold the stop between the thumb and the opponens bar so one end is just proximal to the thumb I.P. joint and the length of the bar supports the thumb. Mark this position, clamp the parts with a Vise-grip, drill two No. 40 holes, and rivet the parts together with one stainless steel rivet as shown in the illustration.

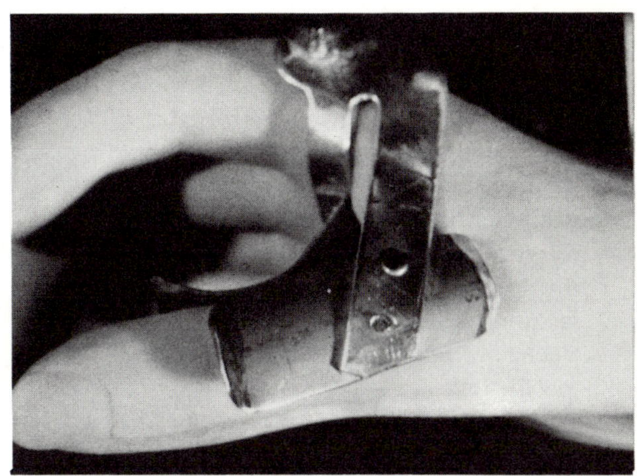

2. Make an assist spring, using a six inch piece of .047 inch stainless steel piano wire, winding three turns on a 3/16 inch mandrel. Bend a 3/16 inch loop close to and at right angles to the coils, then mount the assist spring on the opponens bar, using a stainless steel rivet in the unused rivet hole. Shape the spring into an arch, and bend a hook in the end. The hook should be positioned about 1/2 inch distal to the end of the C-bar.

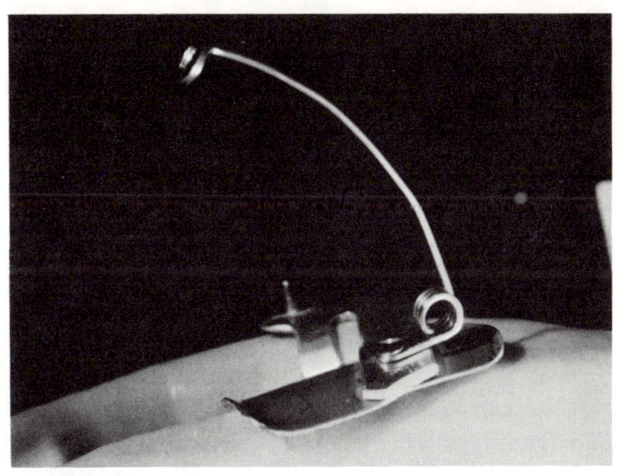

3. Cut a 3/4 inch wide strip of plastic from a piece of plastic tubing, long enough to encircle the thumb and hook onto the assist spring. Punch small holes in the ends of the plastic and hook the thumb piece onto the spring. With the splint on the patient, the spring tension should be just enough to return the thumb against the extension stop after it is flexed. Smooth and polish the stop, and cement a piece of 1/4 inch felt padding onto it.

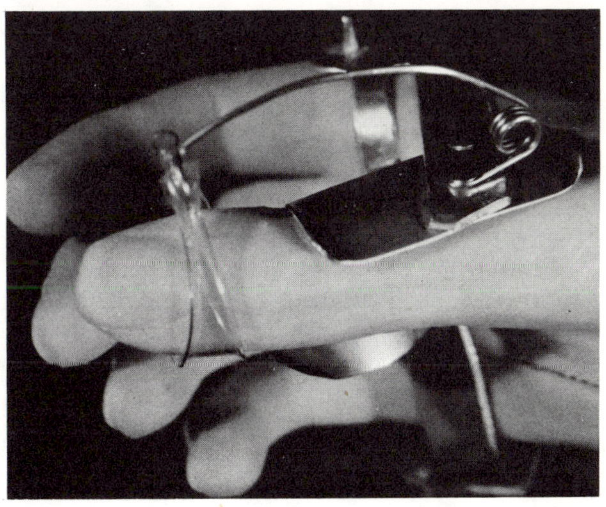

HOW TO MAKE AND FIT AN I.P. EXTENSION ASSIST WITH M.P. EXTENSION STOP

1. Install two attachment studs on the splint as explained in "How to Make and Fit an M.P. Stop". Try an attachment bar on the attachment studs. If the end of the bar strikes the truss stud or billet for the distal strap, trim it with shears as shown in the illustration. If it clears, this is not necessary. Round the corners on the attachment bar, smooth the edges, and polish it.

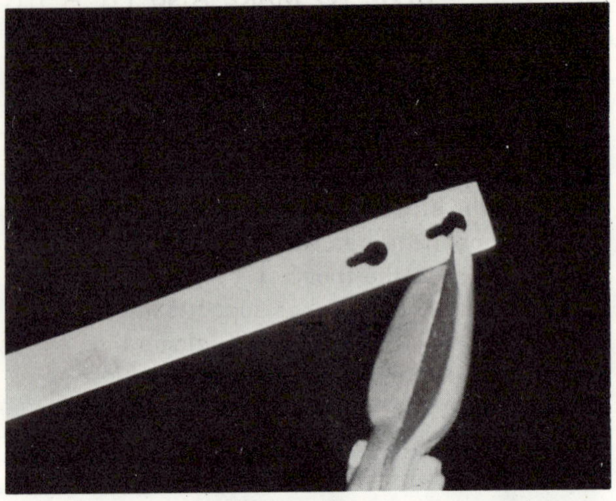

2. Make the M.P. stop out of a curved piece of 3/8 x .064 inch aluminum about 4 inches long. The radius of the curve should be 1-1/2 times the width of the patient's hand. Place the stop on the patient's hand slightly distal of the M.P. joints and mark it in line with the radial and ulnar aspects of the hand. Cut off the stop at these marks, round the corners and smooth the edges.

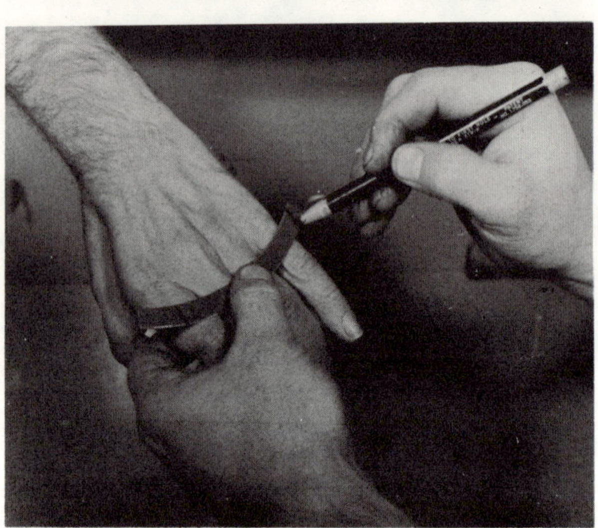

3. With the splint on the patient's hand, install the attachment bar, then bend it down about 10 degrees at the M.P. joints.

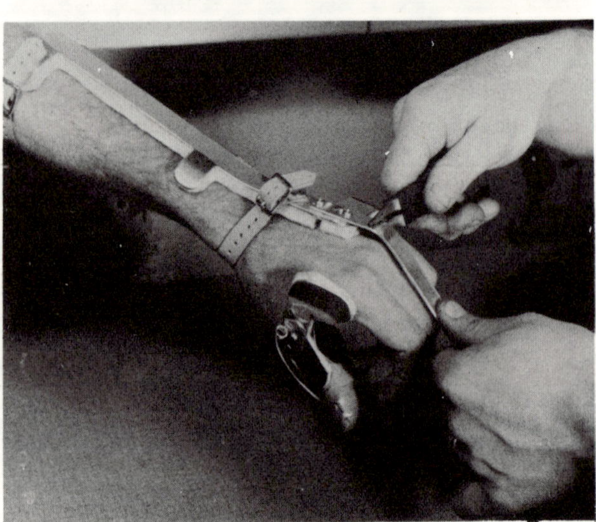

4. Place the M.P. stop under the attachment bar slightly proximal to the P.I.P. joints. Mark the position and clamp the parts together. Drill two No. 40 holes, and rivet the stop to the bar with stainless steel rivets.

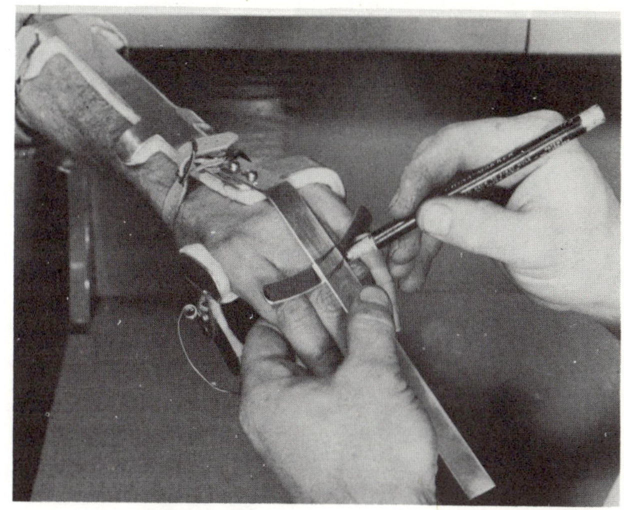

5. Bend the attachment bar up to a 75 degree angle, making the bend slightly distal to the M.P. stop as shown in the illustration.

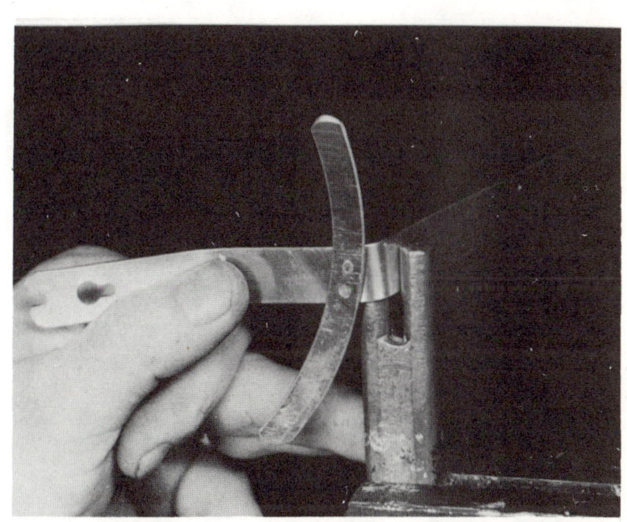

6. Measure 3-1/2 inches out from the bend made in the previous step, and make another bend at that point, bringing the bar parallel to that portion to which the M.P. stop is riveted.

7. Make the extension assist support
    out of a curved piece of 3/8 x 6
    x .064 inch sheet aluminum. The
    radius of the curve should be about
    1-1/2 inches greater than that of
    the M.P. stop.

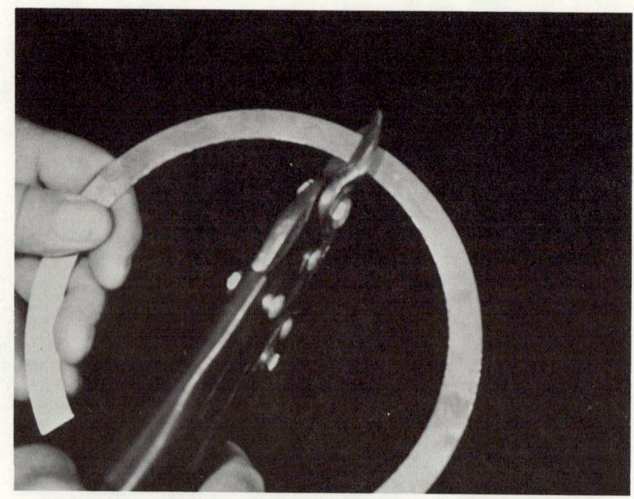

8. Locate the extension assist support
    on the attachment bar centered about
    1/2 inch distal of the last bend made.
    Adjust it so it is concentric with the
    M.P. stop. Drill No. 40 holes and
    rivet it with stainless steel rivets.
    Cut off the excess attachment bar
    stock protruding beyond the exten-
    sion assist support.

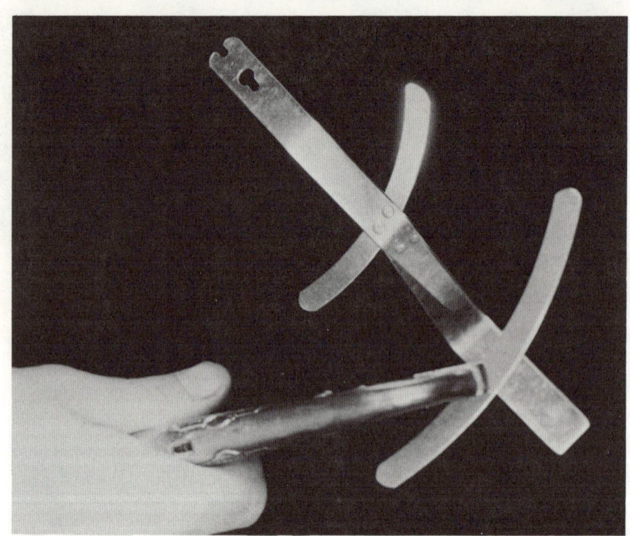

9. With the splint on the patient's hand,
    mark the extension assist support
    directly above each finger, with the
    fingers spread slightly. Drill a No.
    18 hole at each mark for fastening
    rubber bands. Clean out the burrs
    and bevel the sharp edges in each
    hole with a large drill to prevent
    the rubber bands from being cut.

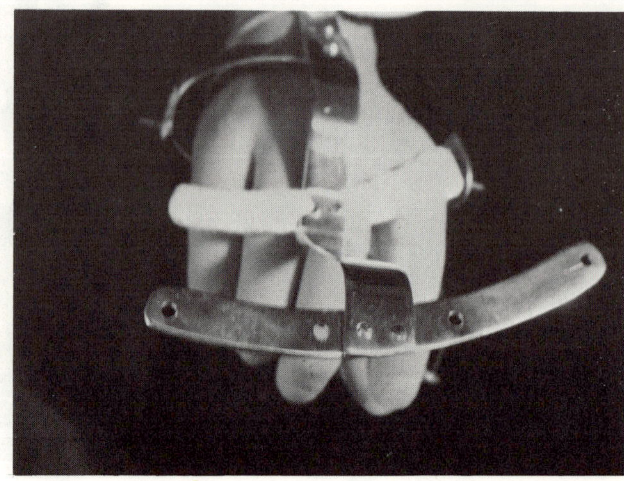

10. Make four plastic finger pieces of 3/8 inch wide strips of plastic, fit them to the fingers, and fasten the ends with grommets or punch No. 30 holes in the ends. Thread one No. 32 rubber band through each finger piece. Thread the rubber bands through the holes in the extension assist support, then try the splint on the patient, with the finger pieces on the distal phalanges. The extension assist support must be slightly distal to the ends of the fingers. The tension of the rubber bands must be sufficient to extend the I.P. joints, but not so great that the flexors cannot stretch them. In some cases the flexors may be so weak that strips of thin "rubber dam" used by dentists must be used until the patient gains enough strength to use regular rubber bands.

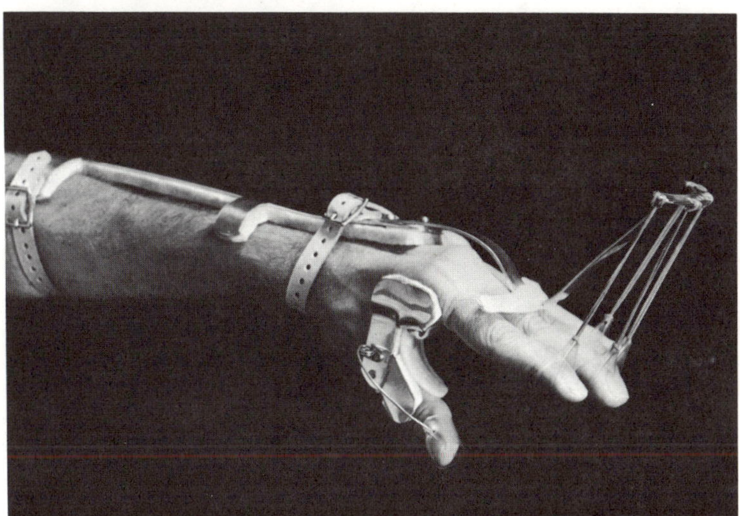

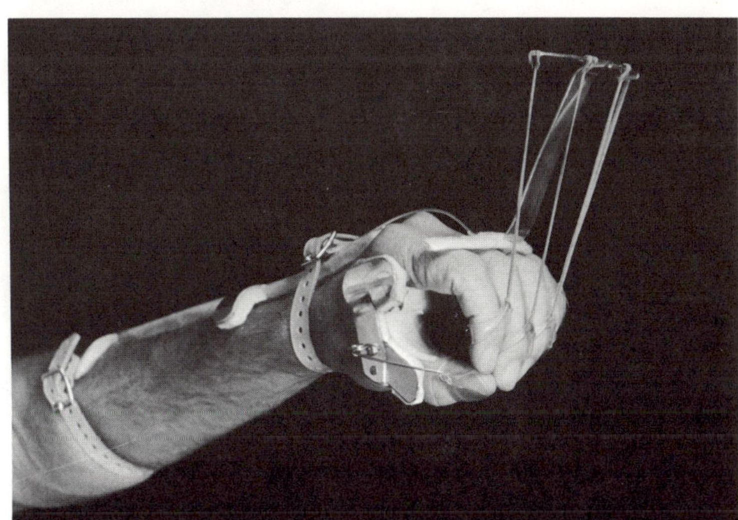

HOW TO MAKE AND FIT AN M.P. EXTENSION ASSIST

1. Install attachment studs on the splint, and fit an attachment bar on them as explained in "How to Make and Fit an I.P. Extension Assist with M.P. Stop". Make a 75 degree bend in the bar 1/2 inch distal of the forward attachment hole.

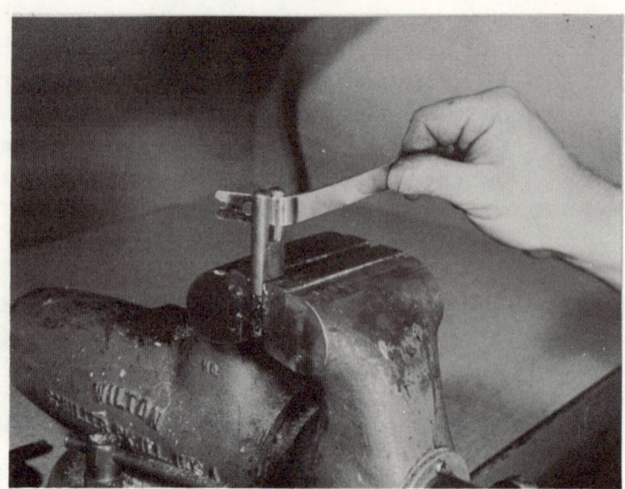

2. Measure 3-1/2 inches from the first bend, then make another at the same angle but the opposite direction, so the proximal and distal arms of the bar are parallel.

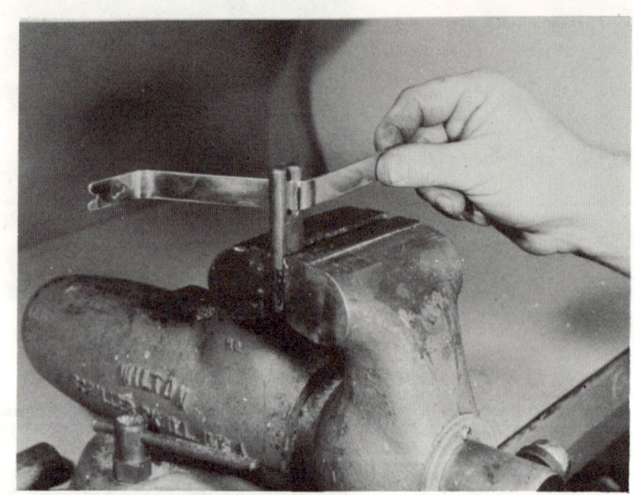

3. Make the extension assist support from a piece of .064 inch sheet aluminum 3/8 x 5 inches, and curved to conform to the patient's proximal I.P. joints. With the splint on the patient, hold the assist support in position on the attachment bar and adjust it so it is directly over the proximal I.P. joints. Clamp it in position, drill No. 40 holes, and rivet the two parts together with stainless steel rivets.

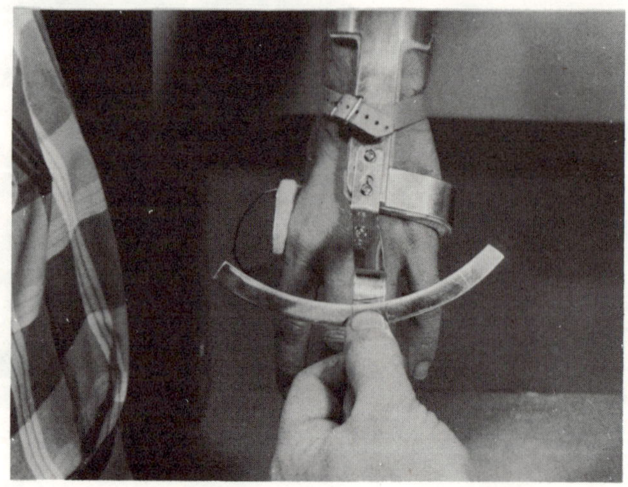

4.   Drill No. 18 holes in the extension
support located directly over the
P.I.P. joints, with the fingers
slightly abducted.   Make and fit
plastic finger pieces and fasten them
to the extension support with No. 32
rubber bands. Put the splint on the
patient and adjust as necessary. The
tension of the rubber bands should be
great enough to allow the I.P. joints
to flex first, then the M.P. joints.
The tension should then help extend
the M.P.'s when the patient relaxes
his grasp.

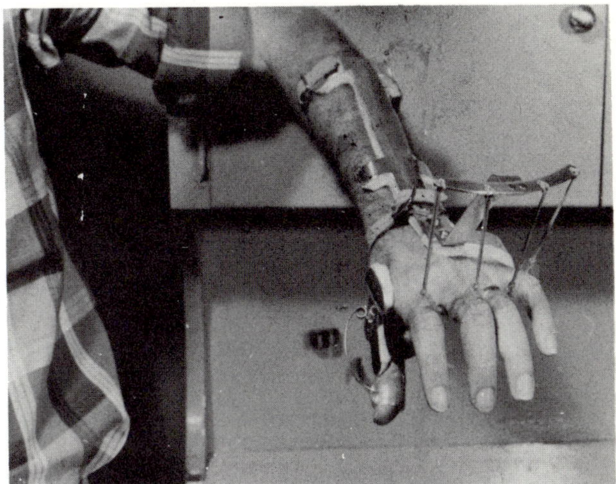

HOW TO MAKE AND FIT AN M.P. SPRING EXTENSION ASSIST

Introduction

The M.P. spring extension assist provides the same function as the rubber band type, but with two differences.

First, the springs lie comparatively close to the hand, so the patient can get his hand into his pocket, and can also perform other acts that would be difficult with the more cumbersome outrigger arrangement required for the rubber band type assist.

Second, the rate of tension build-up per unit of excursion is greater with springs than with rubber bands, making it difficult to adjust the springs to work well, particularly if the patient is quite weak. The basic tension characteristics of the spring are mainly determined by the diameter of the wire of which it is made. It has been found that .047 inch stainless steel piano wire is a good average size for the majority of cases, but if less tension is needed, a smaller diameter wire should be used, and a larger diameter wire should be used if more tension is needed.

The M.P. spring extension assist is a good device for patients who are fairly well stabilized and require few adjustments to compensate for changes. For this reason the rubber band type assist is often used during the time when the patient is changing rapidly and needs numerous adjustments, then the spring type assist is substituted when he stabilizes. It needs few if any adjustments.

1. An attachment bar and studs are fitted to the splint, then a transverse bar is made of 3/8 x .064 strip aluminum cut 1/2 inch shorter than the width of the patient's hand. The transverse bar is riveted to the attachment bar as close to the dorsal extension of the splint as possible. The extension springs are to be mounted on the transverse bar, and since they must arch down to the centers of the proximal phalanges, the bar must be positioned as far proximally as possible to permit the springs to be long enough for good function. The transverse bar should be parallel to the M.P. joints. Drill No. 40 holes and fasten with stainless steel rivets.

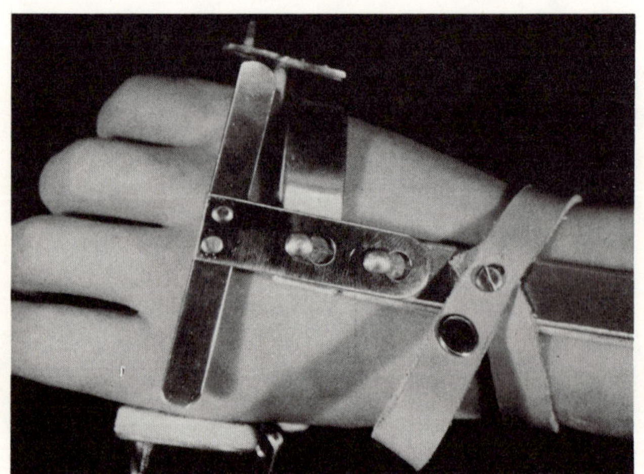

2. With the splint on the patient's hand, mark the transverse bar in line with the M.P. joints, then drill a No. 40 hole at each mark.

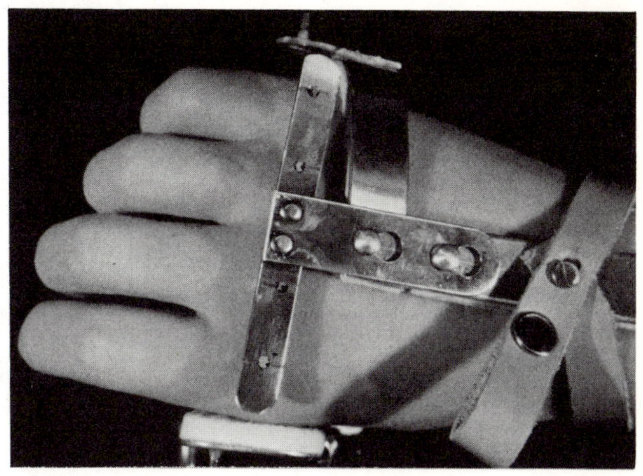

3. Make four assist springs, using six inch lengths of .047 stainless steel piano wire. Wind three coils in each piece about one inch from the end, using a 3/16 inch mandrel, and ending with the one inch part at right angles to the main length of the spring. Bend a 3/32 inch rivet loop in the short end of each spring with the loops at right angles to the coils, as shown in the illustration.

4. Rivet the assist springs to the transverse bar, using stainless steel rivets. The springs must be positioned so they will point distally when bent down.

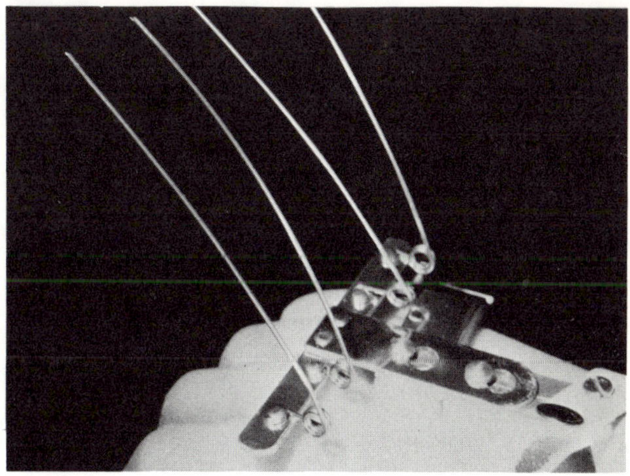

5. Make four finger pieces of strips of 3/8 inch wide plastic cut from plastic surgical tubing. Fit them to the proximal phalanges and fasten the ends with grommets to accommodate the hooks in the ends of the assist springs.

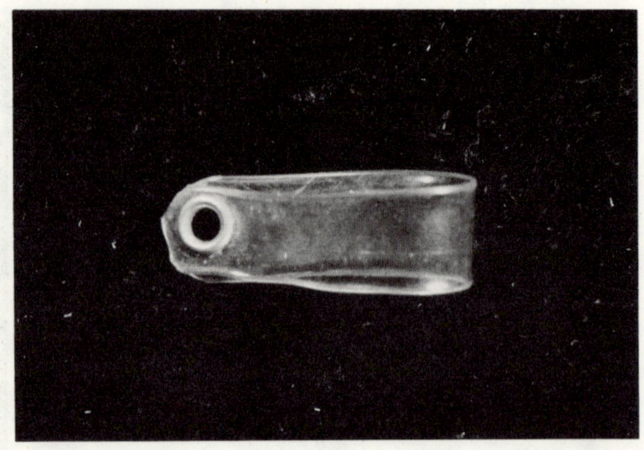

6. Shape the assist springs into arches terminating in hooks directly above the midpoints of each proximal phalanx. Fasten the plastic finger pieces to the hooks, place the splint on the patient's hand and adjust it. The springs should pull directly up on the phalanges, and should be adjusted so the M.P.'s are held in extension enabling the I.P.'s to flex first, then permitting the M.P.'s to flex and bend the assist springs. The springs should extend the M.P.'s when the patient relaxes his grasp. If the springs are too weak, new ones must be made using larger diameter wire; if they are too strong, smaller diameter wire must be used.

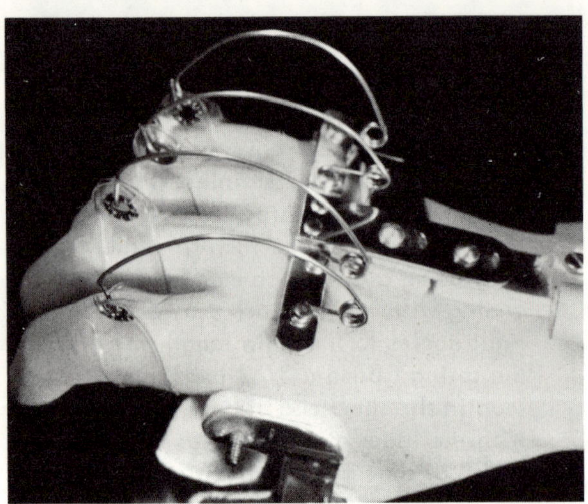

HOW TO MAKE AN ADJUSTABLE M.P. CONTROL

1. Make the ulnar stud mounting extension of a 2 x 5/8 x .064 inch piece of sheet aluminum. Hammer one end concave to match the ulnar extension on the palmar piece, then rivet it on the center line of the extension with stainless steel rivets. Drill a No. 29 hole in it opposite the fourth M.P. joint, then thread the hole with a 6-32 tap. Taper the end of the extension as shown in the illustration.

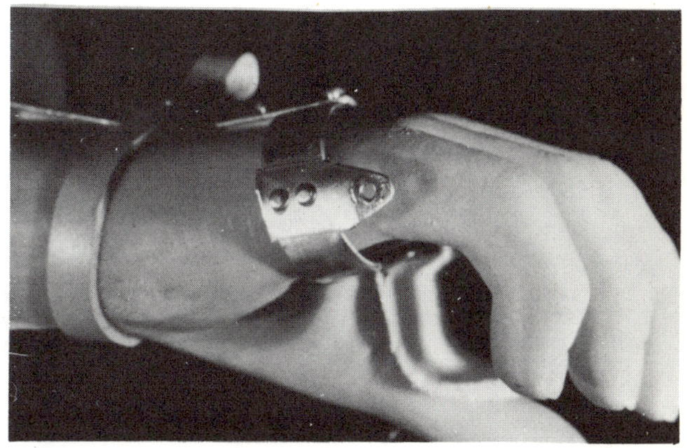

2. Drill a No. 29 hole in the radial extension of the splint, opposite the first M.P. joint, and thread the hole with a 6-32 tap.

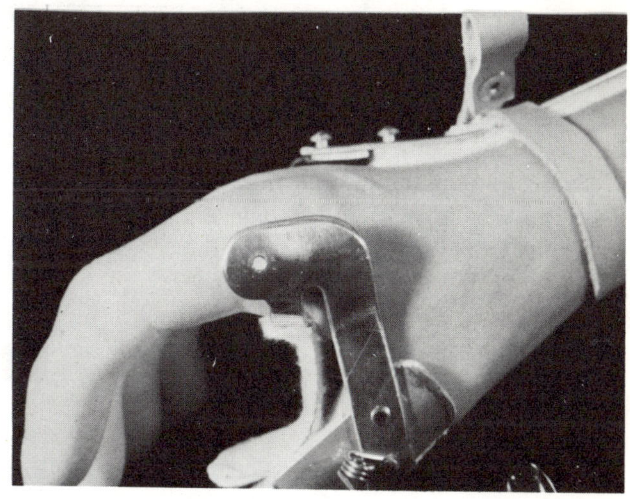

3. Install a 3/8 inch 6-32 round head machine screw in the threaded hole in the ulnar stud mounting extension, then sweat solder around the screw head to prevent it from backing out. Use Alumaweld Solder and Alumaweld Liquid Allmetal Flux, and a torch. Do not get solder on the screw threads. File the screw head flat.

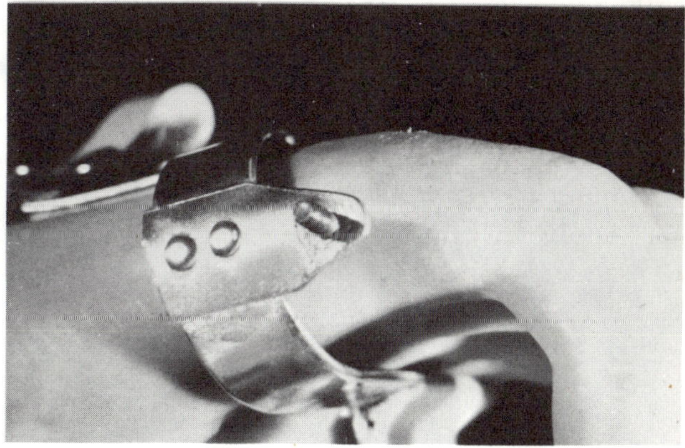

4. Install a 3/8 inch 6-32 round head steel machine screw in the threaded hole in the radial extension, solder it in place, and file the head flat.

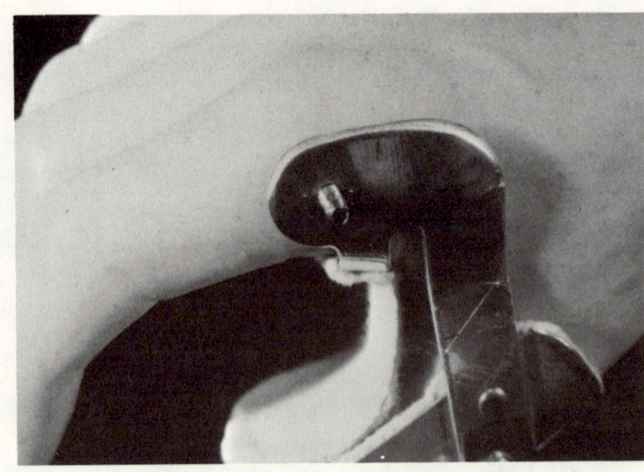

5. Cut two pieces of 2 x 3/8 x .064 inch sheet aluminum for the radial and ulnar links which will hold the swivel bar in position on the proximal phalanges. Drill a No. 29 hole in one end of each piece, then fasten one piece to the ulnar stud with a wing nut. Mark the link 1/4 proximal to the fourth proximal I.P. joint, drill a No. 40 hole at that point, then round the corners. Smooth the edges, and de-burr the holes in the link.

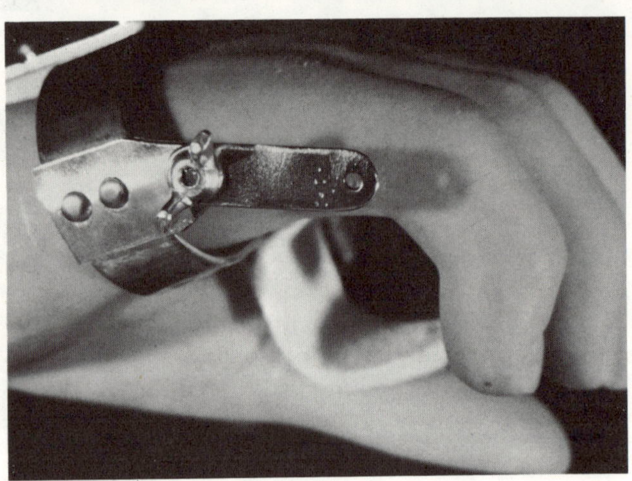

6. Fasten the other link to the radial stud with a wing nut, mark it 1/4 inch proximal to the first proximal I.P. joint. Drill a No. 40 hole at the mark, then round and smooth the link.

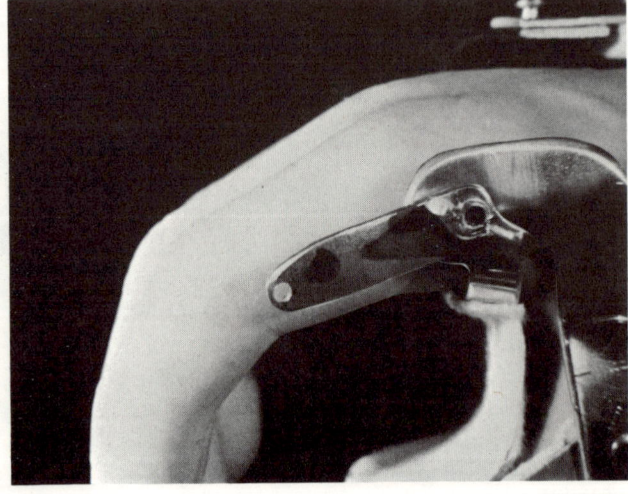

7. Cut a piece of 6 x 5/8 x .064 sheet
    aluminum for the swivel bar. Place
    the piece on the back of the patient's
    proximal phalanges, and make it
    parallel to and even with the radial
    and ulnar aspects of the hand, as
    shown in the illustration.

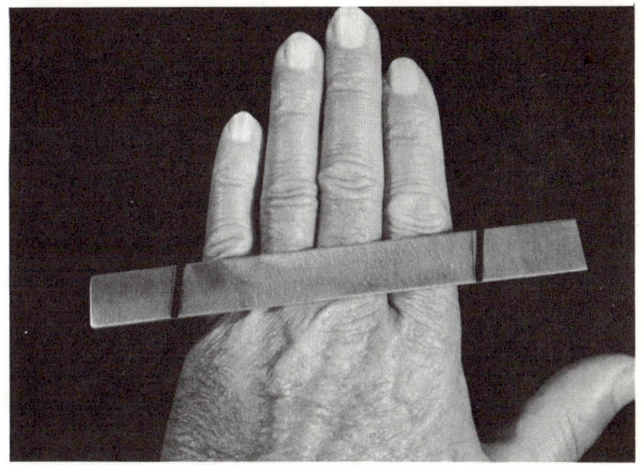

8. Make a 90 degree 1/4 inch radius
    bend in the bar at the marks made
    in the previous step.

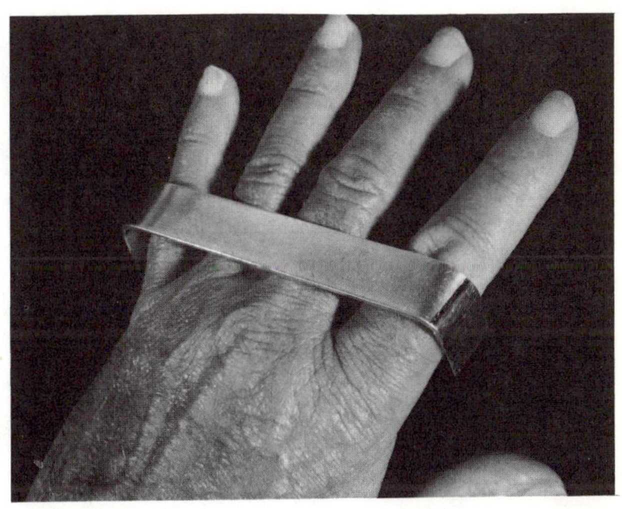

9. To lay out the swivel bar mounting
    holes, make a mark 1/2 inch down
    from the dorsal surface of the bar,
    then make another 3/16 inch distal
    from the proximal edge of the bar
    so the two marks cross.

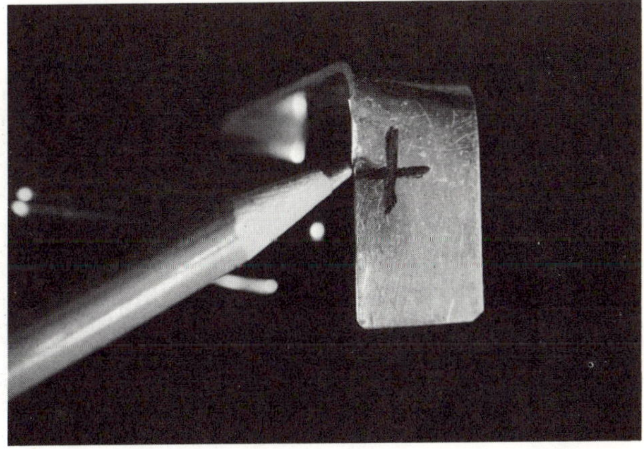

10. Drill No. 40 holes at the inter-
section of the marks made in the
previous step, then trim away
the excess metal around the holes
and smooth the rough edges.

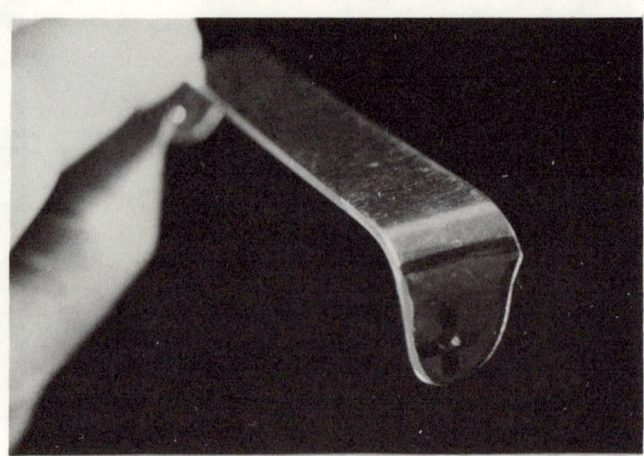

11. Rivet the links to the swivel bar,
using stainless steel rivets, and
placing a Teflon washer between
the links and the bar. The rivets
must be set just tight enough to
hold the bar in position, but not
so tight that its position cannot
be easily adjusted.

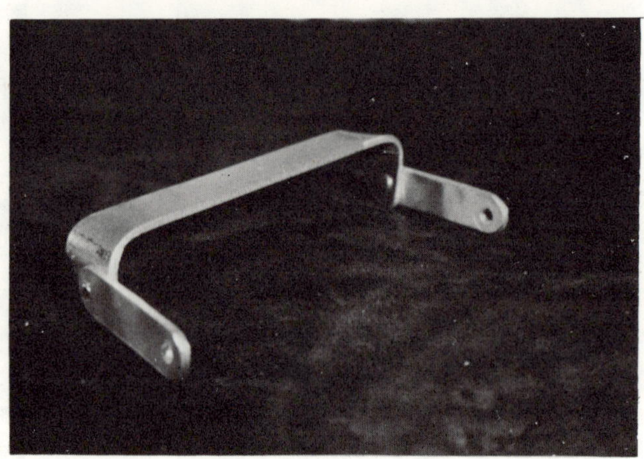

12. Pad the inner surface of the bar
with 1/4 inch felt, then install
it on the splint. Place a star lock
washer between each link and
the splint so the link cannot
change position when the wing
nut is tightened.

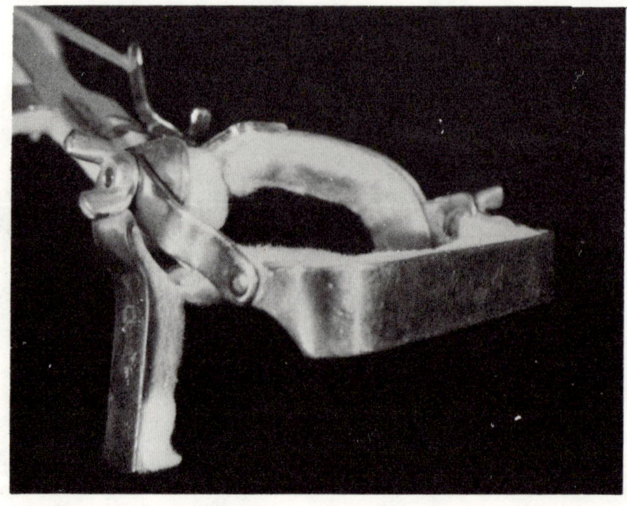

13. Put the splint on the patient's hand and make necessary adjustments. The swivel bar should be adjusted so the desired amount of flexion of the M.P. joints is obtained. The most common use of the M.P. control is to maintain the gains made by the therapist in stretching M.P. contractures, so this adjustment is made after each treatment. The wing nuts must be tightened securely to hold the bar in position.

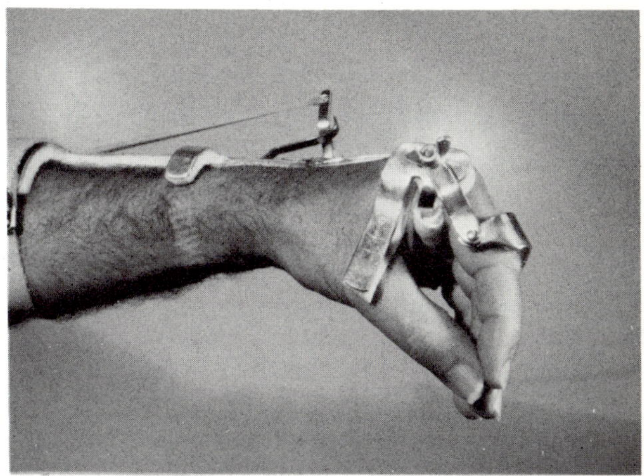

14. The swivel bar must press evenly across the dorsal surfaces of the proximal phalanges. It may be necessary to make slight bends in the links to obtain the best alignment.

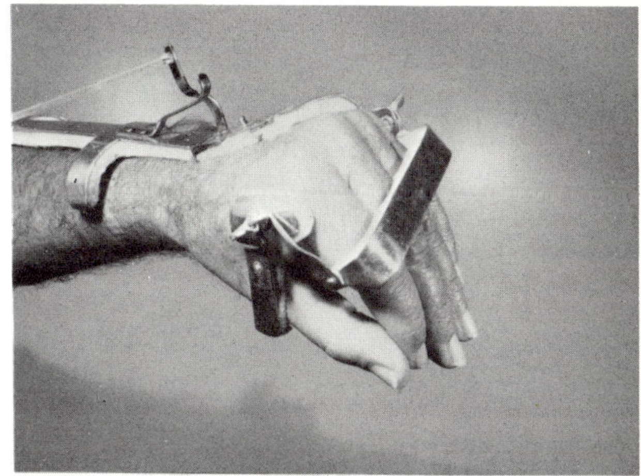

15. An I.P. extension assist is sometimes used with an M.P. flexion control. This is done by following the instructions for making an M.P. extension assist, except the extension assist bar is located farther distally, and the finger pieces are placed on the distal I.P. joints, as in the illustration.

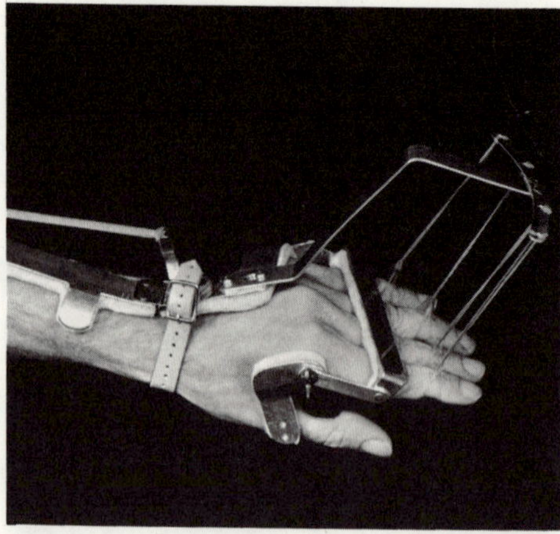

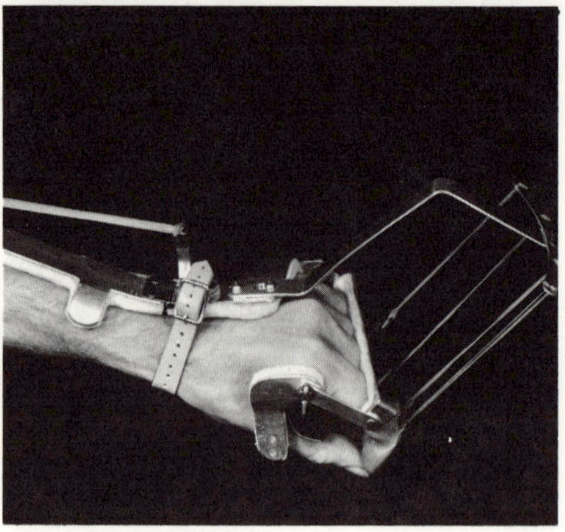

HOW TO MAKE AN I.P. FLEXION CONTROL

1. The I.P. Flexion Control is always used with the M.P. Flexion Control and in making it all the steps required for making the latter are performed up to fitting the radial and ulnar links.   These must be made longer, so they extend an inch distally from the M.P. swivel bar. Make the radial link and assemble it to the swivel bar as illustrated.

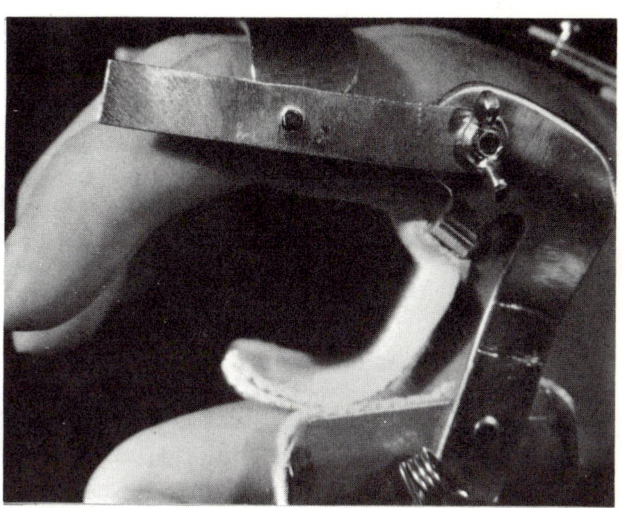

2. Make the ulnar link so it will extend an inch distally from the swivel bar, and assemble it as illustrated.

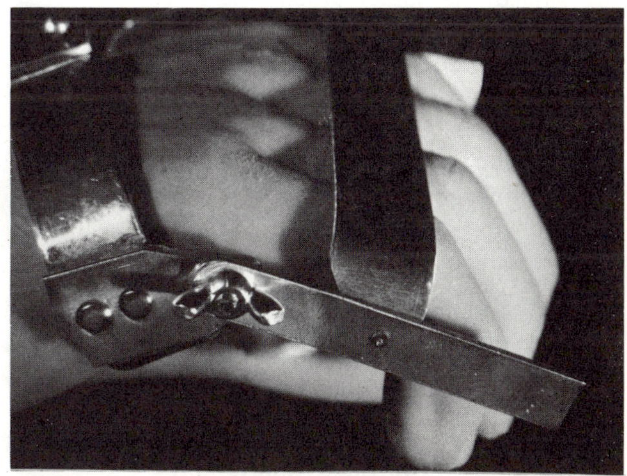

3. Mark the radial and ulnar links
opposite the proximal I.P. joints
of the little and index fingers,
and drill No. 29 holes at the
marks on center. Round the ends
of the links and smooth rough
edges. The ulnar link is illustrated; the radial side is done in
the same manner.

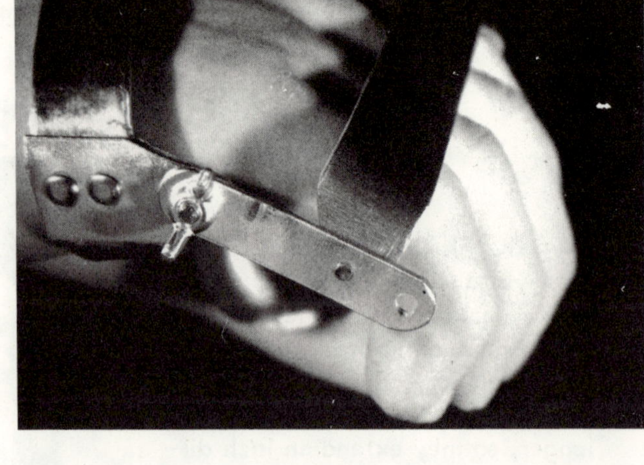

4. Cut two pieces of 1 x 3/8 x .064
inch sheet aluminum to make the
distal links which connect the M.P.
and I.P. swivel bars. Drill a No.
29 hole in one end of each, thread
the hole with a 6-32 tap, then
screw a 3/8 inch 6-32 round head
steel screw into each hole. Sweat
Alumaweld solder around the screw
heads, using Alumaweld liquid flux
and a torch. Fasten the distal links
in place on the links, with 6-32
wing nuts. Use care to see that a
star lock washer is placed between
the links to lock them together when
the wing nut is tightened. This will
prevent the adjustment from slipping.

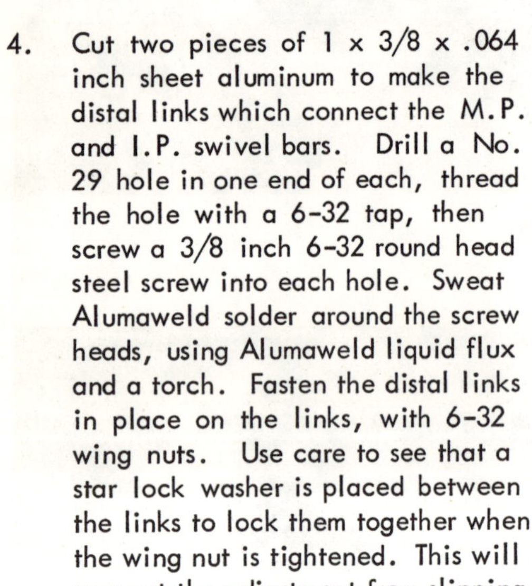

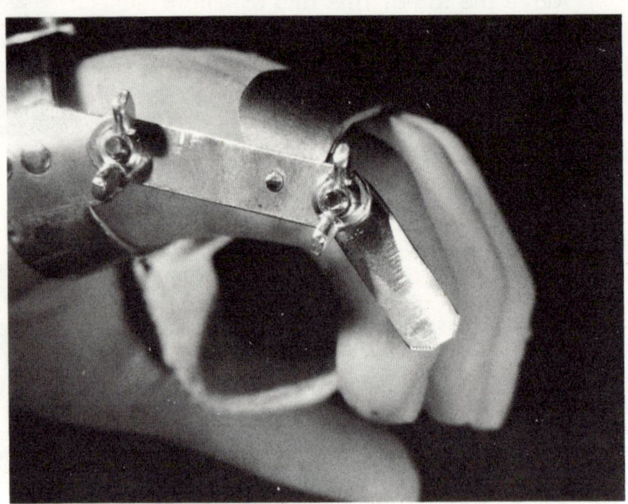

5. Mark the distal links slightly proximal to the distal I.P. joints, then
drill No. 40 holes at the marks.
Make an I.P swivel bar the same
as the M.P. swivel bar, and rivet
it in place between the distal links,
using stainless steel rivets with a
Teflon washer.

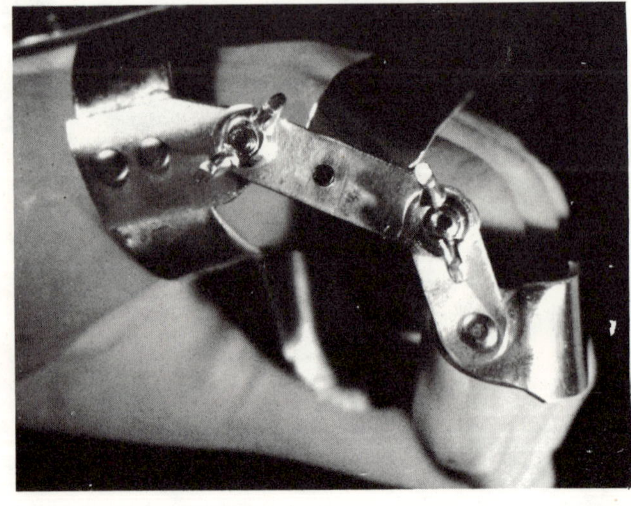

6. Pad the I.P. flexion control swivel bar with 1/4 inch felt, then put the complete splint on the patient and adjust it. The I.P. extension swivel bar should exert equal pressure on all four middle phalanges and should exert it on the same relative location on each one. Major adjustments can be made by loosening the wing nuts and pivoting the links.

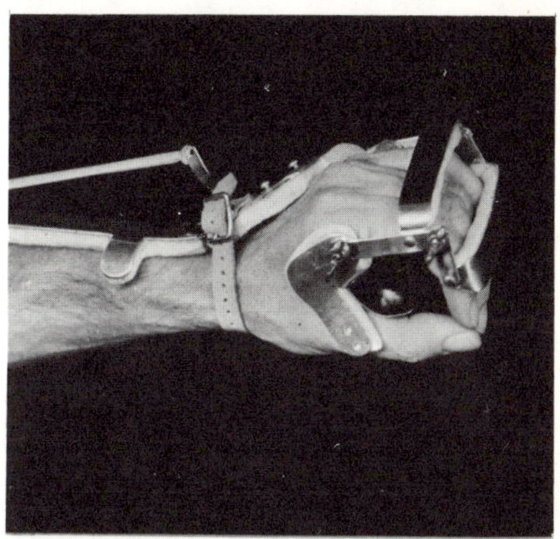

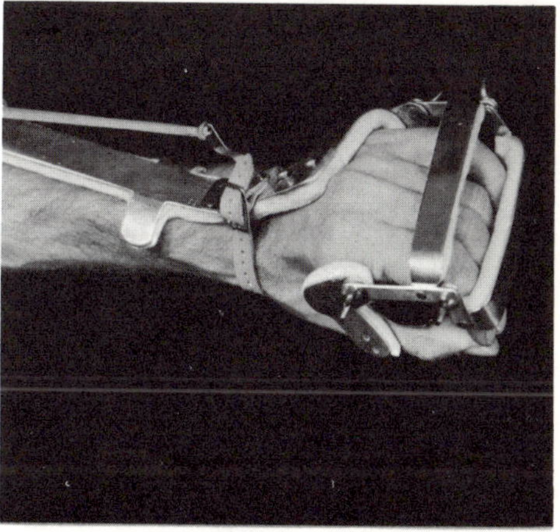

HOW TO MAKE AN ACTION WRIST WITH DORSIFLEXION ASSIST

1. An action wrist is a long opponens hand splint with a hinge that permits the wrist to dorsiflex. The hinge must be located at the wrist joint, so first mark the dorsum of the patient's wrist over the joint, which can be identified as the base of the "V" formed when the wrist is dorsiflexed.

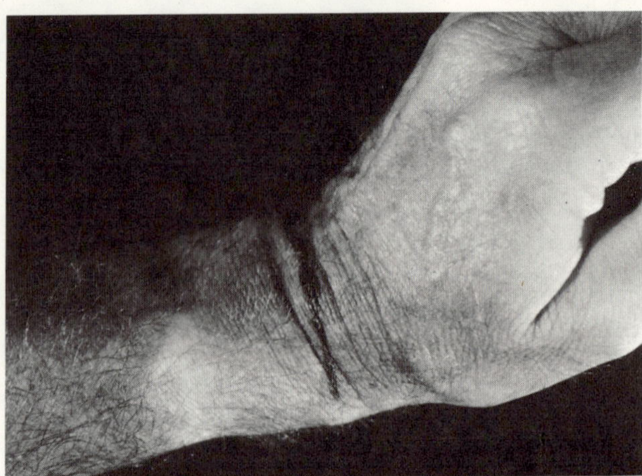

2. Place the splint in position on the patient's arm and mark it over the wrist joint mark made in the previous step. With a hack saw or shears cut the splint at this mark, being careful to make the cut square with the sides of the forearm piece.

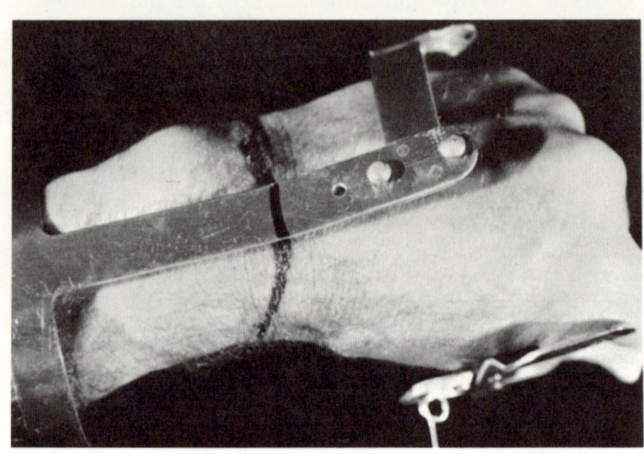

3. A hinge must be installed between the two parts of the splint. A piece of stainless steel 3/4 inch piano hinge is cut from a length of the material, the hinge pin soldered to prevent it from falling out, and 1-3/4 x 1/4 x .064 inch stainless steel operating lever silver soldered to one side of the hinge, as shown in the illustration. The length of the hinge is the same dimension as the width of the splint at the point where it was cut in two. A No. 26 hole is drilled in the end of the operating lever to attach the rubber bands.

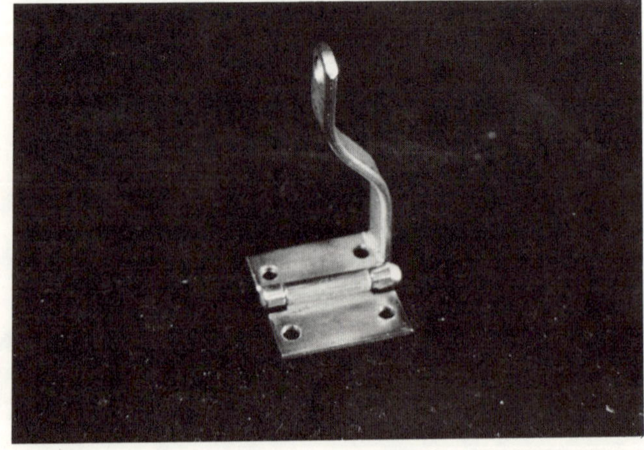

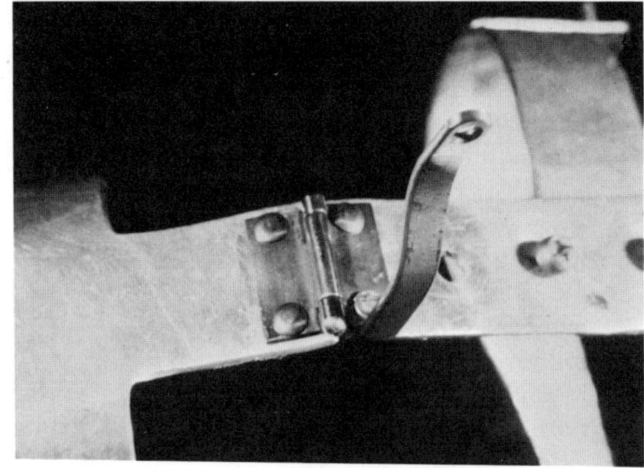

4. The hinge is installed in the splint with stainless steel rivets.

5. Rivet a boot lace hook in the proximal end of the forearm piece for attaching the rubber bands.

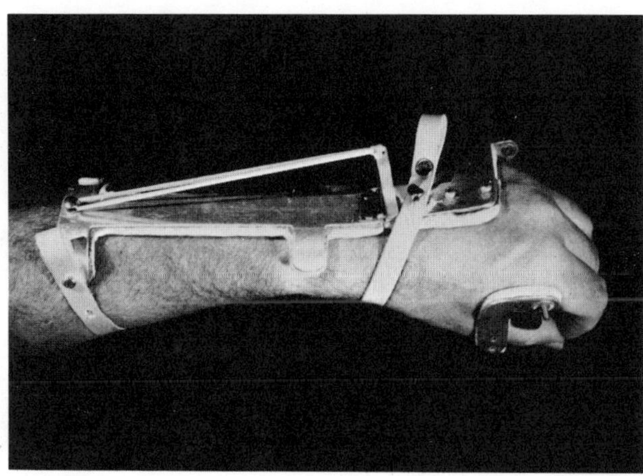

6. Replace the wrist straps, and thread a No. 32 rubber band through the hole in the operating lever and over the boot lace hook. Try the splint on the patient. The rubber band should exert enough force to pull the wrist to about 30 degrees of dorsiflexion after the patient flexes it. It may be necessary to try various weights and numbers of rubber bands to find just the right combination.

7. The hinge type action wrist is usually satisfactory for most patients. However, the axis of the hinge is not in line with the axis of the wrist joint, causing the splint to shift on the arm when wrist movement takes place. If it is necessary to avoid this, the splint can be modified as shown in the illustration. Aluminum bars are extended out from the forearm piece to the palmar piece with pivots adjacent to the wrist joint axis. A stainless steel spring provides the extension assist action.

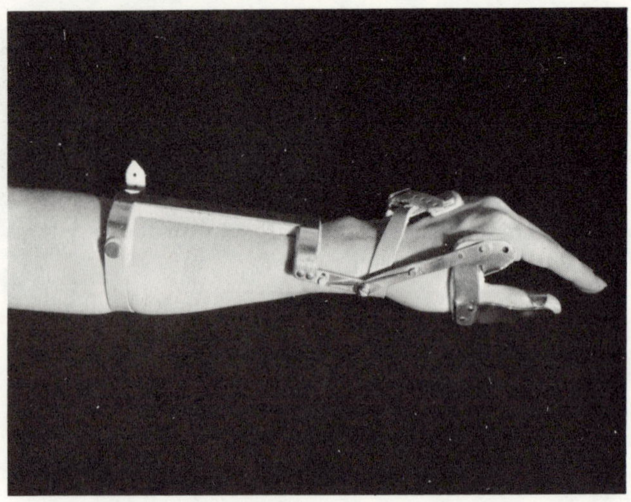

8. Details of the hinge joint construction are shown in the illustration. A 10-32 truss stud is used for the pivot, with a Teflon washer between the bars to serve as a bearing. A stainless steel spring is wound to fit the stud. The arms of the spring may be fastened to the bars with rivets or other convenient means. Sharp wire ends must not be left exposed, as they might hurt the patient.

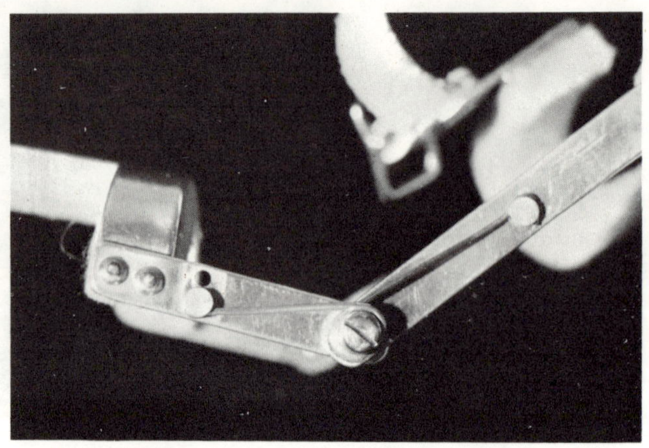

9. In some cases it is desirable to limit wrist flexion to prevent the patient from flexing his wrist beyond the position of maximum finger function. A flexion stop made of 1/4 inch aluminum is riveted to the forearm piece and bent to the shape shown in the illustration so the hinge lever movement is limited.

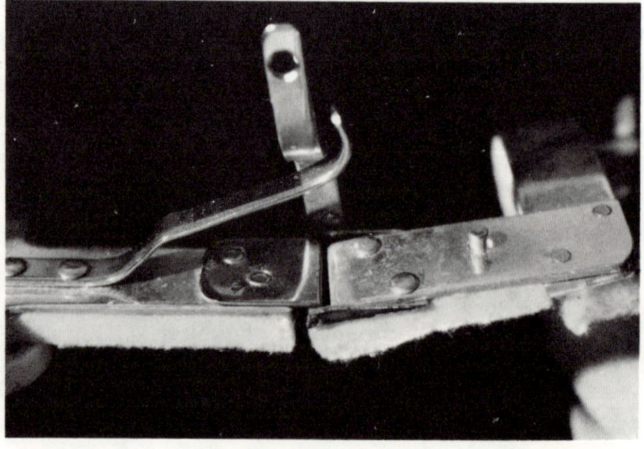

HOW TO MAKE AND INSTALL A HAND SPLINT PROP

1. The hand splint prop is used on the long opponens hand splint only. It supports the patient's hand so the weight of the hand and splint will not be borne by the thumb. Install attachment studs in the palmar extension of the splint. Follow the same procedure as before, but be sure to locate them so the attachment bar, when installed, will be at a right angle to the patient's arm.

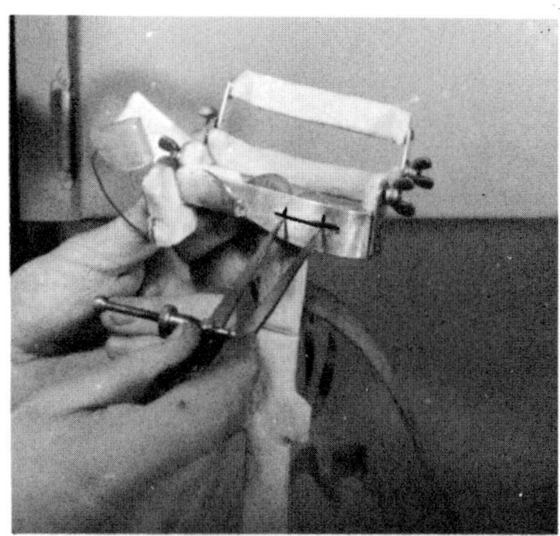

2. Make an 8 inch attachment bar, then bend it into a "U" shape as shown in the illustration.

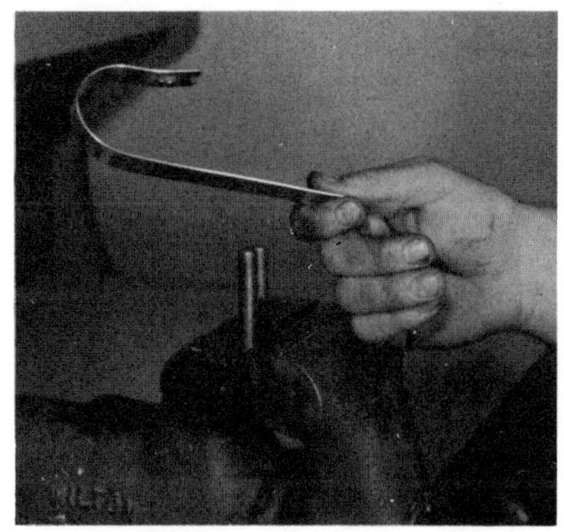

3. Install the prop on the attachment studs and adjust. The lower support arm of the prop should lie flat when the arm is in a relaxed position, and should hold the hand so the thumb bears no weight.

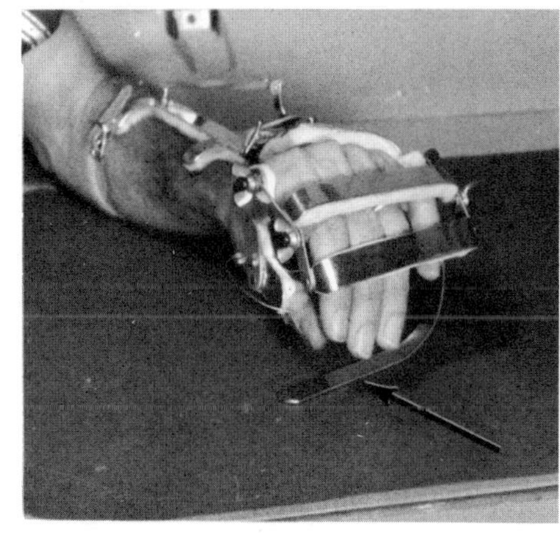

HOW TO MAKE AN I.P. JOINT STABILIZER

1. Cut a piece of .025 inch stainless steel 2 inches by 3-1/2 inches.

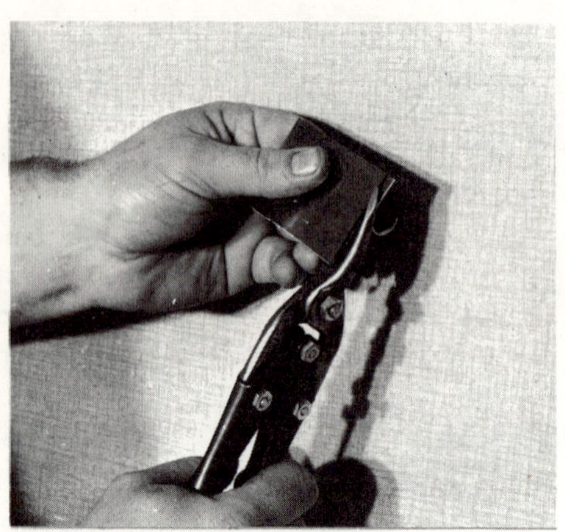

2. Square one corner and sides.

3. Measure the distance from the center of the proximal phalanx to a point just proximal to the distal interphalangeal joint on the patient's finger.

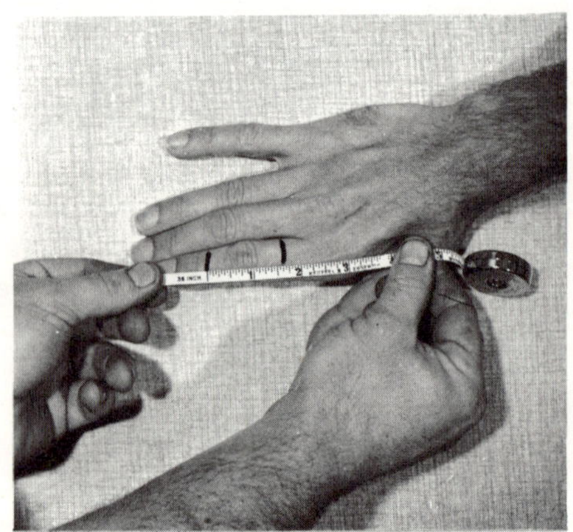

4. Lay out this measurement on one of the
    short sides of the piece of steel, and
    cut it off to this length.

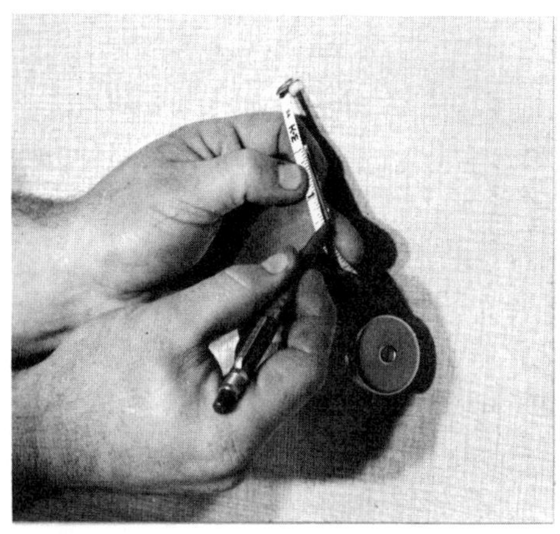

5. Set the dividers to 5/16 inch for adults,
    1/4 inch for children, and scribe along
    both long sides of the piece.

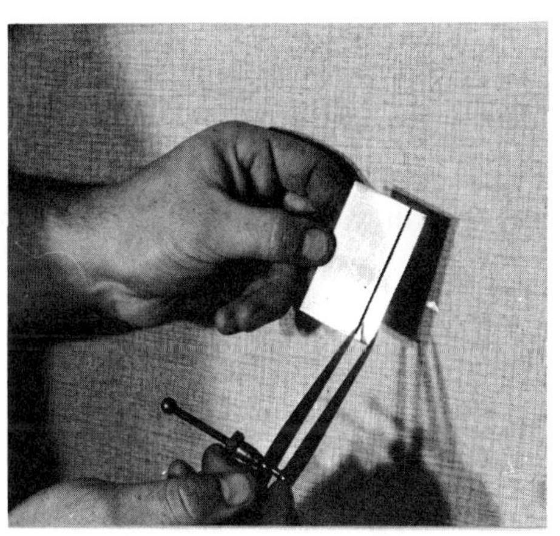

6. Set the dividers at 1/2 inch for adults,
    3/8 inch for children, and scribe across
    one end, connecting the two lines,
    marked in "5".

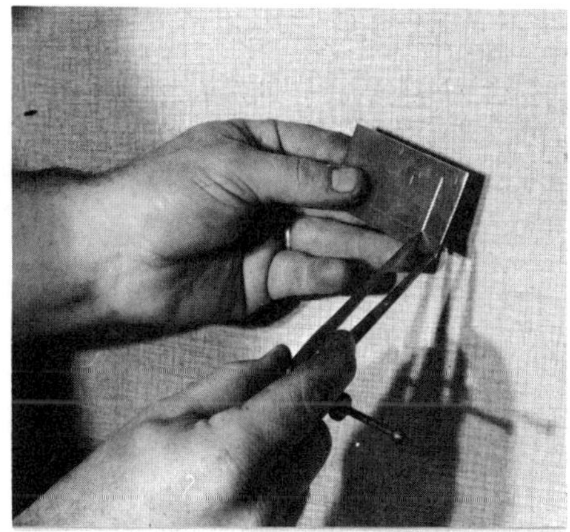

7. With the metal shears, cut along both of the long lines laid out in "5".

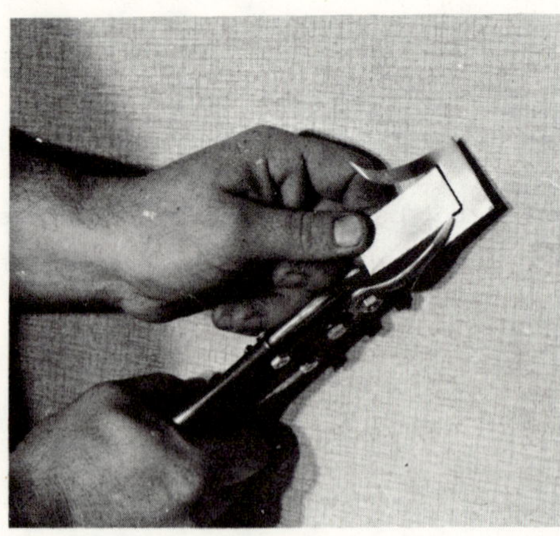

8. With the piece held firmly in the vise, cut the middle portion out with a hammer and chisel, cutting on the line laid out in "6".

9. From the measurement form find the circumference of the patient's proximal and middle phalanges. Lay out these measurements on the long legs of the piece, including the width of the connecting bar, and cut off at the marks. Smooth all edges, and round all corners.

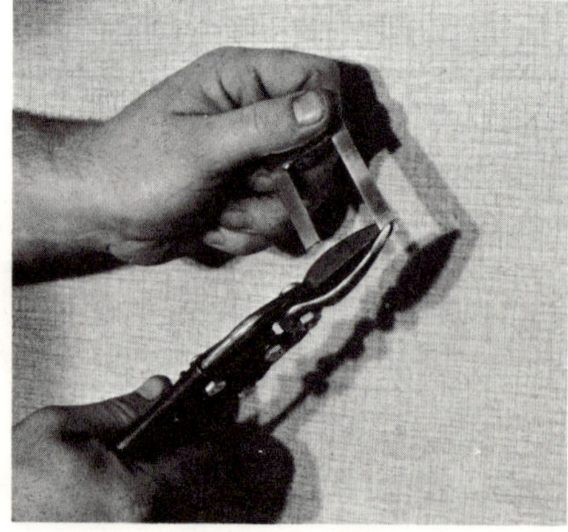

10. Form the connecting bar between the two long legs to the shape of the patient's finger. A 1/2 inch piece of round steel will serve as a mandrel for the average finger. Shape in the vise as shown.

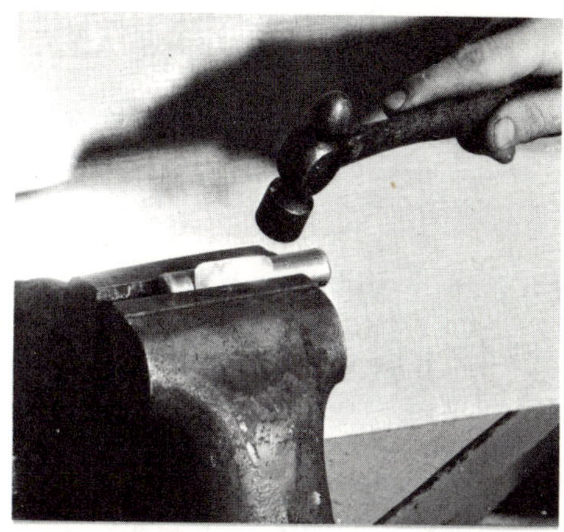

11. Form the two long legs into rings, using the round nose pliers.

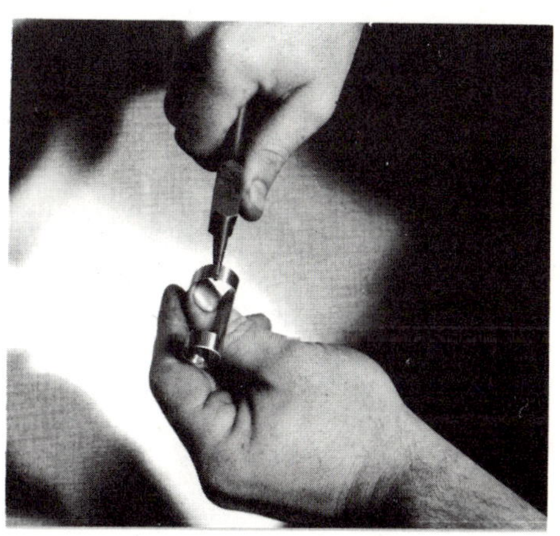

12. Try the stabilizer on the patient's finger and make minor adjustments as needed for snug but pain-free fit.

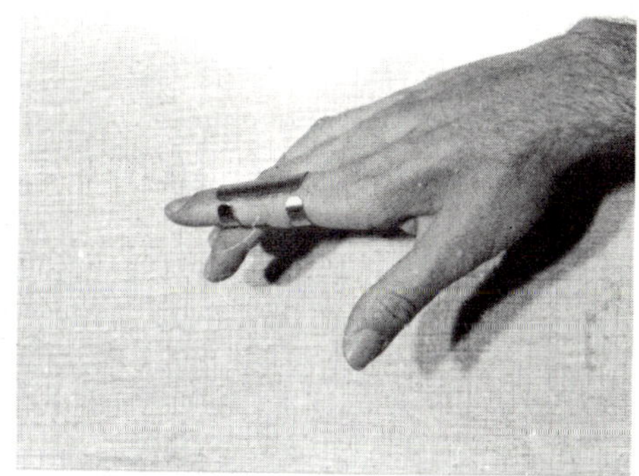

13. If the stabilizer is to be used with an extension assist, a small ring must be soldered to it for attachment to rubber band or spring. Form the ring of 1/16 inch stainless steel wire to a diameter of approximately 3/32 inch.

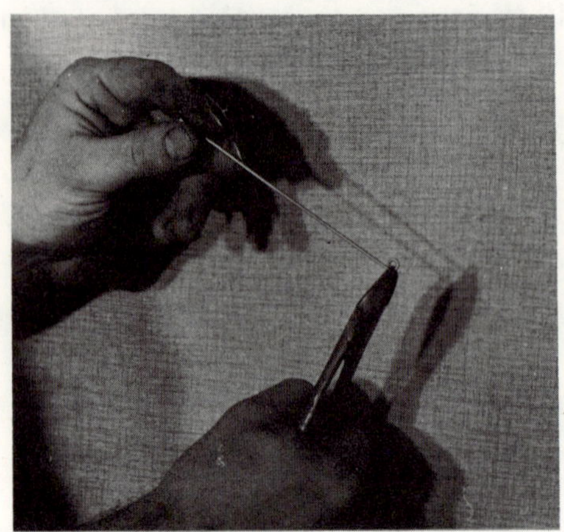

14. Solder the ring to the stabilizer.

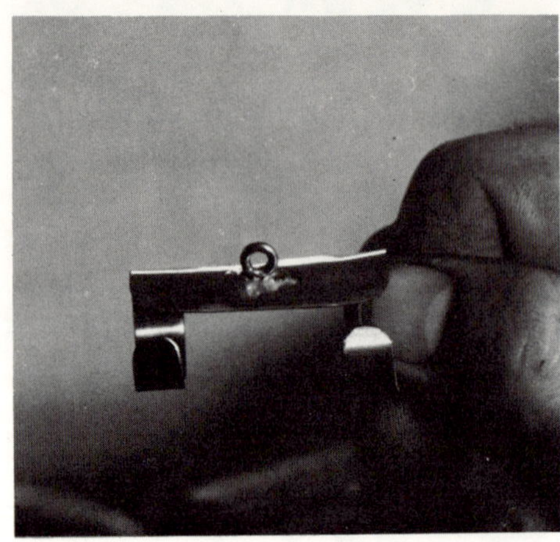

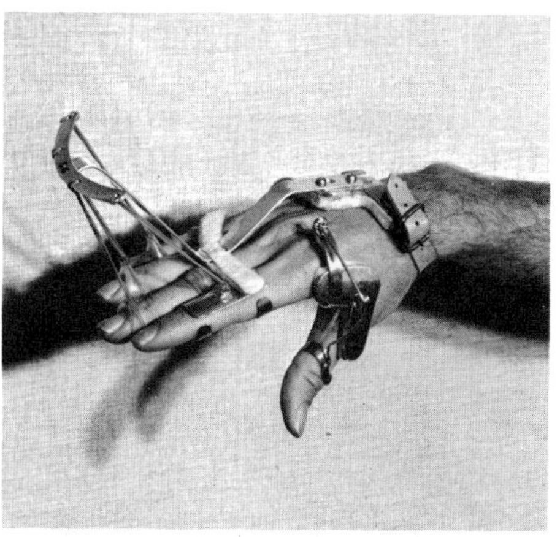

15. Mount the stabilizer on the
    extension assist.

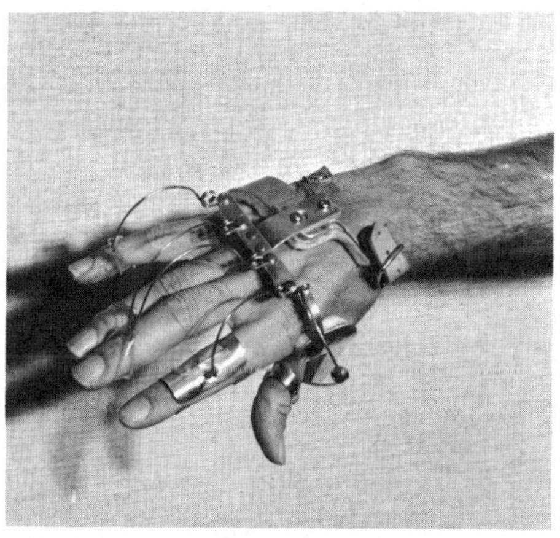

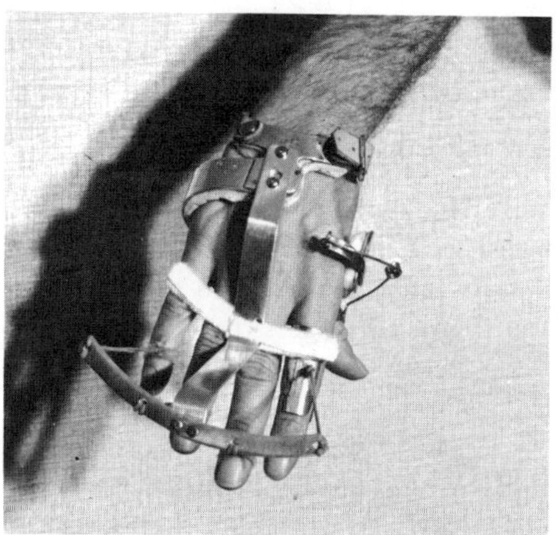

HOW TO MAKE A FINGER-DRIVEN FLEXOR HINGE SPLINT

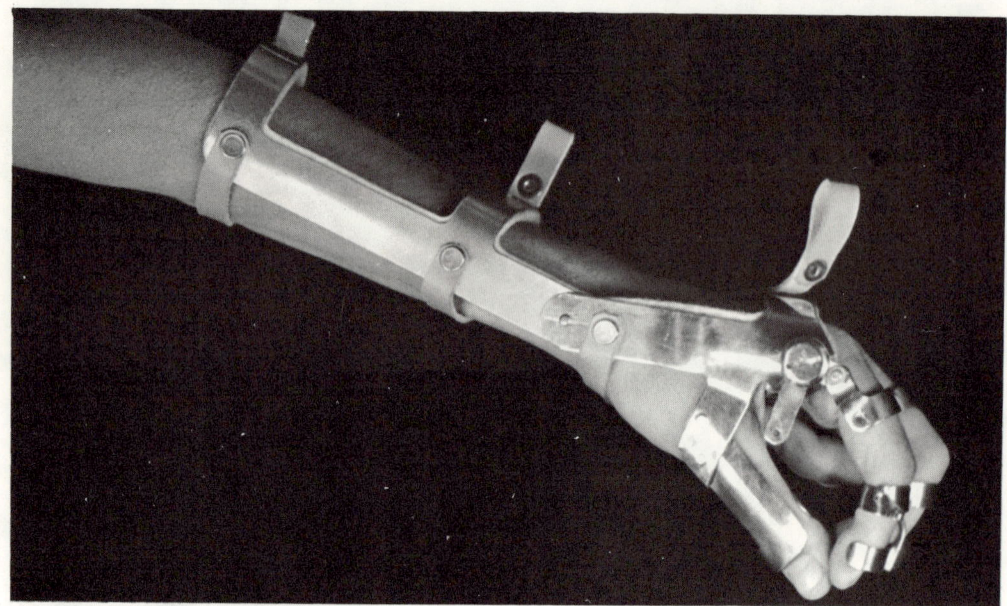

Introduction

The flexor hinge splint is designed to provide a functional "three-jaw chuck" type prehension for the patient who has lost the normal ability to grasp with his fingers. In this type of prehension, the index and long fingers contact the thumb, as shown in the illustration. While it has its limitations, this "three-jaw chuck" grasp is found to be one of the most frequently used in ordinary daily activities, and for this reason was selected for use with this splint.

Of the several types of flexor hinge splints, the finger-driven is the simplest. It is useful for patients who have flexion in one finger and extension in another, or various combinations, as extension in two and flexion in one, and so on. The hinged finger piece harnesses the fingers together, and the thumb piece holds the thumb in a position of opposition to the fingers. The finger or fingers that have good extensor muscles open the grasp, those with good flexors close it, while the finger piece transmits the motion from one to the other.

Modifications of this splint for use where flexion alone, extension alone, or neither is present, will be explained in later sections.

1. Select the correct splint parts kit.
   If the patient's hand width is 2-3
   inches, use No. 2400 Medium; if
   it is 3-5 inches, use No. 2400
   Large. The kits contain all the
   hardware, padding, and straps required. The major stampings are
   shown in the illustration. The
   correct nomenclature is as follows:

   1. Thumb post
       a. Thumb post bar
       b. Thumb post ring

   2. Index finger piece
       a. Lever (1-3/4 inch)
       b. Finger piece (2 inch)

   3. Palmar piece
       a. Palmar extension
       b. Dorsal extension
       c. Radial extension
       d. Center bar

   4. Forearm piece
       a. Distal extension
       b. Proximal extension
       c. Center bar

   5. Joint hinge

All parts may be used to make either
a left or right hand splint.

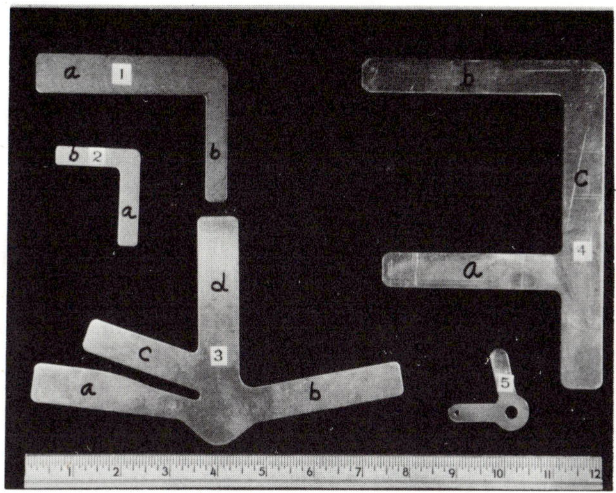

2. Using a lead block with a 1 inch
   channel and a 1 inch steel mandrel,
   form the palmar piece so the center
   bar will fit the curve of the hand at
   the M.P. joint. (In the illustration
   the palmar piece is being made for a
   left hand. For a right hand the stamp-
   ing would be turned over and shaped
   on the other side.)

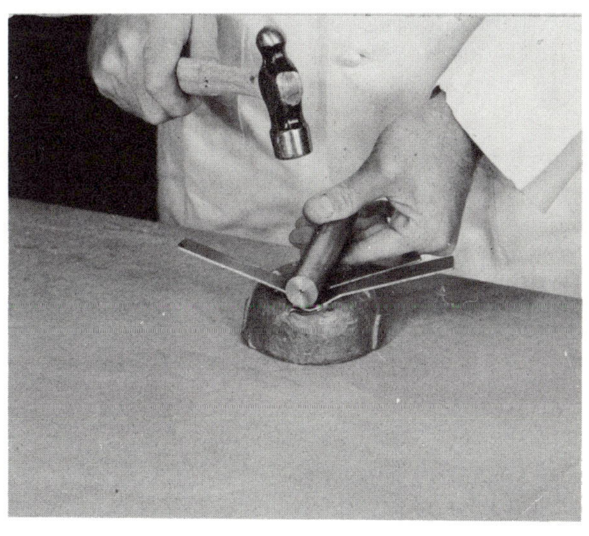

3. When shaped properly, the part should fit the hand as illustrated.

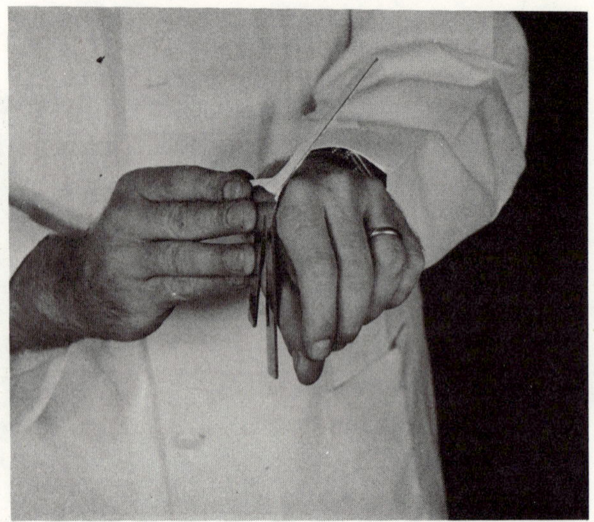

4. Using a small bending fork, shape the dorsal and palmar extensions to follow the contours of the hand.

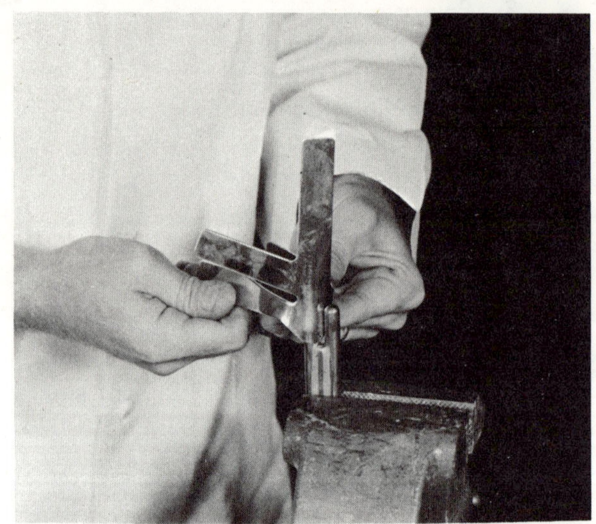

5. The piece should look like this.

6. The palmar and dorsal extensions will be out of alignment laterally, as illustrated, and must be aligned so they are parallel.

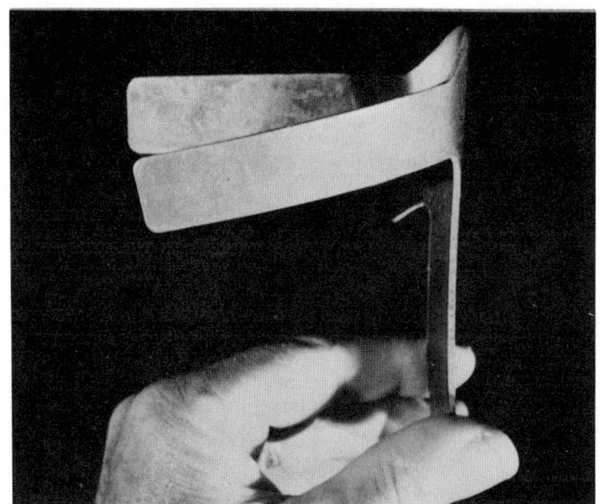

7. Bend the palmar extension parallel with the dorsal extension, as illustrated.

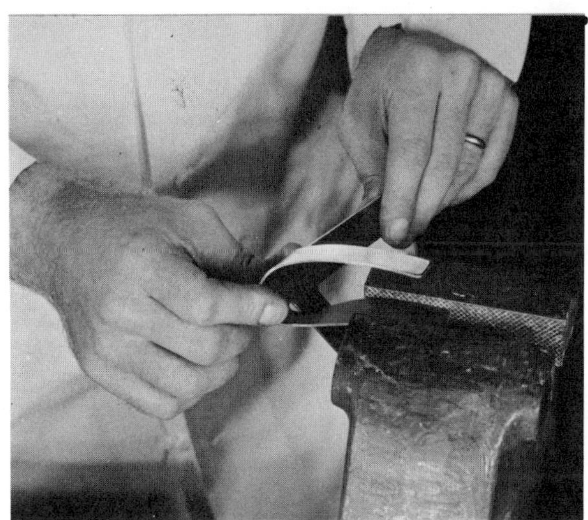

8. They are also usually out of alignment end-wise, as illustrated.

9. Bend the palmar extension into alignment with the dorsal extension, as illustrated. When correctly aligned, the ends will be parallel

10. Bend the radial extension back out of the way.

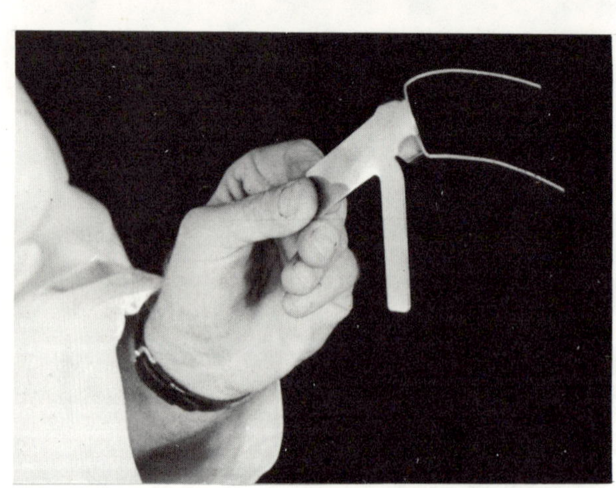

11. Measure the distance between the dorsal and palmar extensions; it should be 1/2 inch more than the thickness of the hand, to allow for the padding.

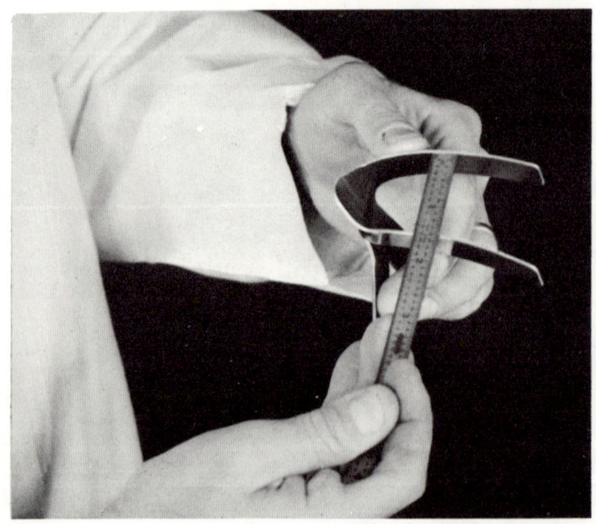

12. With the felt padding temporarily in place, try the splint on the patient's hand to see if it fits properly. It should fit snugly, but not too tight to cause pressure sores or restrict circulation.

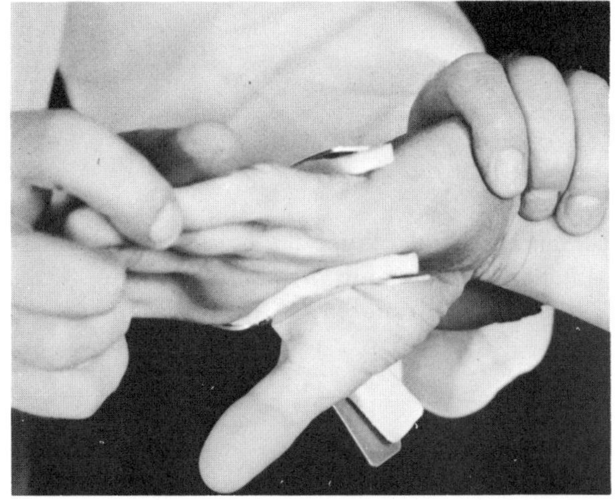

13. Mark the trim line on the dorsal extension at a point corresponding to the midline of the fifth metacarpal.

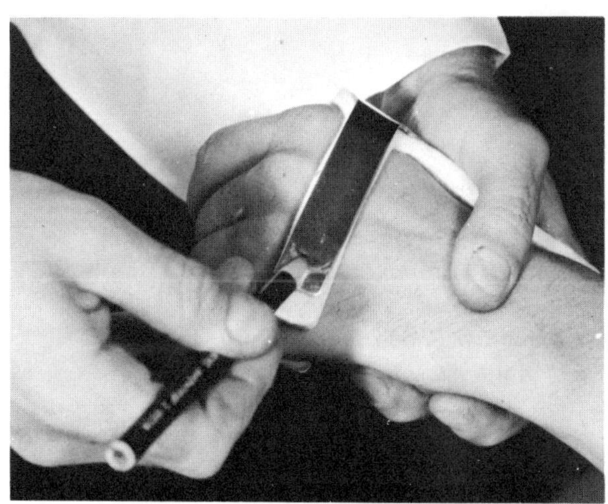

14. Mark a trim line on the palmar extension in the same way. The dorsal and palmar extensions must be trimmed off enough so they will not strike the table or desk top when the patient is eating or working.

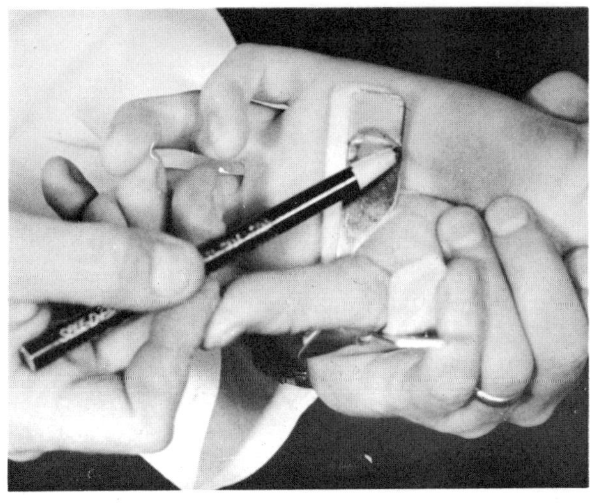

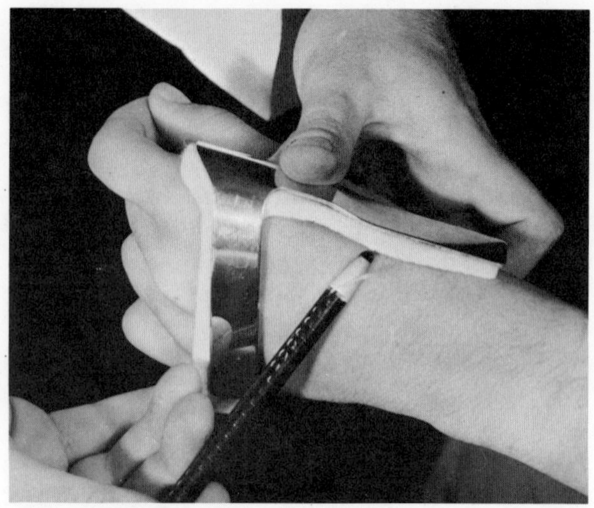

15. If necessary, shape the palmar piece bar to conform to the contours of the wrist as illustrated.

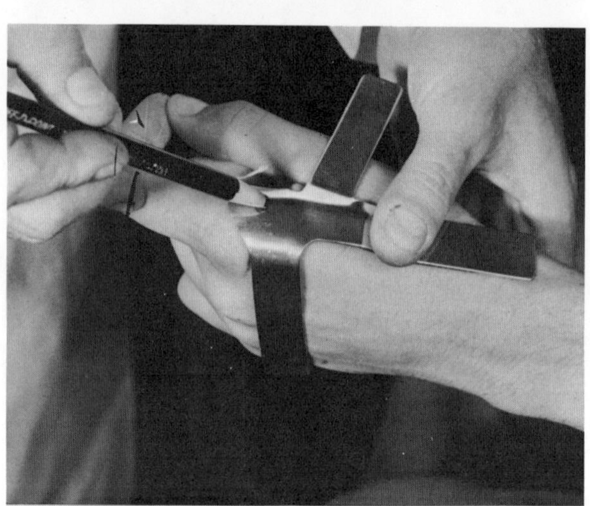

16. Place the splint, without the padding, on the patient's hand, then make a mark on the center bar opposite the M.P. joint of the index finger. This mark will be used to center the joint hinge mounting plate in a later step.

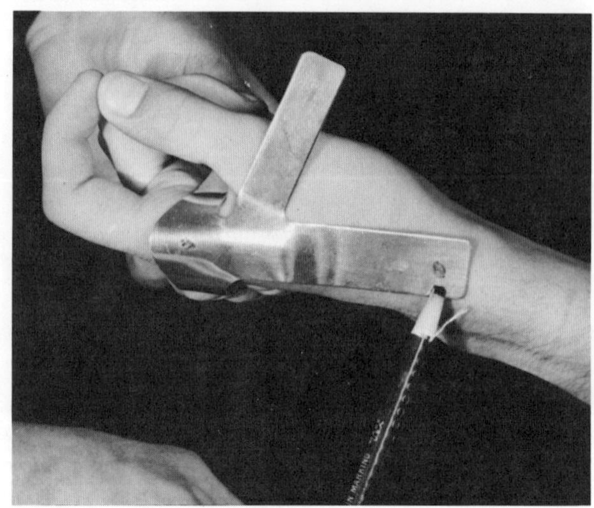

17. Make another mark on the center bar opposite the wrist joint. A hole will be drilled here for riveting the palmar piece to the forearm piece.

18. Scribe the midline of the center bar with a pair of dividers.

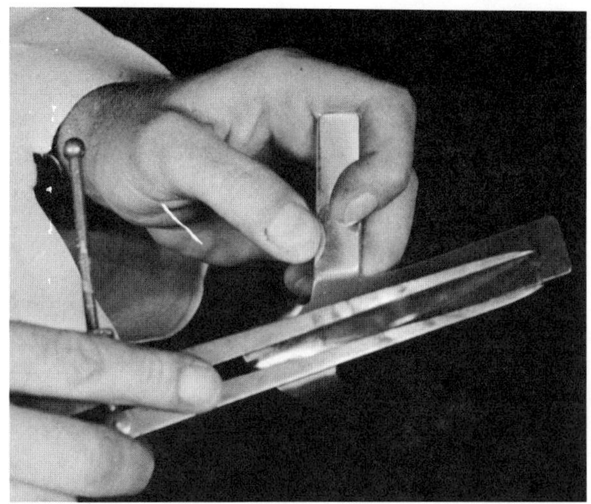

19. Center punch on the scribed line even with the wrist joint mark made in Step 17.

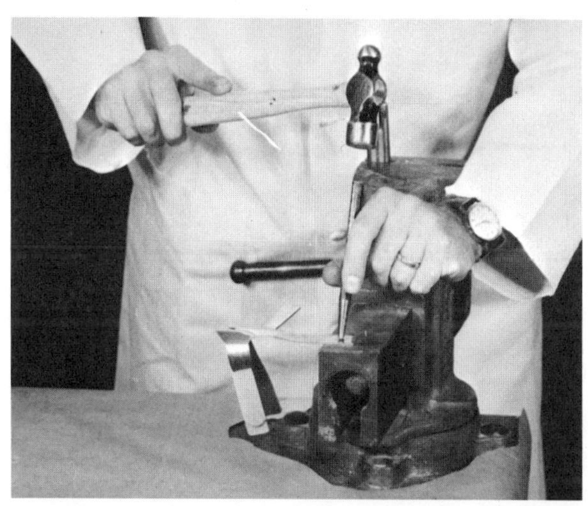

20. Using the center punch mark as a center, scribe a semi-circle tangent with the sides of the center bar, with the dividers.

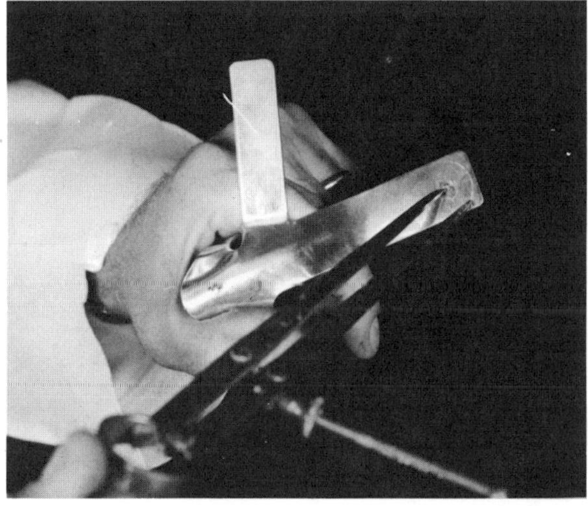

21. Trim the center bar on the mark made in the preceding step.

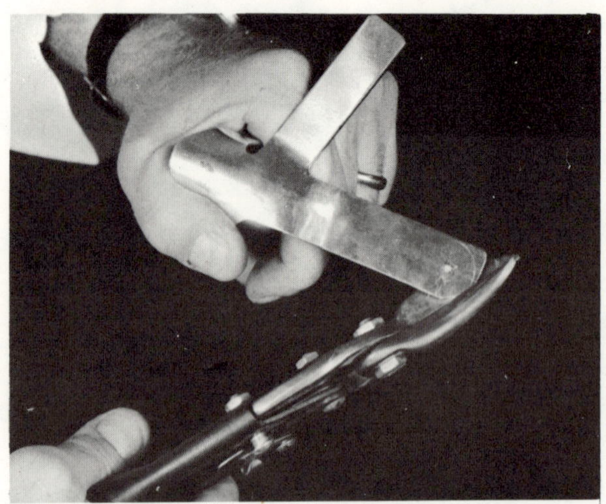

22. Trim the dorsal and palmar extensions on the marks made in Steps 13 and 14.

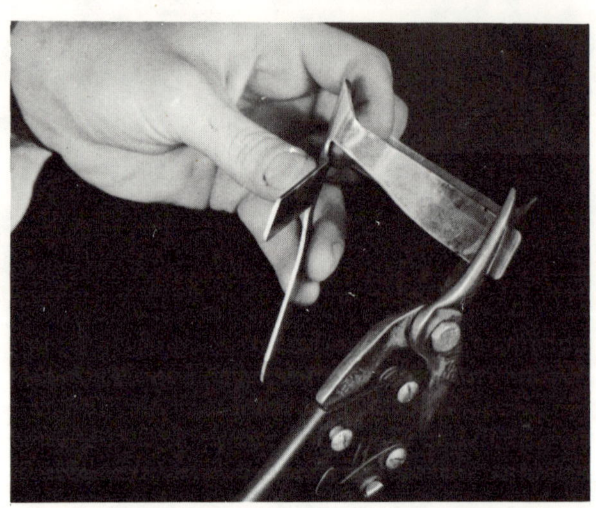

23. Smooth the rough edges on a belt sander or with file and sandpaper.

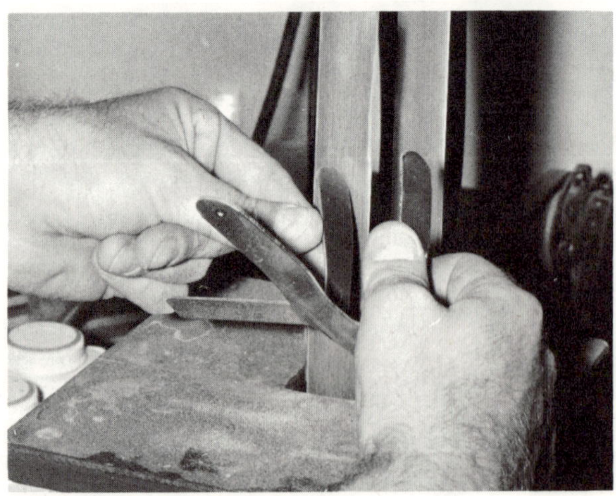

24. Drill a No. 40 hole in the wrist extension at the point center-punched earlier.
Remove the burr with a reamer or a large
drill.

25. The forearm piece must next be fitted
to the forearm. The distal and proximal extensions must be curved to the
shape of the forearm, and the bar must
be shaped to conform to the radial portion of the wrist.

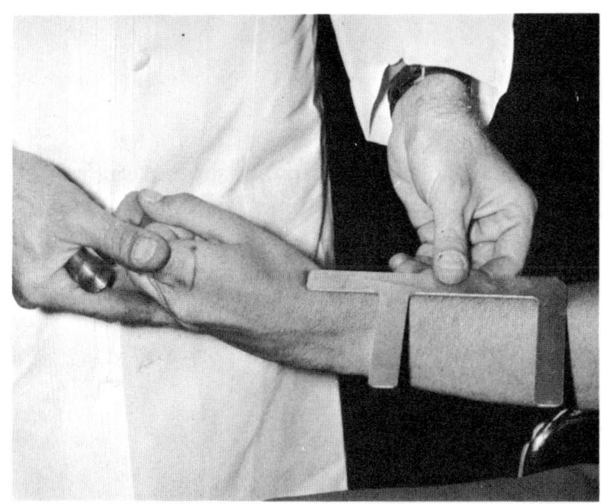

26. Shape the forearm piece center bar with
the lead block and 1 inch mandrel.

27. Shape the forearm piece extensions to conform to the contours of the forearm.

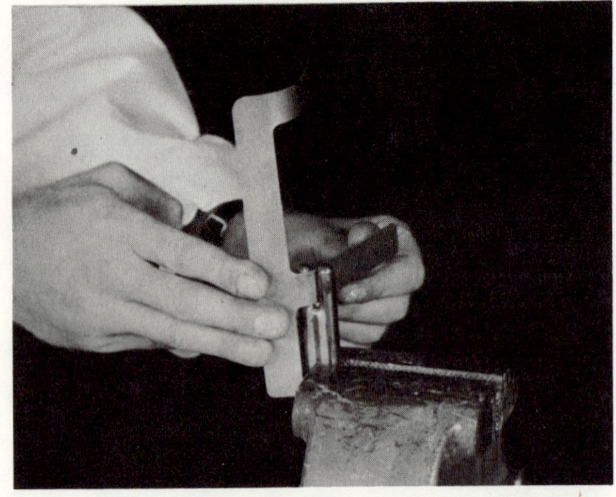

28. Place the felt padding on the patient's arm with the splint over it, and check for proper fit. The extensions and the bar should be contoured to the arm. If gapping occurs as illustrated, re-shape the bar to fit the forearm.

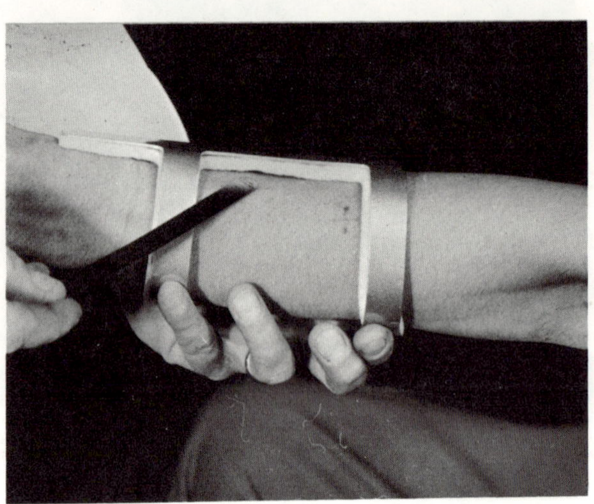

29. Scribe a midline on the distal end of the center bar. Center punch on it at the wrist joint mark. Using the center punch mark as a center, scribe a semi-circle tangent with the sides of the bar. Trim the metal to this line, then drill a No. 40 hole where center punched. Smooth rough edges.

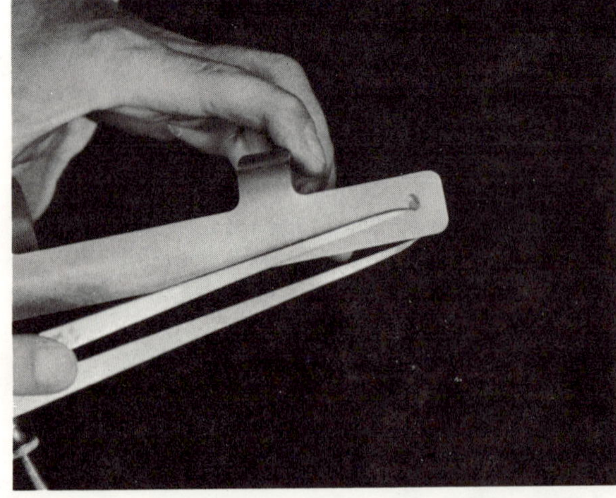

30. Rivet the palmar and forearm pieces together with a stainless steel rivet, being sure to have the palmar piece outside the forearm piece.

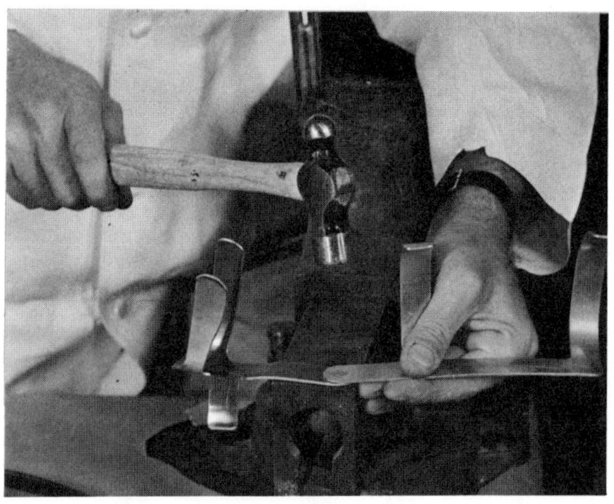

31. Try the wrist strap for location and best fit. It usually works best when the fixed end is about 3/4 inch distal of the riveted joint, and the free end on the dorsal extension at a point on the midline of the dorsum of the hand.

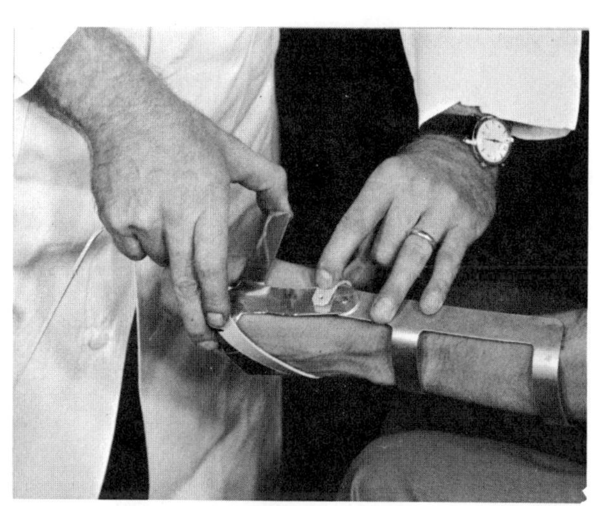

32. Mark both locations with the skin pencil.

33. Mark the locations of the fixed ends of the forearm straps. They go on the radial side of the splint on the midline of the forearm piece center bar.

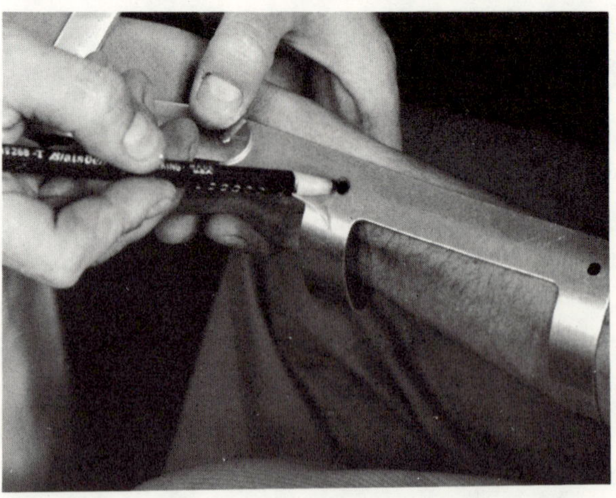

34. Mark the locations of the free ends of the forearm straps. They go on the ulnar side of the splint, at the ends of the distal and proximal extensions.

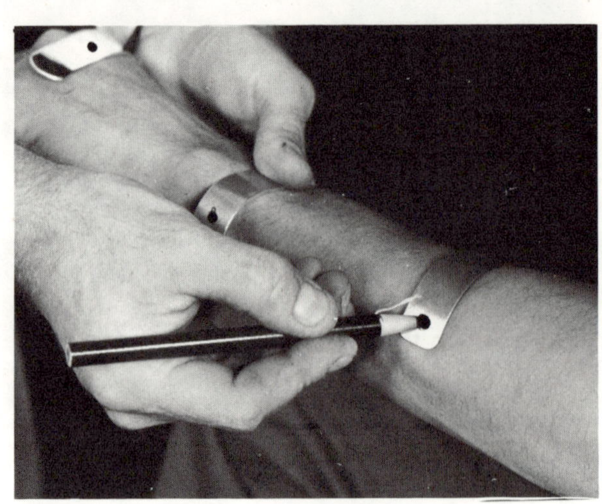

35. Drill three No. 18 holes at the marks on the radial side of the splint for the wrist and forearm straps. Drill three No. 29 holes at the marks on the ulnar side for the truss studs that retain the free ends of the straps.

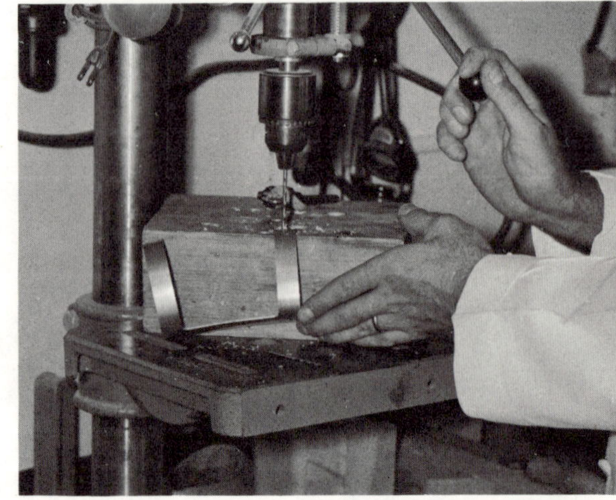

36. Attach the straps to the splint with the Speedy Rivets provided in the kit.

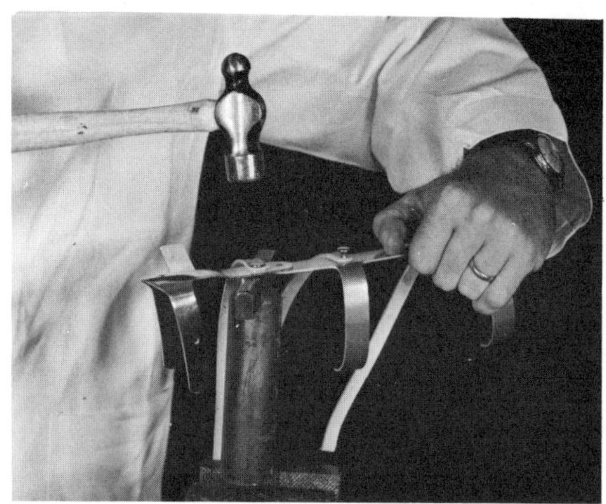

37. Rivet the truss studs into the distal and proximal extensions of the forearm piece, and in the center of the dorsal extension of the palmar piece.

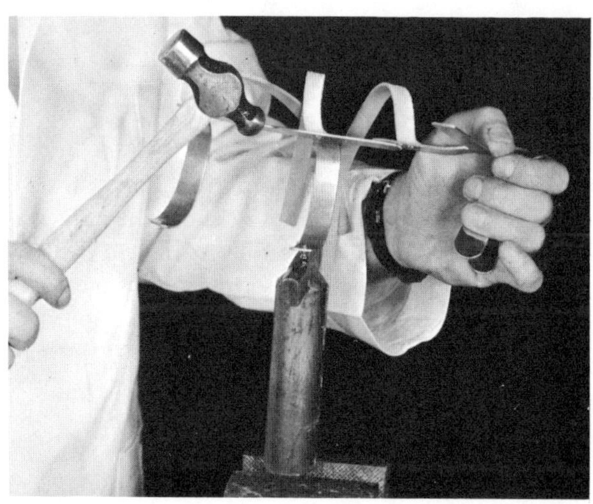

38. Trim the felt padding to proper length so it will fit the splint.

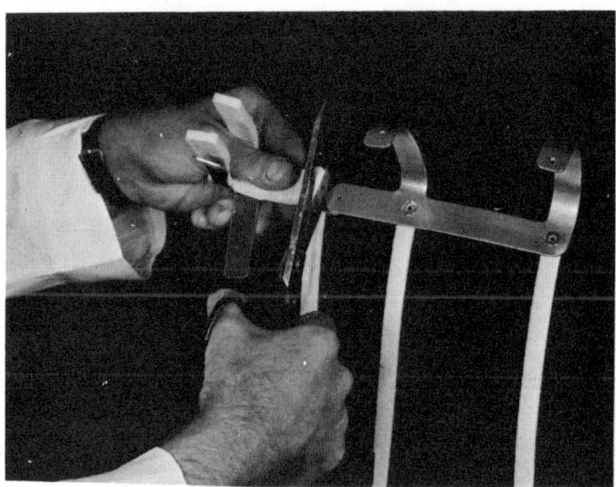

39. Cement the felt padding in place with rubber cement.

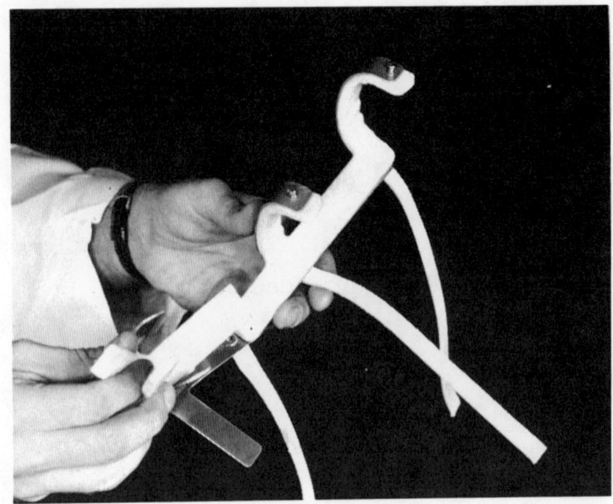

40. With the splint on the patient, mark the free ends of the straps at the points on the truss studs where they will hold the splint firmly but comfortably in place

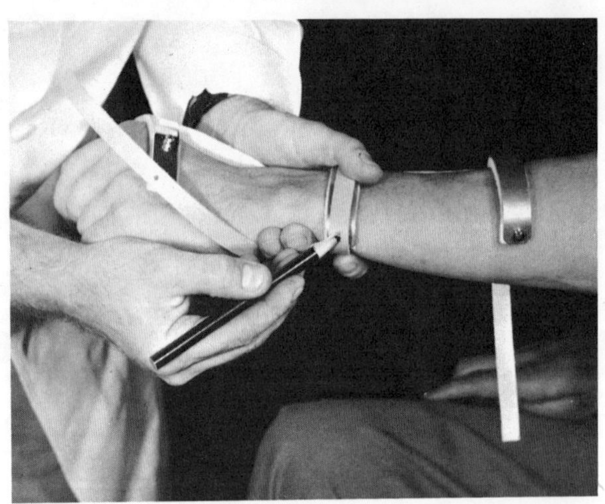

41. Punch No. 6 holes in the straps at the marks, then slit the holes slightly, toward the fixed ends of the straps, so it will be easier to snap the straps over the truss studs.

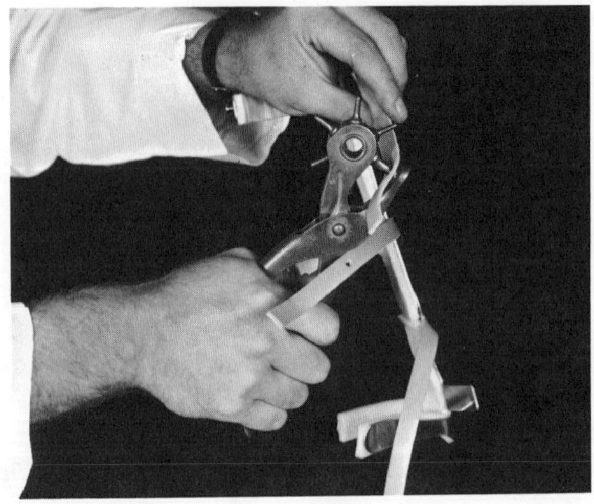

42. Make finger loops in the free ends of the straps, using Speedy Rivets. These loops make it easier for the patient to fasten the straps, as shown in the illustration.

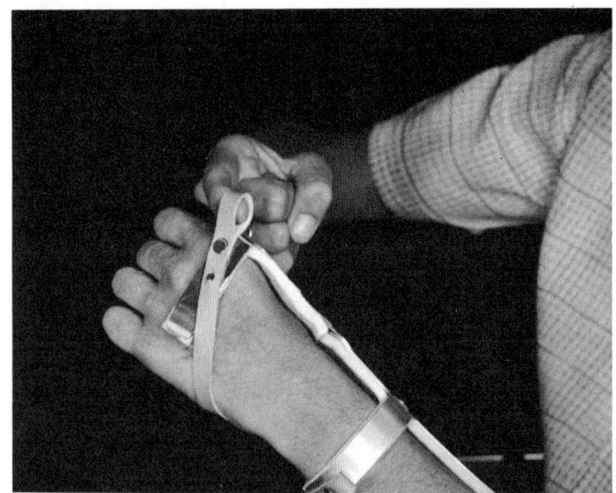

43. Measure and mark the thumb post for preliminary sizing by holding it in place as shown. It should extend along the thumb distally about to the cuticle line of the nail. This is not the final determination of length, but is for the purpose of removing excess metal to make the post easier to shape. Metal proximal to the radial extension can be removed.

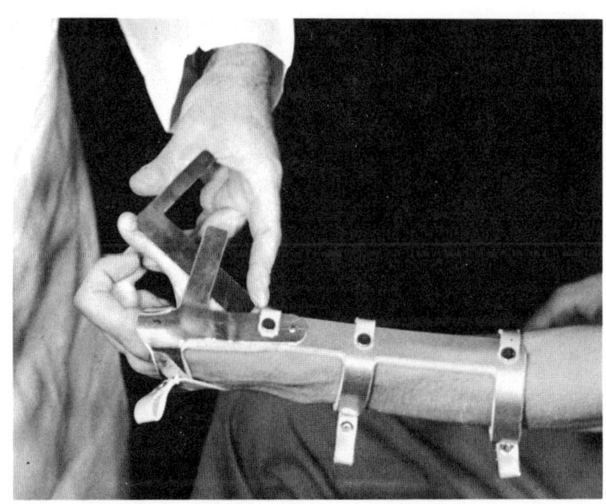

44. Cut the excess metal from the thumb post.

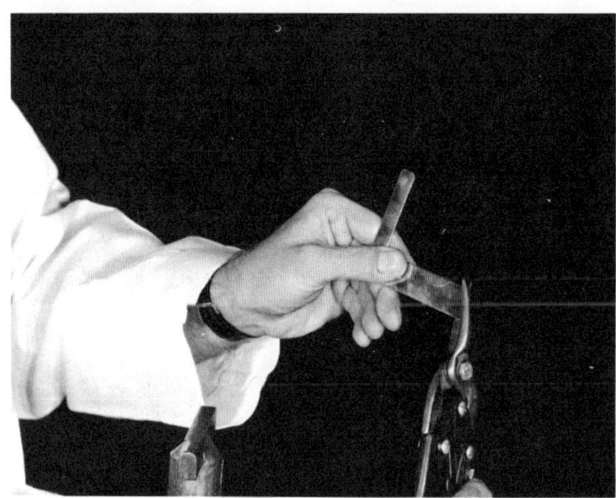

45. With the lead forming block and 3/4 inch steel mandrel, form the thumb post bar into a concave shape that will conform to the curvature of the thumb. Be sure not to reverse the part. For left hands, the thumb post ring is toward you and pointing to your right. For right hands, it is the opposite.

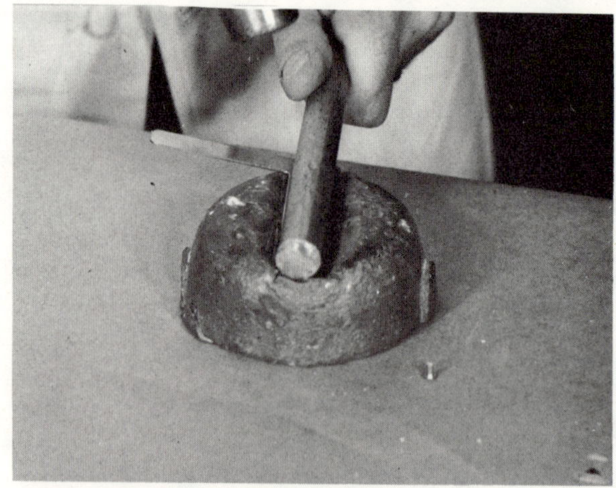

46. Try the thumb post on the patient's thumb, and continue to add to the amount of curvature until it fits perfectly.

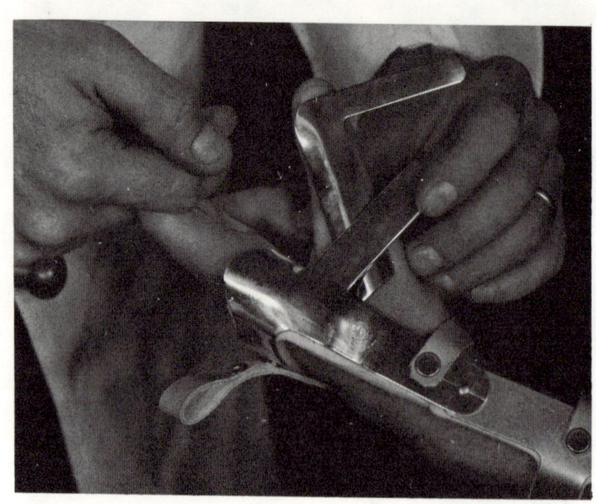

47. Form the thumb post ring to fit around the patient's thumb. Note how it is bent slightly proximally to clear the thumb pad. Trim excess length of the ring with shears.

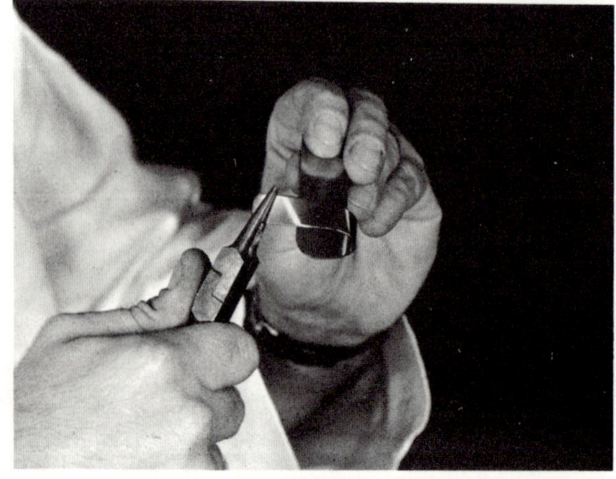

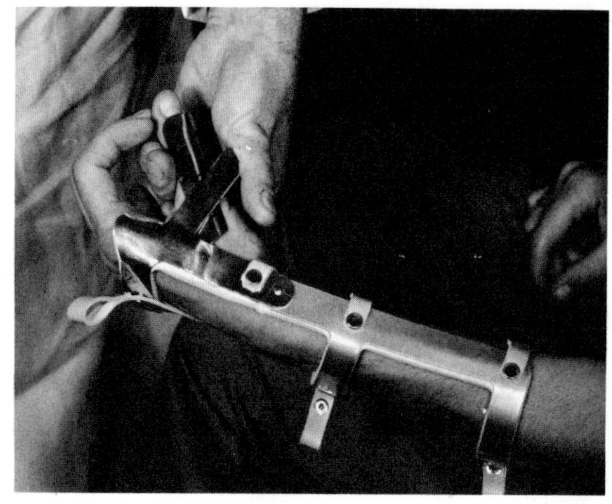

48. Hold the thumb post in place inside
the radial extension and adjust its
position until it holds the thumb in
an opponens position to the index
and middle fingers. Mark both the
thumb post and the radial extension.

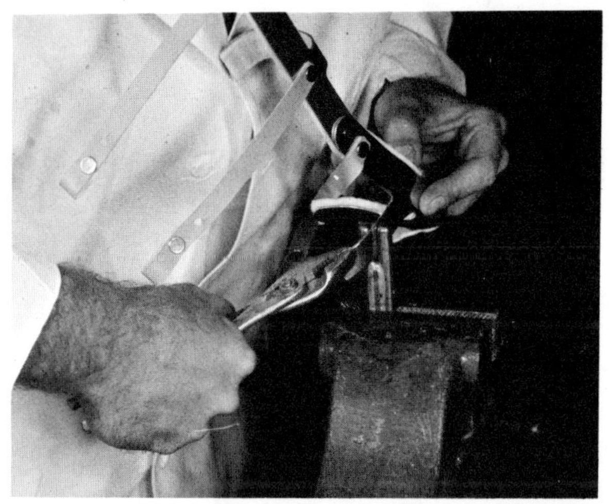

49. With the bending fork and pliers shape
a concavity into the radial extension
into which the thumb post will fit, being careful to locate it between the
marks made in the previous step.

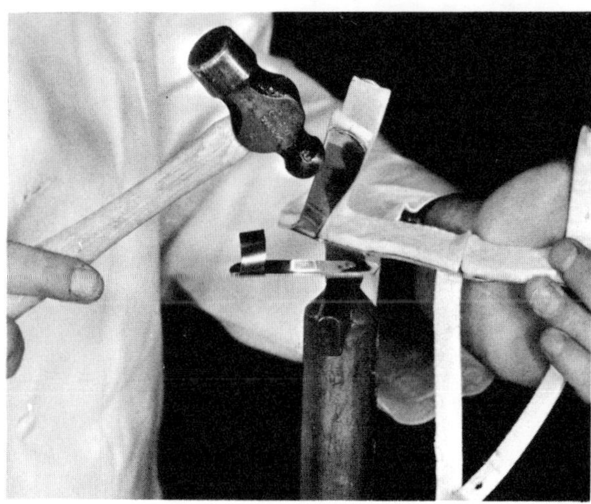

50. Clamp the thumb post into position on
the radial extension, drill a No. 40
hole through both parts, then rivet them
together with one stainless steel rivet.

51. Put the splint on the patient and adjust
the thumb post for comfort and alignment. Align it so the thumb is held in
maximum amount of abduction without
any discomfort. Make sure the thumb
post ring fits snugly but not so tight
as to injure the thumb.

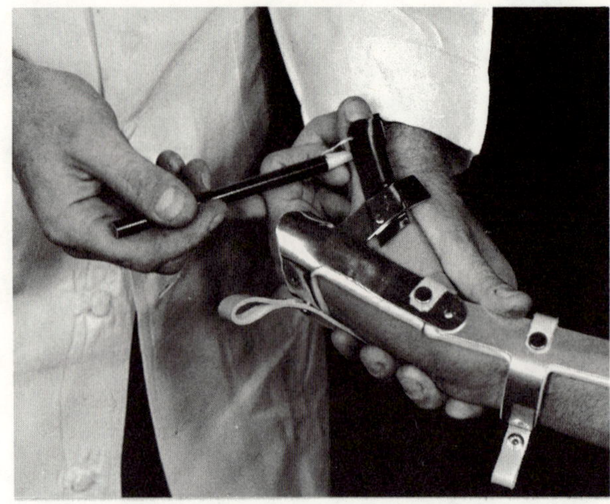

52. Drill another No. 40 hole and install
a second stainless steel rivet to lock
the thumb post permanently in place.
Trim the surplus metal from both the
thumb post and the radial extension,
then smooth and polish all rough
edges.

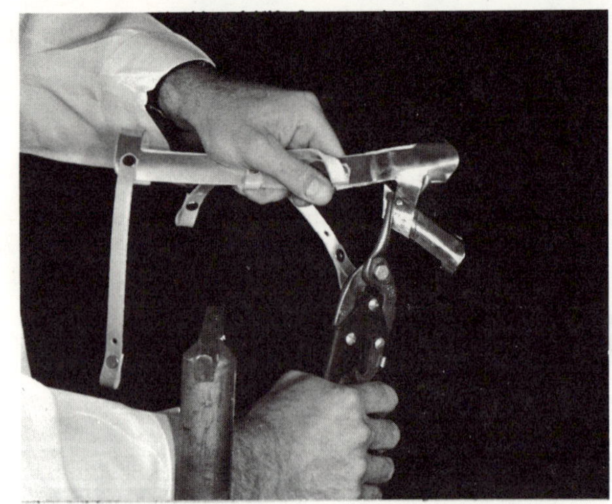

53. Try the splint on the patient and make
any final adjustments necessary.

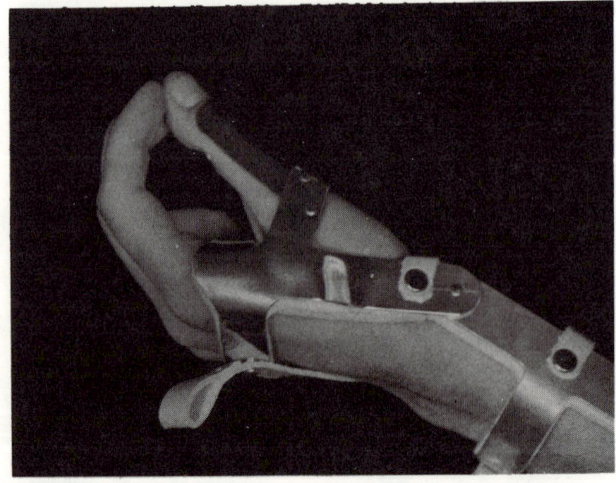

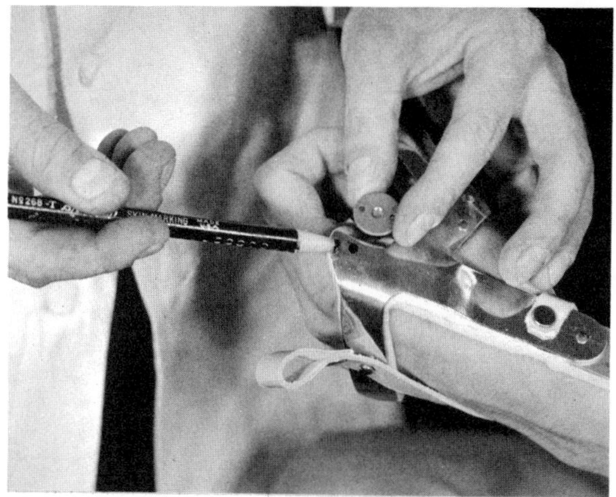

54. Check the mark on the palmar piece made earlier to determine the location for attaching the 3/4 inch joint hinge mounting plate. The plate serves as the base for the finger piece, and its center must be on the center of the M.P. joint of the index finger.

55. Center punch the mark deep enough so it will not be lost when the metal is filed. With a smooth mill file, file a flat on the palmar piece at the mark. The flat must be large enough to accommodate the mounting plate, and must be exactly parallel to the center bar of the splint.

56. Try the mounting plate on the flat to make certain it is exactly parallel to the center bar. Continue filing until it is.

57. With the large center hole exactly on the center punch mark and the small rivet holes centered on the flat, mark through the rivet holes with a scribe.

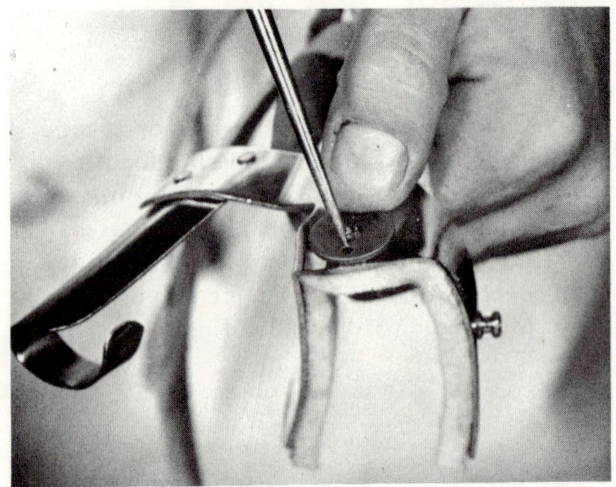

58. Center punch the marks and drill a No. 40 hole at each mark.

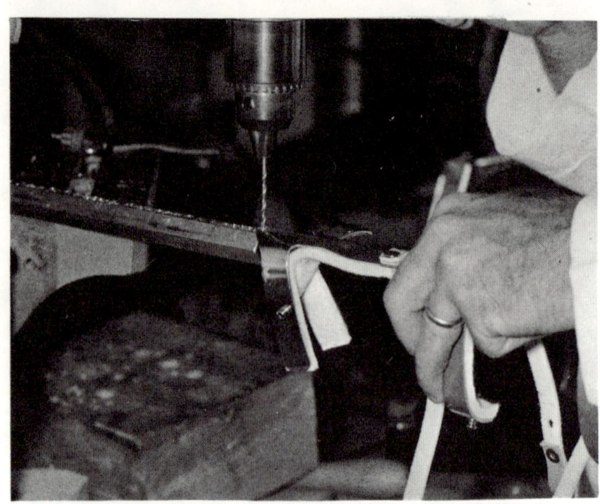

59. Countersink the rivet holes in the mounting plate with a 1/4 inch drill just enough to permit the rivet heads to lie flush with the surface of the plate.

60. Rivet the joint hinge mounting plate to the splint with stainless steel rivets. Use care to avoid denting or bending the plate or the splint while fastening the rivets.

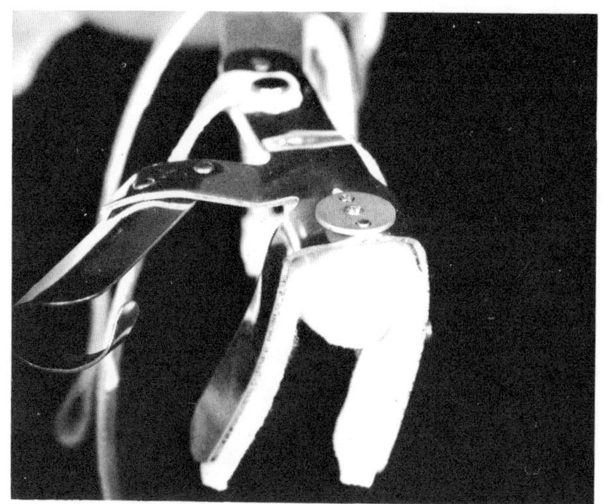

61. Drill the splint through the center hole in the mounting plate, using a No. 21 drill.

62. Thread the No. 21 hole with a 10-32 tap.

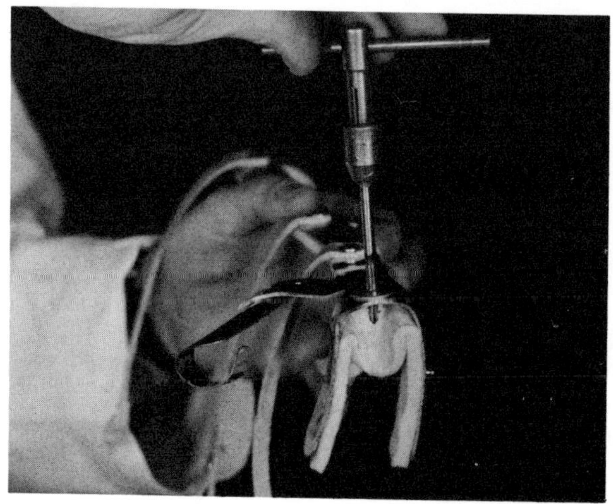

63. Place the small brass joint hinge bushing on the joint screw, then a Teflon
washer, then the joint hinge.

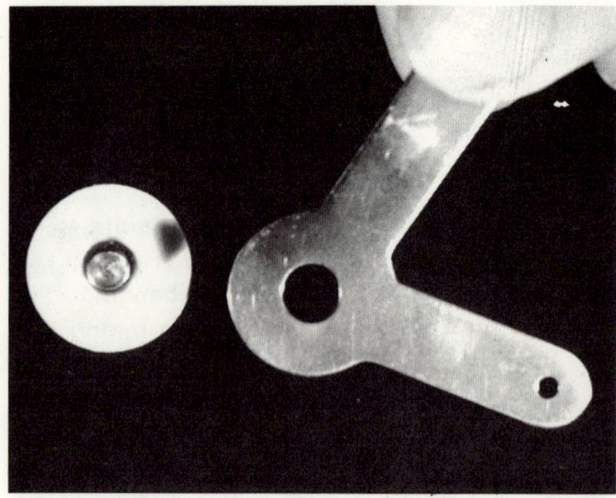

64. The second Teflon washer goes on the
joint hinge.

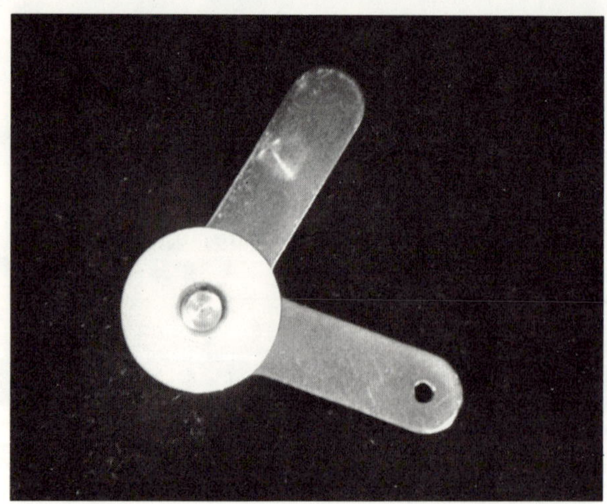

65. Fasten the joint hinge assembly to the
mounting plate on the splint, with the
longer lever with the hole pointing
proximally.

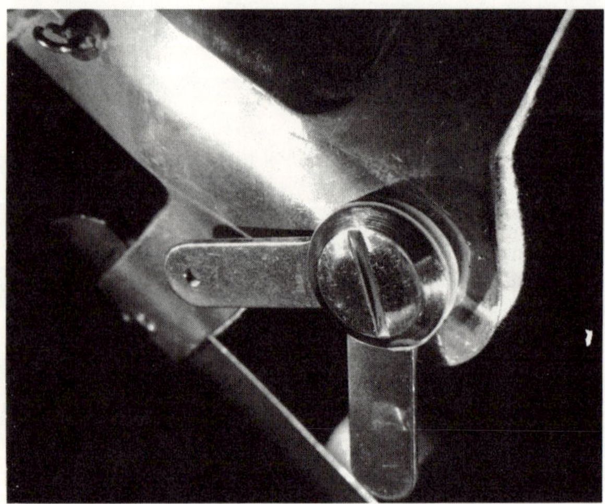

66. To make and fit a finger piece, as illustrated, the index finger piece lever is fastened to the joint hinge and the finger piece formed into a semi-circle to fit the proximal phalanx of the index finger. A half-ring of 3/8 x .037 inch stainless steel is fitted to the proximal phalanx of the middle finger. The middle phalanges of both fingers are fitted with one piece, which is provided with a retaining bar and latch to hold the fingers in place. A single piece of stainless steel is fitted to the distal phalanges, and the entire assembly is silver soldered together onto a shaped piece of 1/8 inch stainless steel rod.

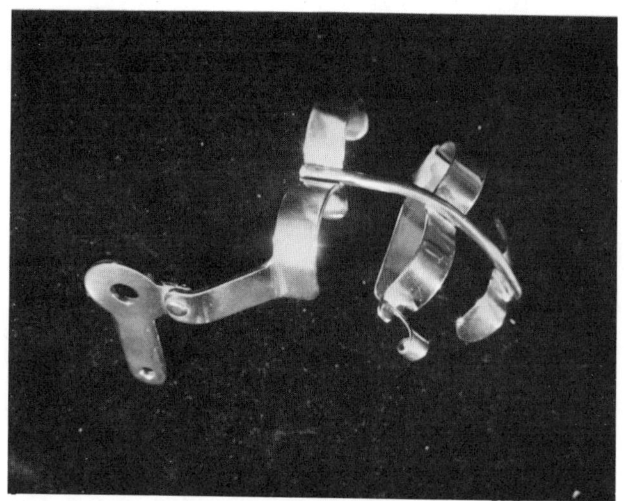

67. The parts needed to make the finger piece are shown left to right: the index finger piece, three 3/8 x .037 x 3 inch stainless steel pieces for making the other finger pieces and retaining bar, 1/8 inch stainless steel rod, and 3/32 inch stainless steel wire for making the retaining bar hinge.

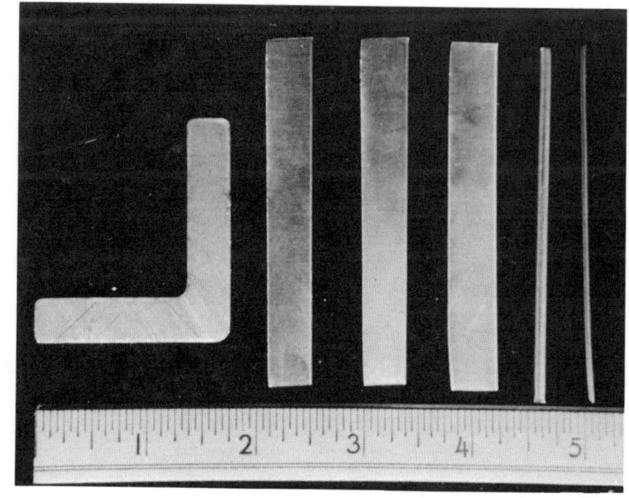

68. Prepare two 3 x 5 inch cards for making the finger tracings by cutting a semi-circle out of one end of each. This permits the cards to fit well down into the space between the fingers.

69. Hold the index and middle fingers in position of opposition on the thumb with one of the cards between them, and trace around the index finger, marking the joints as illustrated. Repeat this procedure for the middle finger.

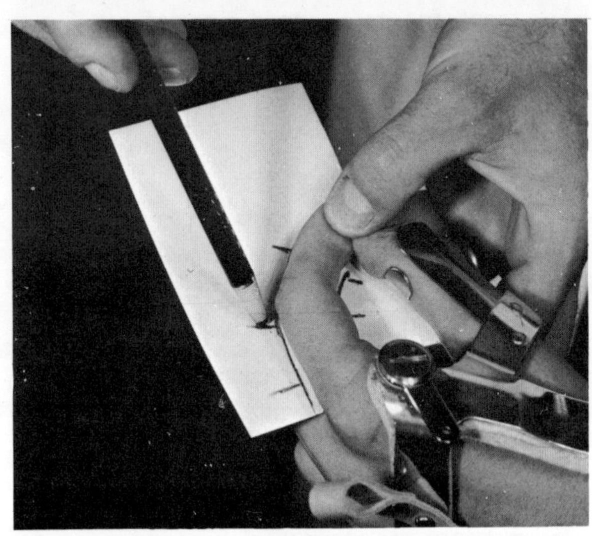

70. Draw lines across the joints on the tracings, then draw lines lengthwise of each phalanx, connecting the centers of the joint lines, for both long and index fingers. Block in the center third of the proximal and middle phalanges where the finger pieces fit.

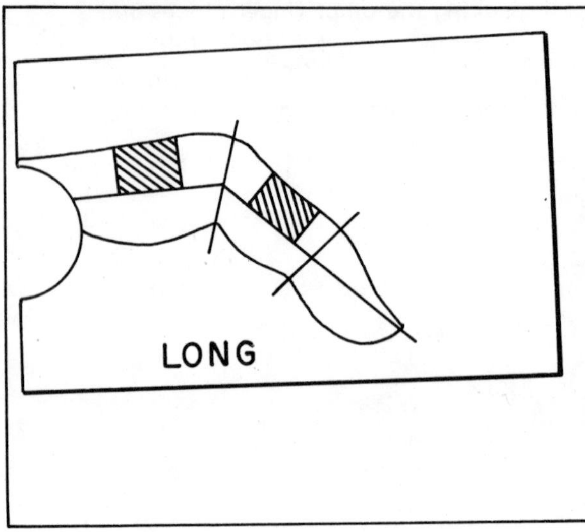

71. Measure the circumferences of the middle and proximal phalanges of the index and middle fingers, and record the measurements on the tracing cards. Use jeweler's ring gauges to make the measurements.

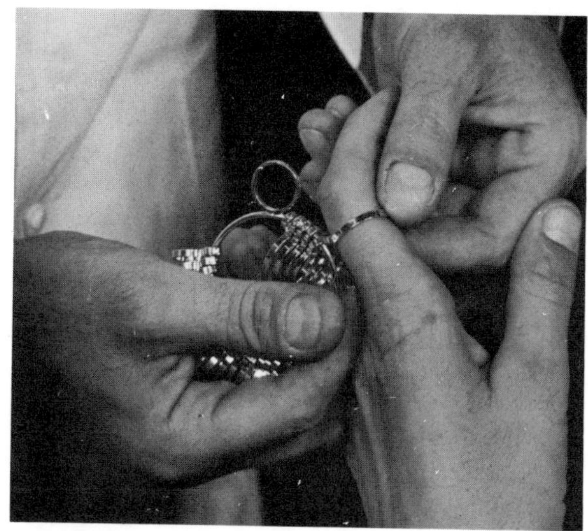

72. Shape the proximal ring for the long finger, using 3/8 x .037 inch stainless steel. Bend it into "U" shape with the round nose pliers.

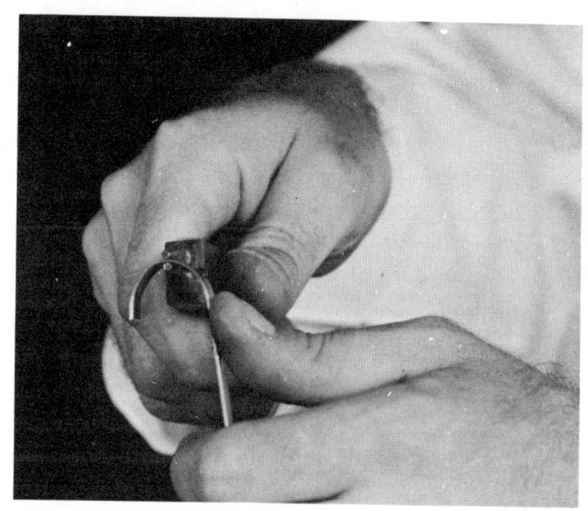

73. Finish shaping the ring on the jeweler's mandrel, using a soft mallet. Shape it at the point on the mandrel that corresponds to the size as determined by the jeweler's rings, for the phalanx being fitted.

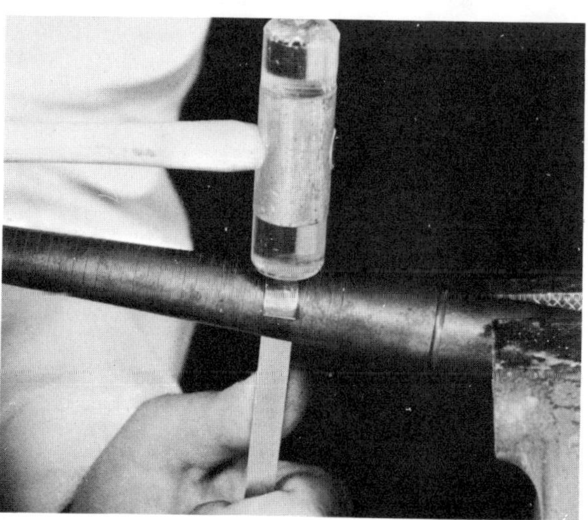

74. Cut the ring from the stock, leaving the cut side a little longer than the other. Round the corners and polish the rough edges.

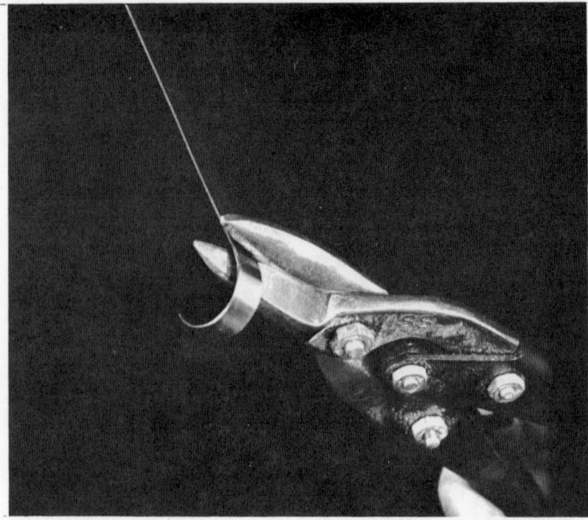

75. Shape the proximal ring for the index finger. Use the "L" shaped index finger piece, and be sure to bend the ring so the lever will lie along the radial side of the finger when it is in place. Use the jeweler's mandrel for final shaping.

76. Check the proximal rings for the middle and index fingers for correct fit.

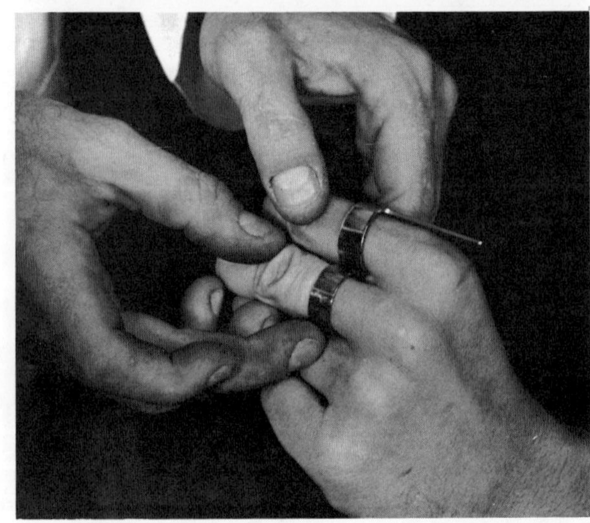

77. Bend the extension on the proximal ring for the index finger at an angle to the rings, as illustrated.

78. Silver solder the two proximal finger pieces together so they will be in the same position when they were tried on the fingers.

79. Silver solder the 1/8 rod in the groove between the two proximal finger pieces.

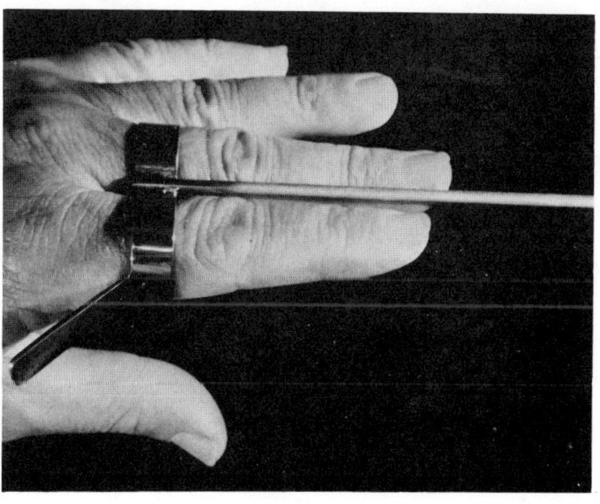

80. Shape the middle finger piece to fit the contours of the middle phalanges of the index and middle fingers, using one of the pieces of 3/8 x .037 inch stainless steel.

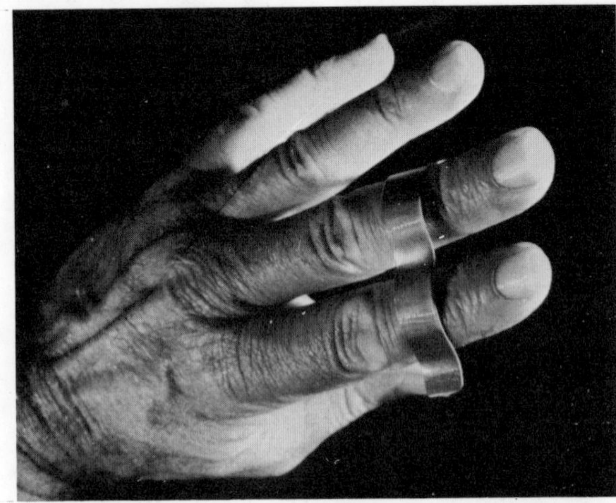

81. The retaining bar is now prepared. It hinges on the lateral end of the middle finger piece and latches to it on the medial end. First trim the corners on the end of a piece of 3/8 x .037 inch stainless steel. Grip it in the vise with a scrap of .064 inch steel or aluminum, with the trimmed end extending about 3/16 inch above it. Bend the end of the bar over the .064 inch material as shown.

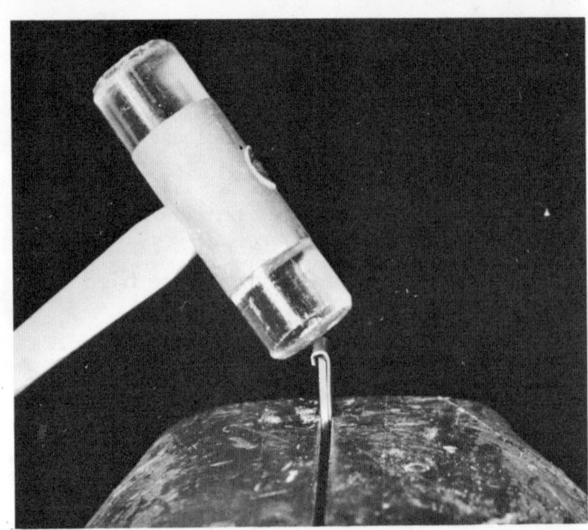

82. Remove the piece of .064 inch material and place a piece of 3/32 inch stainless steel wire in the hook formed in the end of the retaining bar. Squeeze the hook over the wire in the vise.

83. The eye in the end of the bar should resemble the one in the illustration. The wire will be used later to hinge the bar to the middle finger piece. The bar is hinged so the patient can apply and remove the splint more easily. Set the retaining bar aside until the middle finger piece is silver soldered in place.

84. Next, a holding jig will clamp the middle finger piece in place on the rod while it is being silver soldered. This can be easily made by cutting two slots in the end of a piece of 5/16 x 3/4 inch steel. The lengthwise slot is 1/8 x 5/16 inch, the horizontal slot is 3/16 x 1/8. Drill and tap 10-32 threads for the clamp screw, as illustrated.

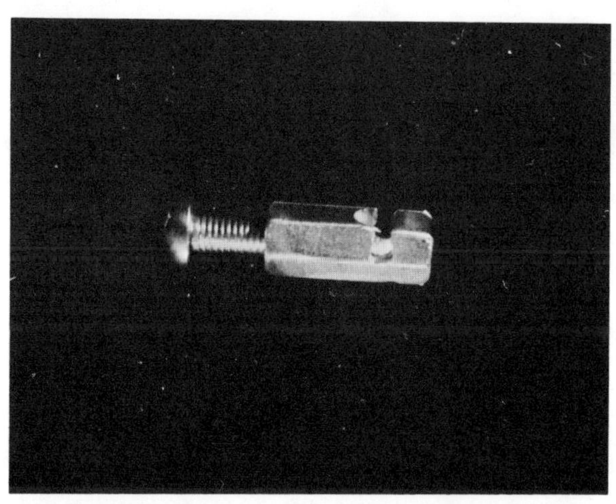

85. Place the middle finger piece under the rod and clamp it in place with the holding jig. Try it on the patient and move the finger piece until it is centered on the middle phalanges of the index and middle fingers. Tighten the holding jig securely.

86. Silver solder the middle finger piece
    in place, then remove the holding jig.

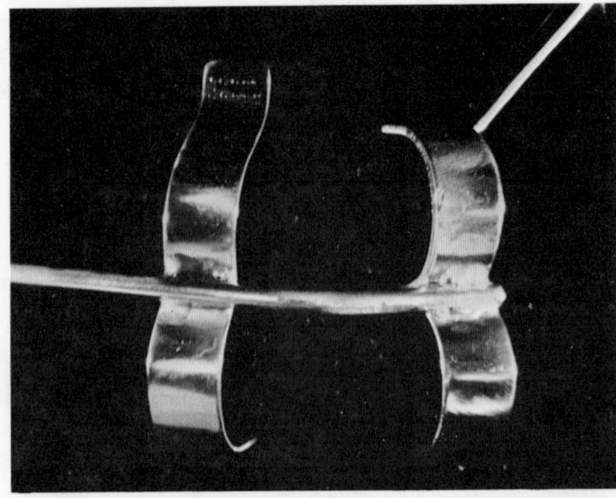

87. Cut off the distal half of the ear on
    the radial end of the middle finger
    piece with shears, as illustrated.

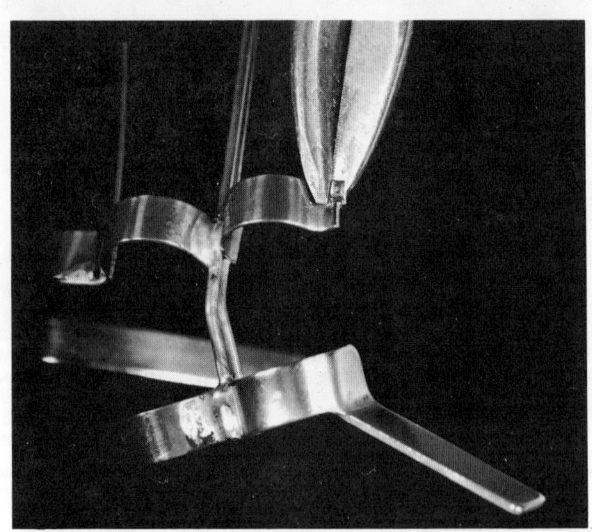

88. File the opening smooth and square
    as shown.

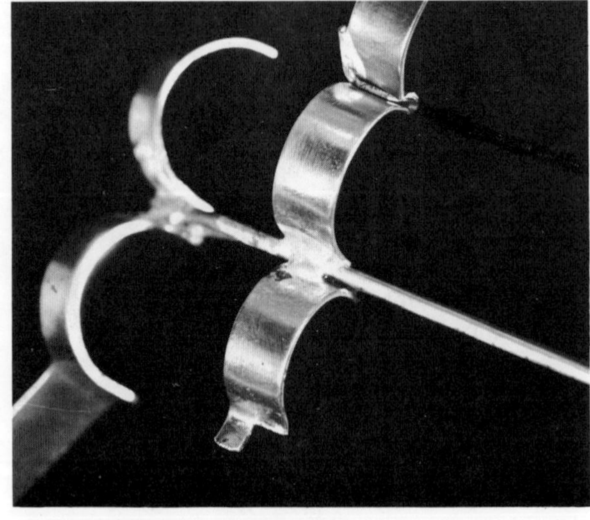

89. With the finger piece on the patient's fingers, hold the retainer bar under the index and middle fingers and mark the radial end of the finger piece where it should be cut off even with the bar. Cut off the excess metal from the bar, smooth the cut edge and round the corners.

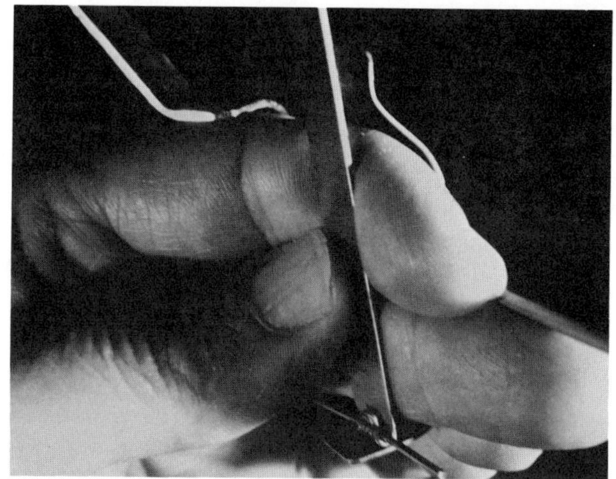

90. Bend the wire in the eye at the end of the retainer bar forming a "U" with the legs parallel. Fit it snugly against the edges of the finger piece. Silver solder the wire to the finger piece in that position to form a hinge, as illustrated.

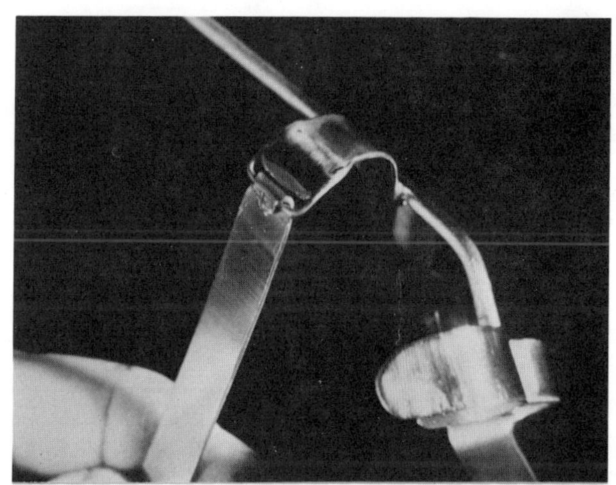

91. With the finger piece on the patient, bend the retaining bar so it keeps the fingers in position but is not uncomfortably tight. The free end of the bar should be bent up by the radial side of the index finger, then marked where it contacts the finger piece. Cut a slot half way through the bar with a hacksaw at this mark, starting the cut from the proximal side of the bar. Trim the bar diagonally from the slot to the end, and bend the end over to form a handle. Smooth and polish.

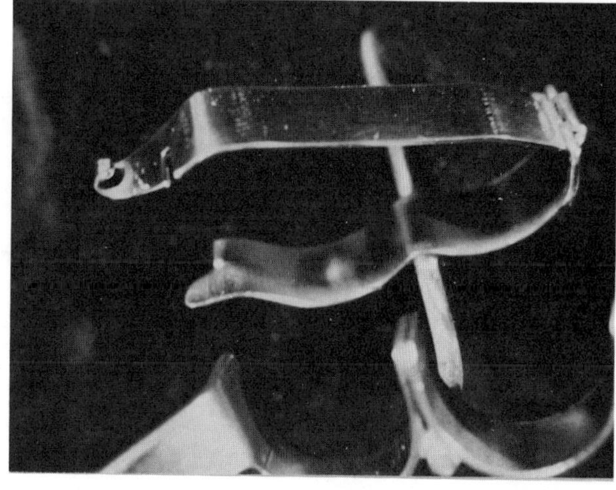

92. Adjust the latch until it closes automatically when the bar is pressed against the tongue on the ring.

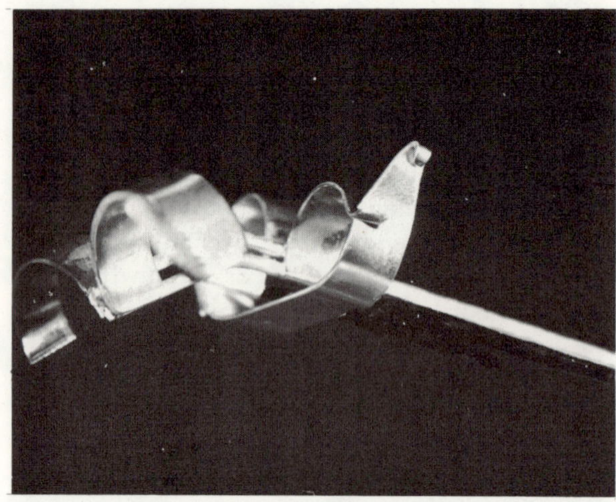

93. The latch must open easily by simply pressing the bar sideways.

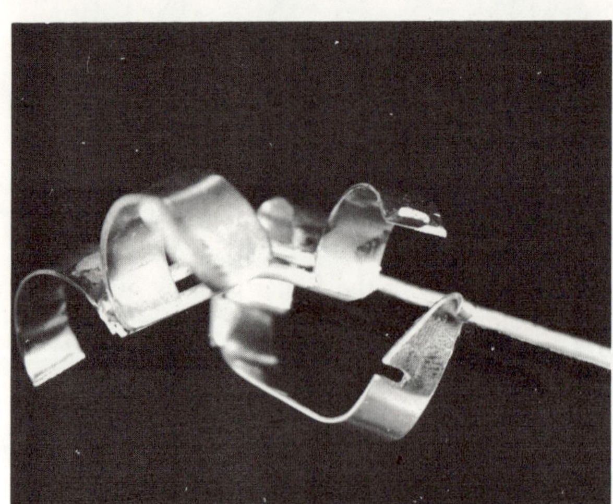

94. Put the splint and the finger piece on the patient. Line up the lever on the index finger piece with the distal lever on the joint hinge, and mark where it will have to be bent to lie flat on the latter.

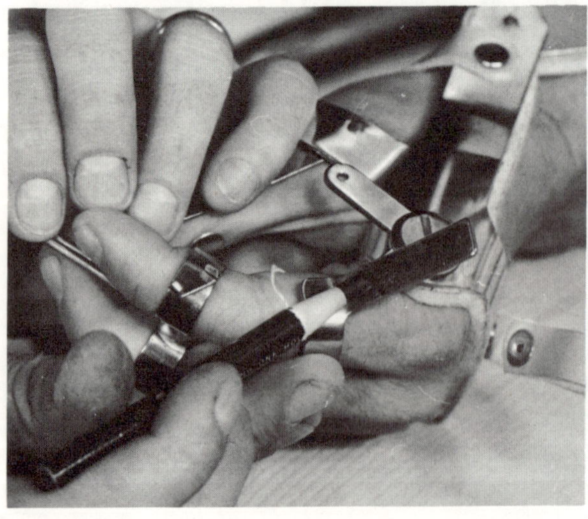

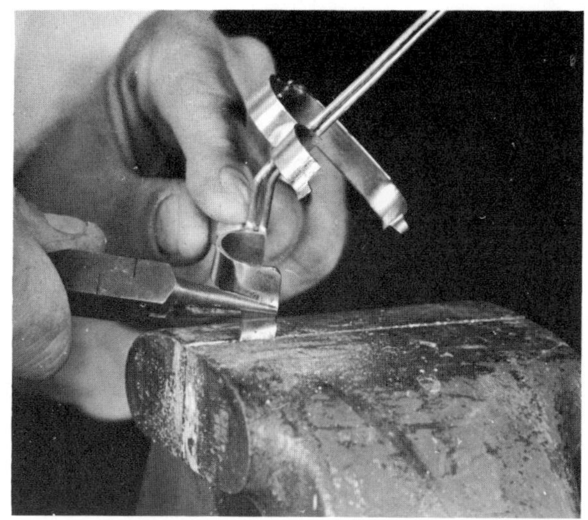

95.   Bend the index finger piece lever with round nose pliers at the mark. Continue adjusting until the two levers are in alignment.

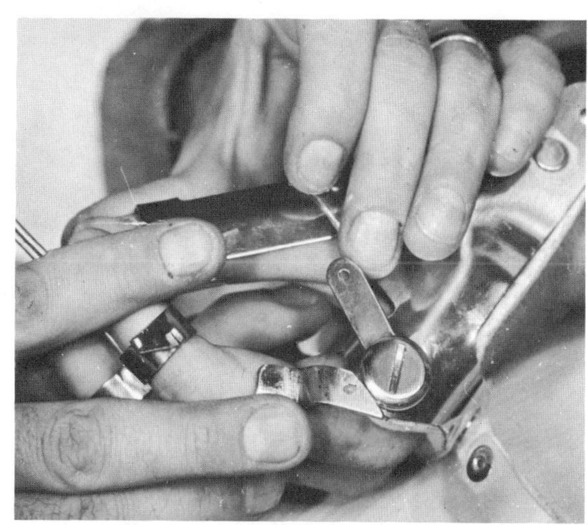

96.   With the parts again in position on the patient's hand, mark the index finger piece lever where it overlaps the short lever on the joint hinge, then use a hacksaw to cut off the excess material at the mark. The levers should line up as shown in the illustration.

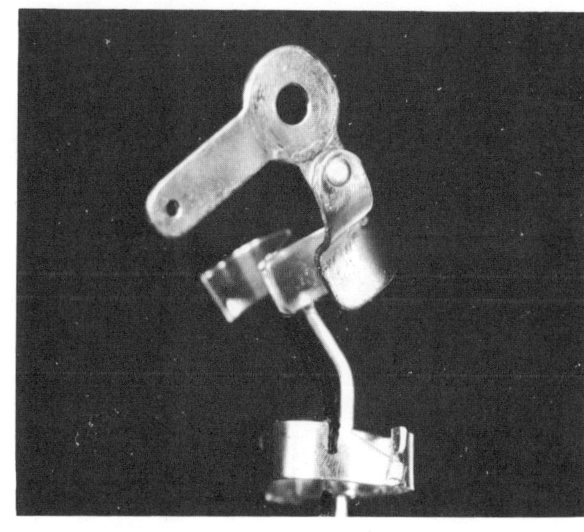

97.   Drill a No. 40 hole through the index finger piece lever, and a corresponding hole through the joint hinge lever, then rivet the two parts together with a stainless steel rivet.

98.   Place the splint on the patient's hand and adjust the relation of the finger piece and the joint hinge at the riveted joint until maximum comfort and smooth function are obtained. Mark the riveted joint, then remove the assembly and silver solder the joint in the position marked.

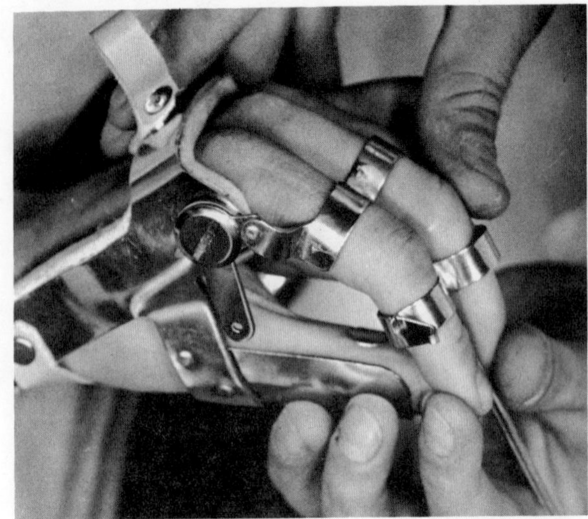

99.   Bend the 1/8 inch rod up to an angle of about 35 degrees.

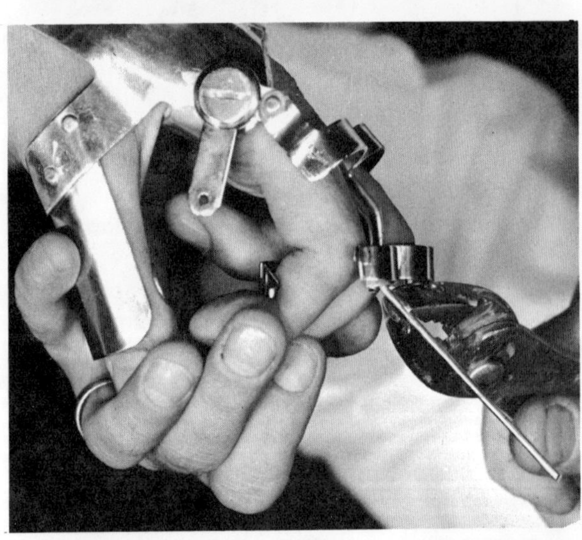

100.   Bend the wire back down again so an arch is formed to clear the fingers.

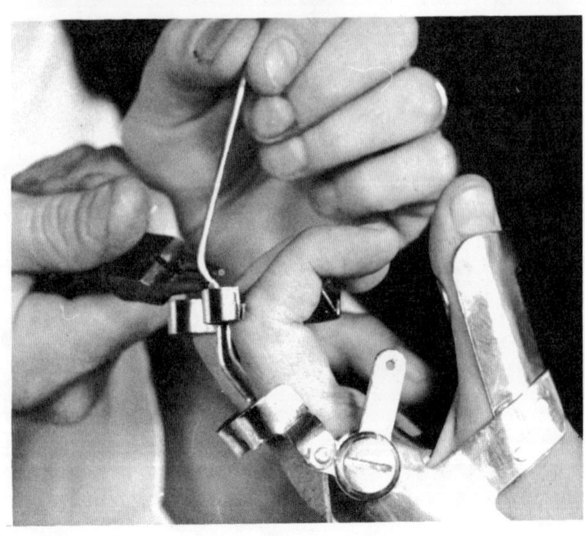

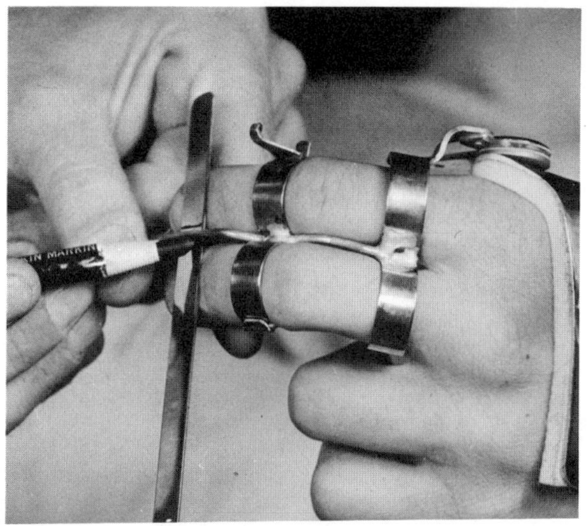

101. Place a piece of 3/8 stainless steel under the wire and across the middle of the distal phalanges of the index and middle fingers. Mark the piece where the wire crosses it.

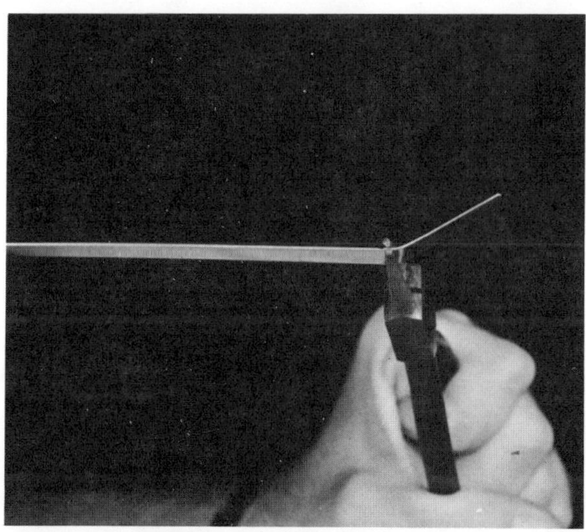

102. Shape the distal finger piece with round nose pliers.

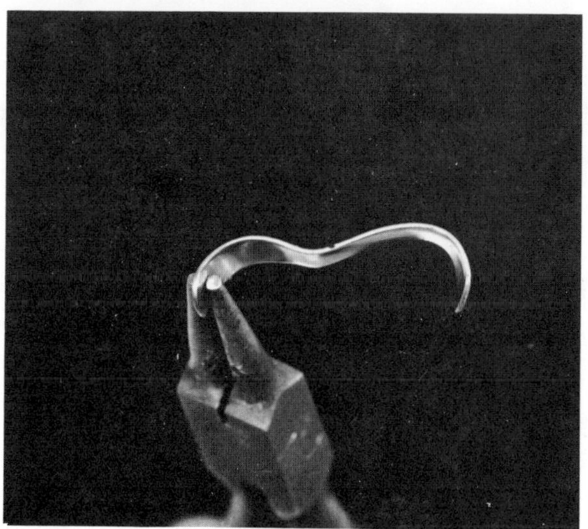

103. Cut off the excess metal, then complete shaping the piece until it is shaped as shown in the illustration.

104. Place the distal piece on the fingers, and mark the position.

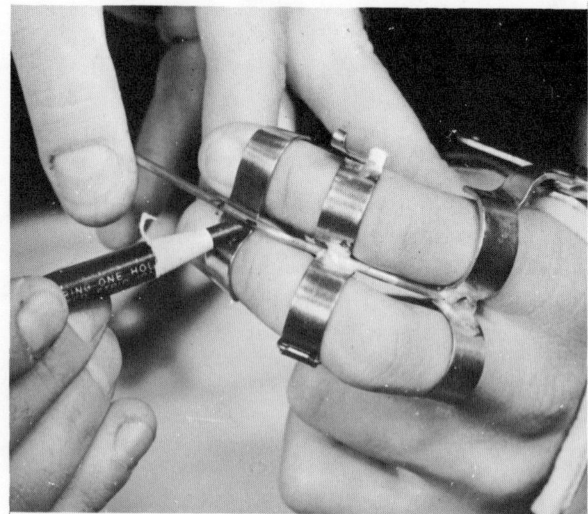

105. Silver solder the distal piece to the mounting wire.

106. Cut off the excess mounting wire, wash off the flux in hot water, and polish the entire assembly.

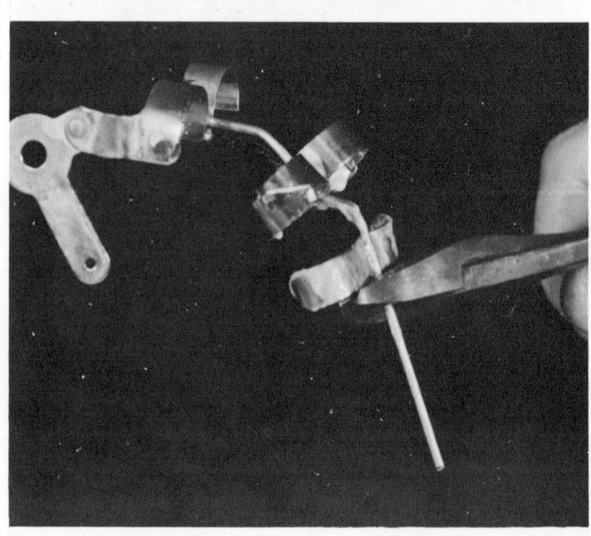

159

107. With the splint on the patient's hand,
     have him operate it. It is usually nec-
     essary to make slight adjustments in the
     rings, wires, thumb post, and retaining
     bar to achieve maximum function and
     comfort. Watch particularly for red
     marks on the skin, indicating excess
     pressure.

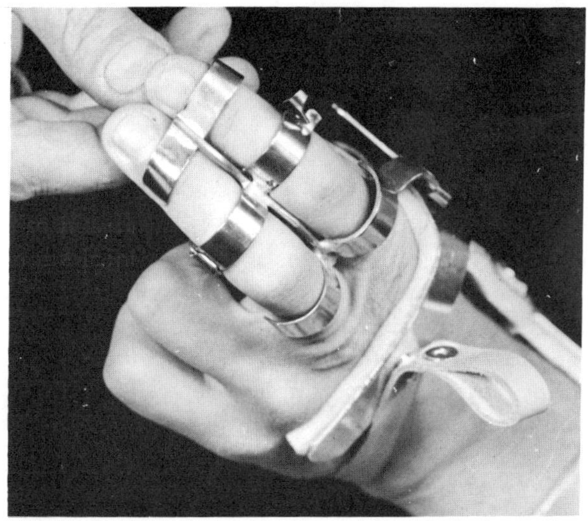

108. The last step is to cement the padding
     in place with Barge or similar material
     replacing the temporary rubber cement.
     The completed splint is shown in the
     illustration.

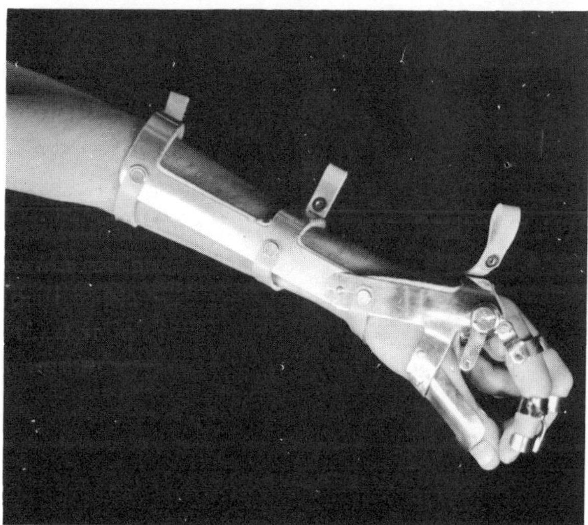

HOW TO ATTACH FLEXION ASSIST TO A FINGER-DRIVEN FLEXOR HINGE SPLINT

Introduction

In those cases where the patient has some fingers with extension power but none with flexion, a flexion assist spring will provide the patient with grasp equal to the force exerted by the spring. The extensors open the grasp and stretch the spring, then the latter closes the fingers on the object to be picked up.

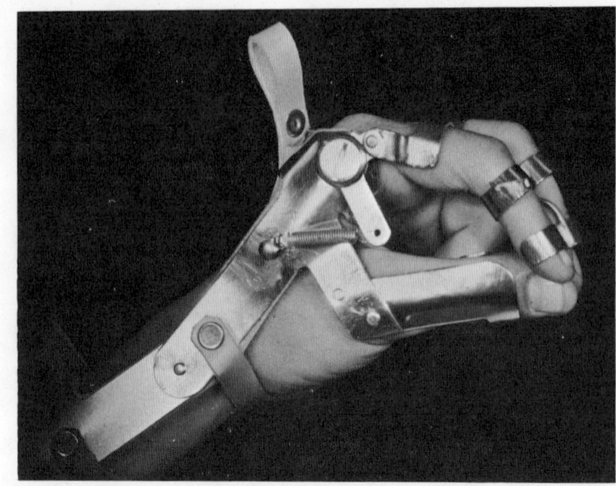

Procedure

1. Punch or drill a No. 18 hole in the the center bar of the splint at the location shown in the illustration.

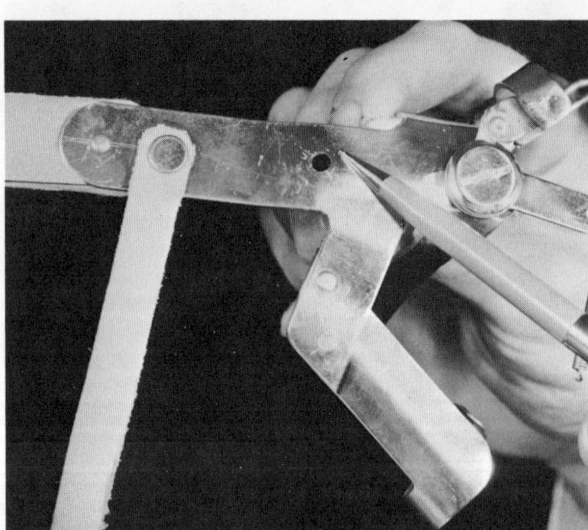

2. Fasten a boothook through the hole.

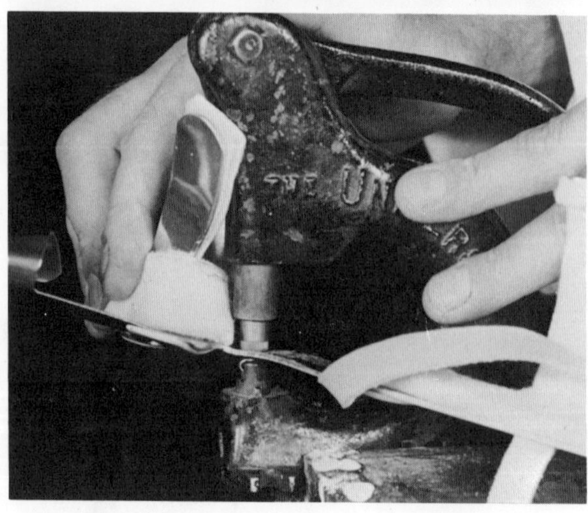

3. The open side of the boothook must face away from the joint hinge lever.

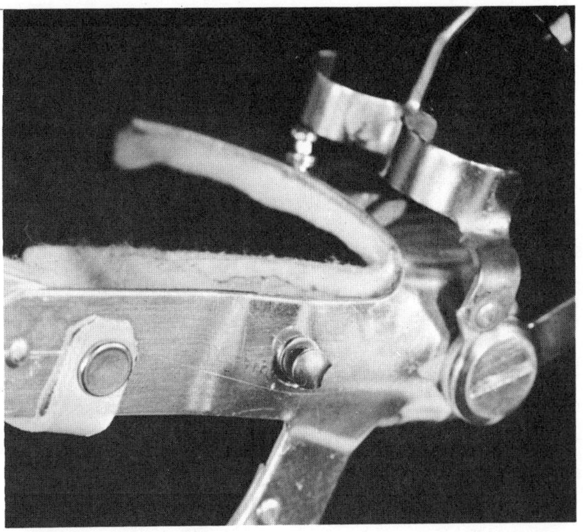

4. Hook one end of a spring into the eye in the joint hinge lever, the other on-to the boothook.

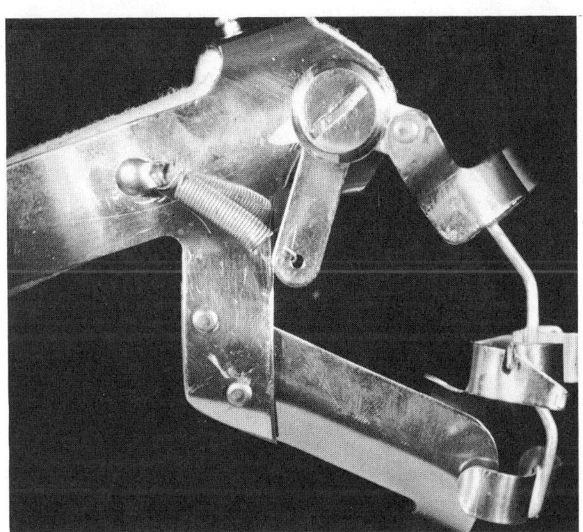

5. Try the splint on the patient. The spring can be made stronger or weaker depending on how strong his extensor muscles are.

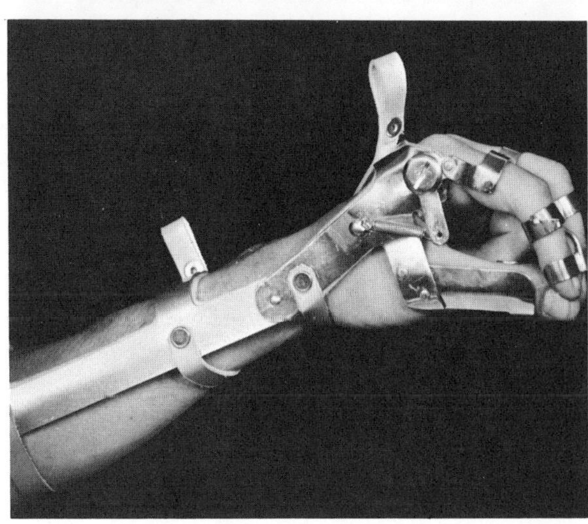

HOW TO ATTACH EXTENSION ASSIST TO A

FINGER-DRIVEN FLEXOR HINGE SPLINT

Introduction

The extension assist works in the opposite way as the flexion assist; in this case the spring substitutes for the lack of extension power. The patient must have some good flexors, so when he grasps he stretches the spring. When he relaxes, the spring separates the fingers.

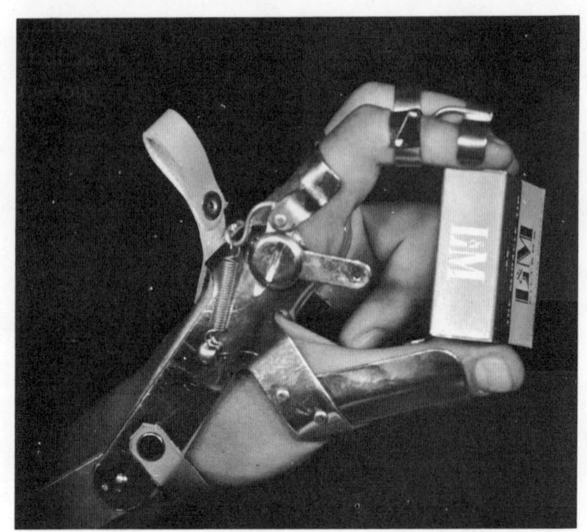

Procedure

1. Shape a hook into the end of a piece of 3/32 inch stainless steel wire. Cut off the wire so the hook is about 1 inch long.

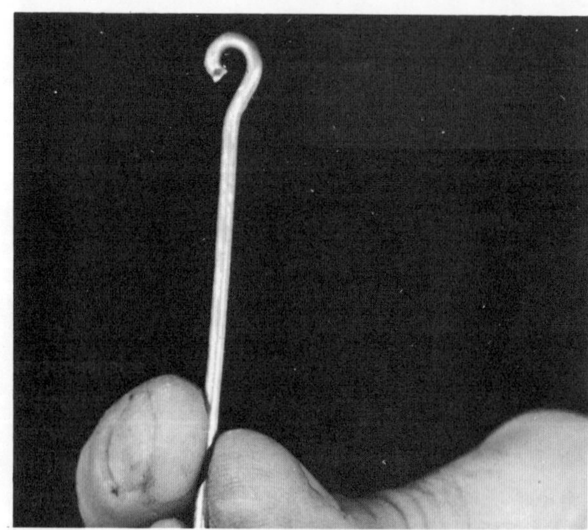

2. Silver solder the hook to the index
   finger ring lever.

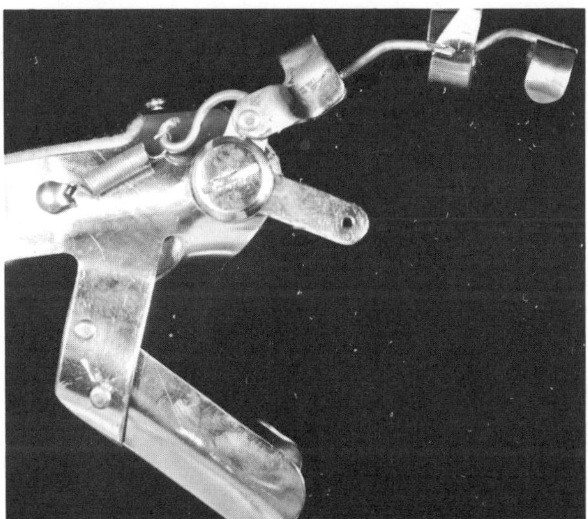

3. Shape the hook so it will not project
   up and catch on clothing, then install a boothook in the same manner
   described for the flexion assist. Install the spring, and check the splint
   on the patient.

HOW TO MAKE A WRIST-DRIVEN FLEXOR HINGE SPLINT

Introduction

It is not unusual to see patients with no finger flexion or extension able to flex and extend their wrists. Many quadriplegics are able to do this. The wrist-driven flexor hinge splint enables patients to turn wrist motion into finger prehension through the differential action of the drive levers on the finger piece and the forearm piece.

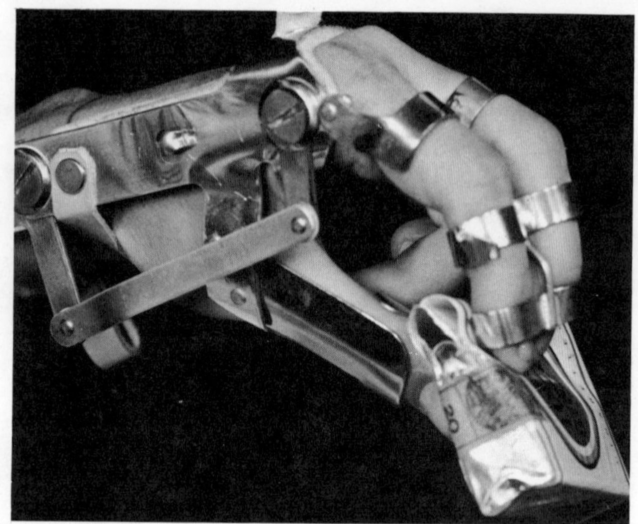

Procedure

1. The palmar piece is shaped in the same manner as for the finger-driven splint. The forearm piece is the same except that the extensions are on opposite sides of the bar, and an operating lever is included. The illustration shows its appearance when properly shaped. To obtain the correct parts, order parts Kit No. 2402, either medium or large size.

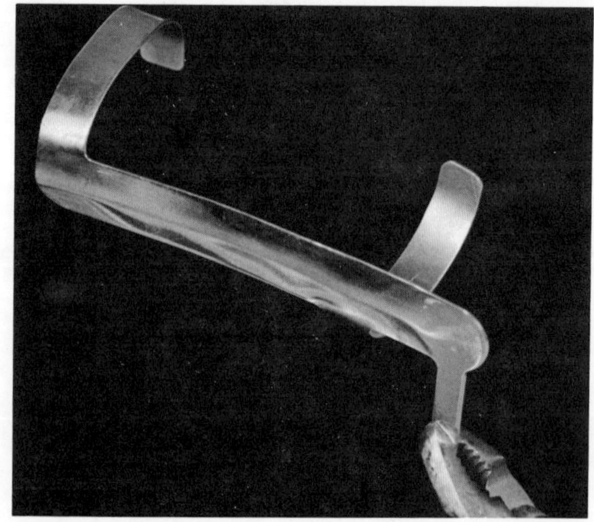

2. Fit the forearm piece on the patient's arm with padding in place, then mark the location of the wrist joint.

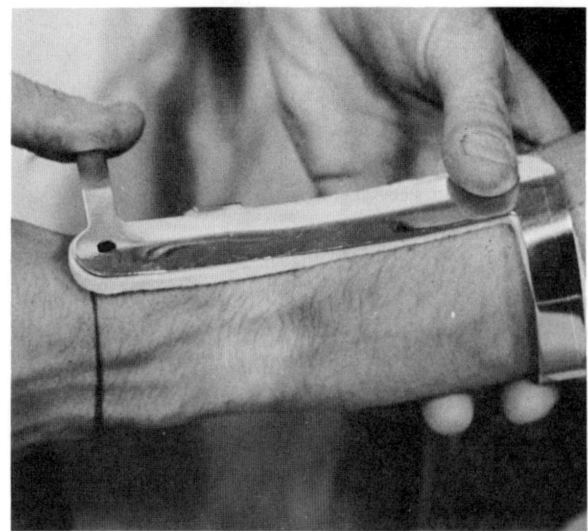

3. Drill a No. 21 hole in the forearm piece at the wrist joint mark, then tap it with a 10-32 tap.

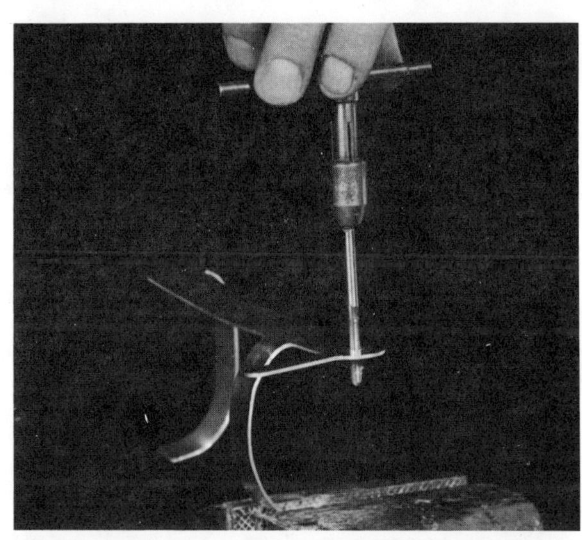

4. Drill a 1/4 inch hole in the wrist joint center of the palmar piece, then assemble the latter to the forearm piece with hinge joint bushing, pivot screw, and two Teflon washers.

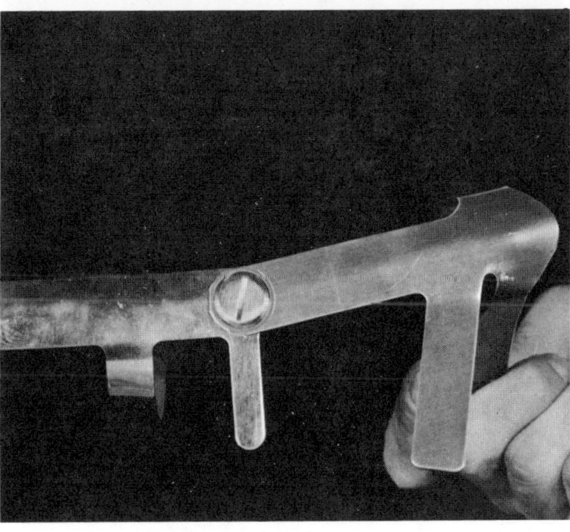

5. The thumb post and finger piece assemblies are made in the same way as
   explained for the finger-driven splint.
   The splint should now look like the illustration.

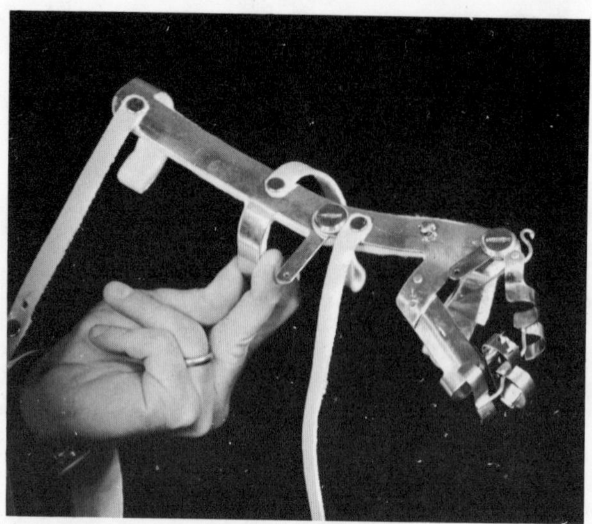

6. Drill a No. 40 hole in one end of the
   3/8 inch aluminum connecting rod,
   round off the corners, smooth and polish. Insert a stainless steel rivet in
   the hole, put a 3/8 inch Teflon
   washer on the rivet.

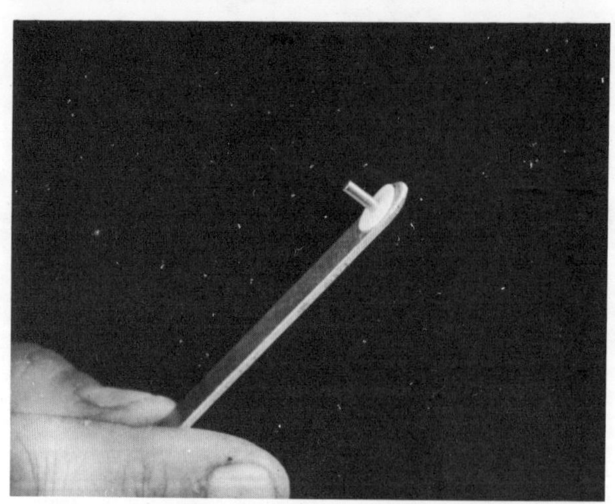

7. Rivet the connecting rod to the forearm piece operating lever. Do not
   rivet it so tightly that it cannot move
   freely.

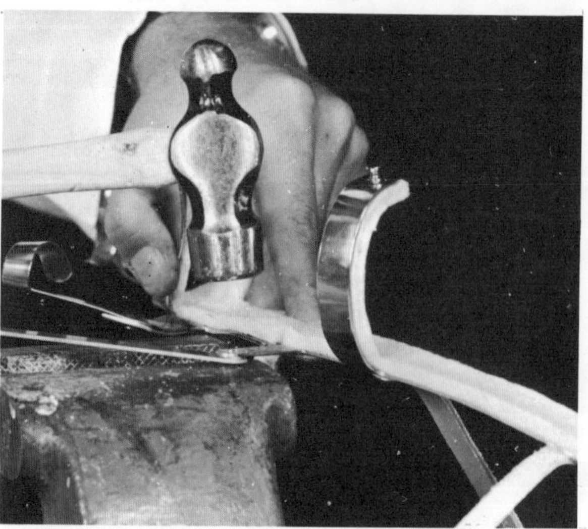

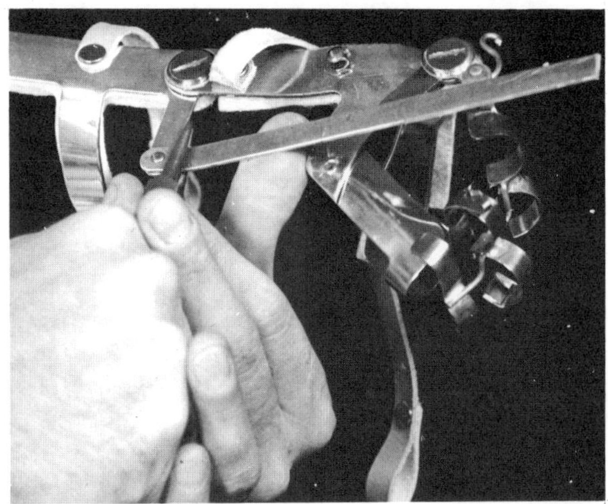

8. Bend the connecting rod so it will line up with the lever on the joint hinge.

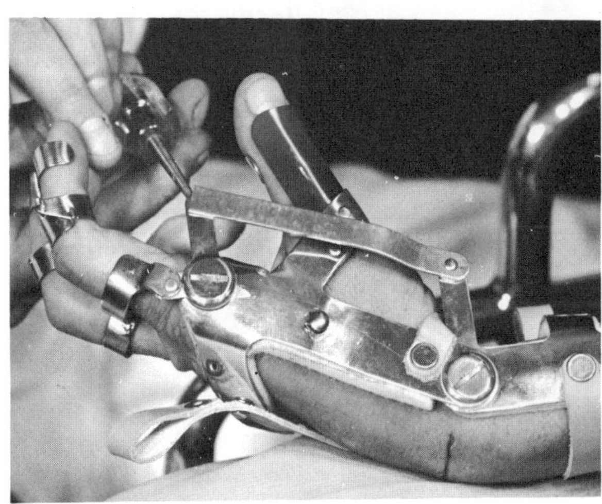

9. With the splint on the patient's hand, direct him to drop his wrist as far as it will go comfortably, then open his grasp as far as it will go without strain. Mark the connecting rod on the lever at this point, drill a No. 40 hole in the rod, and rivet it in place, using the Teflon washer.

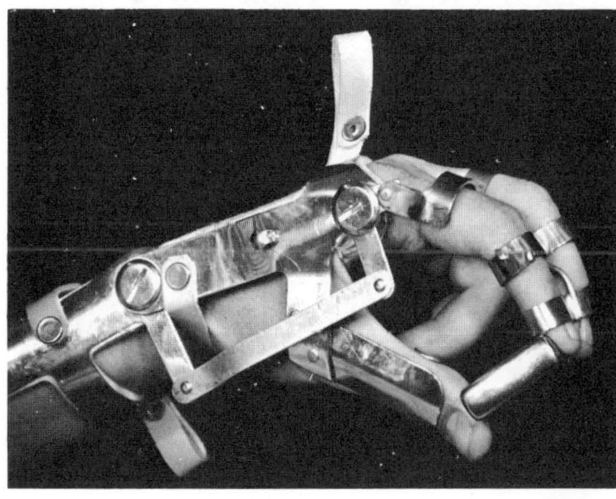

10. Have the patient try the action of the splint. When he flexes his wrist the fingers open; when he extends it, they close. If, after a few tries, he does not have the range to close the fingers completely, shorten the connecting rod a small amount at a time until it is just right.

HOW TO INSTALL THE ARTIFICIAL MUSCLE ON A FLEXOR

HINGE SPLINT TO PROVIDE FLEXION

Introduction

It is not unusual for the patients to have finger muscle loss so great that they have neither flexion nor extension. In these cases, it is necessary to provide outside power enabling them to grasp objects. This can be done with the artificial muscle which is powered by carbon dioxide under pressure that is stored in a small tank. A regulator provided with the artificial muscle allows the gas to flow through a control valve; this in turn permits the gas to flow into the artificial muscle causing it to contract and thereby provide the necessary flexion force. An extension spring extends the fingers when the power is released.

The following instructions show how the artificial muscle is installed on the flexor hinge splint and how the control valve regulator and carbon dioxide supply tank are connected to permit the patient to have finger flexion powered by the artificial muscle.

Procedure

1. Locate the artificial muscle bracket mounting stud on the proximal extension of the forearm piece one inch from the lower edge of the center bar. Drill a No. 40 hole at this point, then rivet the bracket into place. Be sure the flats on the bracket are at a 45° angle to the proximal extension. This makes it possible to remove the artificial muscle without twisting it around too far.

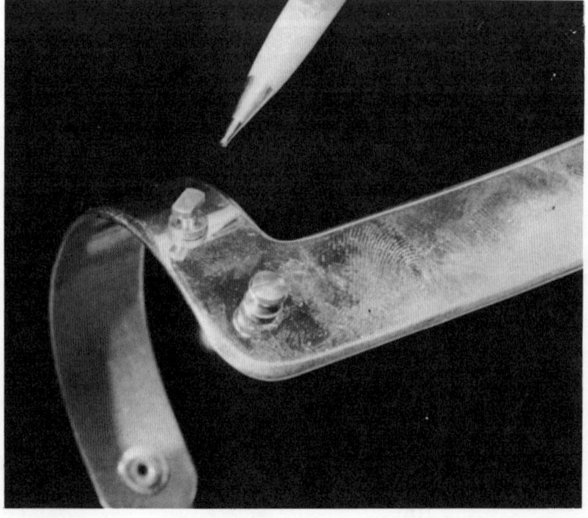

2. In determining what size artificial muscle will be needed, remember standard sizes are six and eight inches in length, but other lengths may be made on order. Insert the artificial muscle attachment rod into the hole in the joint hinge lever. Put the artificial muscle mounting bracket in place on the stud, then use a ruler to measure the approximate length muscle needed. In this case, an eight inch artificial muscle will fit perfectly.

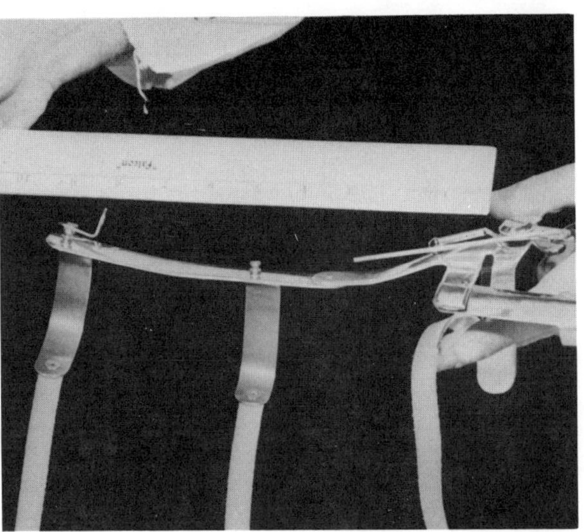

3. Determine the length of the attachment rod. If the artificial muscle attachment bracket is not already in place, attach it to the nipple on one end of the muscle, then put it on the mounting stud. Stretch it full length, open the finger pieces fully, and mark the attachment rod at the point where the end of the artificial muscle comes on it.

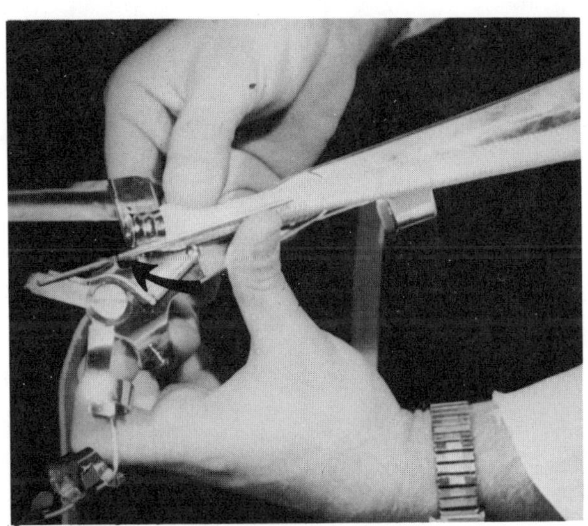

4. Cut the attachment rod off 1/8 inch longer than the point at which it was marked.

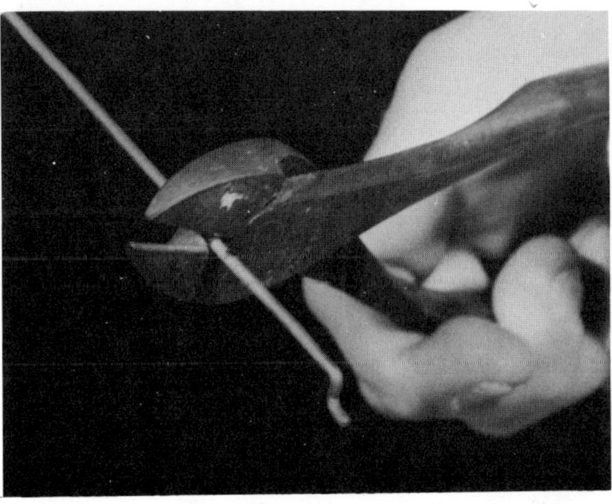

5. Solder the 10-32 headless set screw onto the end of the artificial muscle attachment rod. This provides an adjustment for length as well as a means of attachment. Use a torch and silver solder for this operation.

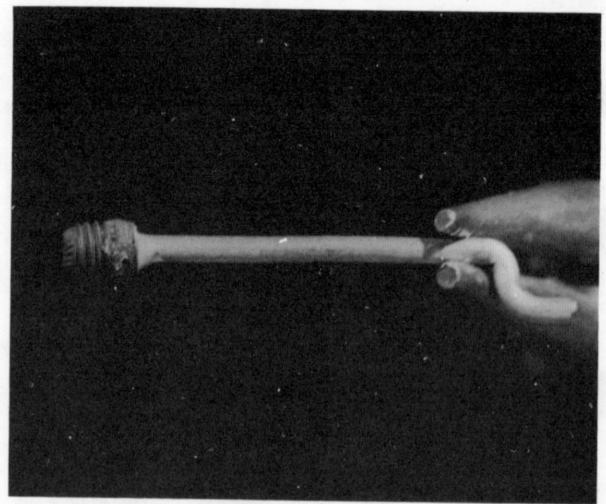

6. Assemble the artificial muscle onto the flexor hinge splint as shown.

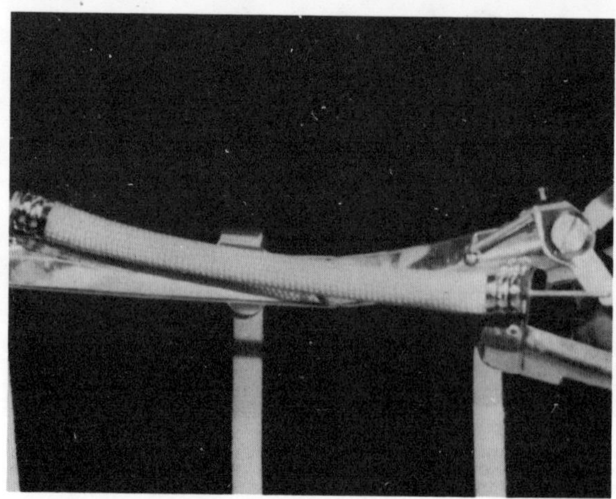

7. Cement the padding into the splint.

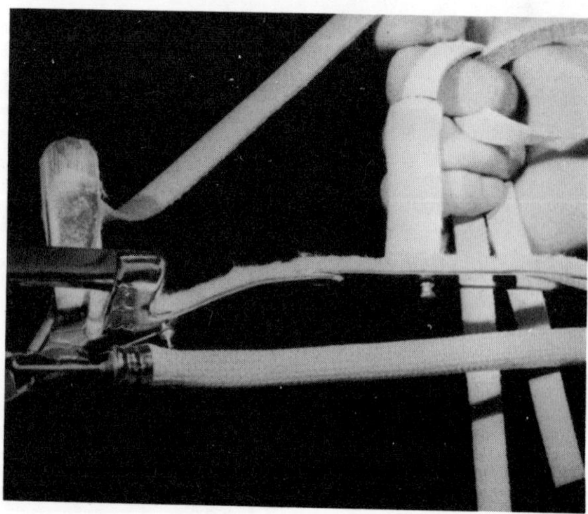

8. Determine where to locate the control valve. The type of control valve and its location is determined by the patient's remaining muscular function which will operate the valve. If he has hip abduction, this movement can be used to operate a control valve mounted on the patient's wheel chair in a position where he can nudge it with his leg using his hip abductors. A lever type valve is used for this application and is clamped to the wheel chair as illustrated. This same type of valve could be mounted in other positions on the wheel chair so that it could be operated by the foot or the elbow if these parts of the body are able to move and operate the valve. If the patient can exert six ounces of force in the form of a push, he can operate this valve.

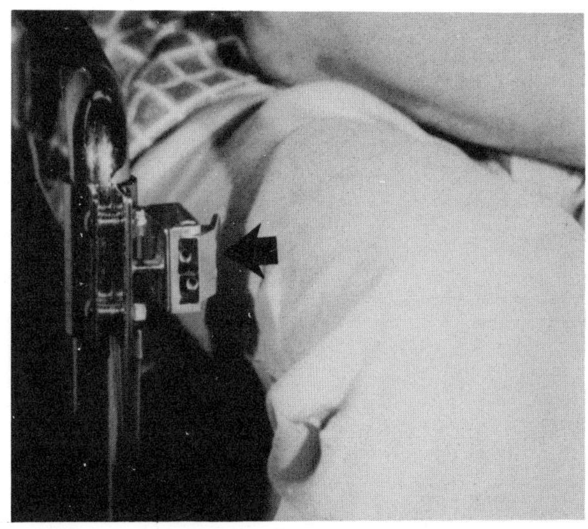

9. Connect the supply tank to the valve and to the artificial muscle on the flexor hinge splint. The supply tank can either be attached to the wheel chair with a bracket or merely laid in the seat along side the patient. In many cases the latter is preferable because of the convenience of being able to remove it without unfastening brackets and clamps. Square the end of a piece of plastic tubing.

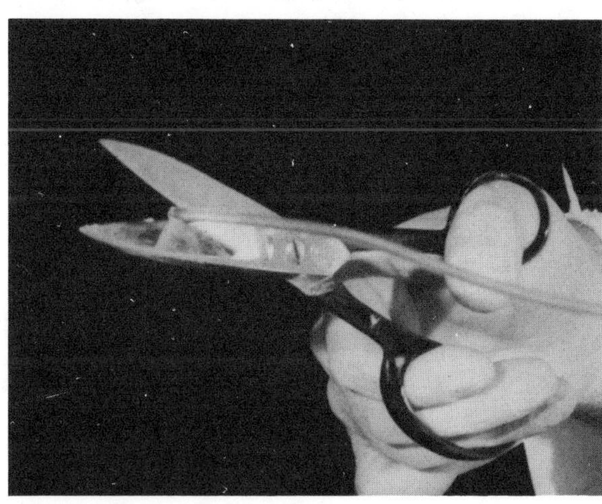

10. Put a ferrule over the end of the tubing, heat the end of it to soften the plastic, then slip the tubing onto the connection on the pressure regulator. Tighten the ferrule finger tight only. NEVER USE A WRENCH.

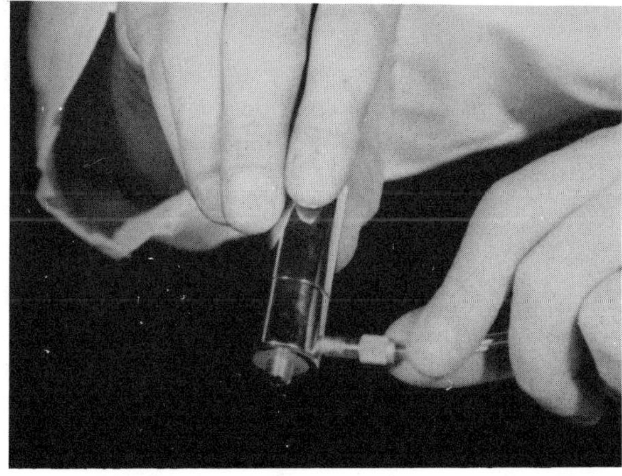

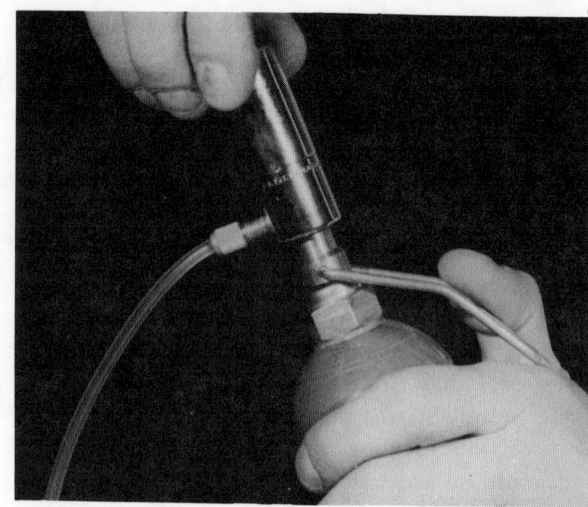

11. Screw the pressure regulator into the
     tank.

12. The valve has two outlets or connec-
     tions, one stamped "O" for "outlet",
     the other stamped "I" for "inlet".
     The tube from the regulator goes to the
     connection marked "I". The tube from
     the muscle goes to the connection
     marked "O". Square off the end of
     the tube from the regulator. Apply
     the ferrule, heat the end of the tube,
     and attach it to the inlet ("I") side
     of the valve as shown. Again, the
     ferrule need only be finger tight.

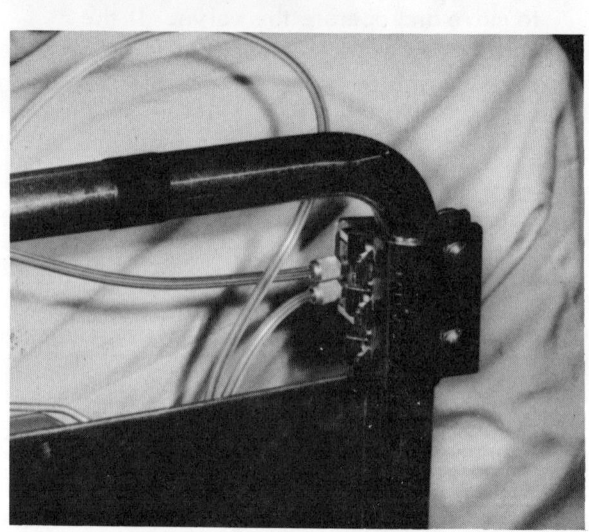

13. Attach another tubing to the outlet
     side of the valve and determine the
     length needed by having the patient
     stretch his arm out as far as it will
     go. The tubing must be long enough
     to accommodate his reach.

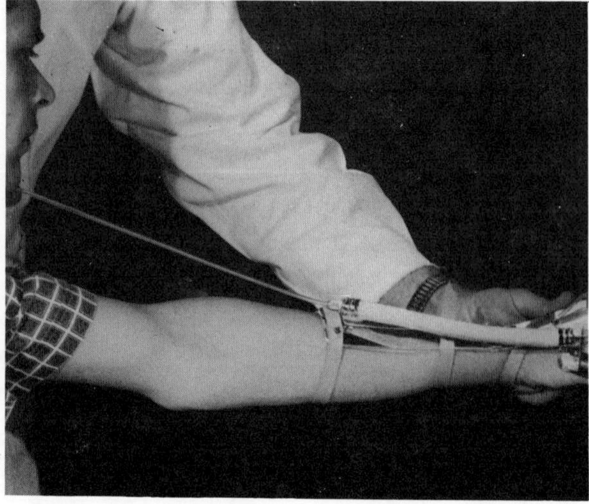

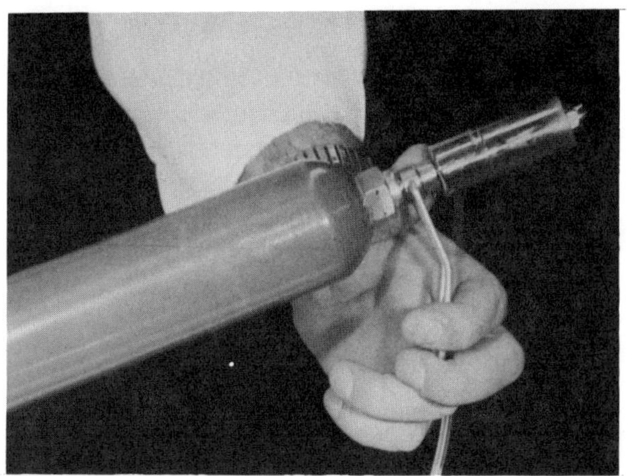

14. Turn on the gas at the tank by turning
the hinged handle counter clockwise.
One quarter turn open is enough.

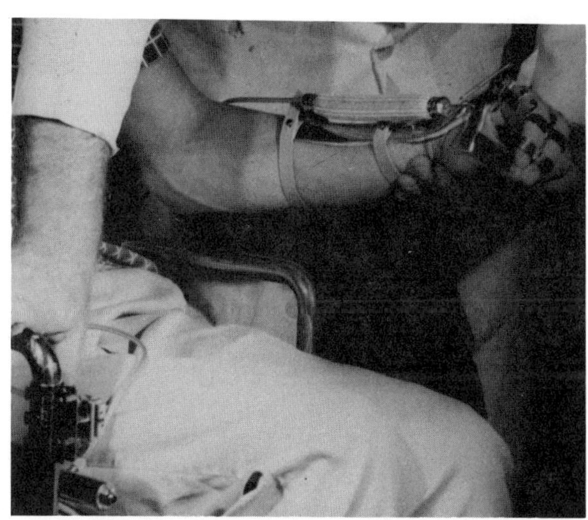

15. Test the action of the artificial muscle
by operating the valve carefully and
observing with how much force the fingers are brought together. It is important that they should not be brought
together with so much force that it will
hurt or injure the patient's fingers.

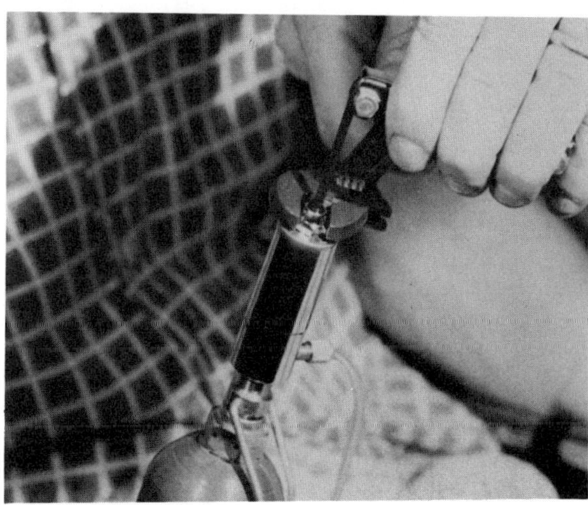

16. To regulate the amount of pressure,
loosen the lock nut on the top of the
regulator and use an Allen wrench to
turn the adjusting screw. Turn the
adjusting screw clockwise to lower
the pressure, counter-clockwise to
raise it. At first, provide only enough
pressure to provide a pinch of one or
two pounds. After the pressure has
been set, tighten the lock screw securely. After the patient has been
trained and can control the valve
himself, the pressure can be raised
to whatever amount he needs to function most efficiently, usually about
five pounds. Use a pinch gauge to
check finger pressure.

17. Test the system for leaks by opening the valve and filling the muscle with gas. Close the valve and allow the muscle to remain contracted. Observe it over a period of time to see if it relaxes, indicating a leak. If it does leak, and you are unable to find where it is leaking, remove the artificial muscle from the splint, immerse the valve, tank, tubing, and muscle in water and watch for bubbles when it operates.

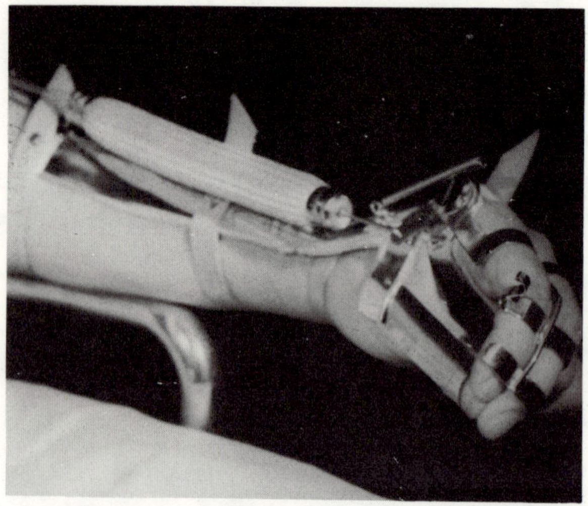

18. In those cases where the patient has shoulder motion, the valve can be mounted on a butterfly type harness as illustrated.

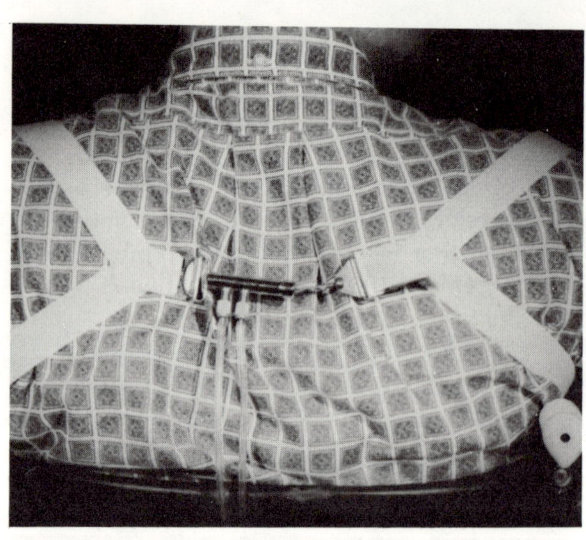

19. A microswitch control can be mounted on the flexor hinge splint finger piece as illustrated, and operated by the patient's own fingers, assuming that they retain some motion. The microswitch, in turn, operates a relay.

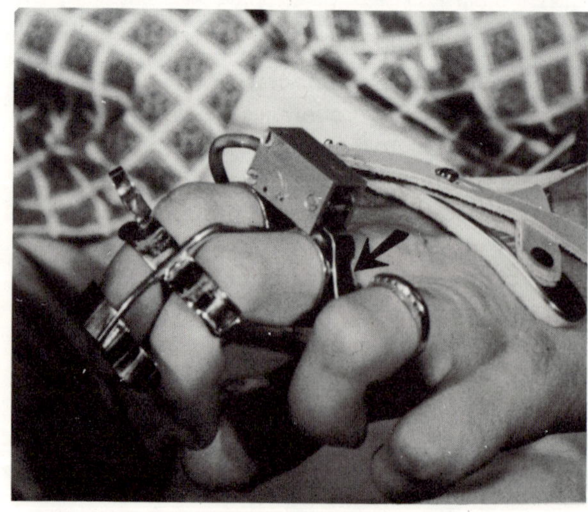

20. The battery-powered relays illustrated are operated by the microswitch. The relays open and close the control valves which control the flow of gas from the cylinder to the artificial muscle.

21. The tubing connections are specially designed so they can be simply plugged into the relay box as illustrated.

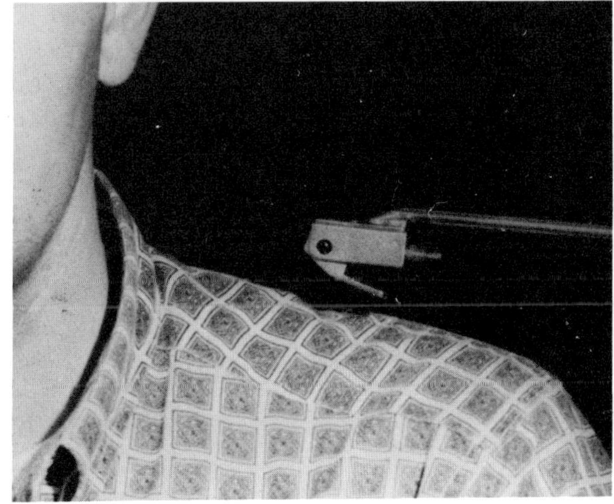

22. If the patient has shoulder elevation, a rod bent at right angles can be attached to his wheel chair and the control valve fastened to it. Raising the shoulder operates the valve.

Lever Type Control Valve

Slide Valve

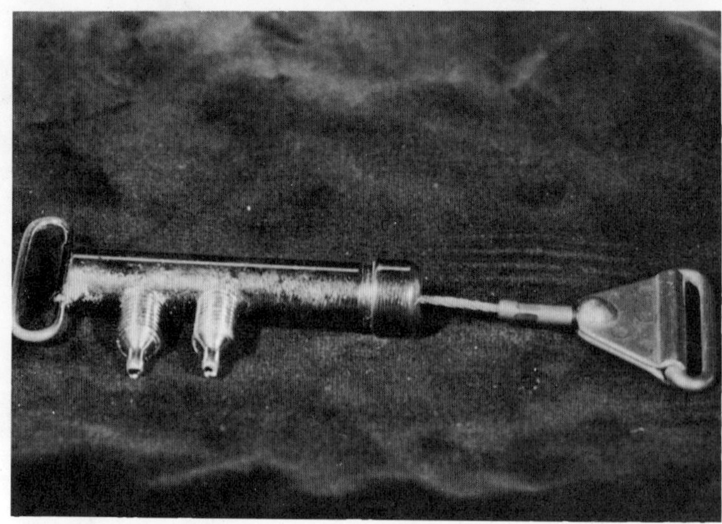

Lever Valve on Bracket
for Shoulder Elevation
Operation

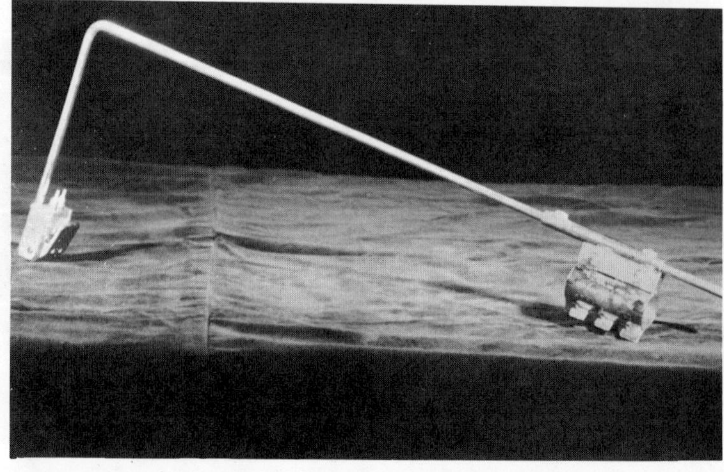

HOW TO INSTALL ELECTRIC POWER ON A FLEXOR HINGE SPLINT

Introduction

With the availability of nickel cadmium rechargeable batteries, electrical power is practical to use to operate the flexor hinge splint. The major components are shown in the illustration. The small 12 volt direct current motor (lower center) drives an enclosed pulley on which is wound the control cable. The cable housing is anchored to a standard Hosmer base plate retainer which is riveted to the splint. The cable runs through a sheave which is pivoted to the hinge joint lever and is anchored to the spring hook. The action of the sheave gives the motor a 2 to 1 mechanical advantage, thus providing a strong grasp and slowing down the action so the patient can control it easily. A paddle type dual micro-switch control enables the patient to close the grasp by pressing the paddle all the way down. He can reverse the motor to release the grasp by pressing the paddle half way down. A Jones plug and fuse make it possible to disconnect the units readily. An 8 volt nickel cadmium battery with a built-in charger suitable for the electric powered flexor hinge splint is made to use with the Arriflex motion picture camera, (see illustration), and is good for wheel chair installation. Smaller sizes are available from electronic supply firms.

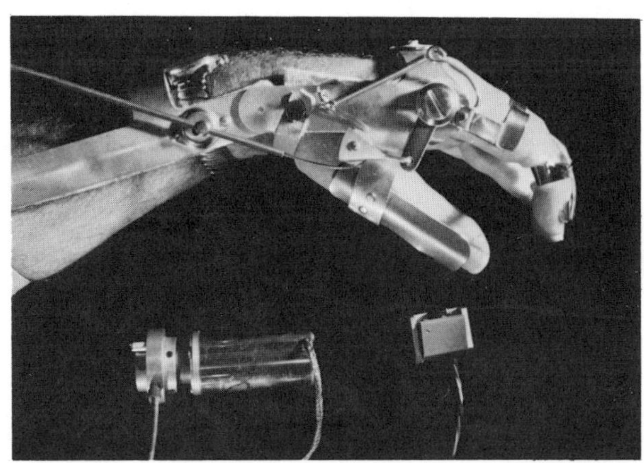

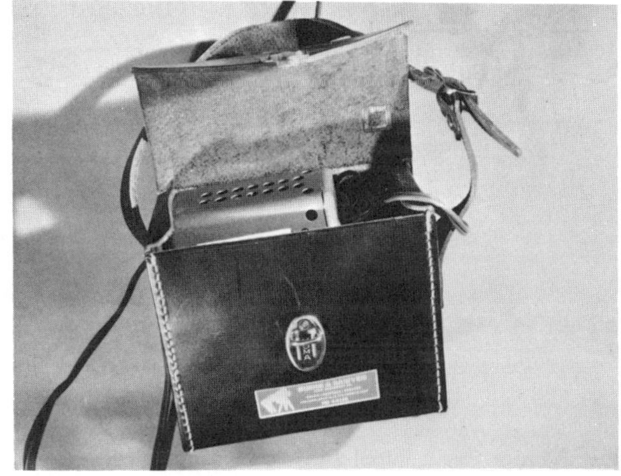

1. Rivet a Hosmer base plate retainer to the center bar of the splint slightly proximal to the riveted joint between the forearm and palmar pieces.

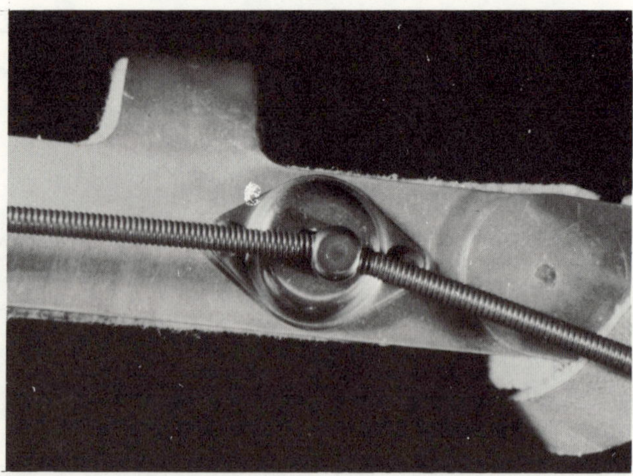

2. Install the sheave in the joint hinge, thread the cable through the sheave, then through the spring eye. Clamp a set screw cable end block onto the cable to prevent it from slipping through the spring eye.

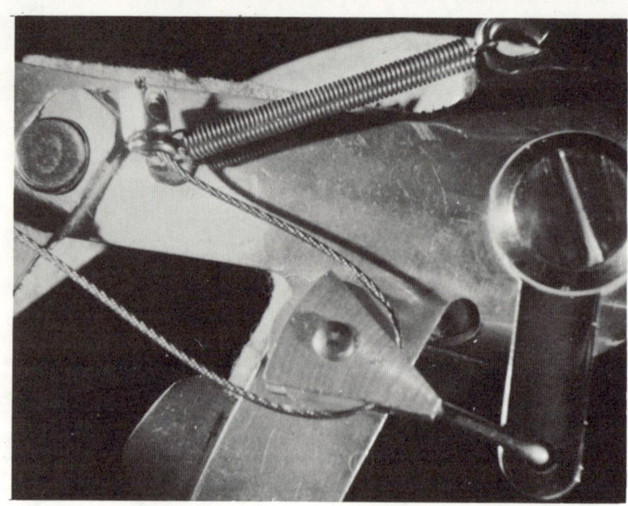

3. Mount the control switch in a position easiest for the patient to operate it, as over the shoulder, near the foot, by the side of the knee, or alongside the head. The action of the switch is progressive. In the illustration, the ball stud attached to the paddle presses the white button of the first micro-switch when it is depressed. Pressing it down all the way causes the ball to close the second switch, opening the first one.

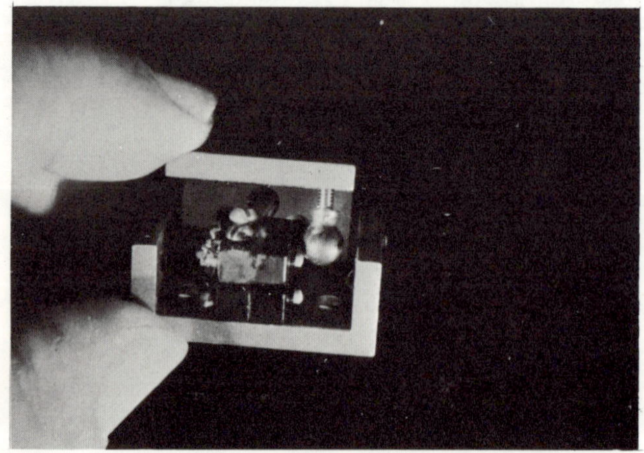

4. Mount the motor in a convenient position so there will be no sharp bends in the control cable. In the case of a wheel chair patient it can be clamped to part of the chair.

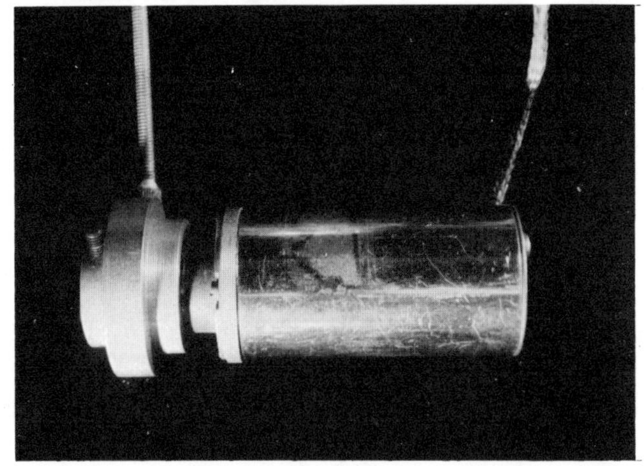

5. Measure the wires from the switch to the motor and from the battery to the motor. Cut them to length, and solder them to the Jones plug. See the circuit diagram (Figure 89) on the next page for the correct connections for the various colored wires. The Jones plug cover is removed by punching out the drive pin on the side of the plug.

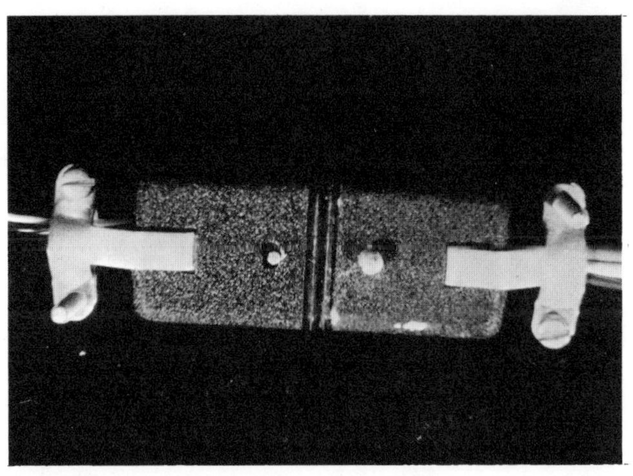

6. Connect the battery wires to the battery. Unwind the take-up spool on the motor until it hits the stop, by pressing the control switch all the way down. Adjust the cable at the spring eye so the slack is taken up with the fingers all the way open.

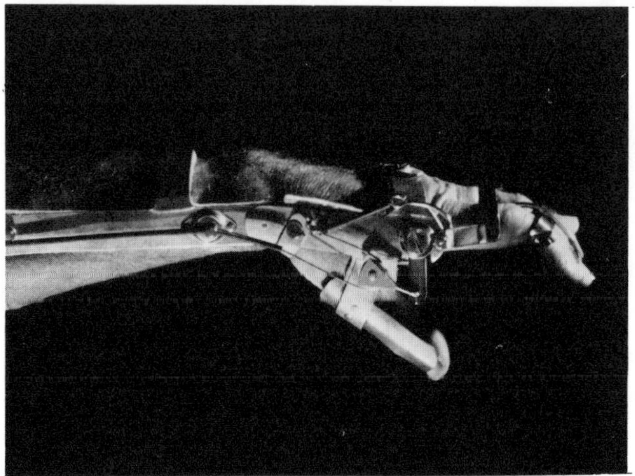

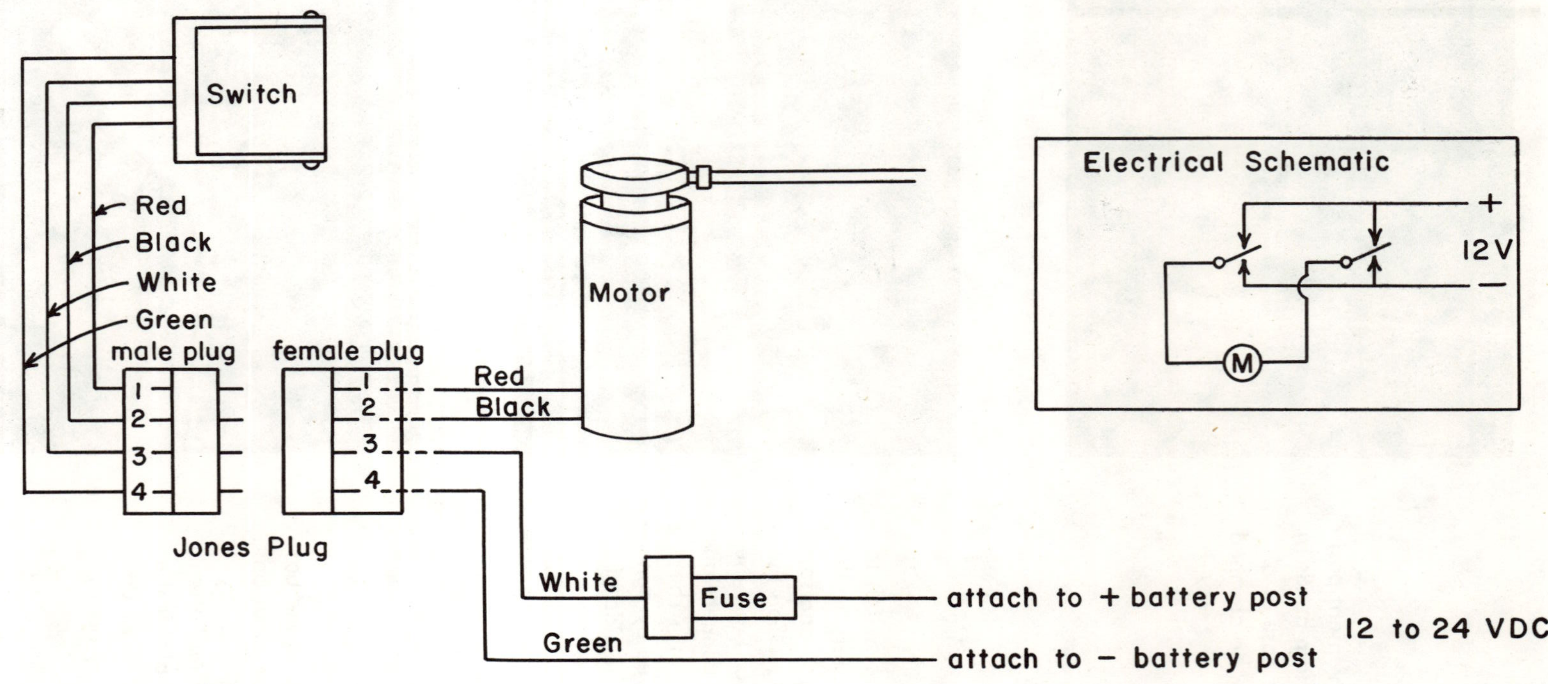

Figure 89. Circuit Diagram for Connecting Electric Powered Flexor Hinge Splint

7. Press the switch to the first position to close the fingers. (If you have to push the switch all the way down to close the fingers, the connections at the battery are reversed.)

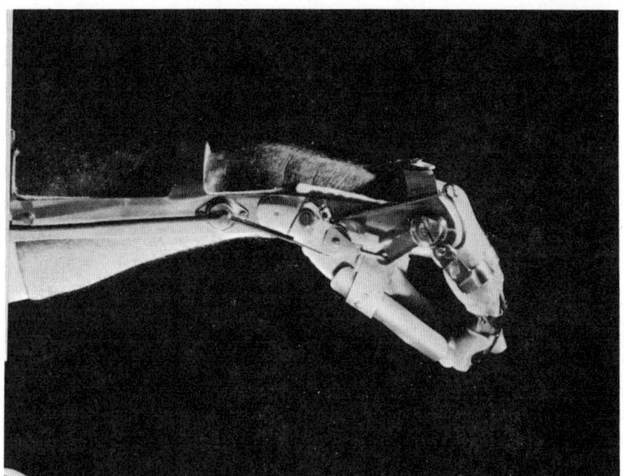

HOW TO FIT A CABLE-CONTROLLED HOOK

Introduction

The cable-controlled hook is a device to replace lost hand function, and has been known to many as the "handy hook". It is quite simple, and comparatively easy to fit. The hook used is a standard voluntary-opening type terminal device of the kind used on artificial arms. It is held in the palm of the patient's hand by a stainless steel receptacle, and is opened and closed by a shoulder loop harness.

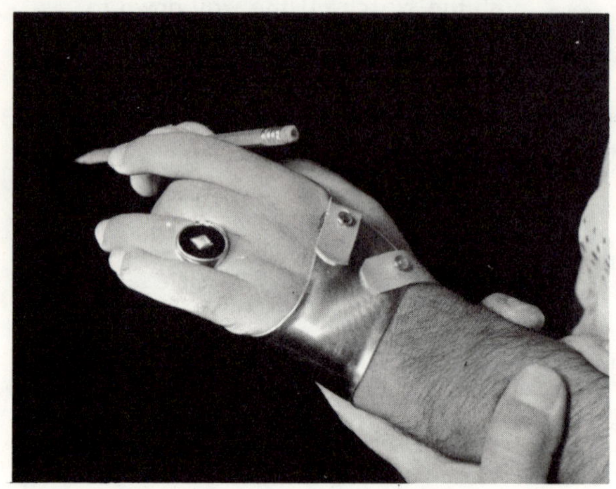

This device may be used in cases where the fingers have no sensation, where the finger joints are so stiff and deformed that movement is difficult or painful or a combination of these problems. If the fingers have sensation and the joints are in reasonably good condition, a functional hand splint is recommended.

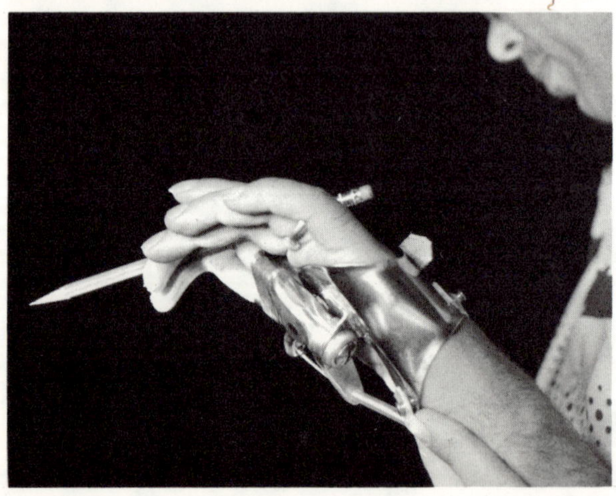

1. Obtain the correct kit of parts to properly fit the patient. Kit No. 208 can be ordered from the Hosmer Corp. complete with harness and cable, and will fit if the following information is furnished:

   a. Right or left hand?

   b. Circumference of hand at meta-carpophalangeal joints __________.

   c. Type of hook wanted __________.

   d. Tracing of hand, palm down flat, indicate metacarpophalangeal joints and styloid.

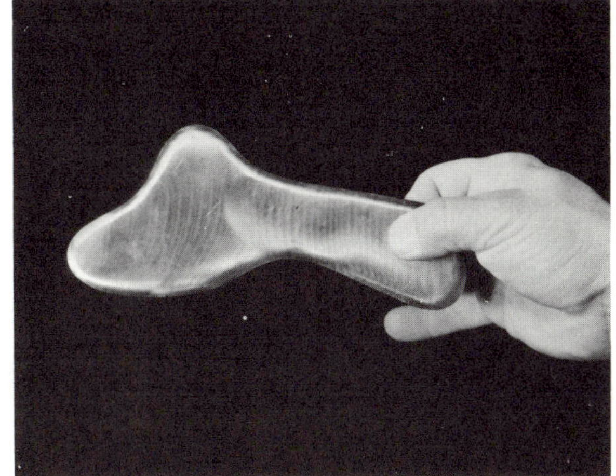

2. Shape the stainless steel receptacle to fit the patient's hand. Check the receptacle for correctness of size by holding it on the patient's hand as illustrated. It should cover the palm, but not extend onto the fingers or wrist area.

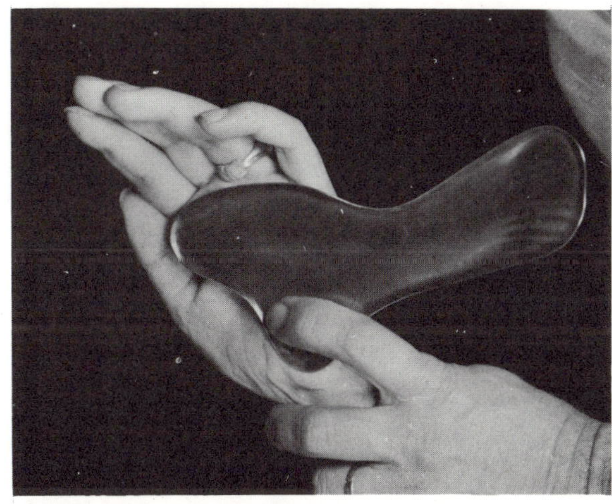

3. Mark the receptacle for the bends needed to shape the receptacle to the hollow of the palm of the hand. The top line follows the metacarpo-phalangeal joints, the lower one follows the carpometacarpal joints.

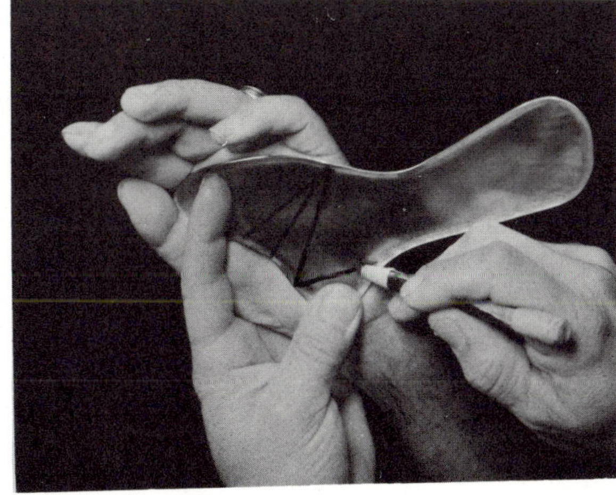

4. Bend the receptacle at the first marks made, as shown.

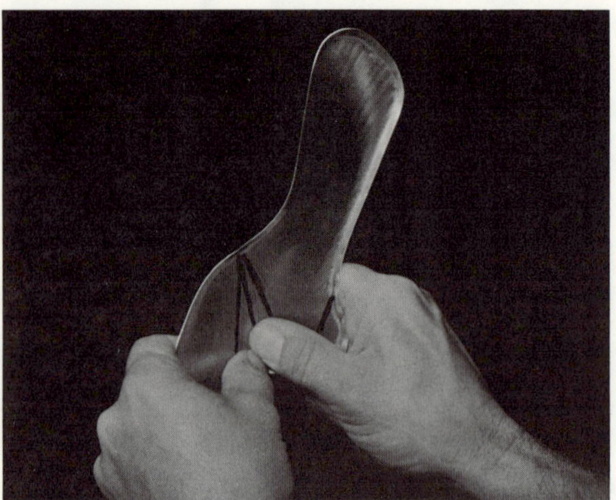

5. Bend the receptacle at the second marks.

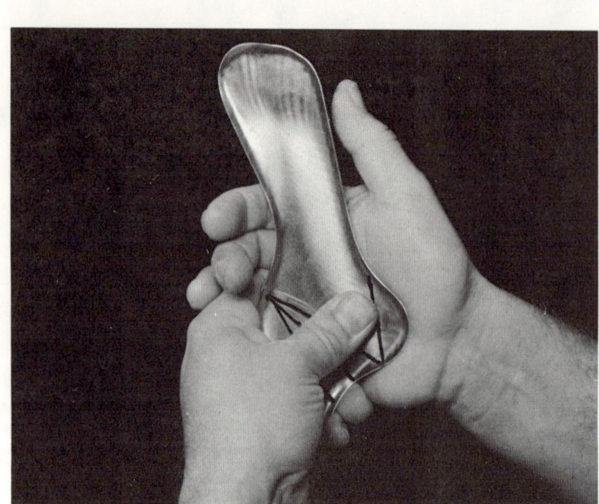

6. Check it on the patient's hand and increase or decrease the amount of bending at each point until a perfect fit is obtained.

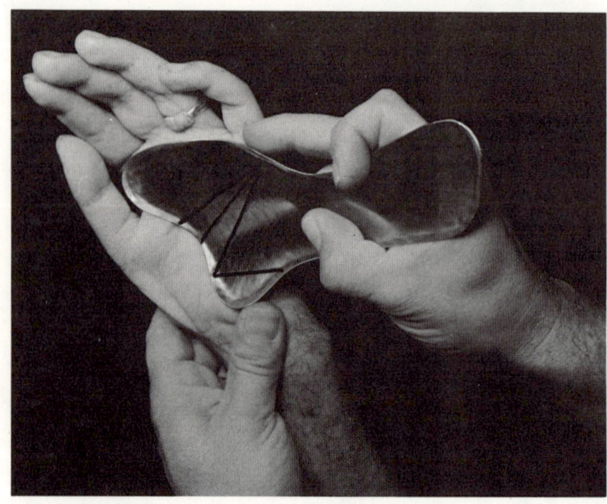

7. With the receptacle in proper position in the palm of the patient's hand, mark it along the ulnar aspect of the hand as illustrated.

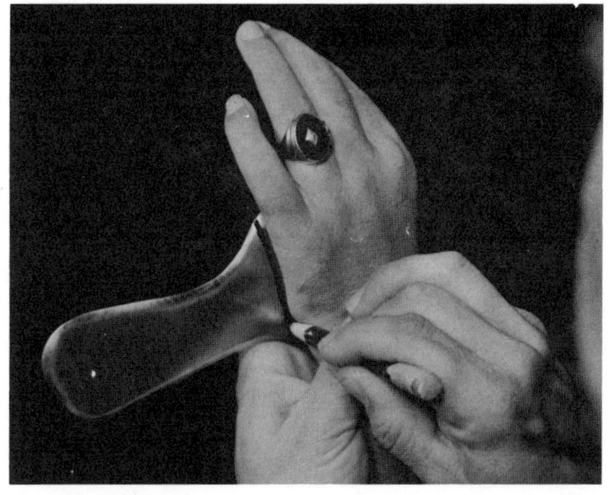

8. Bend the receptacle to a 90 degree angle at this mark. Be sure the flared edges are up when the bend is made.

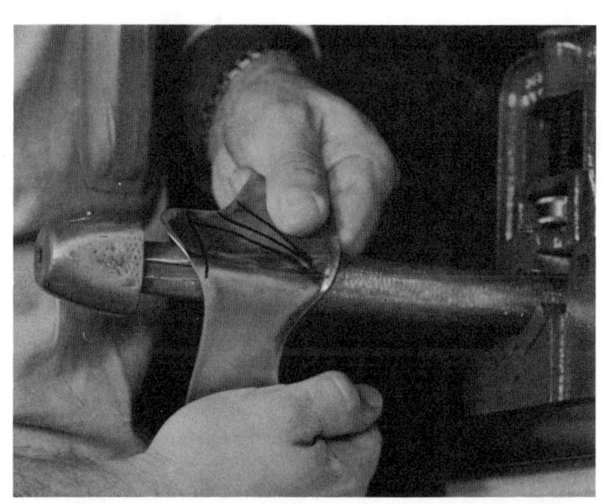

9. With the receptacle again in position on the patient's hand, mark a line on it parallel to the dorsal aspect of the hand as shown.

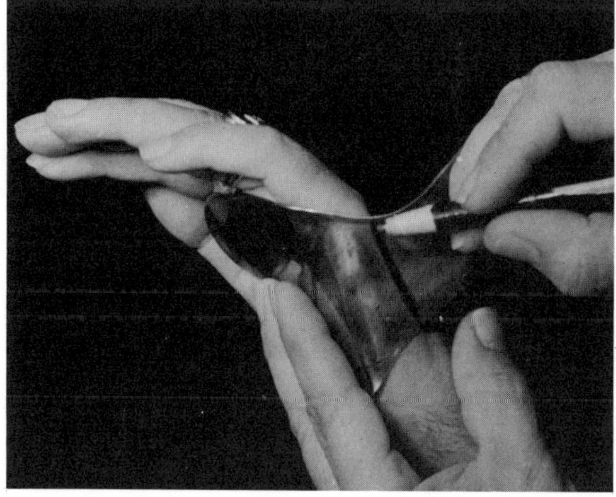

10. Bend the receptacle at this mark, forming it into a "U" shape. Try it on the patient's hand, and vary the bend until it fits well.

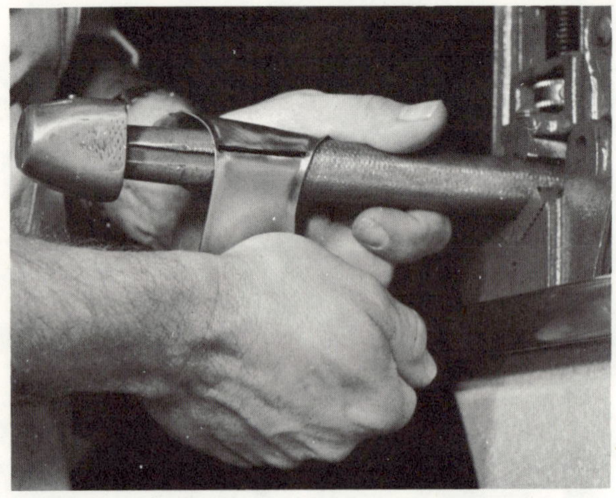

11. With the receptacle on the patient's hand, mark it in line with the radial side of the metacarpophalangeal joint of the index finger, as shown.

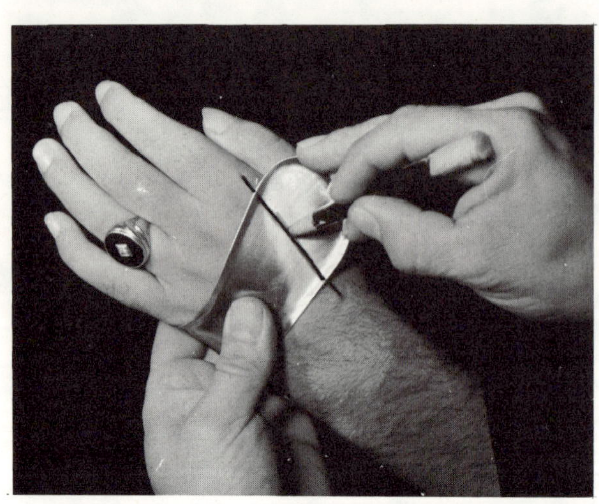

12. Measure over 1/4 inch from this line and sketch a curved line to intersect that point.

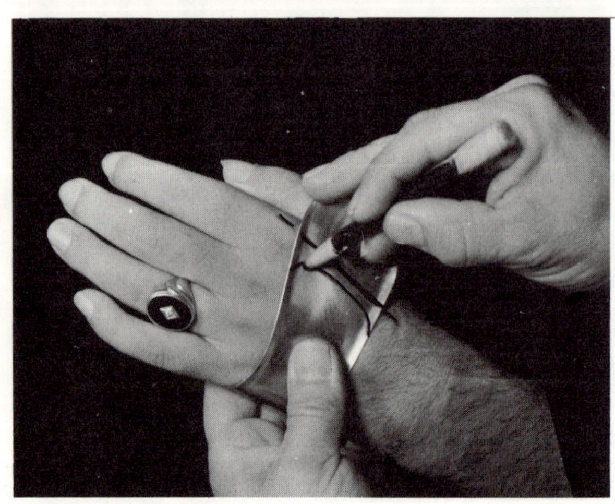

13. Cut off the receptacle at the curved line.

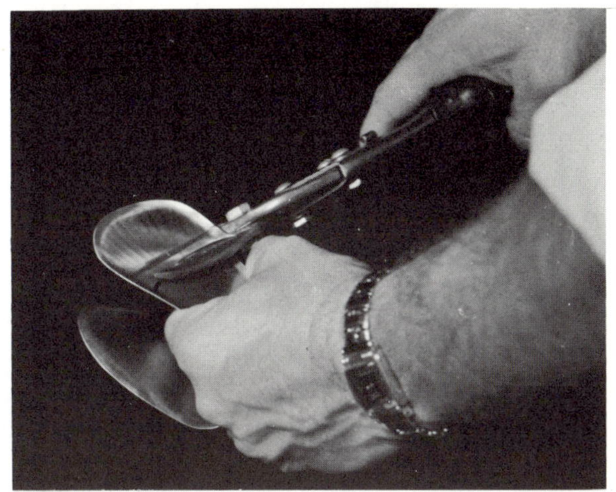

14. Grind the rough edge smooth and then polish it so it will not cut into the back of the patient's hand.

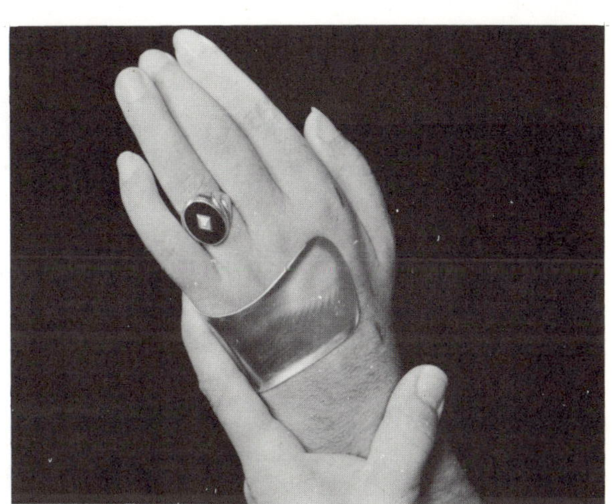

15. Install the strap mounting studs. Mark the location of the studs on the dorsal aspect of the receptacle, as shown.

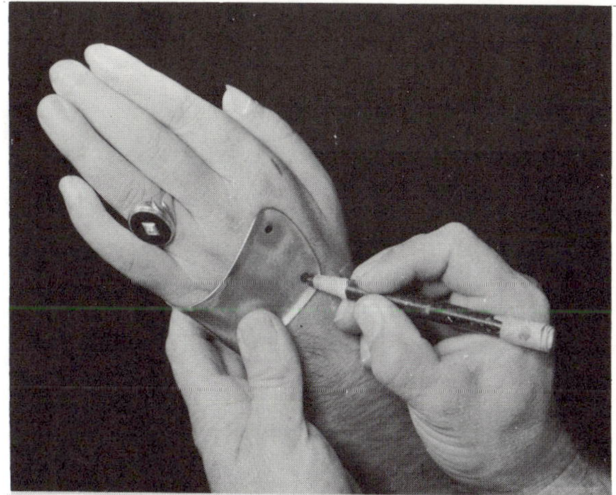

16. Mark the location of the studs for the palmar side of the receptacle.

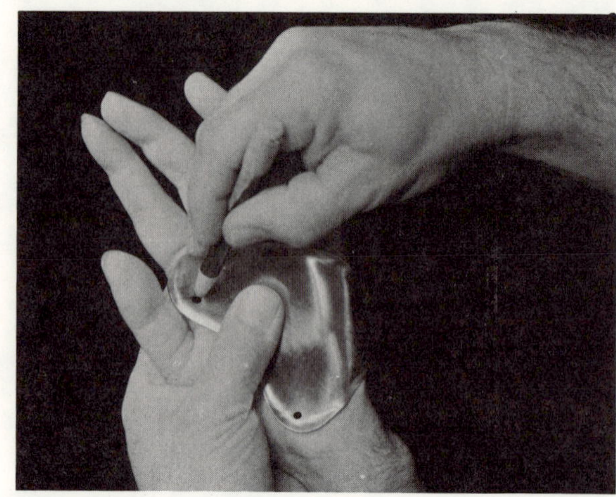

17. Punch or drill 1/8 inch holes at the marks.

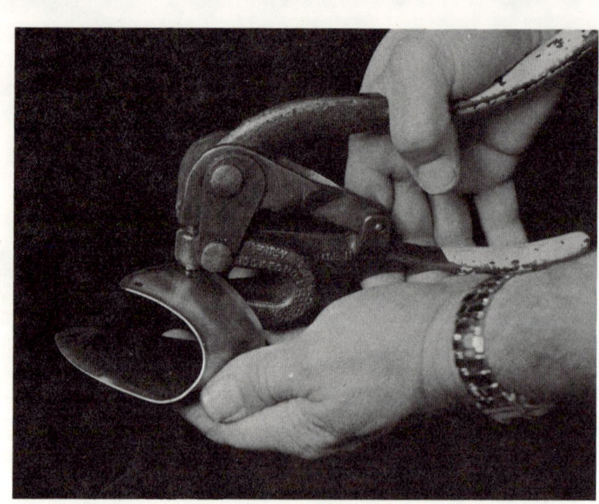

18. Rivet the studs in place. It is easier to do if the receptacle is bent open. If this is done, be sure to bend it back exactly the way it was.

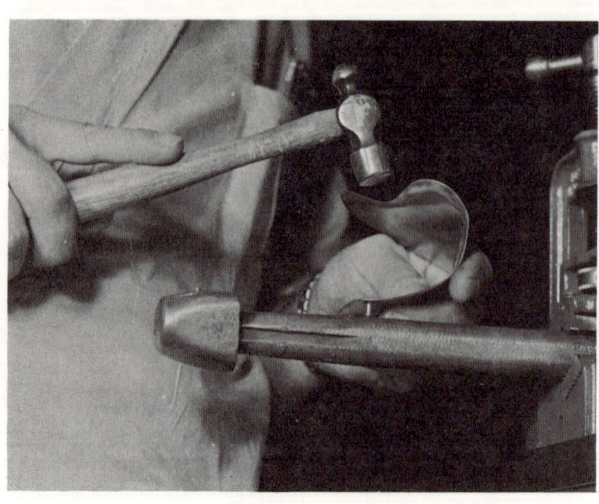

19. Locate the hook mounting bracket. Place the hook mounting bracket on the hook stud, then hold it in various positions on the receptacle until one is found that the patient likes. The illustration shows the average, but some patients want the hook to show hardly at all, while others prefer it projecting out even more than shown.

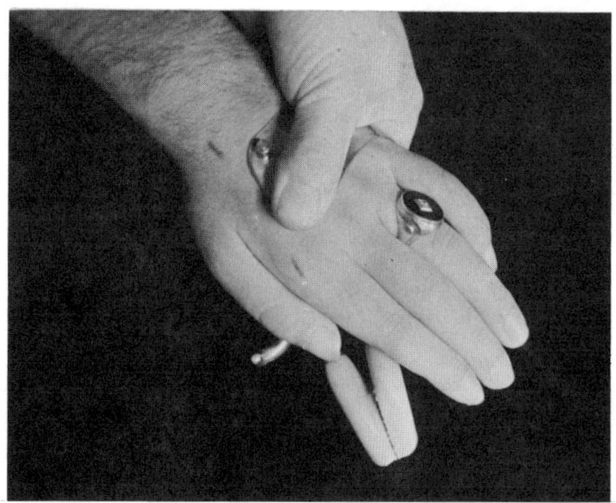

20. Mark the hook mounting bracket location on the receptacle.

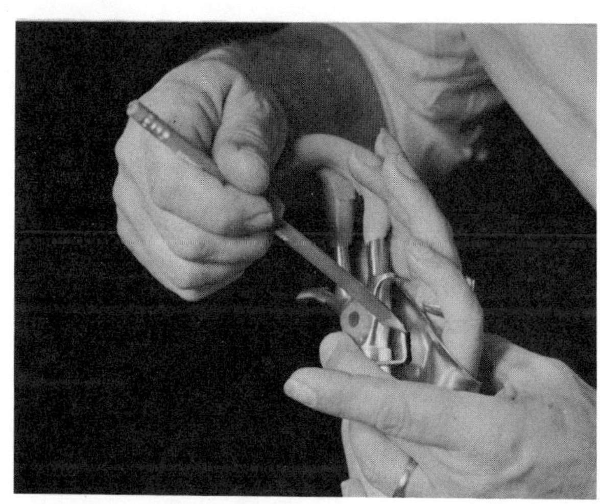

21. Solder the hook mounting bracket on the receptacle in the location marked. Regular stainless steel solder is satisfactory unless the patient will be doing heavy work with the hook in which case silver solder should be used.

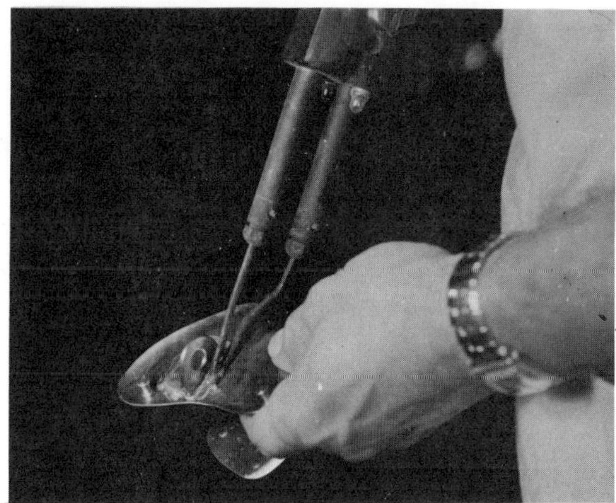

22. Install the hook on the bracket, being
     sure to get the housing thumb in place
     as shown. Tighten securely.

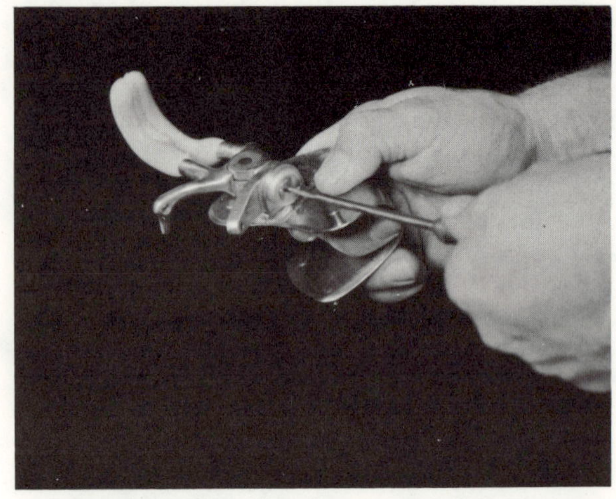

23. Fit and install the wrist straps. They
     should fit snugly enough to keep the
     receptacle from slipping on the hand,
     but not so tightly that the patient's
     circulation will be cut off.

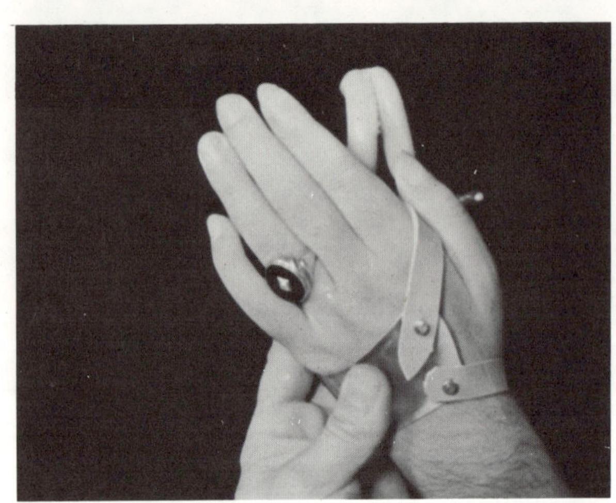

24. Install the cable housing and cable.
     Screw the housing into the housing
     thumb as shown in the illustration.

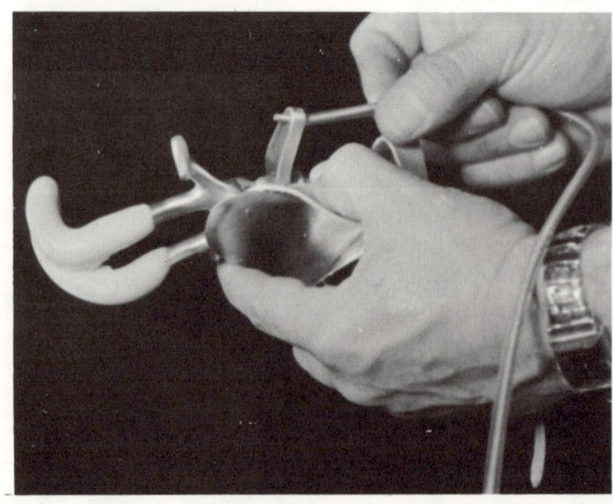

25. Thread the cable through the hole in the hook thumb and then all the way into the cable housing.

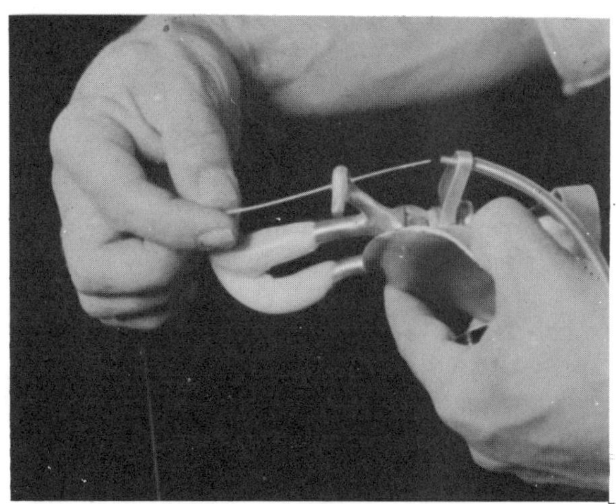

26. Seat the ball terminal on the cable into the hollow provided for it in the hook thumb.

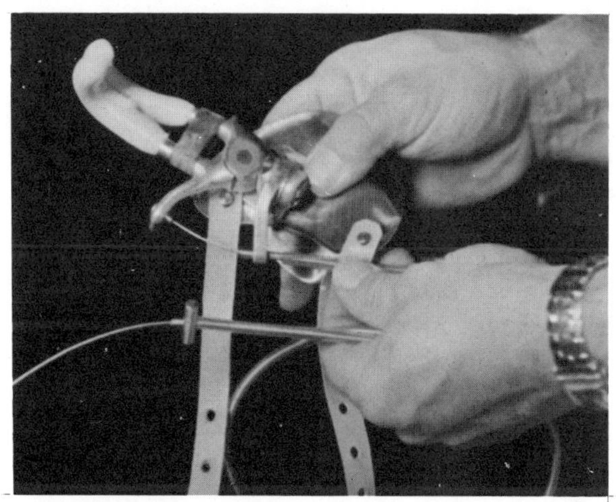

27. Install the housing cross bar into the leather retainer on the harness loop by inserting one end first.

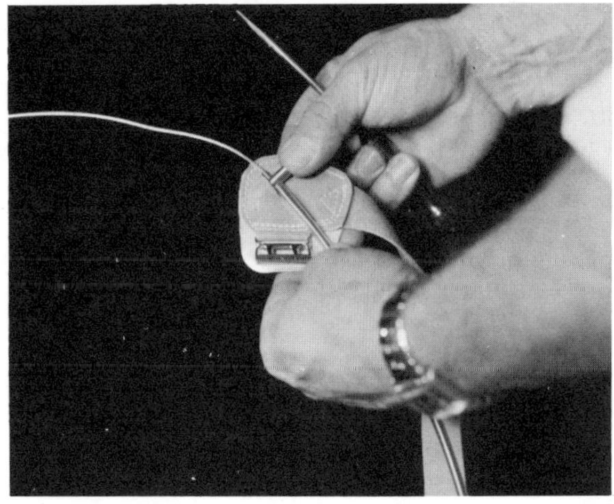

28. Work the other end into the retainer and smooth down the leather.

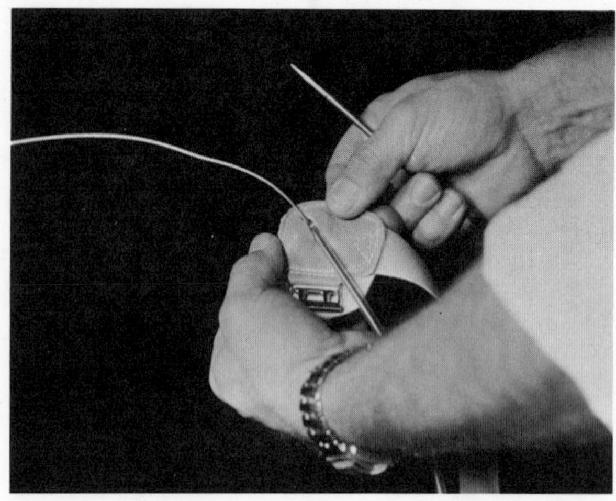

29. Install and adjust the harness on the patient. Place the harness loop to which the cable and housing has been attached on the patient's shoulder. Put it on the same side as the hand that is being fitted. Buckle it in place and put the free end of the cable around the rivet in the harness loop as shown. Be sure the cable retainer is on the cable.

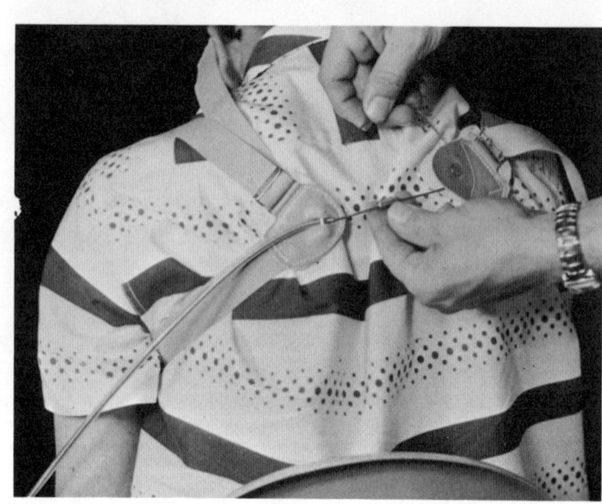

30. Pass the free end of the cable through the cable retainer, pull the cable tight so that all the slack is removed with the patient's arm hanging in normal position, then tighten the cable retainer set screw with an Allen wrench.

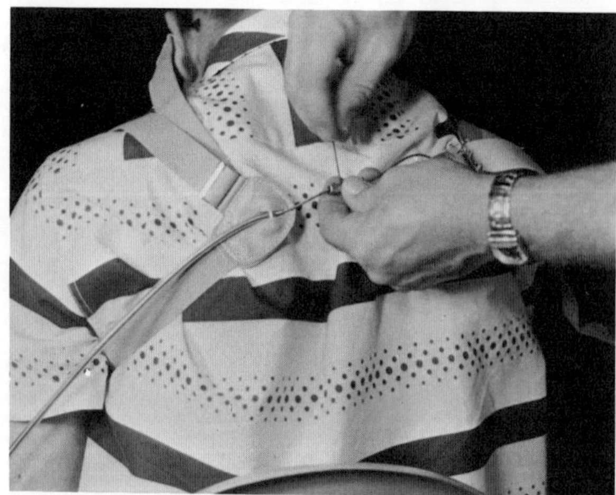

31. Have the patient operate the hook by abducting his shoulders. If the patient has good shoulder motion, he should be able to open and close the hook with no difficulty.

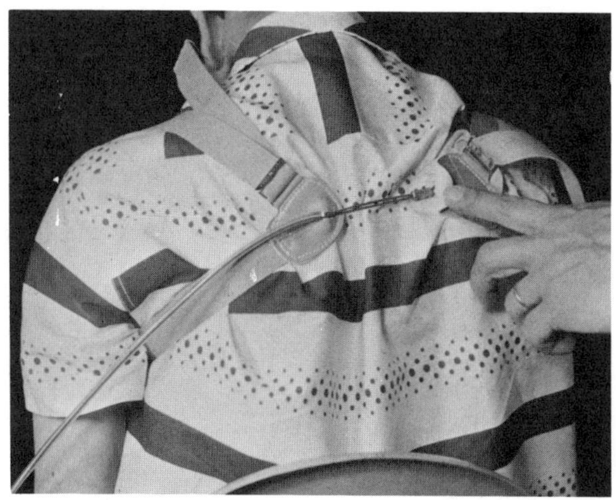

32. If the patient cannot open and close the hook because of limited shoulder excursion, obtain more leverage by fitting the straps down over the upper arms as shown in the illustration. In many cases this will correct the deficiency.

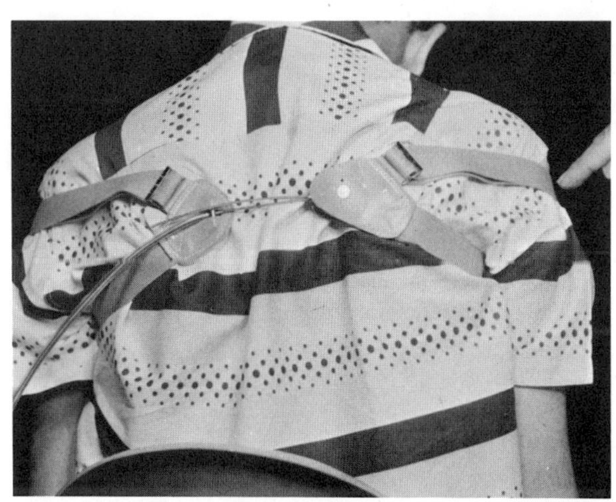

33. If the patient cannot operate the hook with his shoulders, shoulder elevation will usually work. The loop that was on the left shoulder is fitted to the left leg; the loop that was on the right shoulder is fitted to the left shoulder as shown. A strap is run from it over to the right shoulder to keep it from slipping down.

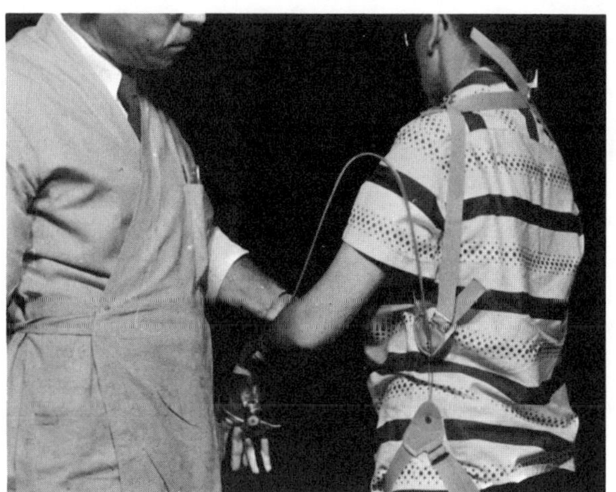

34. Whichever harness variation is used, always cut off the excess cable when the fitting is completed, then tuck the cable retainer under the leather flap on the loop.

35. Have the patient try the hook by picking up several objects, such as a pencil. The hook should be reasonably inconspicuous as in the top view at the right,

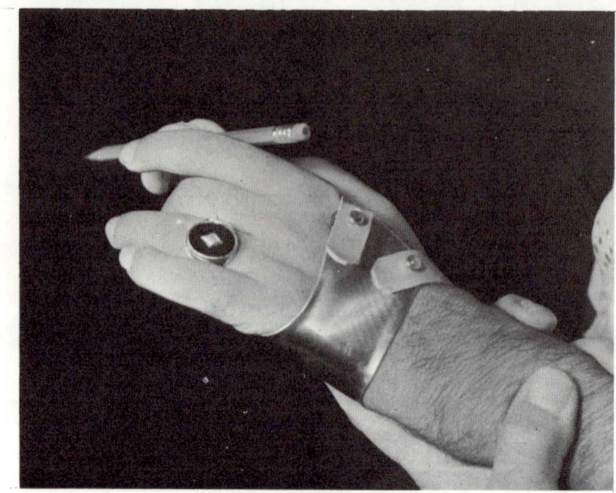

and the side view shown here.

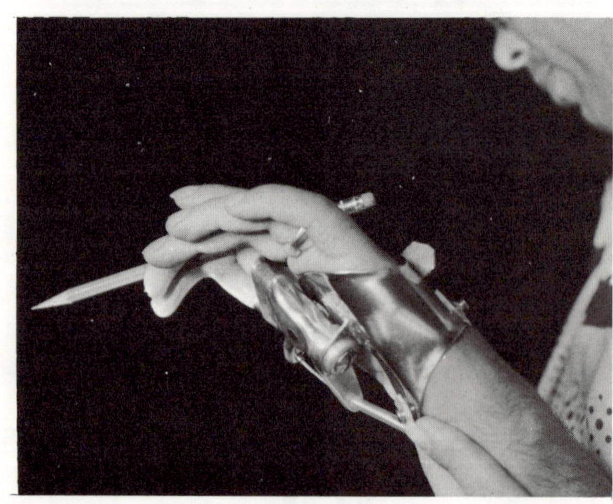

HOW TO FIT A DORSAL WRIST SPLINT TO A CABLE-CONTROLLED HOOK

Introduction

In many cases the patient will not only suffer from lack of sensitivity in the fingers and immobility of the finger joints, but in addition will have paralysis of the muscles of the wrist. The first deficiency calls for the cable controlled hook, the second requires the addition of a wrist splint to this device. The splint stabilizes the wrist joint in a position of function.

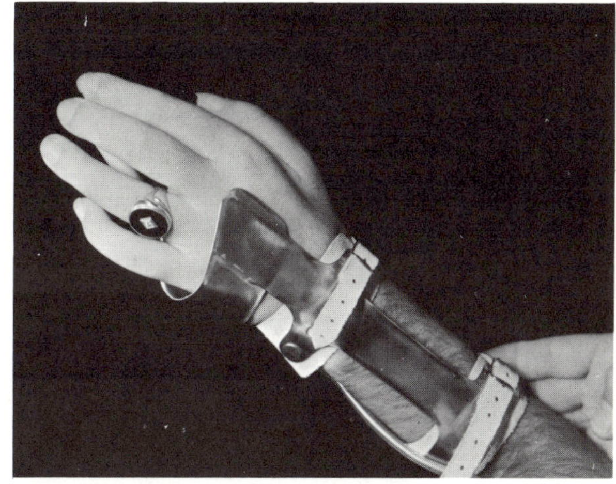

The wrist splint can be fitted as either a dorsal or volar type. In the first instance the splint lies on the extensor aspect of the forearm, in the second it lies on the flexor aspect.

It is common to use a functional arm brace of one kind or other in combination with the cable controlled hook and wrist splint. If the brace is fitted with a pronation stabilizer or assist, it is necessary to use a dorsal wrist splint. If the brace is fitted with a supination stabilizer or assist, it is necessary to use a volar wrist splint. Sometimes a patient who is not fitted with a brace will prefer the volar splint because it is less noticeable.

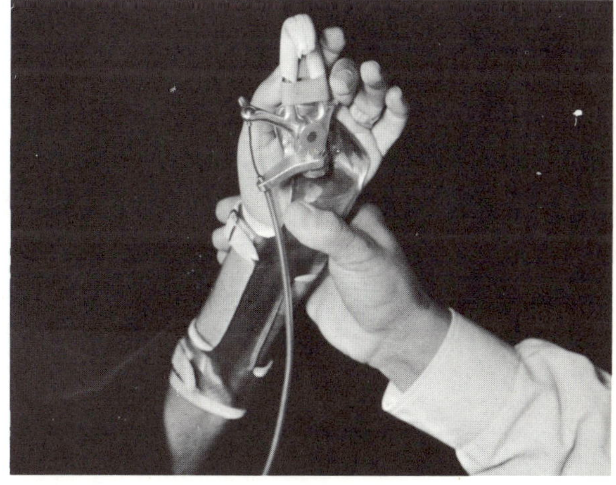

1. Select the correct size splint for the patient. Females and children require the small size, men and large females the large size. Check it on the patient's arm as shown.

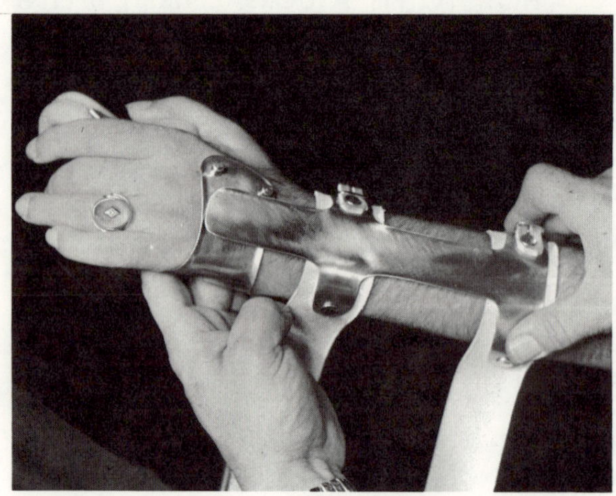

2. Holding the splint in place, as illustrated, mark it at the proximal edge of the receptacle.

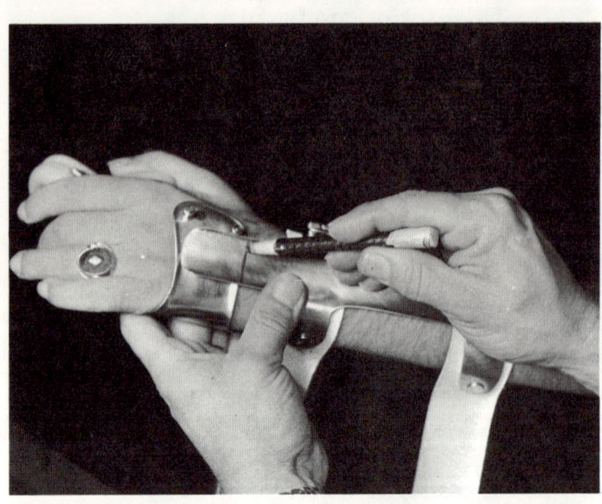

3. Bend the splint at the mark, to an angle of about 15 degrees.

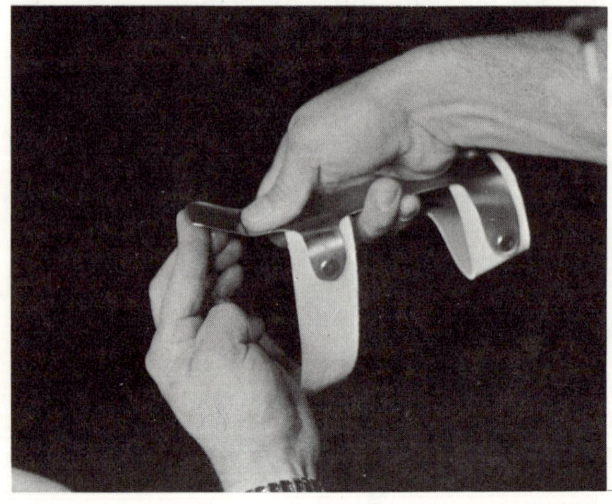

4. Mark the straps the right length to encircle the patient's arm.

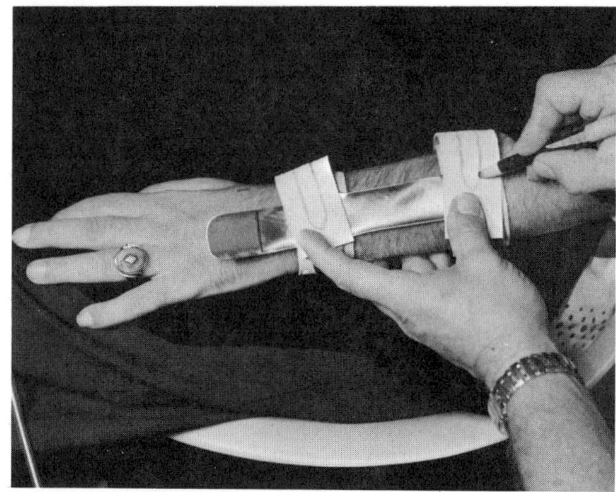

5. Trim the straps to the marks.

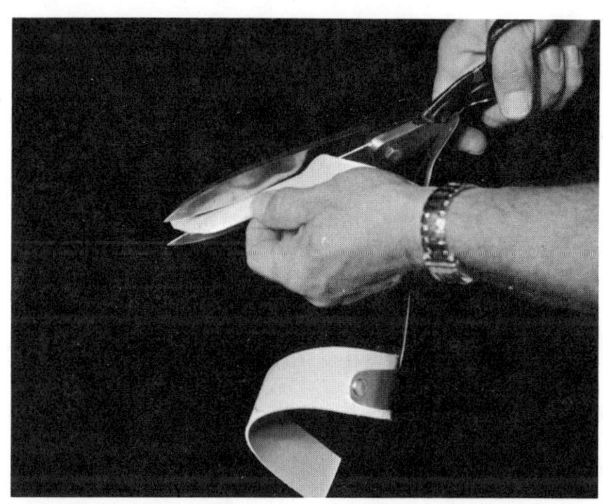

6. Punch holes in the strap billets.

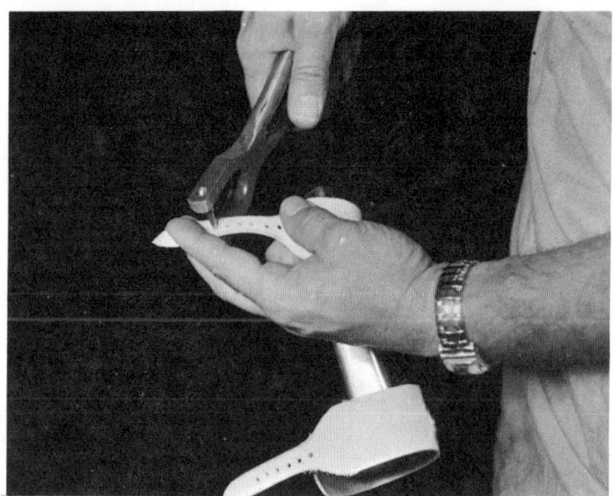

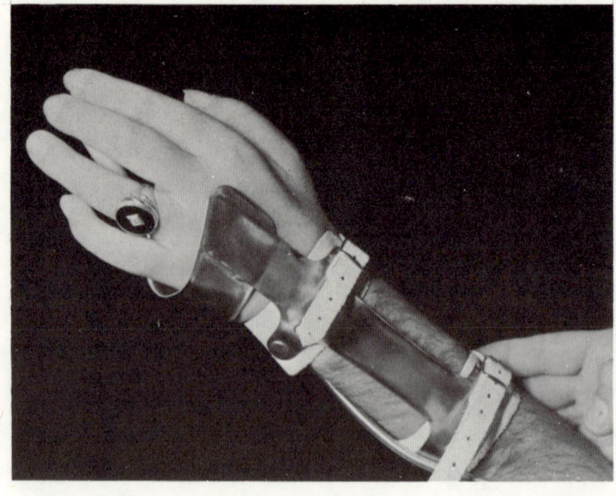

7. Solder, silver solder, or rivet the splint to the receptacle in the position shown.

INSTRUCTIONS FOR FITTING A VOLAR WRIST SPLINT TO A CABLE-CONTROLLED HOOK

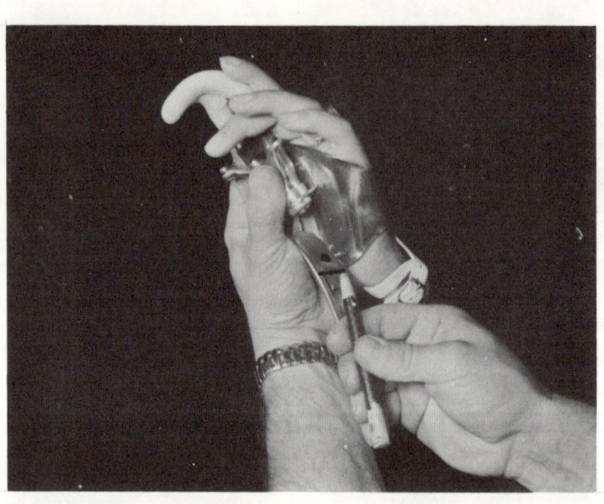

1. With the splint in position on the volar aspect of the patient's arm, mark the splint where it rests on the strap mounting stud, and at the proximal edge of the receptacle; drill a 5/16 inch hole at the stud mark.

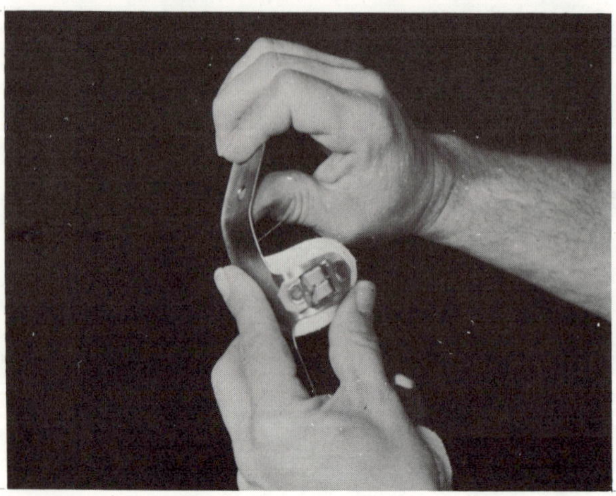

2. Bend the end of the splint at the other mark to about a 15 degree angle.

3. Trim and punch the straps as with the dorsal splint, then try it on the patient. The stud on the receptacle should fit through the hole in the splint, and the bend should hold the patient's wrist in a functional position, as illustrated.

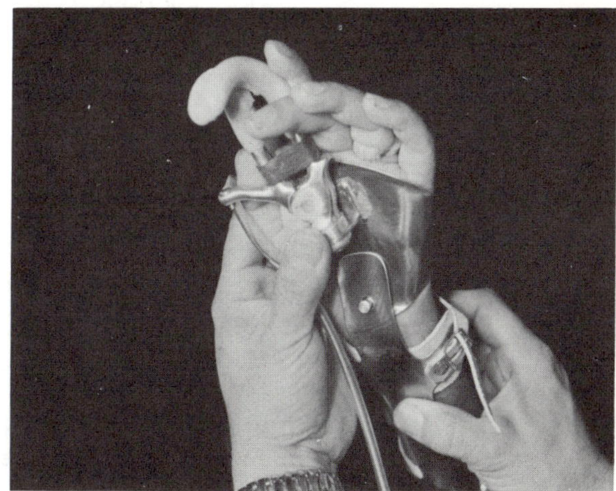

4. Solder, silver solder, or rivet the splint to the receptacle.

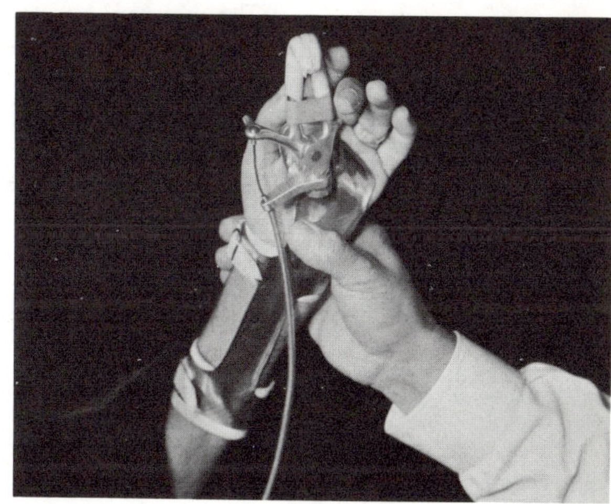

HAND SPLINTS FOR TRAUMATIC LESIONS*

In some cases, traumatic lesions of the hand cause permanent disability that can not be avoided. In other cases, however, the hand can be restored to function through the proper use of splints. We shall deal here with those splints or braces used in temporary disability of the hand with expected return to partial or full functional activity, excluding functional braces for the hand and arm that are intended for cases of permanent paralysis or deformity.

The surgeon frequently requires an active hand splint in the management of a hand injury, but may not know that the required splint is available in an inexpensive, prefabricated form, easily modified to the individual patient's needs. It is the aim of this article to provide a compendium of the various types of hand splints available from manufacturers and list indications and contraindications for their use. The splints to be discussed have one or more of the following objectives:

1. To immobilize the hand for better healing,

2. To increase flexion or extension of the joints,

3. To maintain certain positions of the wrist or fingers.

The two main types of splints in use today may be classified as either "rigid" or "dynamic".

* Marmor, Leonard and Sollars, Raymond: "Hand Splints for Traumatic Lesions," The Journal of Trauma, 3:6:551–562, November, 1963. Copyright ©1963. The Williams & Wilkins Co., Baltimore 2, Md., U.S.A.

Purpose of Splinting

The entire purpose of splinting may be summed up in one phrase--to maintain the position of function in the hand. This simple purpose is sometimes forgotten in the surgeon's attempt to achieve union of a fracture, or to prevent deformity in an injured or postoperative hand. After an injury or operation, the hand tends to drift into a nonfunctional position. The wrist becomes flexed and the extensor tendons tighten and flatten the arch of the hand. The thumb is drawn to the side of the hand, the metacarpal-phalangeal joints are extended, and their tendons tighten and produce a "claw" effect (Figure 90).

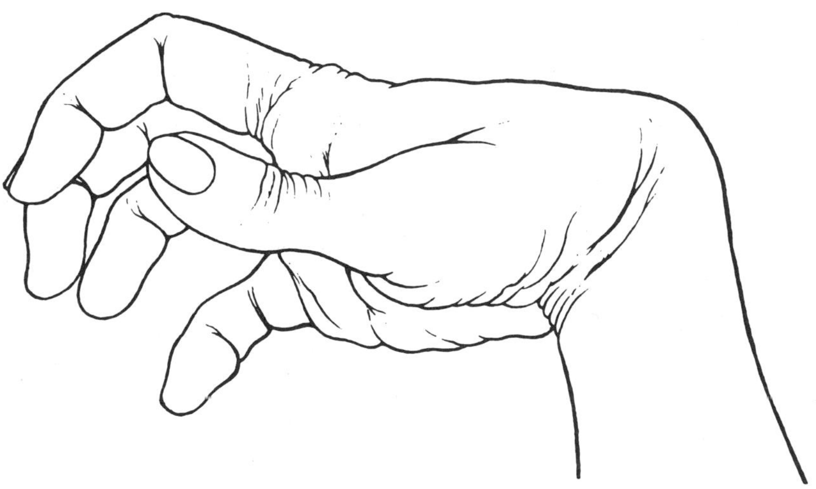

Figure 90.

The "position of function" of the hand is as follows: the wrist is dorsiflexed 30 degrees, which puts proper tension on the flexor and extensor tendons and relaxes the intrinsic muscles of the hand. The normal palmer arch is maintained, the thumb is in abduction and opposition, the metacarpal joints are flexed approximately 45 degrees, and the proximal interphalangeal joints are flexed 45 degrees. This position may be simulated by grasping a baseball in the hand (Figure 91). The most important joint in maintaining the position of function of the hand is the wrist joint. Dorsiflexion of the wrist to 30 degrees, as stated above, balances the muscles of flexion and extension so that neither overpowers the other. The key joints in the fingers are the meta-carpal-phalangeal joints. At 45 degrees of flexion, the collateral ligaments are at their longest. If the fingers are allowed to remain in extension, and contracture of the ligaments occurs follow-ing injury, flexion of the fingers will not be possible and the hand will be seriously impaired.

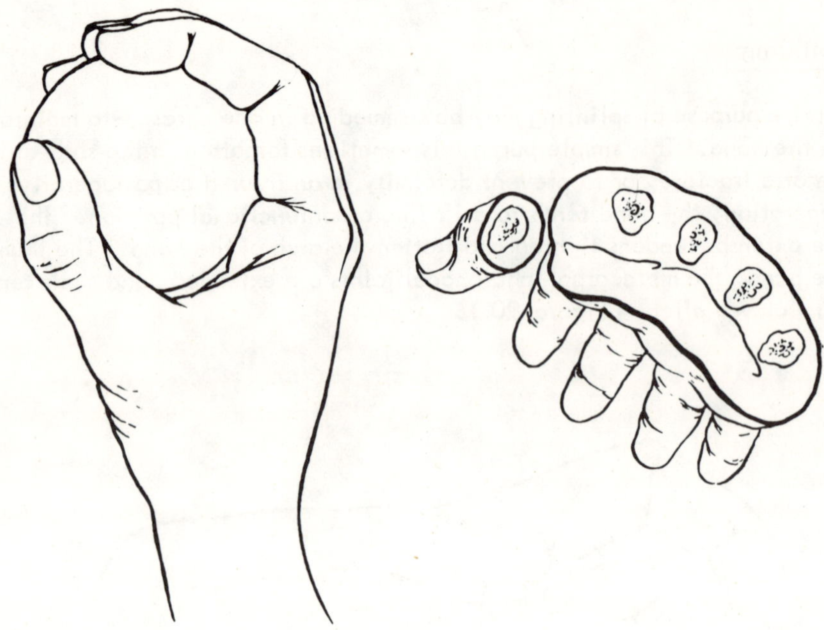

Figure 91.

Rigid Splints

The function of the rigid splint is to maintain the wrist, hand or fingers in a fixed positon determined by the surgeon. The indications for use of this type of splint are fracture, postoperative care following tendon or nerve repair, and infections.

Plaster of Paris is the most commonly used rigid splint at the present time. It should be noted, however, that the use of plaster on the hand differs somewhat from its use elsewhere on the body. It is advised that flat slabs or splints be used whenever possible rather than plaster casts which completely encircle the hand and fingers. The plaster should either be wrapped on with gauze bandage that does not shrink when wet or should be replaced later by dry gauze when the plaster has set.

The palmar arch should be carefully molded into the plaster. If the arch of the hand is allowed to flatten, disability of the hand will result.

The digits should be left exposed whenever possible so that the adequacy of the circulation may be easily determined. Only the involved fingers should be immobilized, in order to allow the other fingers to be freely active and to prevent their stiffening from disuse. Active motion tends to pump the edematous fluid out of the hand and fingers. This fluid would otherwise prevent motion and contribute to fibrous tissue formation. It is important that the thumb be left as free as possible and placed in a functional position, not held to the side of the hand.

Volkmann's Contracture

One of the most dire complications which may result from rigid splinting is development of a Volkmann's ischemic contracture with severe loss of hand function. It is not clearly understood whether a tight cast with obstruction of arterial flow or interference with venous return is the etiology of this condition, but the best treatment is prevention.

The patient with an impending Volkmann's contracture will exhibit one or all of several signs and symptoms, pain--severe burning, lack of pulse, pallor of the hand and fingers, and paresthesia--numbness. If any one of these occurs it should be regarded as an emergency situation requiring immediate treatment. The pain is usually very severe and does not respond well to narcotics. It is often described as a burning feeling or a pins-and-needles sensation. The cast or splint should be removed immediately; there is no time to waste. The gauze and dressing should be split down to the skin, since a blood soaked dressing can be as hard as plaster. If there is no rapid improvement in the condition a stellate or axillary block may be indicated to relieve the vascular spasm and improve the situation. If there is still no improvement, surgical intervention is necessary.

Contraindications for Rigid Splinting

Long immobilization of the hand and fingers should be carefully avoided to prevent stiffness in the joints. Patients with Dupuytren's contracture or osteoarthritis should not be held in rigid splints if possible, since stiffness develops very rapidly in these cases.

The use of the so-called "banjo" splint should be condemned vigorously. This splint produces an imbalance of the muscles by holding the fingers out straight and tends to produce contractures of the collateral ligaments. It also interferes with maintaining the reduction in finger fractures. The practice of strapping fingers to a board or tongue blade is detrimental for the same reasons.

In the past some physicians have treated finger fractures by placing a hard roller gauze bandage or hard ball in the palm of the hand and wrapping the fingers about it to form a rigid splint. This method is poor, since it tends to compress the flexor tendons against the fracture and leads to adhesion of the tendon, and compression of the palmar arch.

In summary, rigid splinting has an important role in the treatment of hand trauma, but rigid splints must be used judiciously and carefully to avoid more injury than benefit. The late Sterling Bunnell has summarized the dangers of the technique, "Rigid splinting makes rigid hands".

Dynamic Hand Splints

The dynamic hand splint differs from a rigid splint in that it allows active motion and provides dynamic power from spring action or elastic bands. Dynamic splints have a definite place in the treatment of hand injuries but seem to have been neglected in the past because their indications and functions have not been well understood.

Indications for the use of dynamic splints are to mobilize joints that are stiffening, to prevent development of deformities, to substitute for the function of peripheral nerves that have been injured, and to prevent stiffness.

Wrist Splints

There are three basic splints available for use at the wrist. The <u>volar cock-up splint</u> (Figure 92) is a basic splint which may be used independently or to which other active splints may be attached. It is indicated when the surgeon wishes to obtain or maintain wrist extension. These splints are very light and are interchangeable for the left or right wrist. The function of the splint is to dorsiflex the wrist by spring action and to resist flexion but allow it to some degree. The oblique strap passing across the neck of the splint and around the hand is necessary to prevent displacement of the splint.

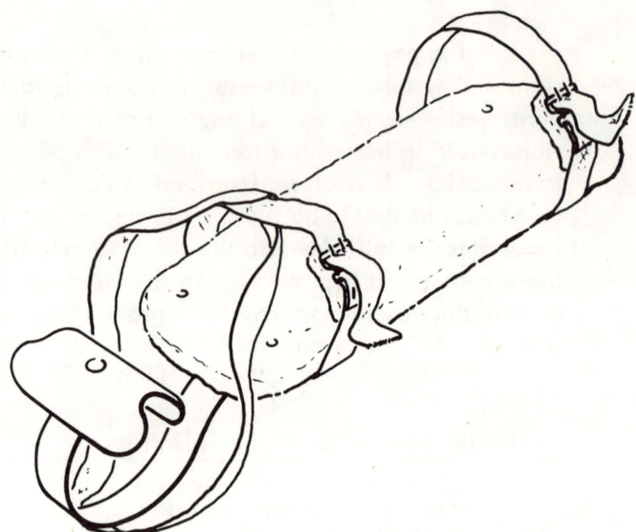

Figure 92.

The <u>dorsal wrist splint</u> (Figure 93) is applied to the back of the wrist as the name indicates and is used to maintain flexion of the wrist and prevent extension. There is no true spring action. The surgeon may bend the splint and set it in the desired degree of flexion. Its use is indicated in cases where constant flexion must be maintained, as in postoperative care after repair of the flexor tendons or the median or ulnar nerve.

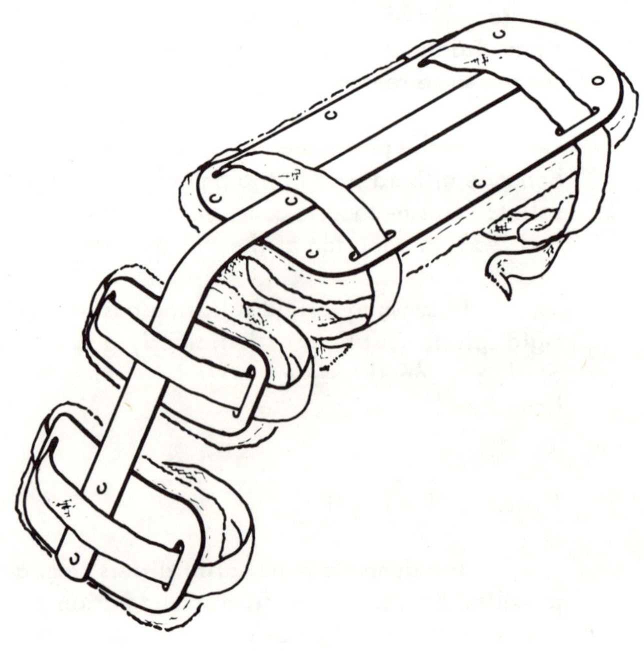

Figure 93.

The <u>Thomas suspension splint</u> (Figure 94) is an excellent splint for patients with a radial nerve palsy. It is especially well suited for use in fractures of the humerus in which contusion of the radial nerve has occurred and full return of nerve function is expected. The function of this splint is to extend the wrist, the metacarpal-phalangeal joints, and the thumb, without interfering with function of the hand in flexion. It aids in preventing overstretching of the extensor muscles by the strong flexors and restores muscle balance.

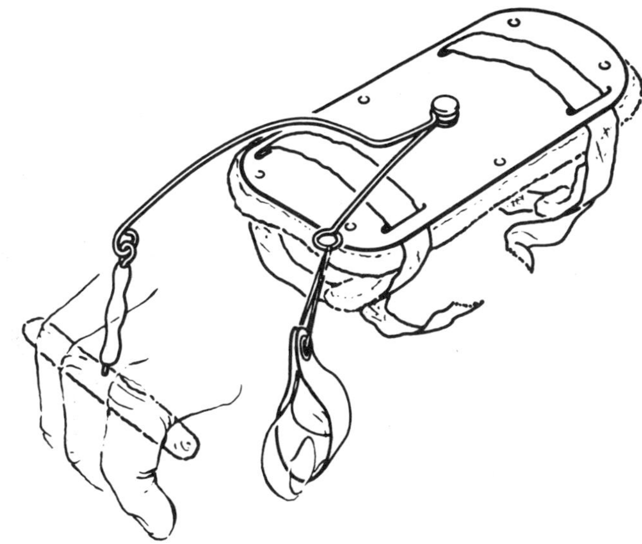

Figure 94.

Hand Splints

The <u>Bunnell knuckle-bender splint</u> (Figure 95), designed by the late Sterling Bunnell, is a basic splint to which other attachments may be added. It is indicated for the correction of extension stiffness in metacarpal-phalangeal joints, such as the "claw" hand or extension contractures after removal of a cast. The contra-extension or flexion power of the splint may be gradually increased by adding rubber bands to the splint. It is advisable that the patient wear the splint for only brief periods in the beginning to prevent problems from undue pressure of the skin. This is especially true in rheumatoid arthritis where the skin is thin and atrophic. In fitting the splint it is very important that the hinge of the splint fall at the axis of the joint, the center of the head of the metacarpal.

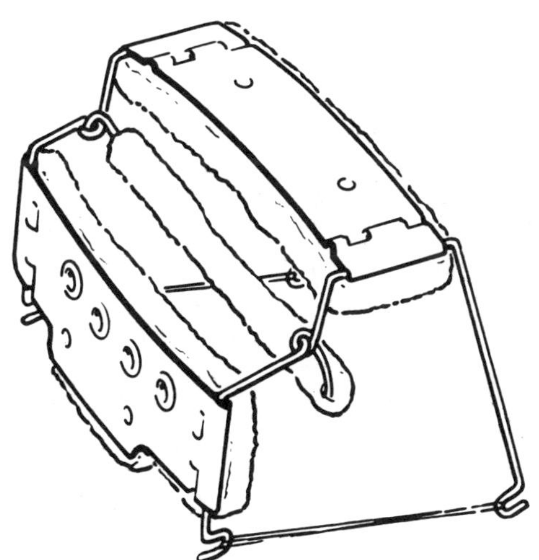

Figure 95.

The _reverse_ _knuckle-bender_ (Figure 96) is indicated for the patient with flexion contractures of the metacarpal-phalangeal joints. This splint is used to gain extension in these joints. It has been of.great value following operation on rheumatoid hands to repair ulnar drift by repositioning the extensor tendons. The splint aids in obtaining better extension of the fingers in these cases.

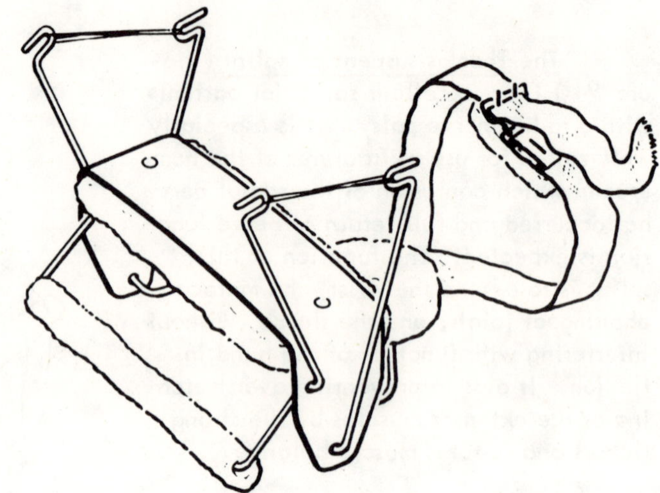

Figure 96.

Finger Splints

In many cases of hand trauma, the proximal interphalangeal joint will be involved as well as the metacarpal-phalangeal joints. When the patient has limitation of motion in both joints, the surgeon will find the following attachments to be of great value in treatment of the proximal interphalangeal joint problem. These are referred to as attachments since they are used along with the hand splints presented in the preceding paragraphs.

An _extra_ _flexion_ _segment_ (Figure 97) is available and may be added to the knuckle-bender hand splint for the cases in which the proximal interphalangeal joints are stiffened in extension and the physician desires to promote flexion.

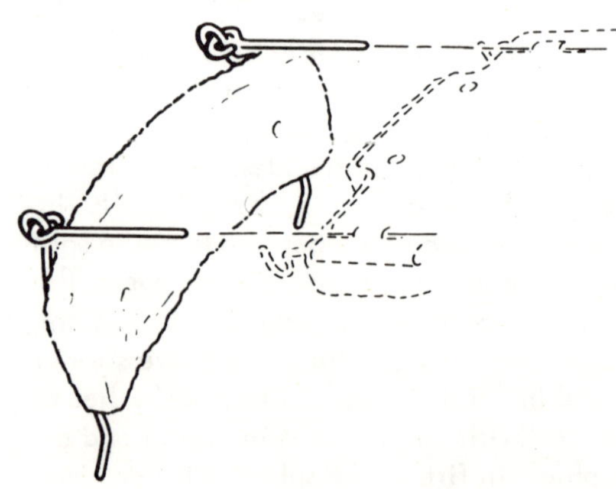

Figure 97.

An <u>extension</u> <u>segment</u> <u>or</u> <u>out-</u><u>rigger</u> (Figure 98) is available for attachment to the reverse knuckle-bender type of splint. By means of finger loops or stirrups and rubber bands, the fingers are pulled into extension to reduce flexion contractures.

In those cases in which only the proximal interphalangeal joint is involved and all other joints are free, a simple splint may be used on the one finger alone. Four types of splints are available for one-finger splinting.

The <u>miniature</u> <u>knuckle-bender</u> is used in those cases in which there is limitation of flexion and the finger has stiffened in a position of extension. The miniature knuckle-bender splint is little more than a reduced model of the standard knuckle-bender. It utilizes rubber bands in the same way to provide gradual flexion force.

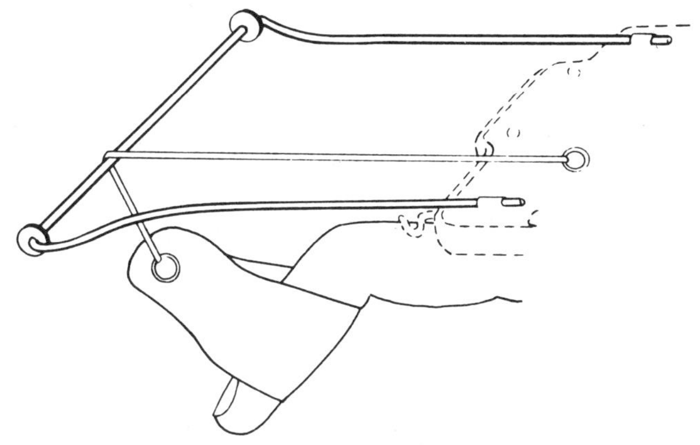

Figure 98.

In the same way the <u>miniature</u> <u>reverse</u> <u>knuckle-bender</u> is utilized when the proximal interphalangeal joint of one finger exhibits a flexion contracture and the physician desires to promote extension. Again, it is little more than a miniaturized version of the standard reverse knuckle-bender splint.

The <u>safety</u> <u>pin</u> <u>splint</u> (Figure 99) may be used on the finger that has a moderate amount of flexion contracture, in order to stretch the joint into extension. It is less cumbersome than the previous two splints but also exerts less force. It is activated by the spring of the metal it is made from and acts upon the proximal and distal finger joints.

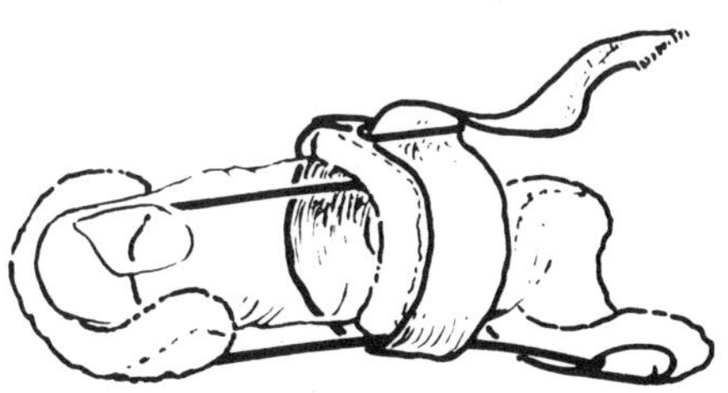

Figure 99.

The <u>clock spring splint</u> (Figure 100)
is utilized only in the case of the strongly
flexed finger. It has a very strong dynamic
spring action. As its name implies, its extension force is derived from the fact that
it is made from clock spring metal.

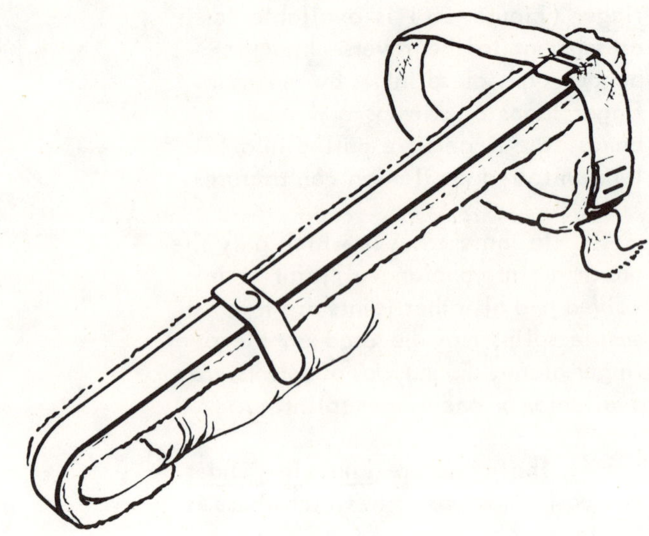

Figure 100.

Special Problems and Complications

A spring action to promote extension of the wrist may be obtained along
with flexion of the metacarpal-phalangeal
joints as provided by the Bunnell knuckle-
bender splint. The attachment used for
this purpose is called an <u>Oppenheimer
attachment</u> (Figure 101). It is ideal for
combined median and ulnar nerve palsy
producing a "claw" hand. Its use will
change a hand from the classification of
"intrinsic minus" to "intrinsic plus" and
will also aid in bringing the thumb into opposition.

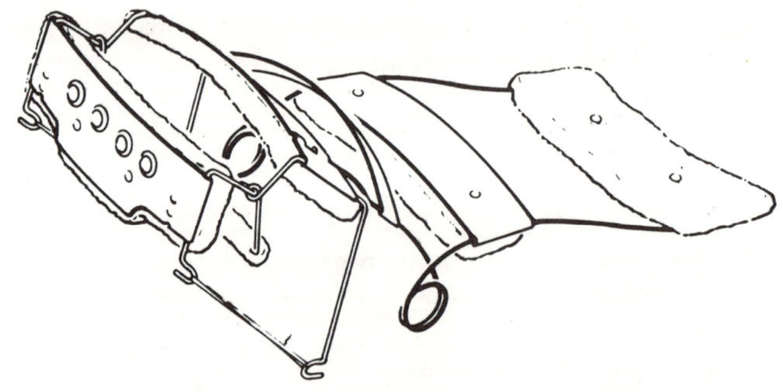

Figure 101.

A <u>Dupuytren's contracture splint</u> (Figure 102) is available to provide a dorsal rigid splinting action of value in treating the patient with postoperative palmar fasciectomy. In this type of surgery it is necessary to immobilize the palmar incision to prevent it from separating. However, early mobilization is necessary because of the tendency of these patients to develop stiff fingers. The Dupuytren's contracture splint provides an ideal remedy for this problem since it immobilizes the palm but does not interfere with the motion of the distal finger joints.

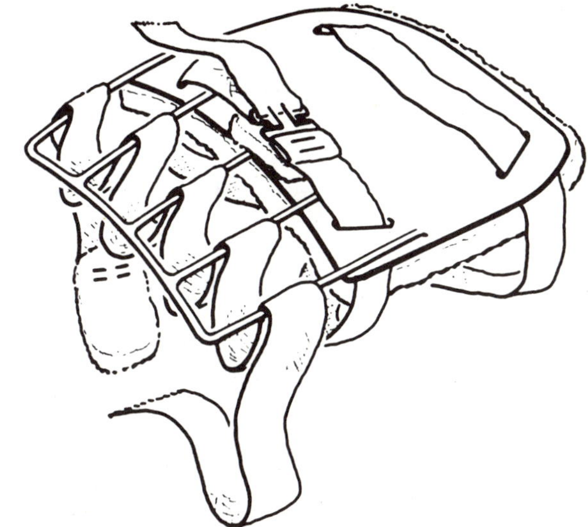

Figure 102.

Conclusion

An armamentarium of light-weight inexpensive hand splints is available for the treatment of traumatic hand lesions. They have been classified, described and presented in a concise and comprehensive manner for ready reference by the surgeons.

These splints are available in prefabricated form and may be obtained from suppliers on either coast. Printed catalogue sheets and price lists may be obtained on request from the suppliers. (The authors will supply information concerning the suppliers on receipt of a written query.) These splints are sufficiently inexpensive that a small stock may be maintained by the surgeon or, in any case, may be obtained by mail order rapidly enough to apply to the patient within a few days following prescription. Many orthotists maintain a small supply or will gladly assist in ordering for the surgeon.

CHAPTER III.   FUNCTIONAL ARM BRACES

BASIC ANATOMY OF THE ARM AND SHOULDER

Introduction

It is necessary to understand some of the basic anatomy of the arm and shoulder before an intelligent approach can be made to fitting functional arm braces. In this section we will cover the major movements of the arm and shoulder, relating the movements to the principal muscles that cause them.

For a more complete discussion of the complexities of the arm and shoulder the student is urged to study Rasch, P. J. and Burke, R. K., Kinesiology and Applied Anatomy. Philadelphia: Lea and Febiger, 1963, pages 167 through 217. This book covers the subject very thoroughly, and should be in the library of every prosthetist-orthotist.

Movements of the Shoulder Girdle

Adduction: The shoulder moves back. In the illustration the movement is shown looking down at the figure from above. The chief adductor of the shoulder girdle is the trapezius.

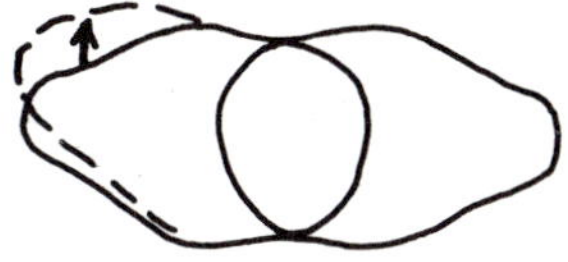

Abduction: The shoulder moves forward. The chief abductors of the shoulder girdle are the serratus anterior and the pectoralis minor.

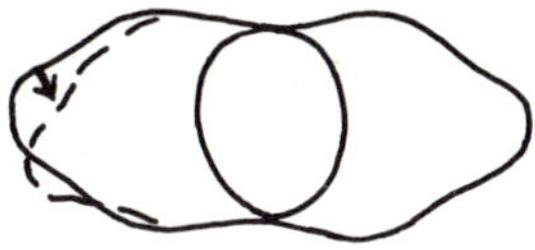

Elevation: The shoulder moves upward, as in "hunching". The chief elevators of the shoulder girdle are the trapezius and levator scapulae.

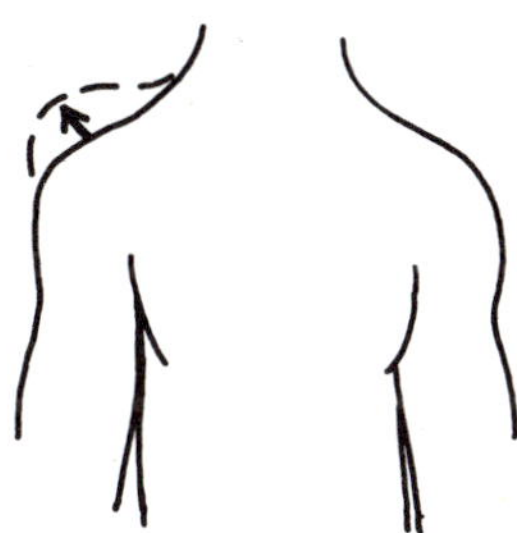

Depression: The shoulder moves downward. The chief depressors of the shoulder girdle are the subclavius, rhomboids, and pectoralis minimus.

Subluxation: Partial dislocation of the gleno-humeral joint which occurs because of the weakness of the supporting musculature.   Not to be confused with shoulder depression, subluxation may require the support of an arm brace to prevent excessive strain on muscles and ligaments.

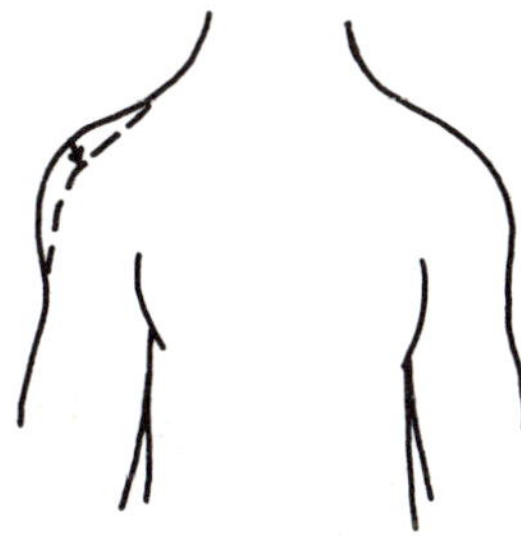

Movements of the Shoulder Joint

Abduction: The arm moves laterally away from the body. The normal individual can continue the movement until the arm is pointing straight up, a total range of approximately 180 degrees. The shoulder abductors are the deltoid, supraspinatus, infraspinatus, and teres minor.

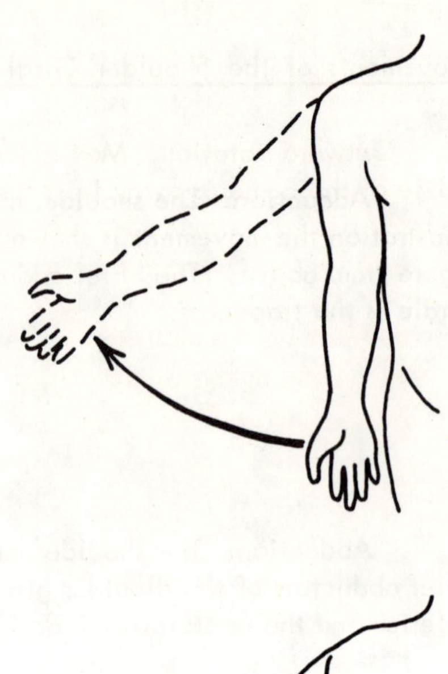

Adduction: The arm moves from the abducted position back to the normal position of rest at the side. The adductors of the shoulder are the pectoralis major, latissimus dorsi, teres major, and coracobrachialis.

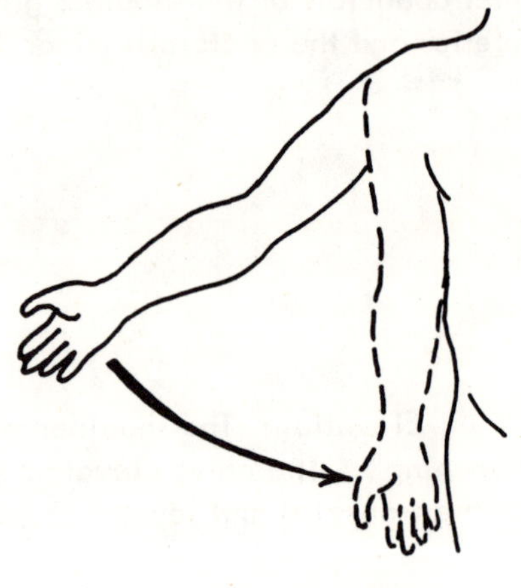

Flexion: Forward movement of the arm at the shoulder joint. The shoulder flexors are the deltoid, pectoralis major, and coracobrachialis.

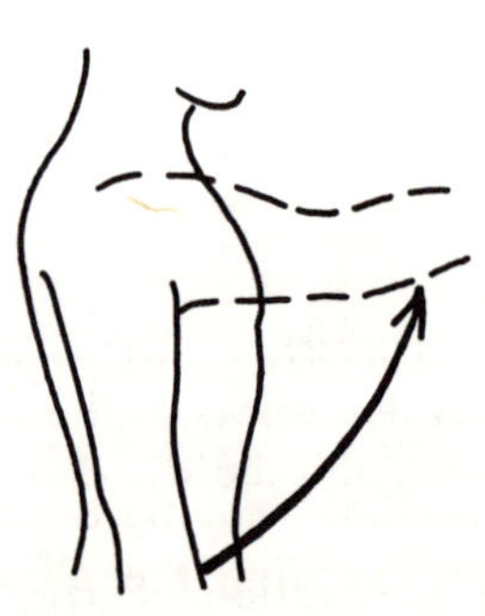

Extension: Backward movement of the arm at the shoulder joint in returning the arm to the normal position after the joint has been flexed. The shoulder extensors are the latissimus dorsi and the teres major. Movement of the arm back of the position of rest may be described as "hyperextension"

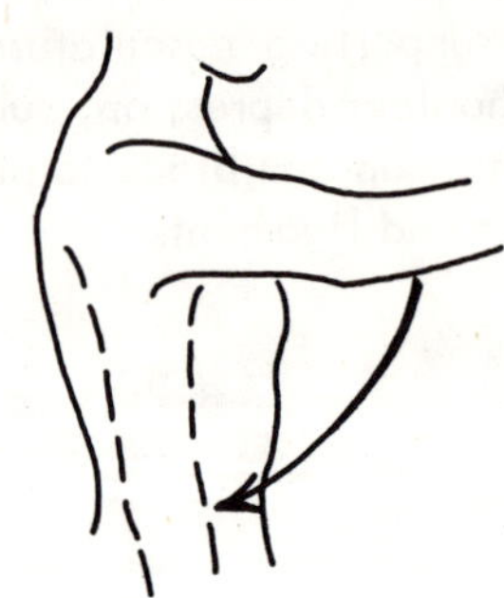

Outward rotation: Movement of the inner condyle of the humerus away from the body. With the elbow flexed, outward rotation causes the hand to be moved in an arc laterally away from the body. The outward rotators of the shoulder joint are the posterior fibers of the deltoid, the infraspinatus, teres minor, and coracobrachialis.

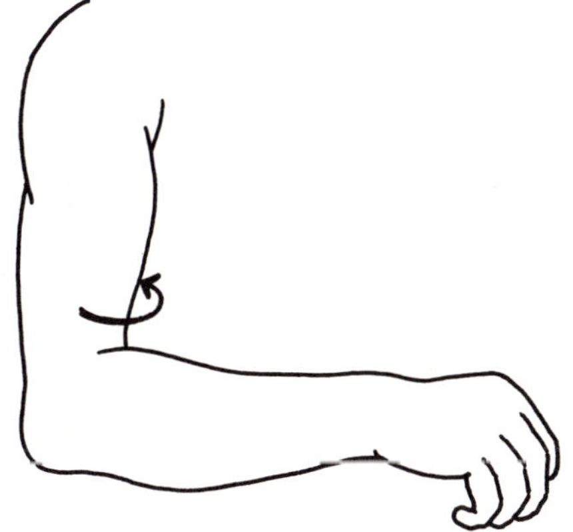

Inward rotation: Movement of the inner condyle back to a position close to the body from a position of outward rotation. The inward rotators of the shoulder joint are the anterior fibers of the deltoid, the latissimus dorsi, teres major, subscapularis, and pectoralis major.

Movements of the Elbow Joint

Flexion: Moving the hand toward the shoulder by bending the elbow joint. The flexors of the elbow are the biceps brachialis, pronator teres, brachioradialis, and brachialis.

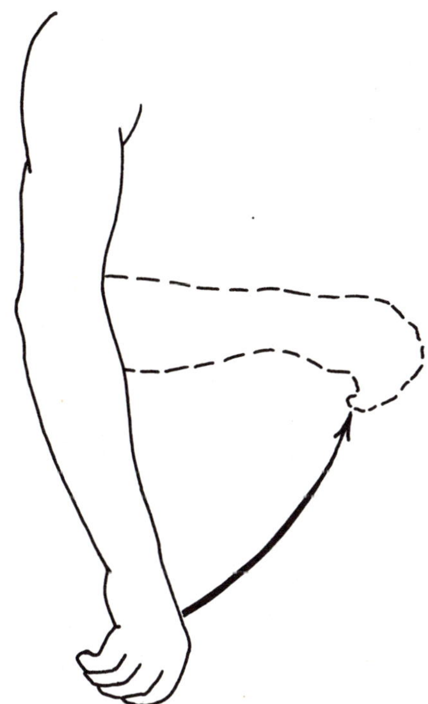

Extension: Straightening the arm after it has been flexed at the elbow. The elbow extensors are the triceps brachii and the anconeus.

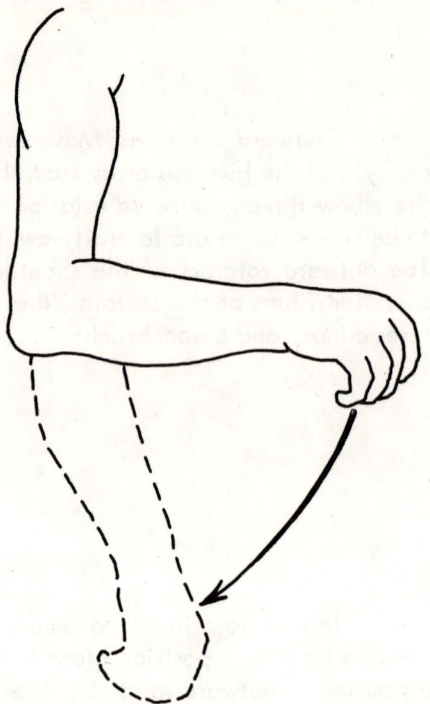

Pronation: Inward rotation of the forearm, moving the hand from palm up to palm down. The pronators are the pronator teres, brachioradialis, and pronator quadratus.

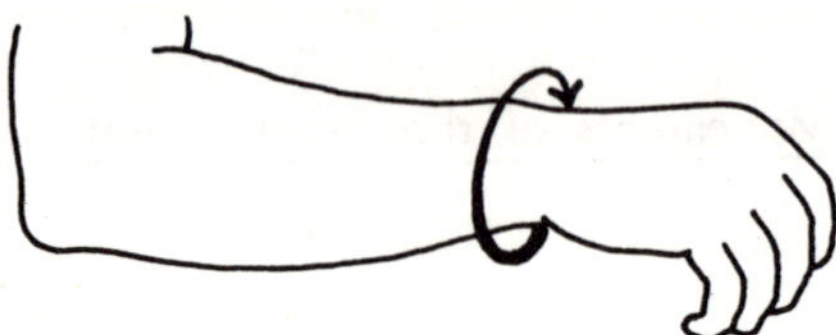

Supination: Outward rotation of the forearm, moving the hand from palm down to palm up. The supinators are the biceps brachialis, brachioradialis, and supinator.

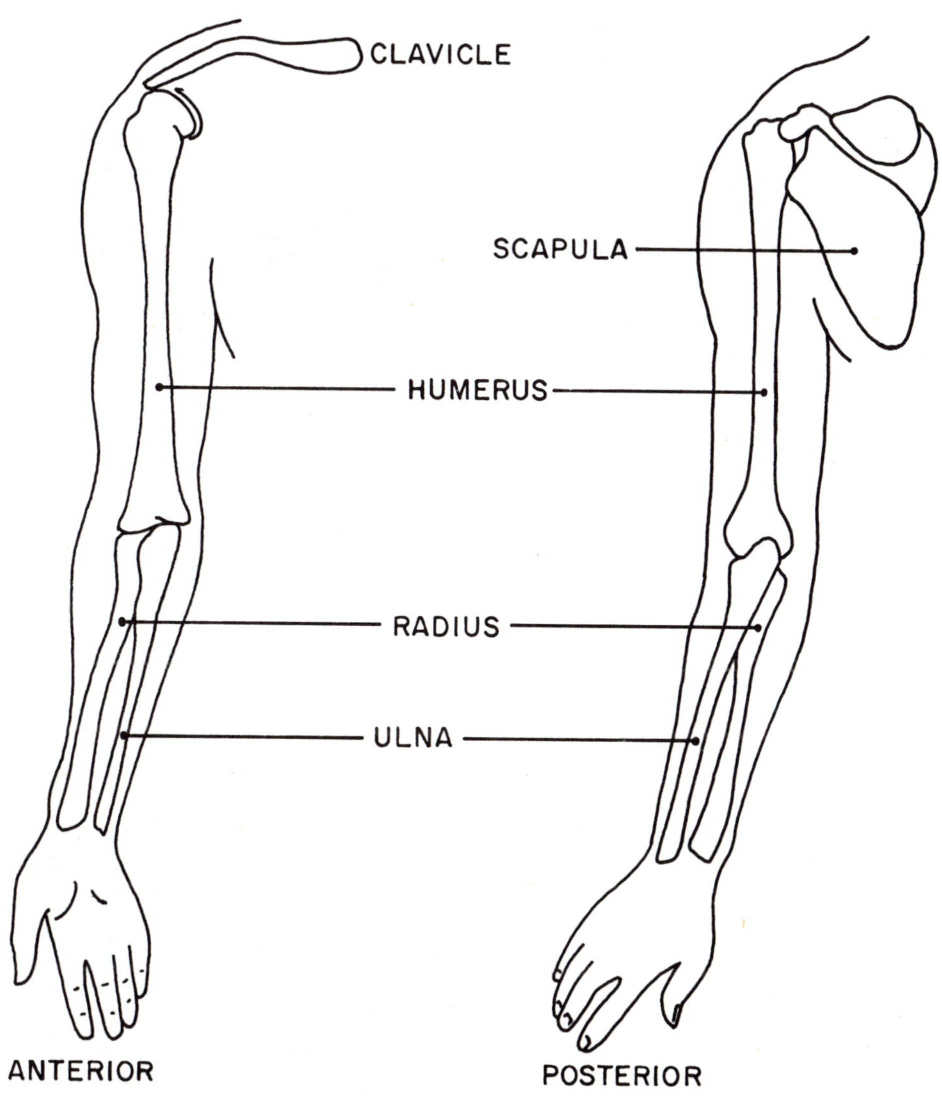

Figure 103. Bones of the Arm and Shoulder

HOW TO MAKE ORTHOTIC MEASUREMENTS FOR A FUNCTIONAL ARM BRACE

1. Measure and record the distance from the ulnar styloid to the medial elbow condyle of the affected arm, using a tape measure. Record all data on an "Orthotic Information – Functional Arm Bracing" form. A sample form is shown on the following page.

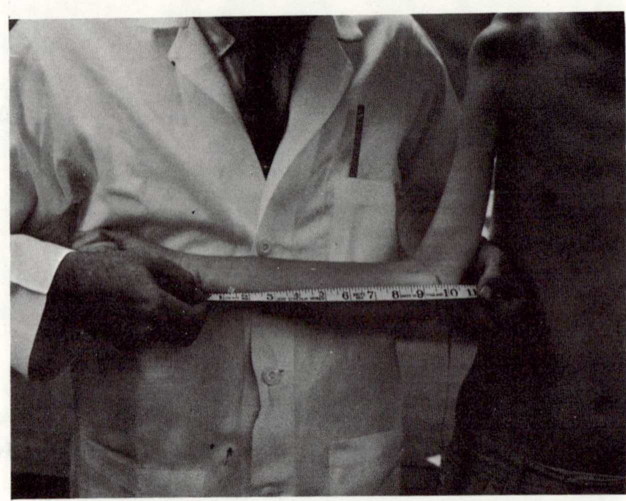

2. Measure and record the distance from the medial condyle to the axilla using a tape measure.

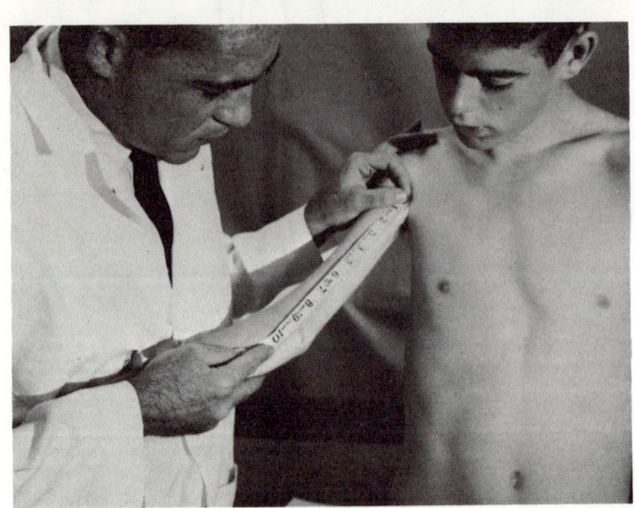

3. Measure and record the distance from the lateral condyle to the acromion, using a tape measure. If there is sub-luxation of the shoulder joint, it is important to support the upper arm so the head of the humerus is in its normal position in the glenoid cavity.

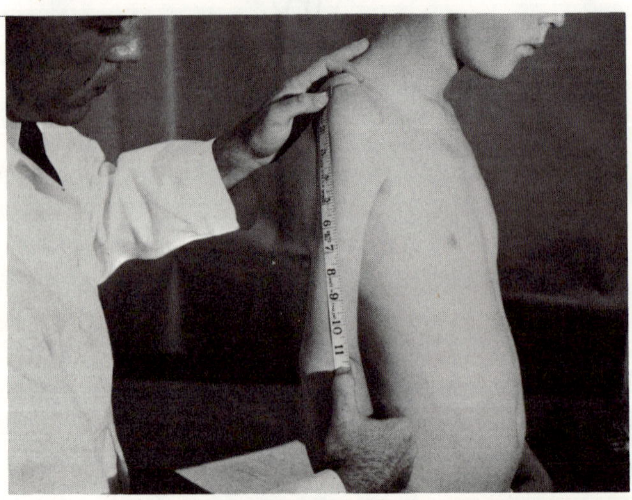

ORTHOTIC INFORMATION – FUNCTIONAL ARM BRACING
MEASUREMENT FORM

Name _________________________ Sex ______ Age ________ Height __________ Weight ________

Address ____

LENGTH MEASUREMENTS

Left _________ Right _________

Ulnar styloid to medial elbow condyle . . . . L _____ R _____

Medial condyle to axilla . . . . L _____ R _____

Lateral condyle to acromion . . . . L _____ R _____

Lateral condyle to top of shoulder . . . . L _____ R _____

CIRCUMFERENCE MEASUREMENTS

Around wrist over ulnar styloid . . . . L _____ R _____

Around wrist just above ulnar styloid . . . . L _____ R _____

Around upper arm just above fold of elbow . . . . L _____ R _____

Around upper arm at axillary fold . . . . L _____ R _____

Around chest at the axillary fold . . . . L _____ R _____

BILATERAL MEASUREMENTS

Across back from left to right axillary fold __________

Across chest from left to right axillary fold __________

4. Measure and record the distance from the lateral condyle to the top of the shoulder with the arm hanging in a neutral position. Place a straight edge horizontally across the top of the shoulder, then measure from this edge to the center of the styloid. Again be sure that the head of the humerus is in the glenoid cavity.

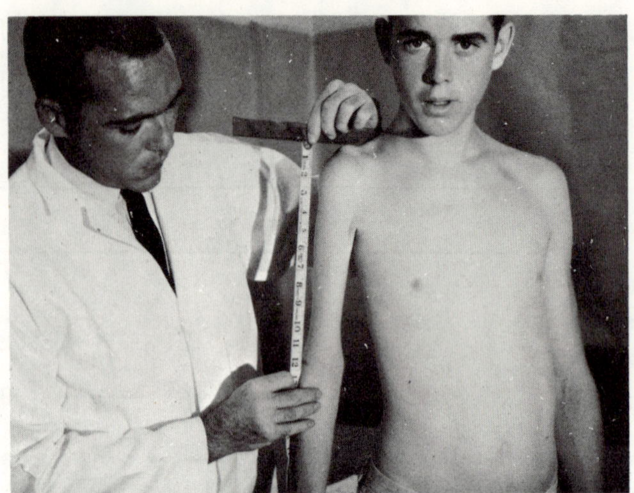

5. Measure and record the circumference of the wrist over the ulnar styloid, using a tape measure.

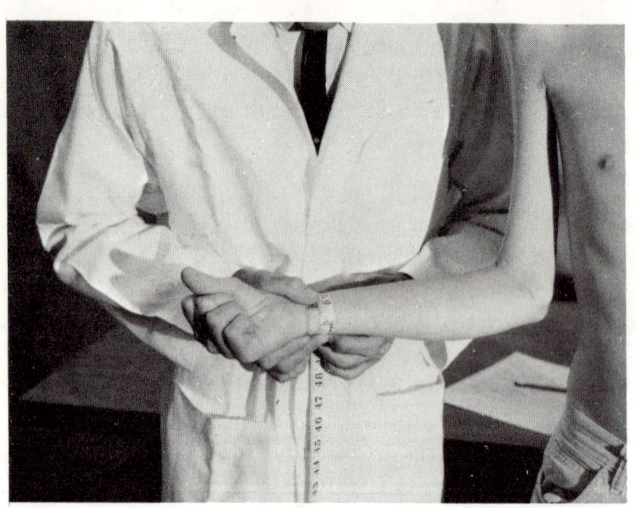

6. Measure and record the circumference of the wrist slightly proximal to the ulnar styloid, using a tape measure.

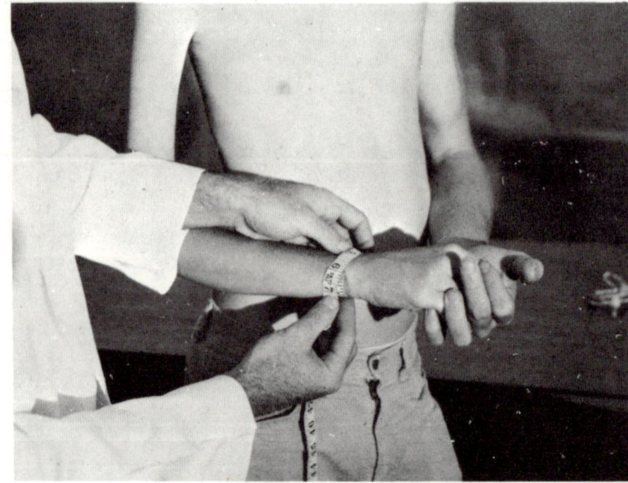

7. Measure and record the circumfer-
    ence of the upper arm just above
    the fold of the elbow using a tape
    measure.

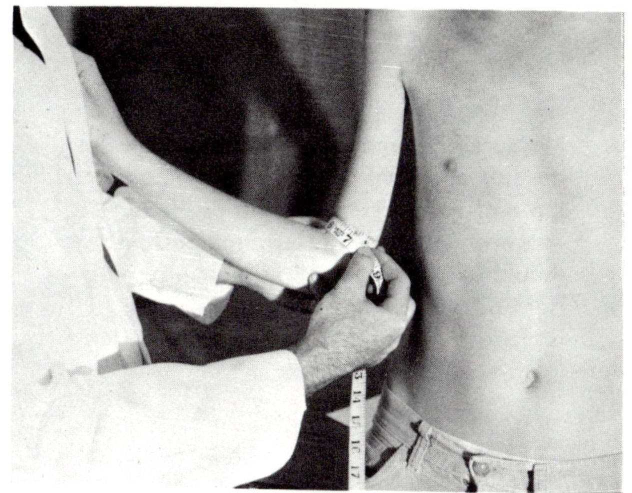

8. Measure and record the circumfer-
    ence of the upper arm at the axillary
    fold using a tape measure.

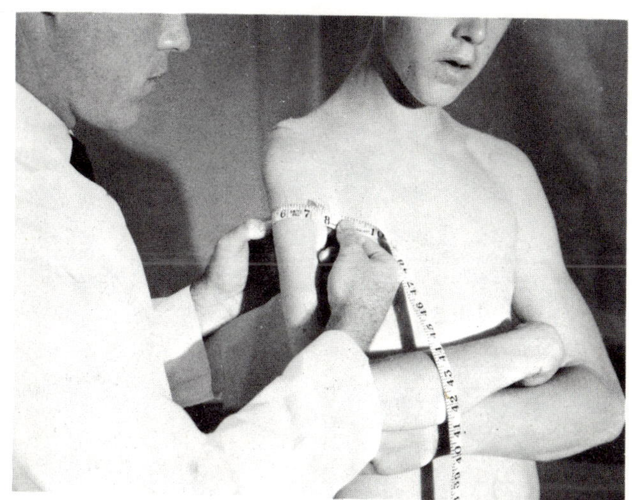

9. Measure and record the distance be-
    tween the left and right axillary folds
    across the patient's back, using a tape
    measure, with the patient relaxed and
    in a normal upright position.

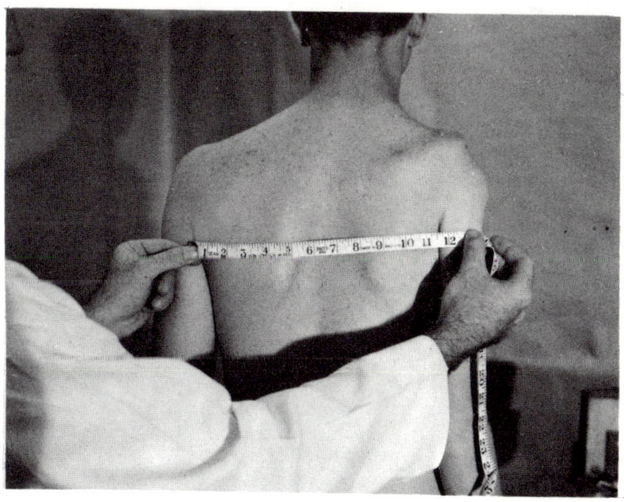

10. Measure and record the distance between the left and right axillary folds across the patient's chest using a tape measure. As in the preceding step, the patient should be relaxed and in a normal upright position when this measurement is made.

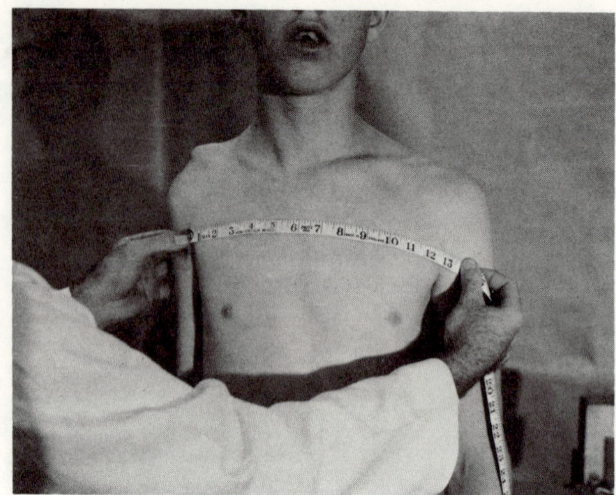

HOW TO MAKE AND MODIFY THE MODEL FOR
THE FUNCTIONAL ARM BRACE SHOULDER CAP

1.   Cut off a length of 12" cotton stockinette long enough to reach from the patient's shoulder to his iliac crest.

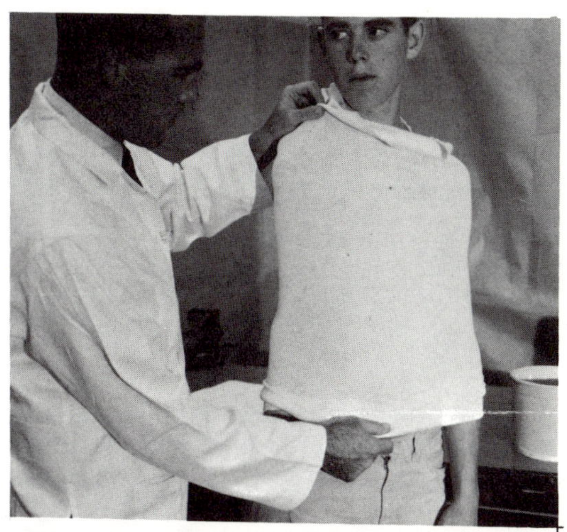

2.   Mark the stockinette on the patient's arm opposite the axillary fold to serve as a guide for cutting an arm hole.

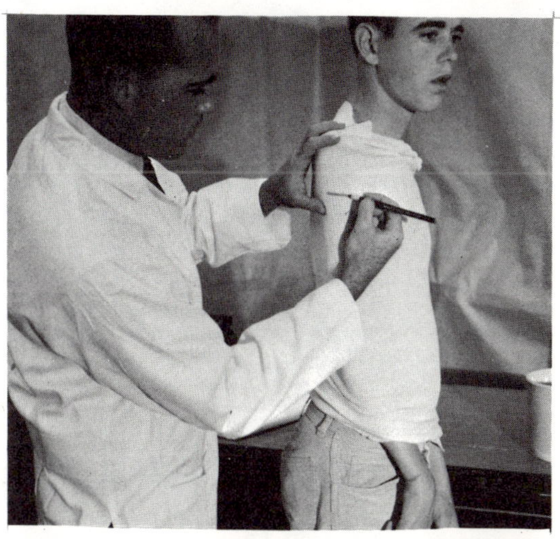

3.   Remove the stockinette and make a 2" slit at the point where the mark was made.

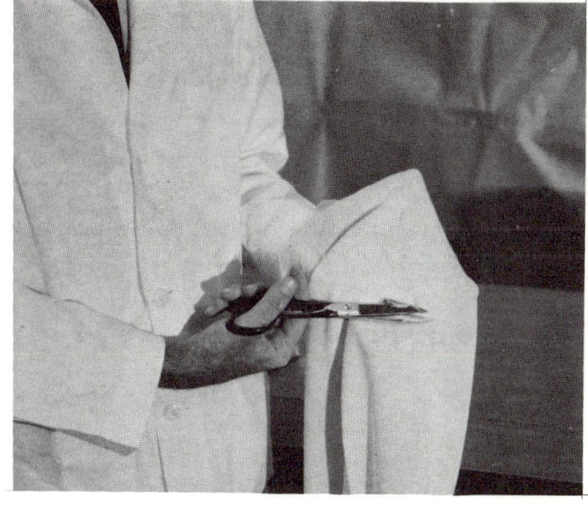

4. Pull the stockinette down over the patient's body and insert his arm into the slit prepared in the previous step.

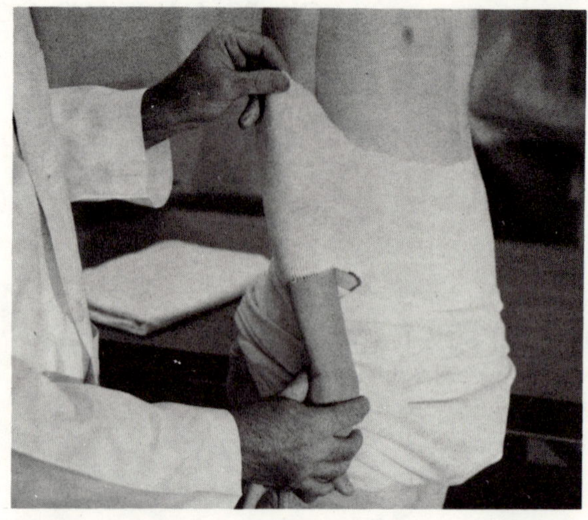

5. Pull the stockinette up around the patient's neck; use Yates clamps to hold it snug and free of wrinkles. It is particularly important to be sure that there are no wrinkles in the stockinette over the shoulder as this will be a weight bearing area in the final socket.

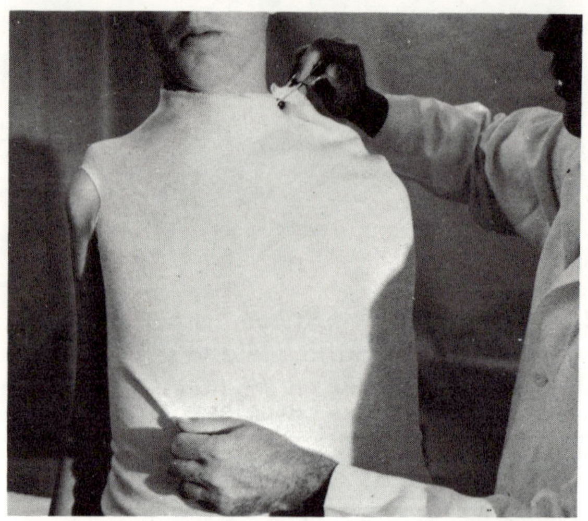

6. In those cases where the patient does not have sufficient strength to support his shoulder, it is necessary to extend the shoulder cap down the side of the body and mold it over the crest of the ilium, so that weight can be borne on that part. In such cases it is necessary to pull the stockinette down over the crest of the ilium.

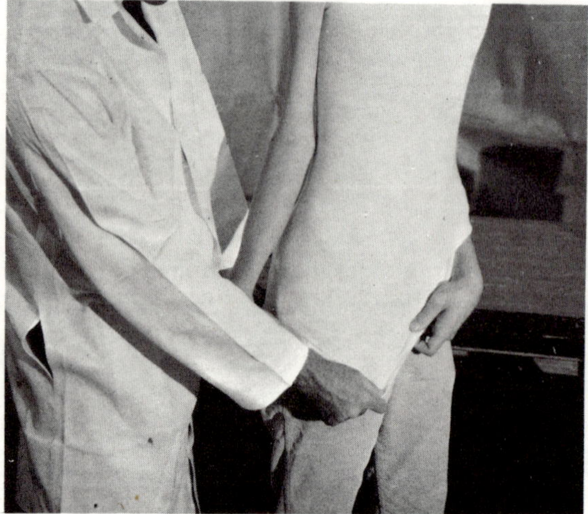

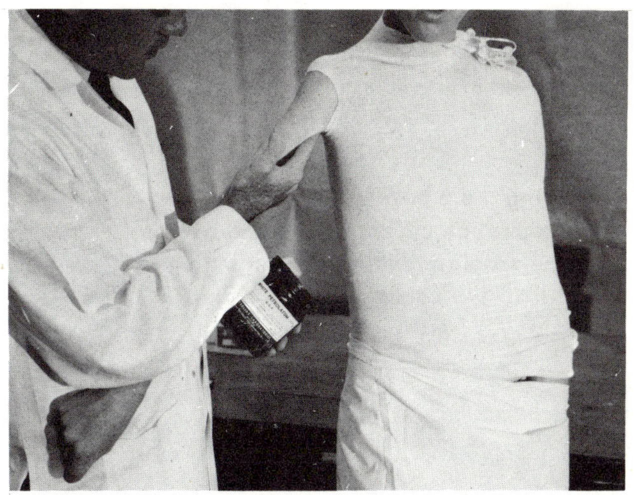

7. Before applying the plaster bandage, apply vaseline to the axilla area to prevent the plaster from sticking in the axillary hair.

8. Submerge the two rolls of elastic plaster bandage on end in a bucket of cold water for one minute. (Use 12 c.m. "Ruhrstern Elastik" elastic plaster bandage, supplied by Fillauer Surgical Supplies, Chattanooga, Tennessee.)

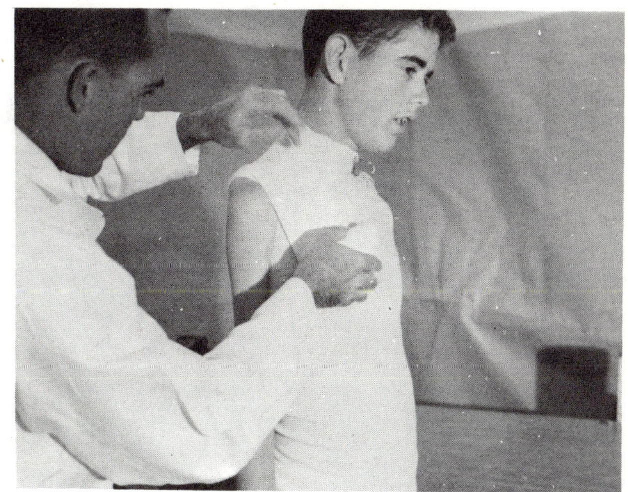

9. Apply the first roll of elastic plaster bandage, starting posterially, then bringing the bandage over and down the anterior portion of the shoulder. Do not stretch the plaster bandage in applying it; simply lay it in place.

10. Bring the bandage horizontally around the chest then up and over the shoulder again, continuing until all the bandage is rolled into place.

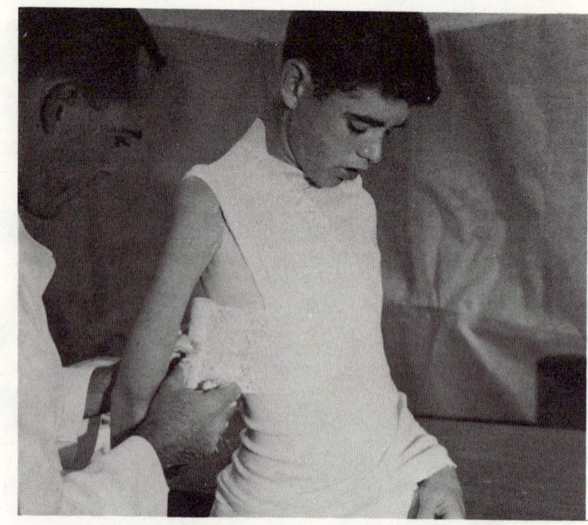

11. Apply the second roll of elastic plaster bandage, following the same pattern as before, building up the thickness and bringing the material up to the level of the axilla area.

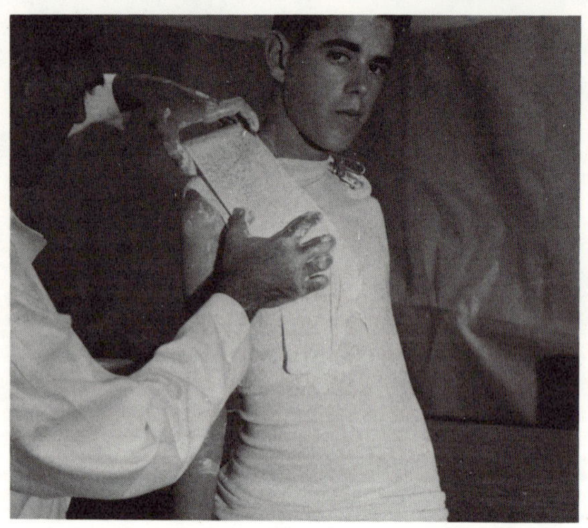

12. Smooth the plaster bandage with the hands, making certain that it conforms to all the bony prominences, and into all of the hollows.

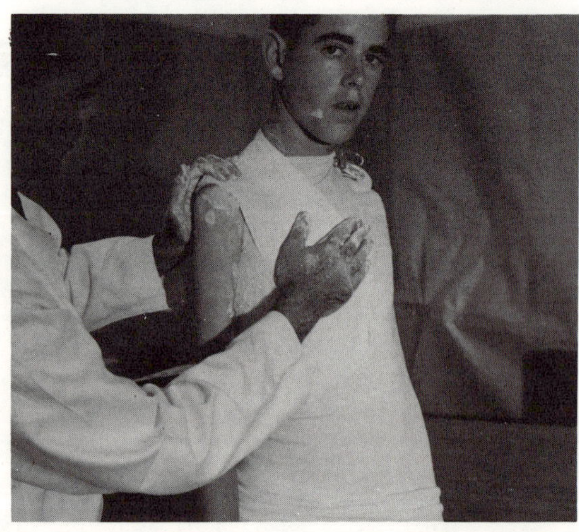

13. Apply two rolls of regular orthopedic plaster bandage to reinforce the elastic material so the cast will hold its shape later when the model is poured.

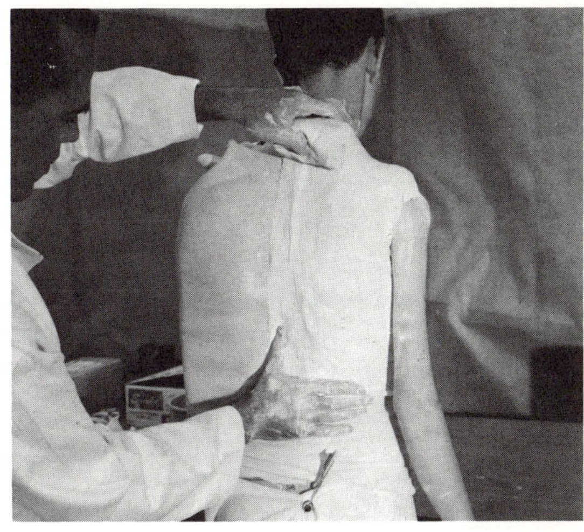

14. In some cases it is desirable to bring several wraps of the regular plaster bandage around the body on the side opposite that on which the cast is being made, to hold the plaster wrap in place. This is commonly done where there is not sufficient prominent bony structure to key the wrap in place, and prevent it from changing position as the material sets up. It is also very helpful where the orthotist is taking the cast by himself, and doesn't have anyone else to help him hold the material.

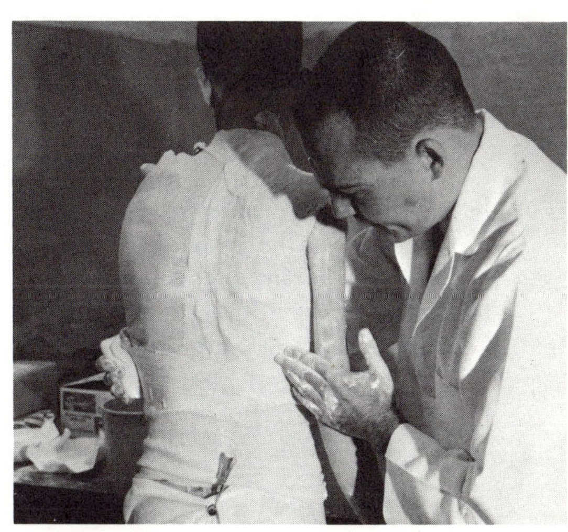

15. Slit the plaster bandage with shears at the axilla area as shown, and fold under the excess material to form a smooth rounded surface.

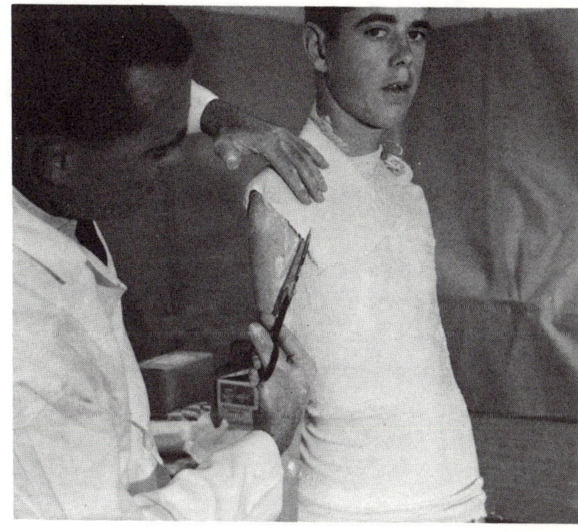

16. With one hand, hold the patient's upper arm so the head of the humerus is kept in position in the glenoid cavity. At the same time use the other hand to shape the plaster bandage over the top of the shoulder so it will conform to the bony prominences and hollow areas.

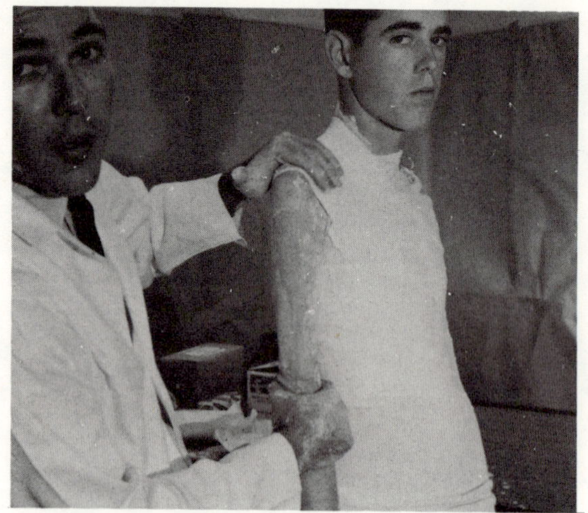

17. In case of subluxation where support is required for the shoulder, it is necessary to extend the wrap down to, and over the crest of the ilium. To accomplish this, continue the wrap on down well over the crest of the ilium, as illustrated.

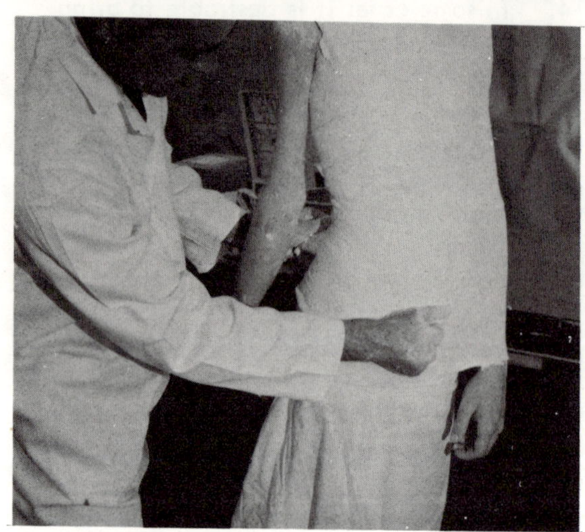

18. As the plaster bandage sets up, shape it over the crest of the ilium, using the thumbs and fingers of both hands as illustrated. The purpose of this is to give a good fit so that weight can be supported without discomfort.

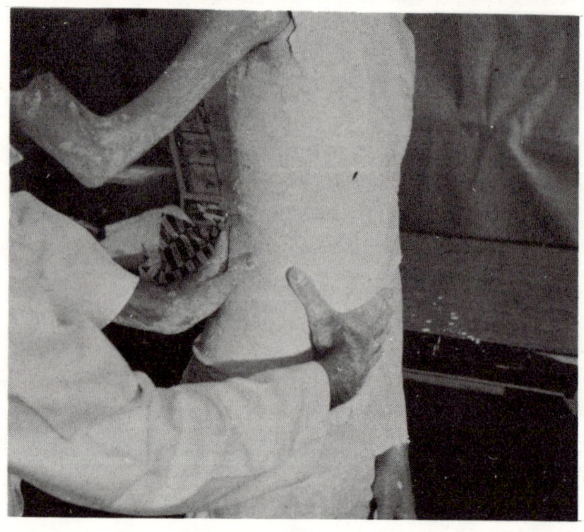

19. With a pair of plaster shears make a vertical slit in the wrap cast and stockinette opposite the affected side.

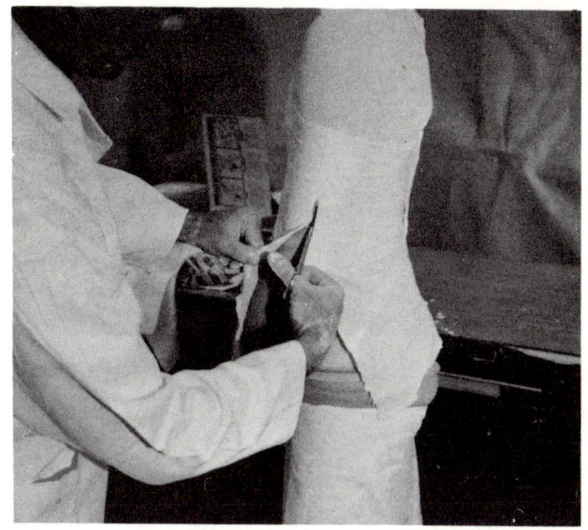

20. Remove the wrap cast from the patient. Take a wet towel and clean all the plaster and vaseline from the patient's shoulder, arm, and axilla area.

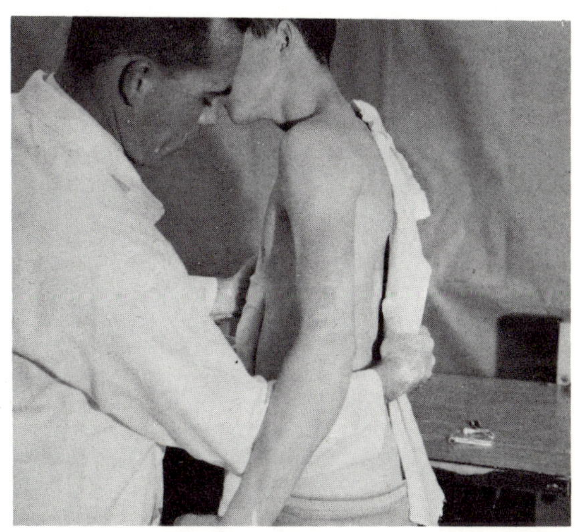

21. Inspect the wrap cast. It should be free of wrinkles, and should reflect the shape of all the bony prominences and hollows in the patient's shoulder and upper arm area. Illustrated in the photograph is a wrap cast made to extend down over the crest of the ilium.

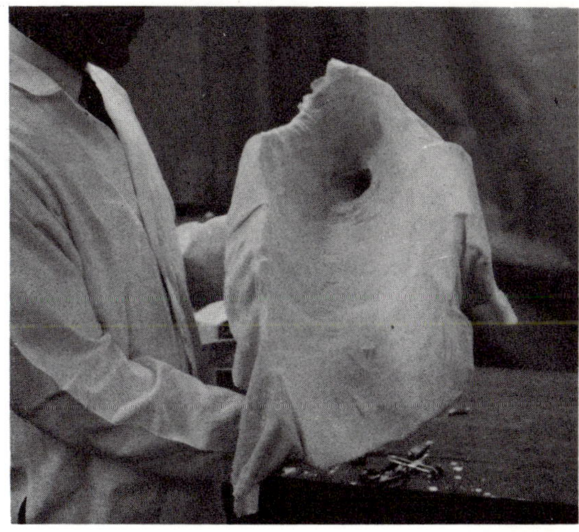

22. Illustrated in this photograph is a standard wrap cast where it is not necessary for the shoulder cap to extend down over the crest of the ilium.

23. Cut a piece of cardboard shaped to cover the arm hole in the wrap cast.

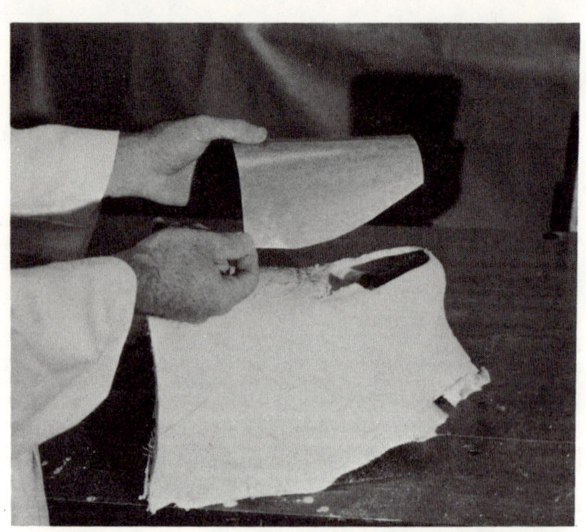

24. With a roll of plaster bandage, seal the cardboard over the arm hole.

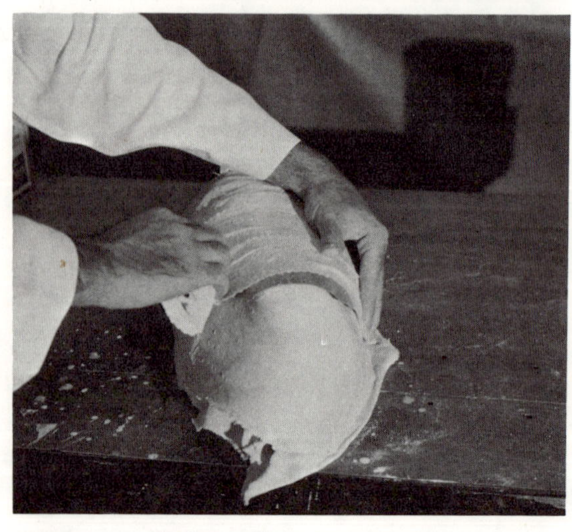

25. Continue the plaster bandage around the open end of the wrap cast to form a bulkhead to seal it so that the plaster will not leak when the mold is poured.

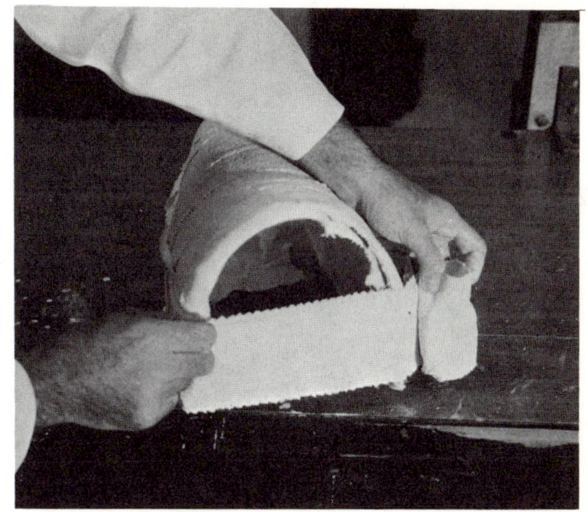

26. Smooth the bulkhead so that it will form a flat, smooth surface when the mold is poured.

27. Cut a hole in the bulkhead for the mandrel. The mandrel is made of a piece of 3/4 inch pipe with a bolt in the end to prevent it from turning in the plaster.

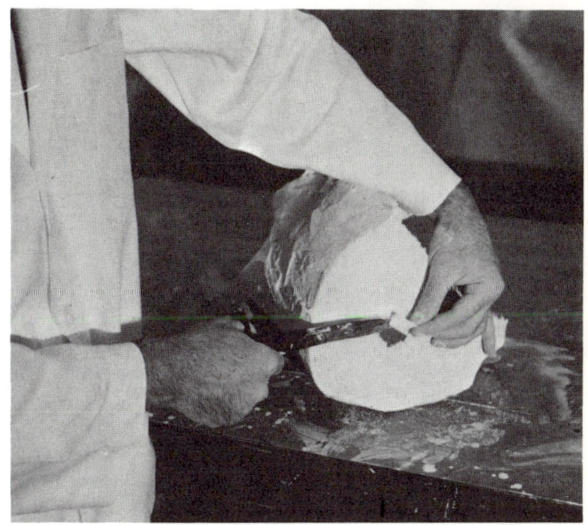

28. Insert the mandrel through the hole and up into the wrap cast, then fill the cast with plaster. When the plaster has hardened remove the wrap cast.

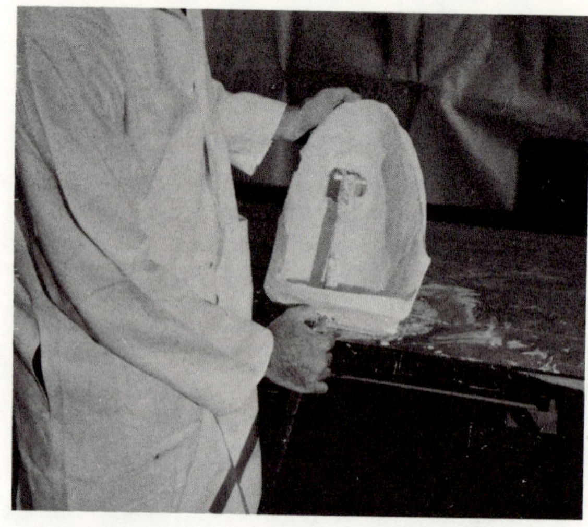

29. Use Durite on the model to smooth it and to remove all irregularities and surface roughness.

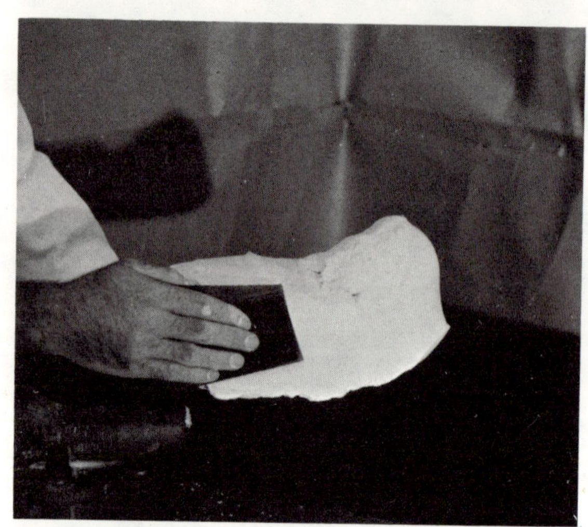

30. Outline the location of all bony prominences on the model with a skin pencil. These areas include the clavicle, acromion, and the spine of the scapula.

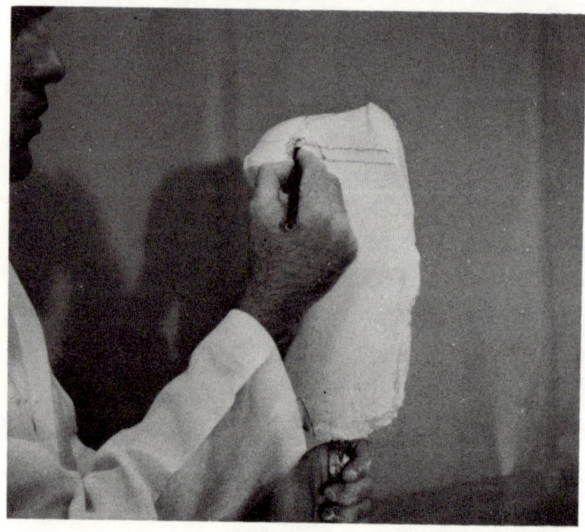

31. Use 3/4 inch number 18 brads to form a guide for building up the relief over the bony prominences. Drive the brads in far enough to give the correct amount of thickness, which will vary between 3/16 and 1/4 of an inch.

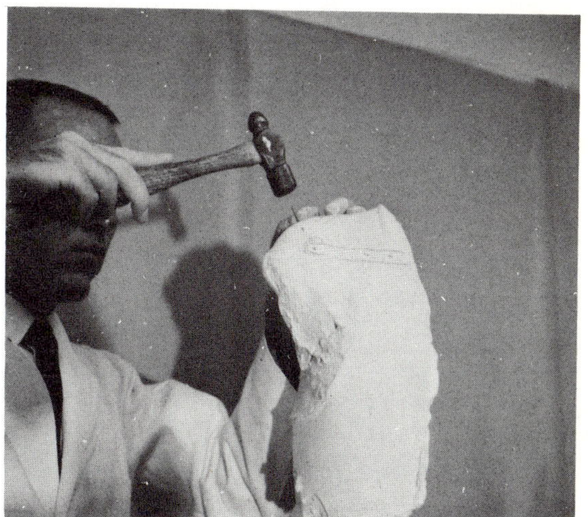

32. Apply plaster to the areas to be built up, using a spatula and keeping the material within the outlines previously marked, and being sure that the brads are thoroughly covered.

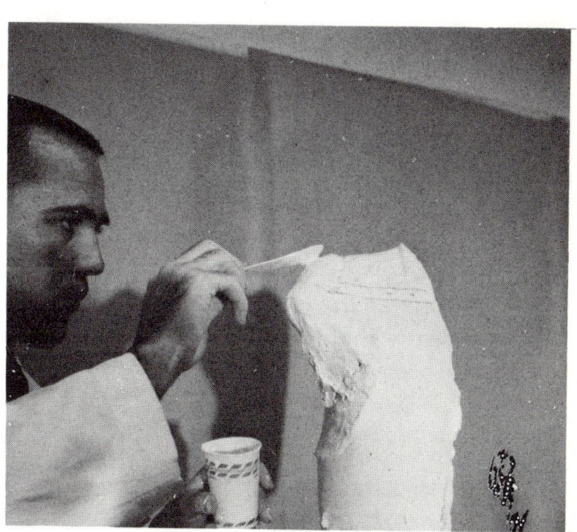

33. Smooth the reliefs with Durite, down to the heads of the brads, and fairing into the outlines previously marked.

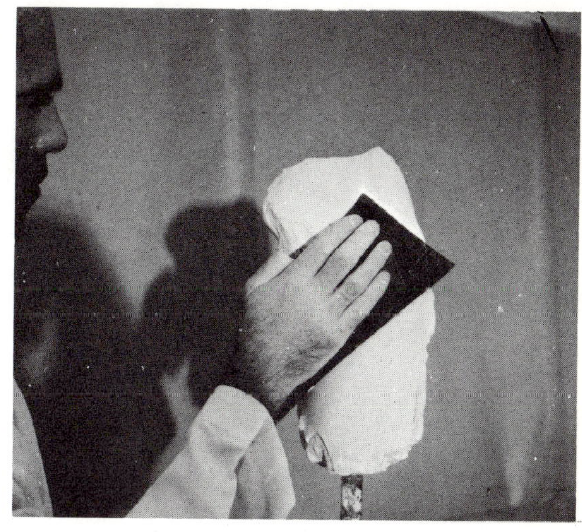

34.   Apply three or four coats of Hi-Glo
      parting lacquer to the cast.

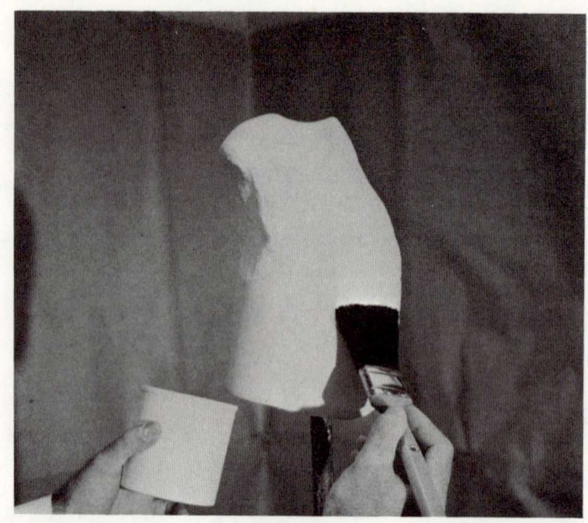

35.   To make the wax check socket, pull
      four layers of six-inch cotton stocki-
      nette over the modified model of the
      patient's shoulder.

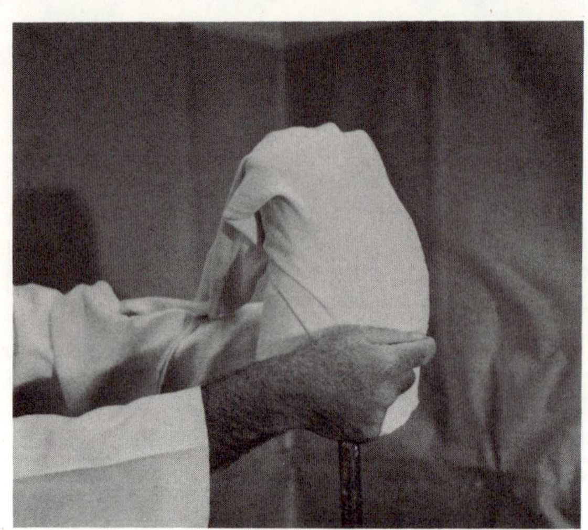

36.   Tie the layers of stockinette securely
      at the base with a piece of string.

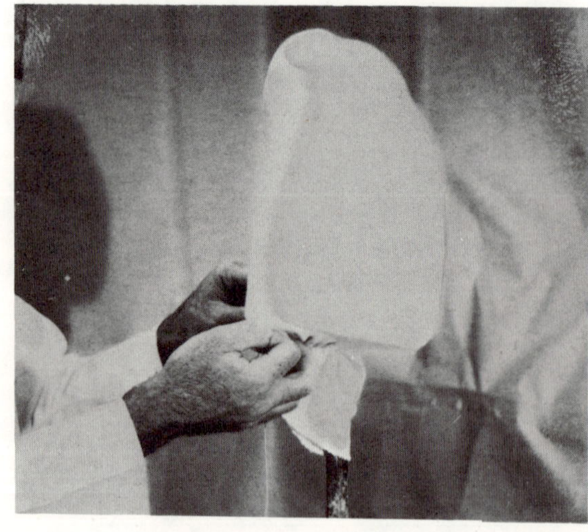

37. Melt two pounds of Welco Number 180
white wax in a container. Bring it to
a temperature of about 350 degrees
Fahrenheit, then saturate the layers of
stockinette thoroughly with the wax as
illustrated. Use great care not to spill
any hot wax on the skin as serious burns
will result.

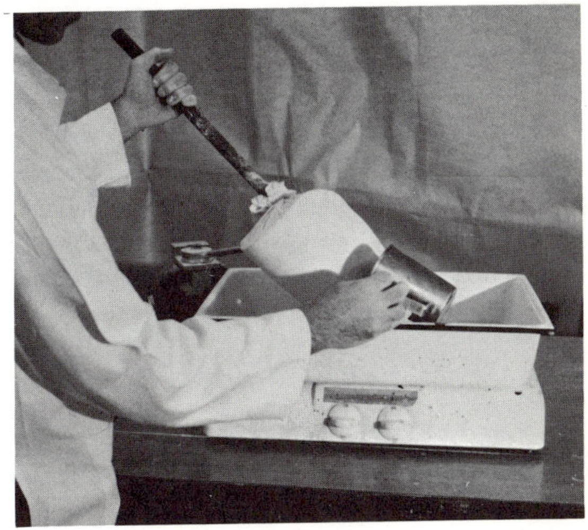

38. With the stockinette thoroughly saturated
with the molten wax quickly dip the
model into a bucket of cold water so the
wax will set. Use the hands to mold the
check socket into any hollow spots that
might have been bridged over by the
stockinette. After doing this hold the
model with the check socket under a
stream of cold water until it is thoroughly chilled.

39. Outline the trim lines on the check
socket. Plan to make an opening just
barely large enough for the patient to
slip his arm through to get the check
socket on for the first fitting.

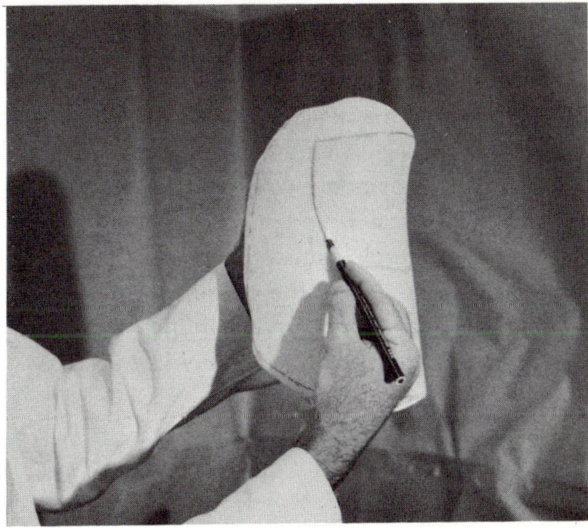

40. Cut the wax check socket along the trim line with a cast cutter or sharp knife, and remove it from the cast.

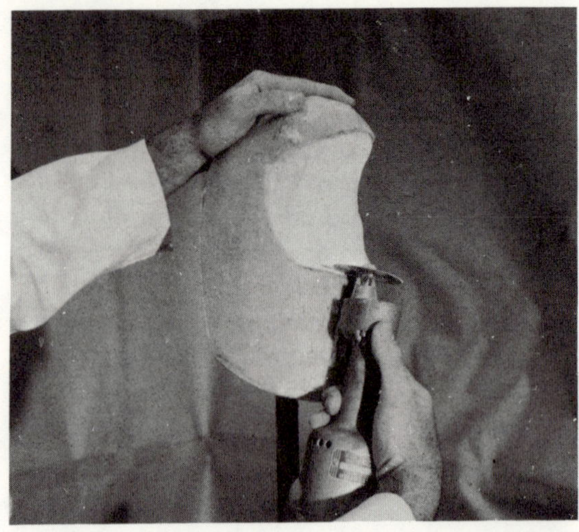

41. With a hammer handle or other round smooth piece of material, smooth off all the rough edges on the check socket where it was trimmed, to prevent the patient from being injured.

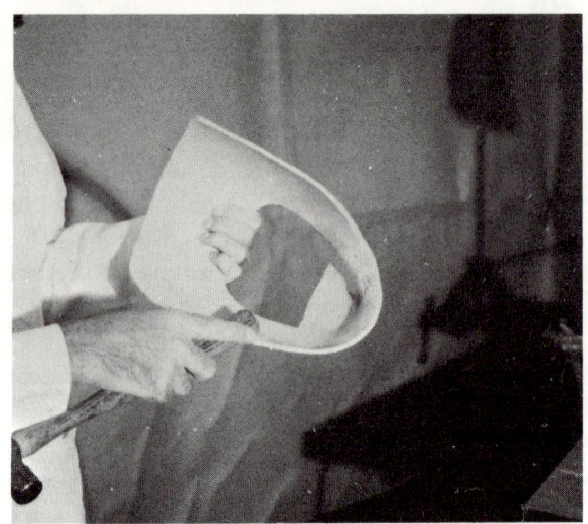

42. Put the check socket on the patient, then flex and extend his humerus about 90 degrees each way. Check to see if any part of the edges of the arm hole in the check socket cut into the flesh. If any do, mark the amount of material that must be removed to provide adequate range of motion without discomfort.

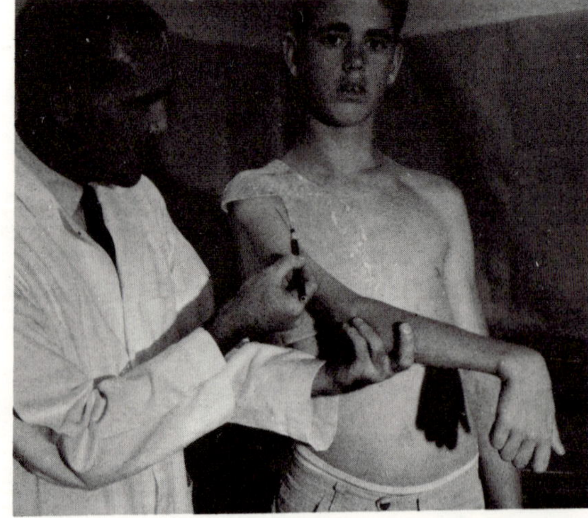

43. Check the trim line on the posterior aspect of the check socket to make sure that the vertical edge is parallel to the spine and that no part of the check socket impinges on bony areas or soft tissues.

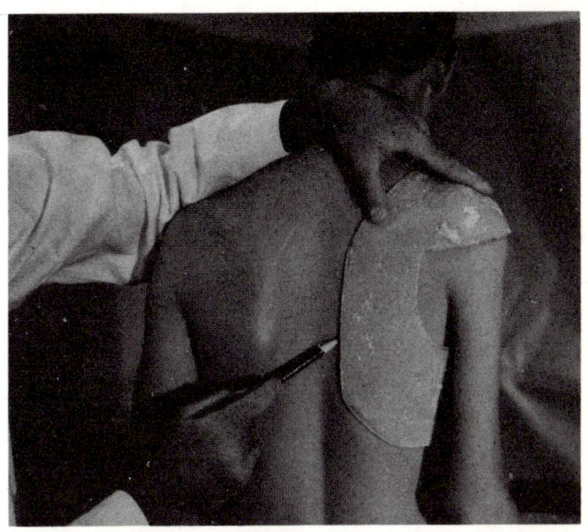

44. Check the trim where it is close to the neck to make sure there is ample clearance, and that the edge does not cut into any of the soft tissues around the area of the neck.

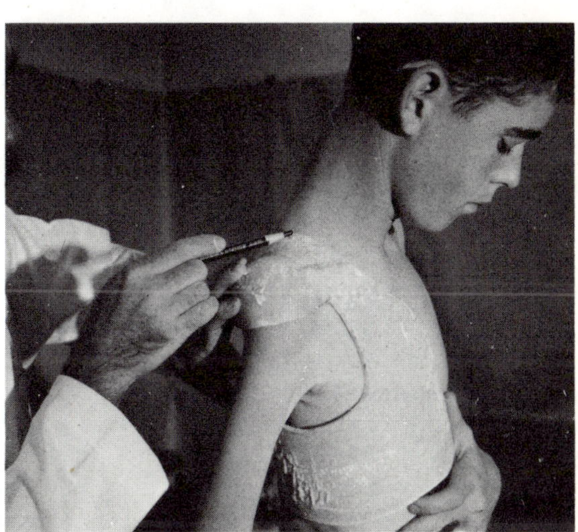

45. Press downward on the check socket so that pressure is applied to the patient's shoulder to see if there are any painful pressure points.

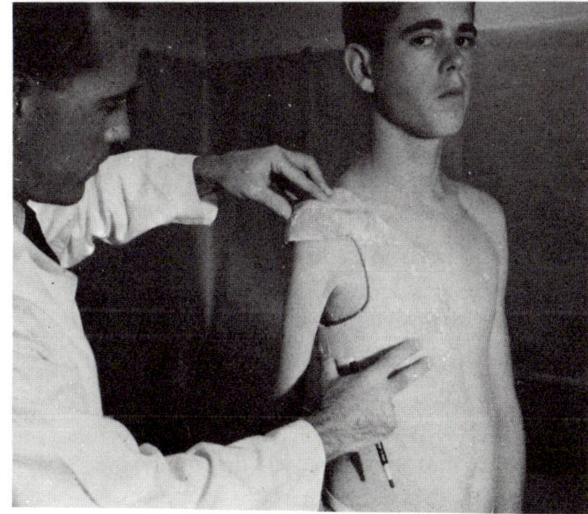

236

46. If the shoulder cap is the type that
extends down to bear weight on the
crest of the ilium, it is quite important to check the fit at this point.
Make sure that when downward pressure is applied no discomfort is felt
by the patient where the pressure is
applied over the crest of the ilium.

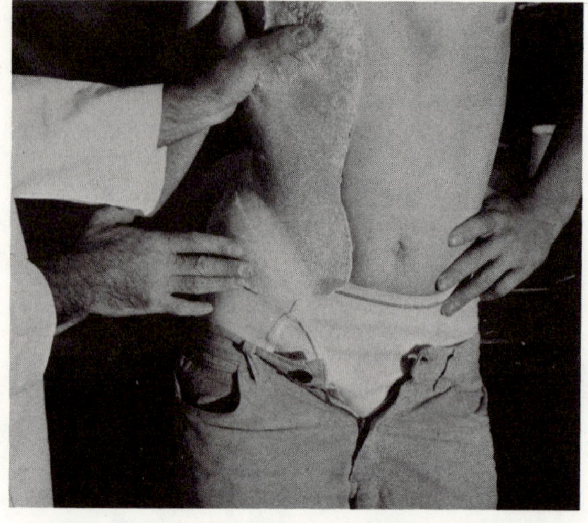

47. If any modifications in the shape
of the check socket are necessary
to correct painful pressure points,
use a heat gun to soften the wax,
make the necessary changes, then
chill the wax to harden it, using
a damp towel.

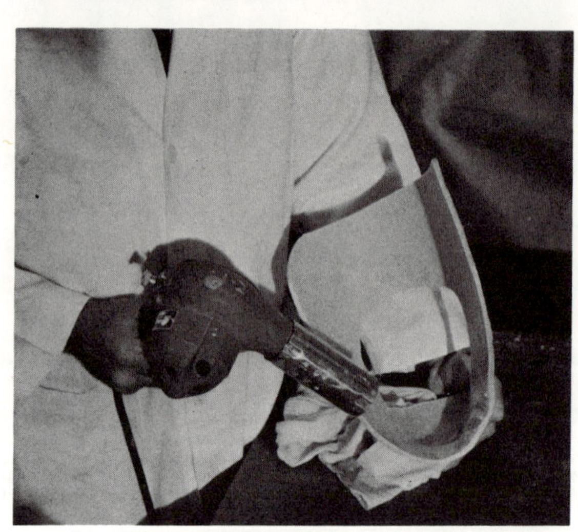

48. The check socket is used to make
the final lamination mold so that
the finished shoulder cap will reflect all the modifications made
in the check socket. Cover the
arm hole of the check socket
with cardboard, seal off the arm
hole and the end of the socket
with the plaster bandage, then
insert a mandrel and fill the
check socket with plaster all in
the same way as was done earlier
with the wrap cast.

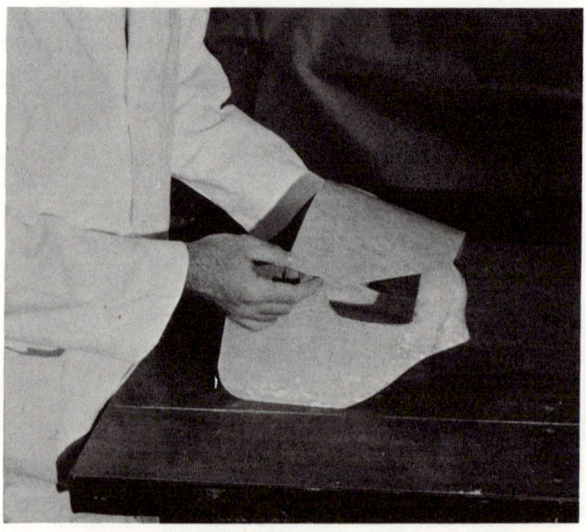

49. Remove the wax check socket from the new mold, then smooth it as was done earlier with the mold made from the wrap cast. Apply three coats of Hi-Glo parting lacquer. The mold is now ready to lay up the lamination.

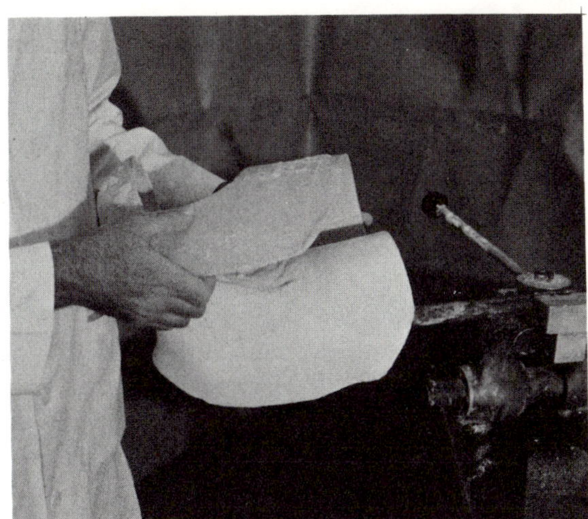

HOW TO LAY UP THE SHOULDER CAP MODEL FOR LAMINATION

1. Apply a piece of horse hide large enough to cover the entire cast, with enough extra material to allow for clamping the edges together to hold them for sewing. Be sure to apply the material so the "stretch" of the hide is in the horizontal plane so that it will follow the contours of the cast without bridging. Staple the horse hide in place with a few staples applied to the medial or "rough" side of the model. The leather liner will serve to cover the "Kemblo" sponge rubber liner to form a soft inner surface on the shoulder cap to provide maximum comfort for the patient.

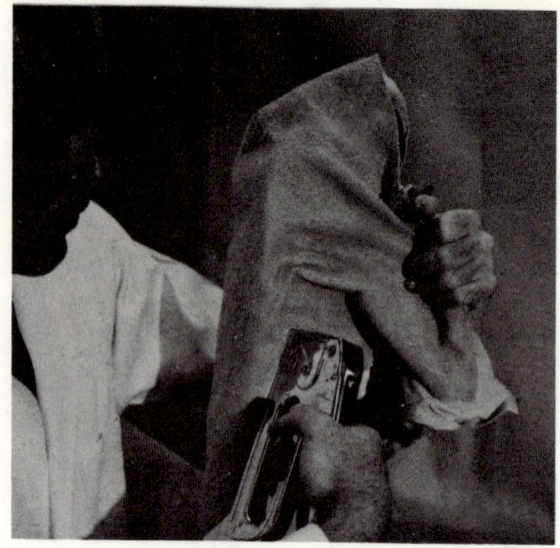

2. Stretch the horse hide smoothly over the cast, bringing the edges together and holding them in place with Yates clamps. The seam should go across the top of the shoulder and down the medial side of the model.

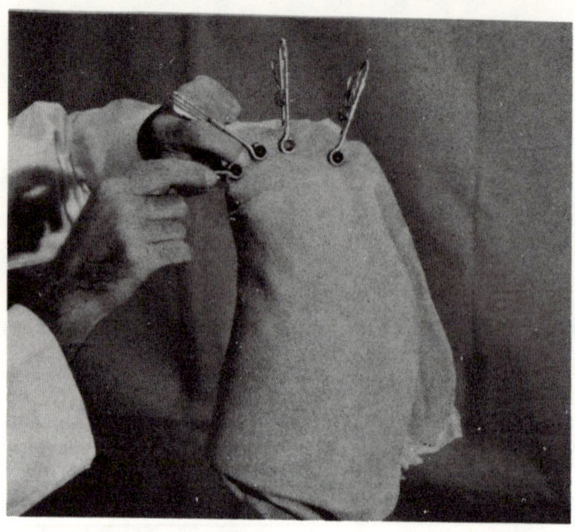

3. With the leather stretched tightly over the model and held in place with the Yates clamps, mark a line along which the leather should be sewn to provide a good tight fit.

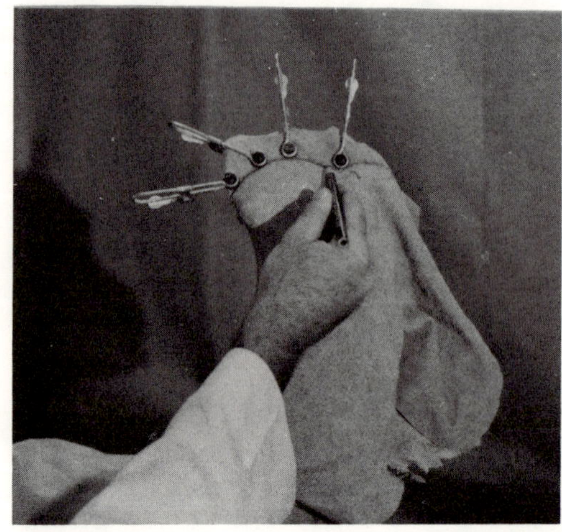

4. Remove the leather from the model and sew the edges together along the line made in the previous step. Trim the excess leather parallel to the seam with a pair of shears.

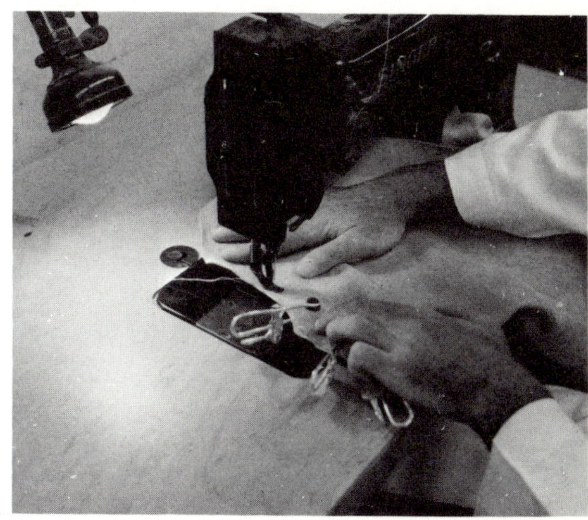

5. Put the horse hide liner back on the model and staple it into place, being sure that it is free of wrinkles and fits tightly over the surface of the model.

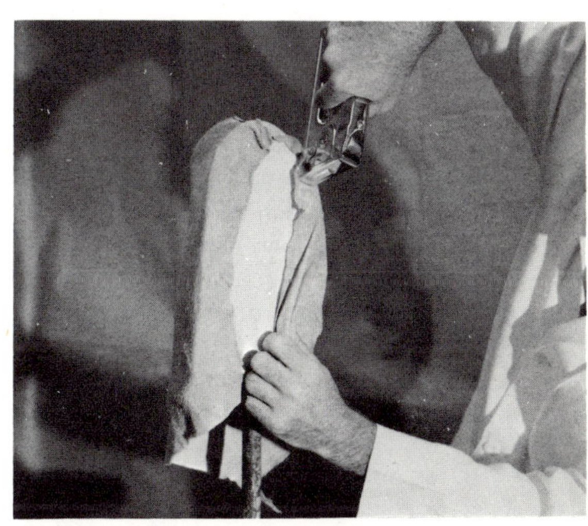

6. Carefully shape the leather into the hollow portion of the model so that there will be no bridging over hollow spots.

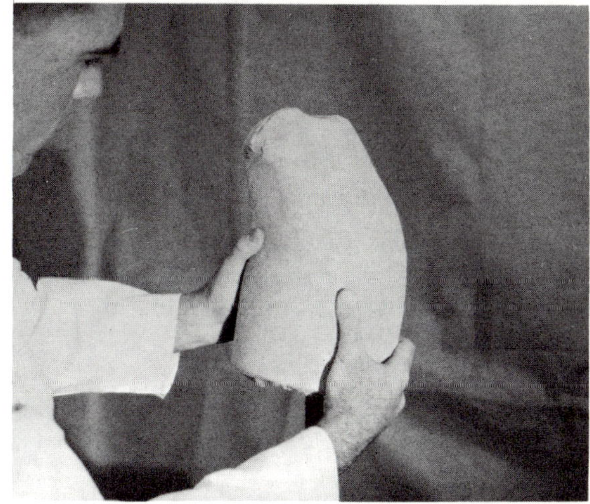

7. Outline the trim lines on the leather, using blue chalk. These will serve as a guide in fitting the pieces of "Kemblo" sponge rubber sheeting to the model.

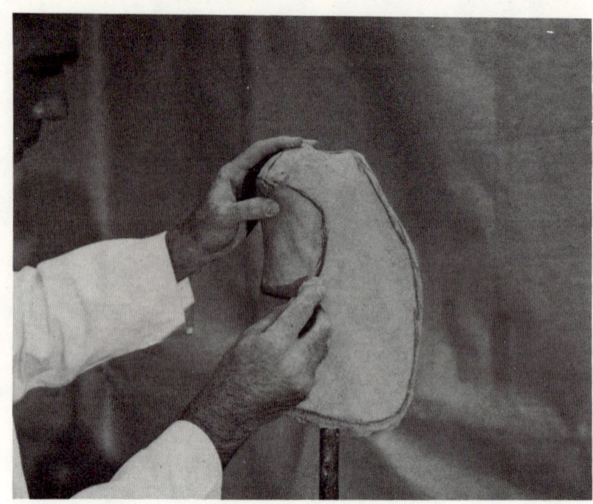

8. Cut a strip of 1/8 inch "Kemblo" 3 inches wide, and long enough to cover the top of the shoulder on the cast. Place this strip of "Kemblo" on the cast as illustrated, so it will pick up the blue chalk marks which were placed on it in the preceding step.

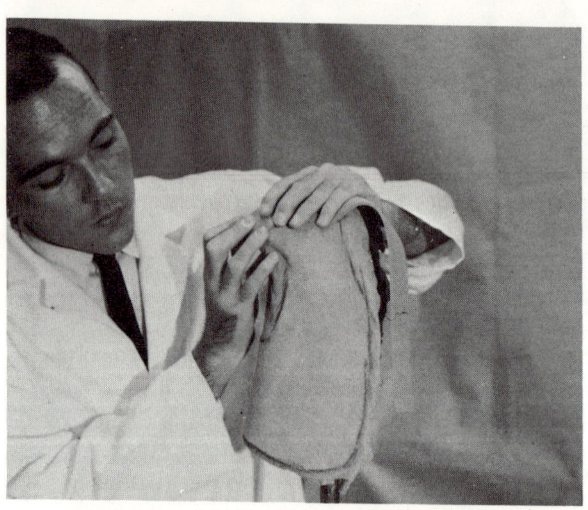

9. Using the chalk marks picked up on the "Kemblo" as a guide, trim the material to shape so that it will fit accurately on top of the shoulder of the model.

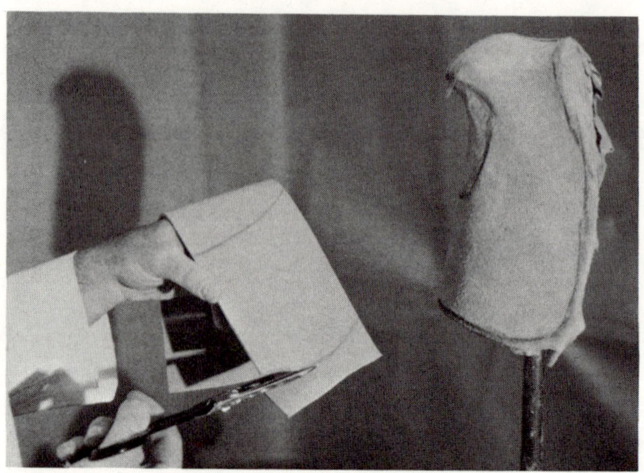

10. Place the piece of "Kemblo" again on top of the shoulder, position it in place, then index it so that the mark goes from the "Kemblo" onto the leather. These index marks will make it possible to accurately reposition the "Kemblo" when cementing it in place.

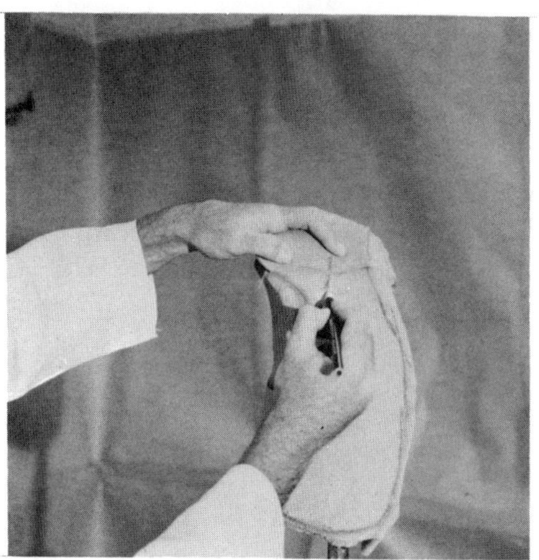

11. Apply a coat of bonding cement (Stabond, Barge, etc.) to the leather in the area where the "Kemblo" is to go.

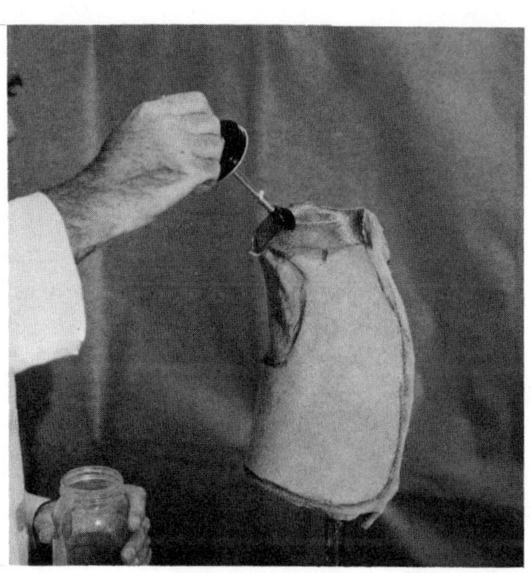

12. Apply a coat of the cement to the side of the "Kemblo" that will go on the leather.

13. After the Stabond cement has dried, apply the piece of "Kemblo" to the shoulder in the proper position as marked, and press it firmly in place.

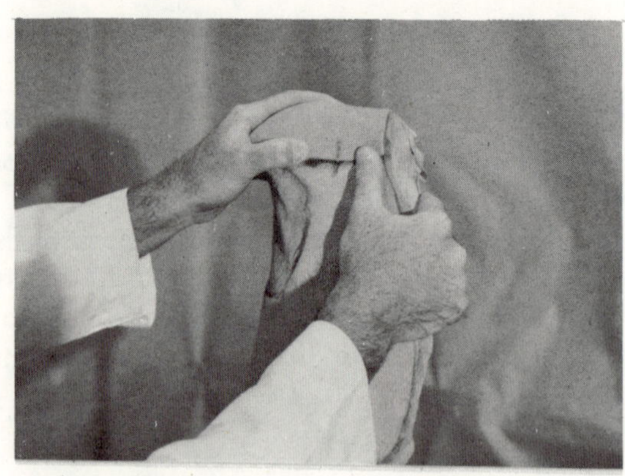

14. Chalk the edges of the "Kemblo" piece that was cemented to the top of the shoulder. Apply the chalk generously so it will "pick up" on the next piece of "Kemblo" to be fitted, thus providing an accurate line for cutting the next piece of material.

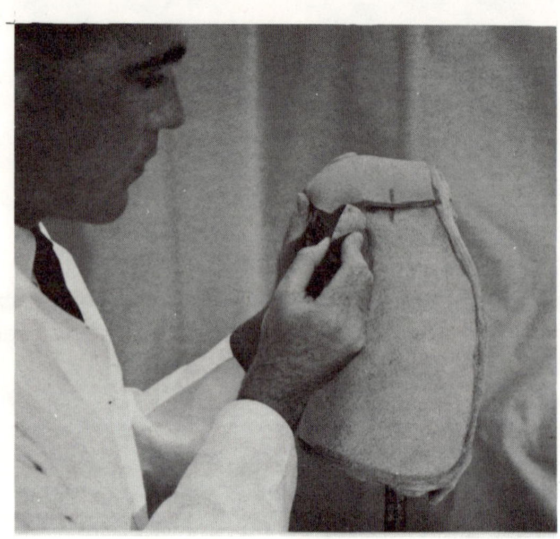

15. Cut a piece of "Kemblo" large enough to cover the anterior aspect of the model, and press it firmly against the chalked edge of the "Kemblo" cemented to the top of the shoulder. Trim to the mark picked up on the new piece of material, then make a further trim down the sides to the other trim lines picked up for the model. Mark an index line so the piece can be accurately repositioned.

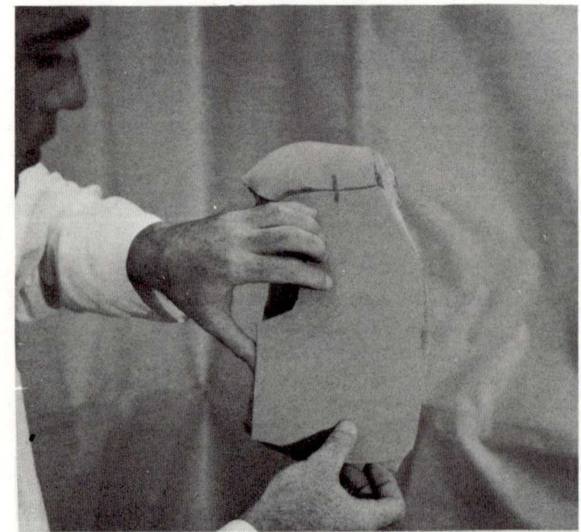

16. Apply bonding cement to the leather where the second piece of "Kemblo" will be applied, and to the "Kemblo", being very careful to get plenty of cement on the edges of the material where they will butt together. Allow the bonding cement to dry.

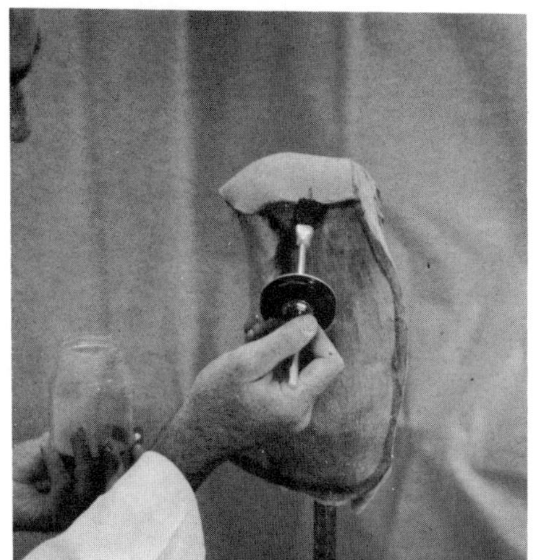

17. Apply the piece of "Kemblo" to the anterior portion of the model, being very careful to match the index lines while butting together the edges of the "Kemblo" so there is no gapping. Smooth the "Kemblo" onto the leather so there are no wrinkles.

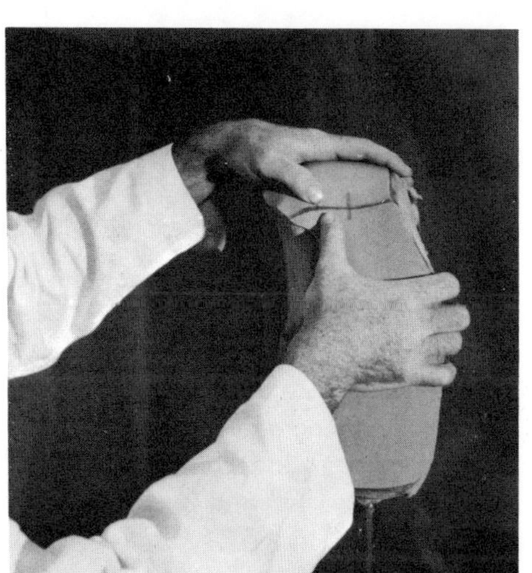

18. Continue to match the edges of additional pieces of "Kemblo" in the manner described in previous steps until the entire surface of the shoulder cap, as outlined in the trim lines, is completely covered with "Kemblo."

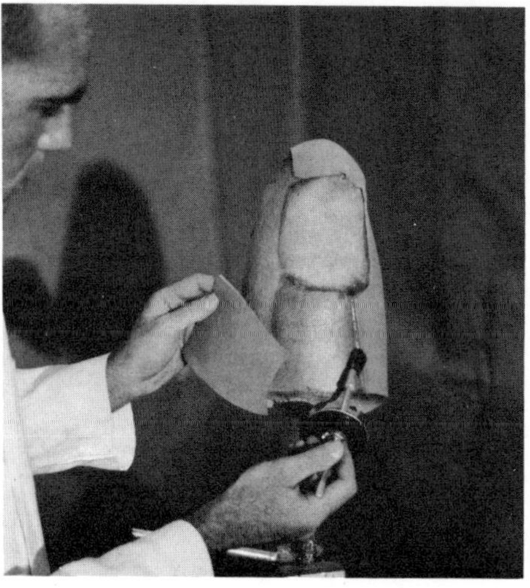

19. Double check to make sure that the "Kemblo" is not bridging any hollow spots, that all of the butt joints are firmly cemented together, and that the entire surface is smooth without any wrinkles or bulges.

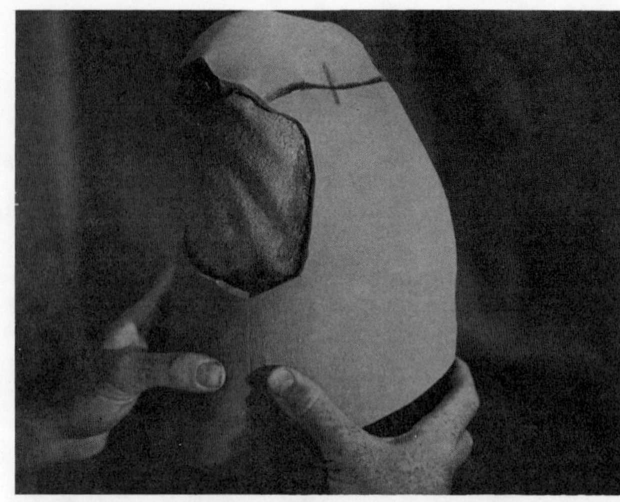

20. Drill a quarter inch hole in the mandrel pipe as close to the plaster model as possible. The purpose of this hole is to enable vacuum to pull the inside P.V.A. (Poly Vinyl Alcohol) bag closely against the liner on the model. In making the shoulder cap it has been found to be good practice to use a P.V.A. bag in this way as a parting agent and to prevent the resin from soaking into the "Kemblo" and leather liner.

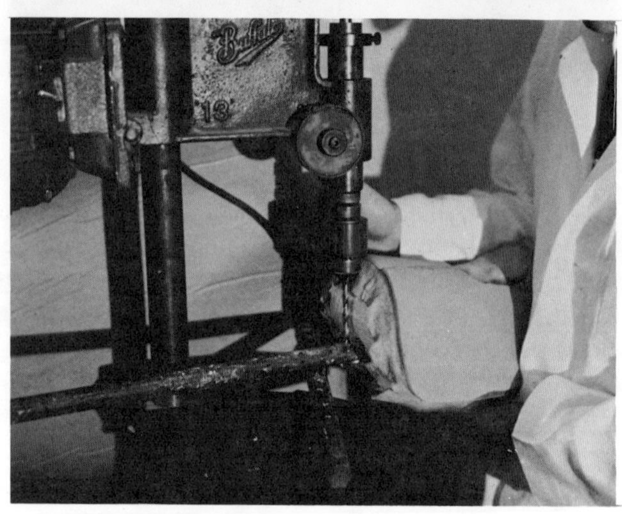

21. Tie a three inch long piece of scrap stockinette over the vacuum feed hole that was drilled in the mandrel. The purpose of this is to prevent the P.V.A. bag from being sucked into the hole when the vacuum is applied, thus stopping it up and shutting off the vacuum.

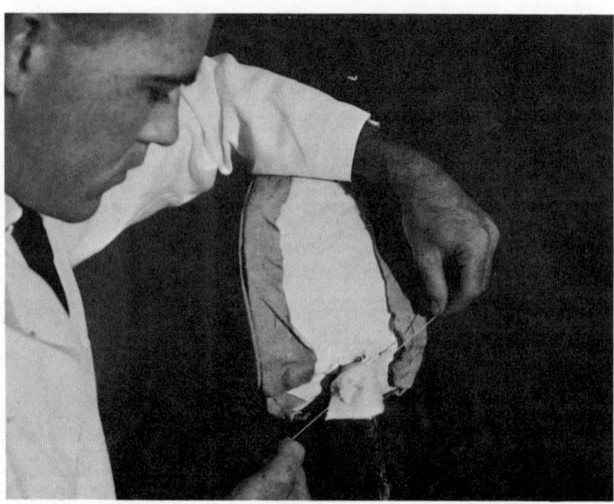

22. Sprinkle talcum powder on the "Kemblo" to make it easier to slide the P.V.A. bag down over the layup

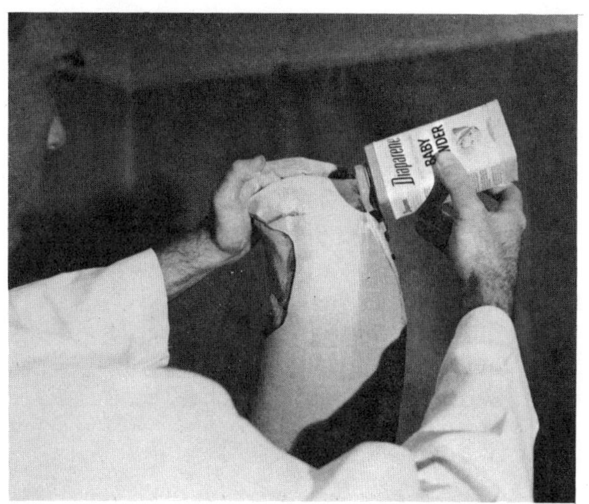

23. To determine the amount of P.V.A. sheeting to cut from the roll for making the inner P.V.A. bag, first measure the circumference of the model at the distal end.

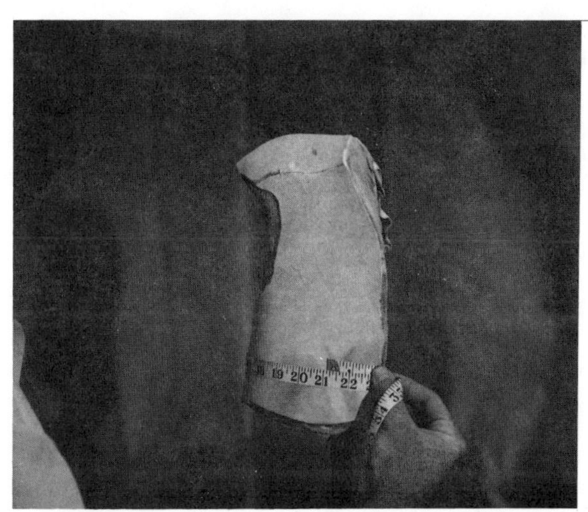

24. Measure off a length of P.V.A. sheeting equal to the circumference measured in the previous step, plus 2 inches for overlap, and cut off the piece of P.V.A. with a pair of shears.

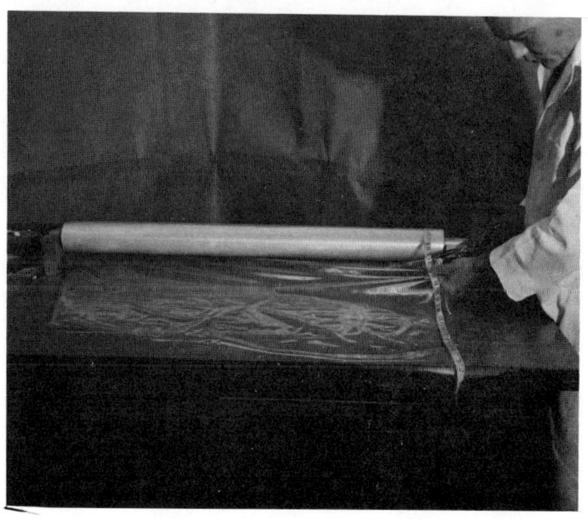

25. Adjust the P.V.A. pressure sleeve
    jig so that the dimension across the
    adjusting bar is half the circum-
    ference measured on the model. In
    this case the model measured 21
    inches in circumference, so the
    bar is set at 10-1/2 inches.

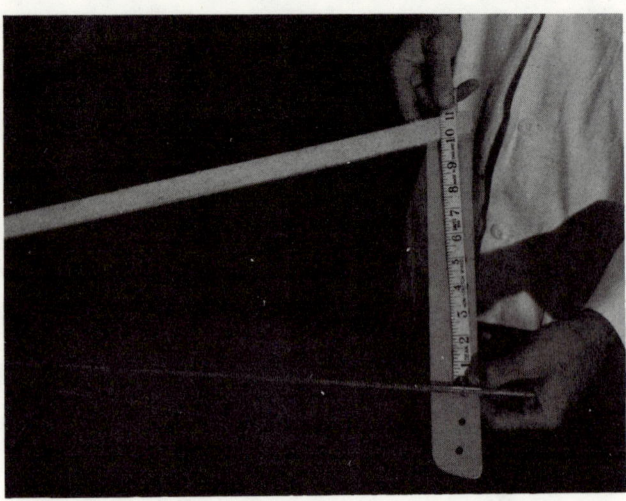

26. Lay the sheet of P.V.A. on a flat
    surface with the long dimension
    left to right, then place the jig on
    the P.V.A. with the fixed arm
    toward you. Fold the edge of the
    P.V.A. material over the padded
    surface of the fixed arm so the
    edge of the P.V.A. is on the upper
    surface of the arm and even with its
    inner edge.

27. Clamp the P.V.A. material to
    the arm of the jig with spring
    clips.

28. Fold the P.V.A. material toward you over the movable bar of the jig, then on toward you and over the fixed bar. Hold the two layers of the P.V.A. onto the fixed bar with the fingers while removing the nearest clamp, fold the P.V.A. over and under the fixed bar, then replace the clamp so the P.V.A. material is overlapped on the padded bar. Repeat this with the other clamp, making sure the P.V.A. is pulled tight and free of wrinkles.

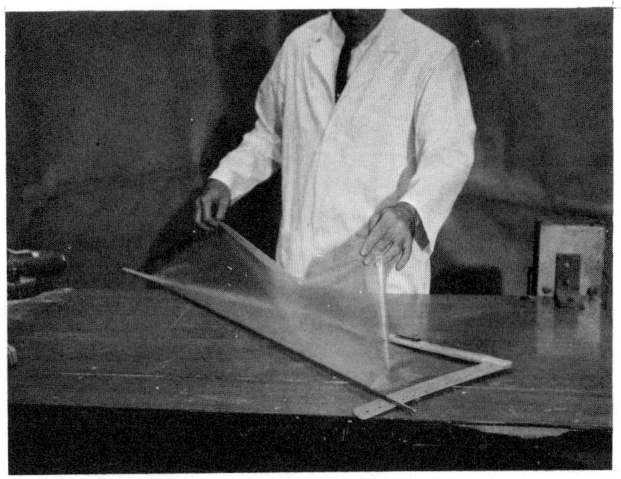

29. Trim the P.V.A. material even with the outer edge of the fixed arm of the jig, using a pair of shears. First trim to one of the clamps, then remove the clamp and attach it to the part already trimmed, then continue to the second clamp, handling it in the same way until the material is trimmed to the edge along its full length. Be sure to save the piece of P.V.A. that was trimmed off for use later to tie the P.V.A. bag at the base around the mandrel to prevent leakage at that point.

30. Apply a thin coating of P.V.A. glue to the facing edges of the seam of the P.V.A. bag between the two clamps. Do not use too much glue as it will weaken the seam. (P.V.A. glue is made by dissolving scraps of P.V.A. in warm water.) Go immediately to the next step.

31. Immediately after the P.V.A. glue is applied, press the two edges together and iron them with a hot sealing iron. Be sure the seam is smooth and is completely bonded.

32. Remove one of the clamps, then apply P.V.A. glue to the seam from the point where the seal left off in the preceding step to the end of the bag.

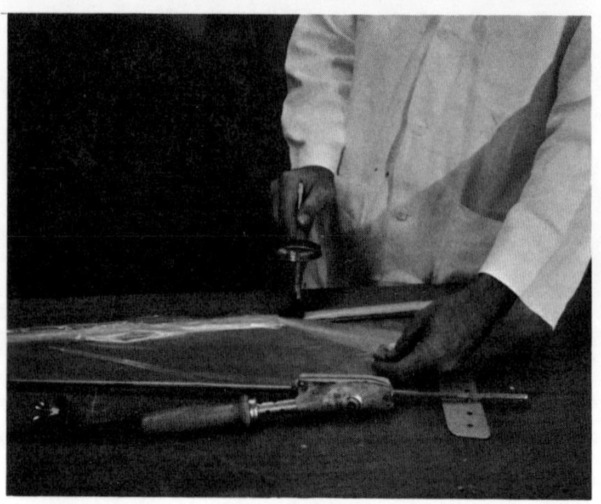

33. Iron the portion of seam just glued with the sealing iron as was done with the center section. Repeat these two steps on the other end, making the P.V.A. bag complete.

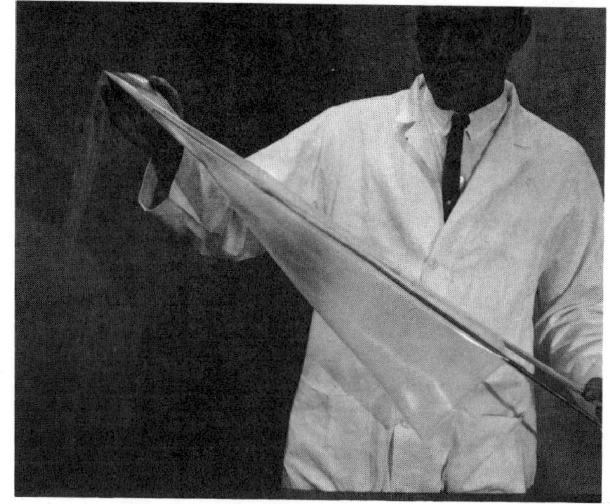

34. Slide the completed P.V.A. bag off the jig and check to see that the seam is evenly and completely sealed. Any ragged edges of the P.V.A. material can be trimmed with a pair of shears or sealed down with additional P.V.A. glue.

35. Wet a turkish towel with cold water and wring it out enough so that it does not drip. Lay the damp towel on a flat surface, then lay the P.V.A. bag full length on the towel. Fold the towel over the P.V.A. bag and roll it up so that it is completely enclosed by the wet cloth. This is to soften the P.V.A. material so it will stretch to conform to the shape of the model without breaking or tearing. The moisture makes the P.V.A. material soft and elastic so that it will more easily conform to irregular shapes. On the other hand, if the P.V.A. bag is over-saturated with moisture, it will be weakened and will split. Five minutes in the damp towel is about average. Give more time in a dry climate, less in a humid one.

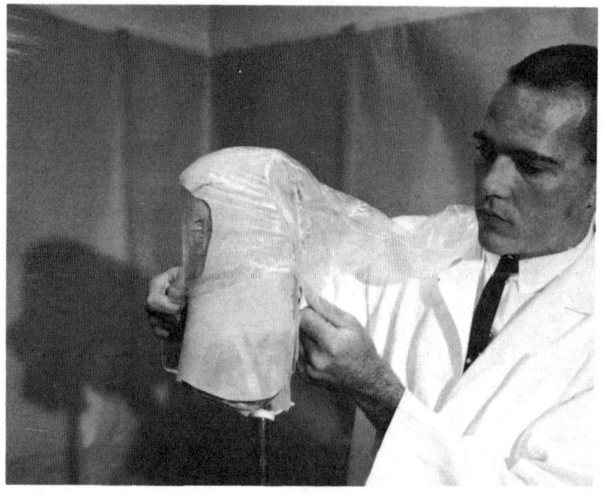

36. Pull the moistened P.V.A. bag down over the model, being sure that the seam is on the medial side of the model where the laminate will be trimmed away later.

37. Pull the P.V.A. bag down snugly over the layup, then tie the bag firmly around the mandrel at the bottom using the scrap of P.V.A. trimmed off in step 29. Be sure it is tied under the stockinette that was placed there earlier to avoid permitting the P.V.A. material to get into the vacuum hole in the mandrel.

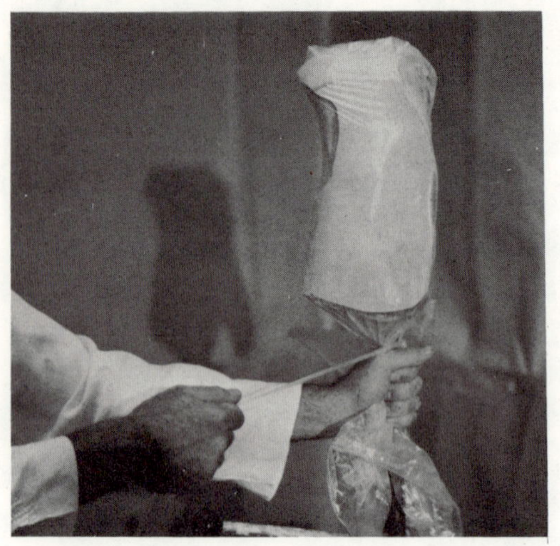

38. Grasp the top of the P.V.A. bag and pull it up and over toward the trim or medial side of the cast where the material will be cut away later. This will prevent wrinkles in the finished shoulder cap.

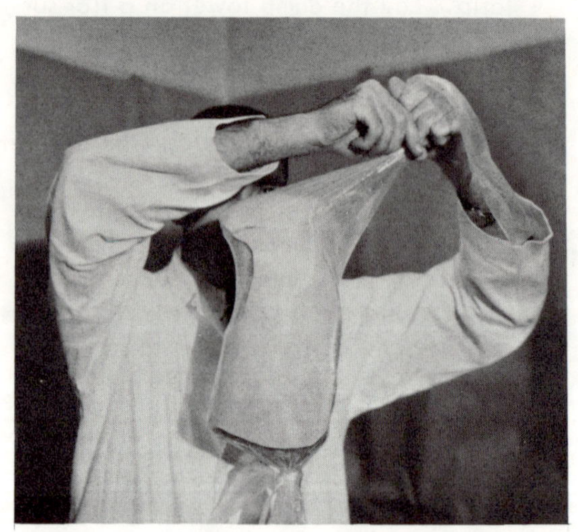

39. Pull the end of the P.V.A. bag down over the medial side of the model tight enough to remove most of the wrinkles from the top of the shoulder area.

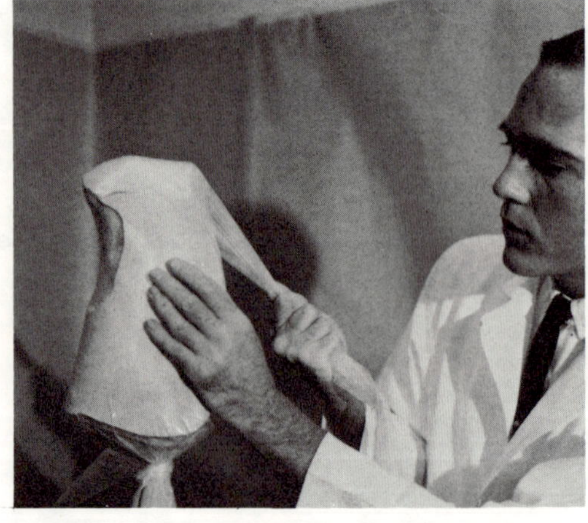

40. Secure the end tightly with a scrap of P.V.A. material as close to the model as possible.

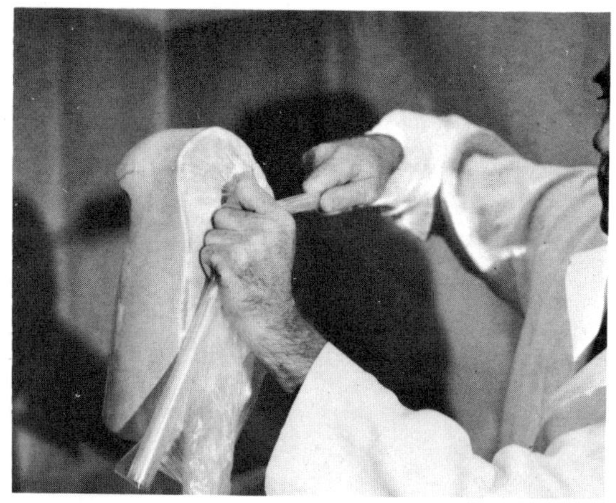

41. Trim away the excess P.V.A. material, both at the shoulder level and around the mandrel.

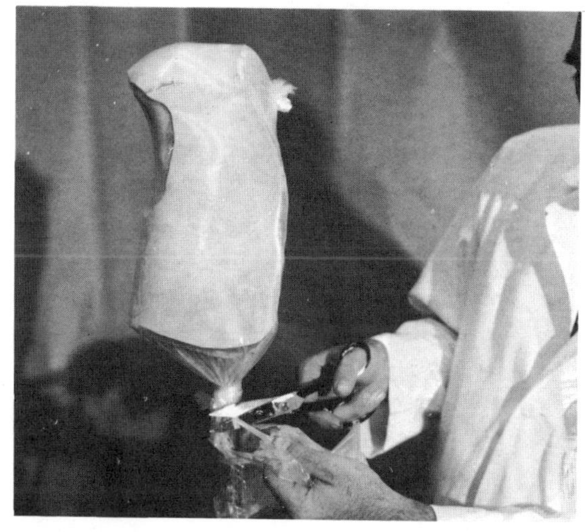

42. If a few wrinkles remain over the shoulder area, as is often the case, these can be removed by application of warm air with a heat gun. Be very careful not to overheat the P.V.A. material, or it will melt and ruin the bag. The heat causes the P.V.A. material to dry and shrink slightly thereby removing the wrinkles, and only enough heat should be applied to accomplish this result.

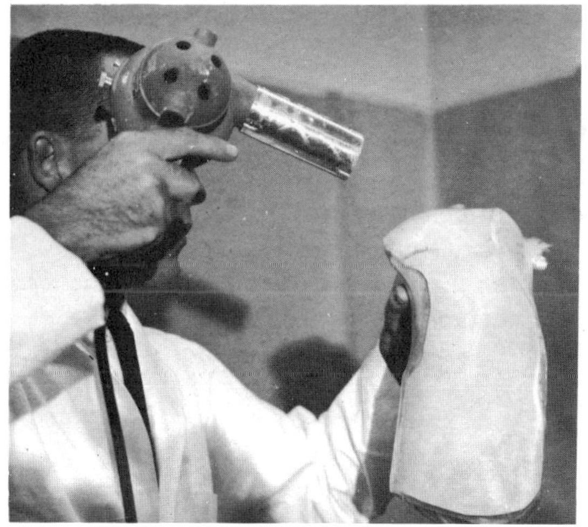

43. Apply vinyl plastic electrical tape over the P.V.A. material at the mandrel, covering the P.V.A. binding and on down over the lower binding and over the lower edges of the P.V.A. bag to make a complete seal to prevent any vacuum leaks.

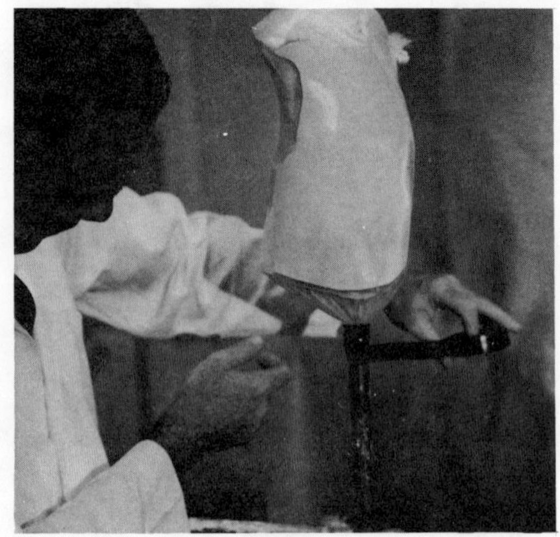

44. Attach the hose from the vacuum plastic trap to the end of the hollow mandrel, and seal the joint with plastic tape to prevent leakage. A plastic trap similar to the one illustrated is always used to prevent plastic from leaking into the source of vacuum and ruining the equipment.

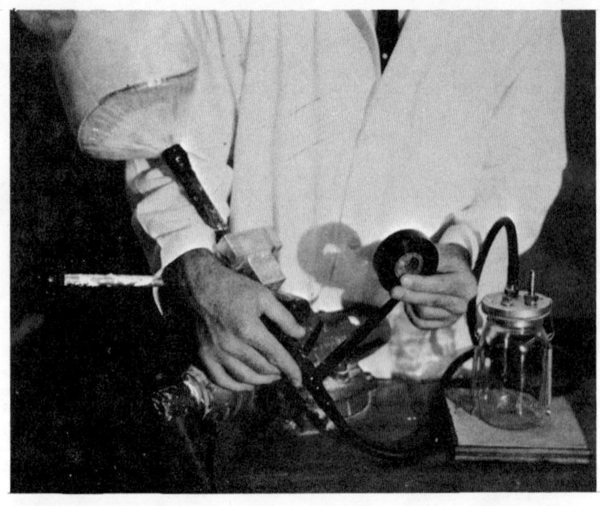

45. Attach the hose from the source of vacuum to the plastic trap and turn on the vacuum. Ten to twenty inches of vacuum is needed to draw the P.V.A. bag tightly against the layup.

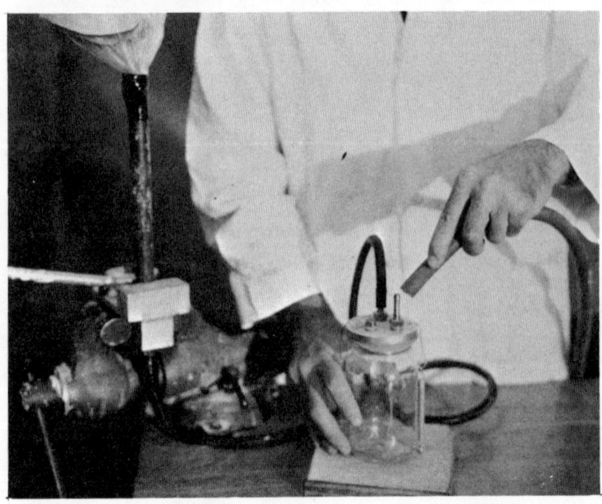

46. As the vacuum pulls the P.V.A. bag in closely to the model, work it carefully with the hand to make sure it follows the hollows, and that all the wrinkles are worked out, resulting in a perfectly smooth glass-like surface.

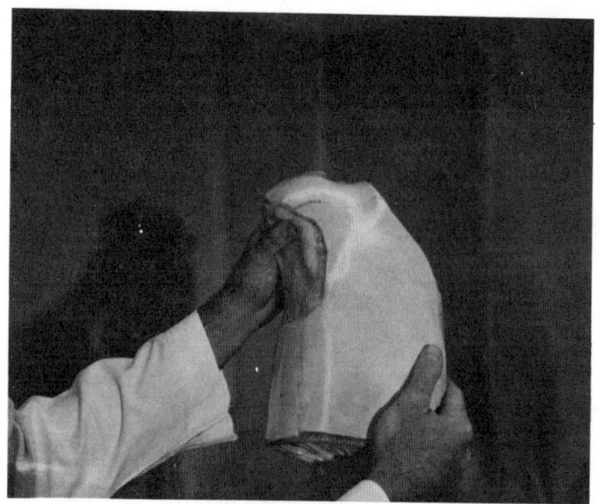

47. Cut a piece of dacron felt large enough to enclose the model, hold the edges together and mark the felt for the seam, as illustrated.

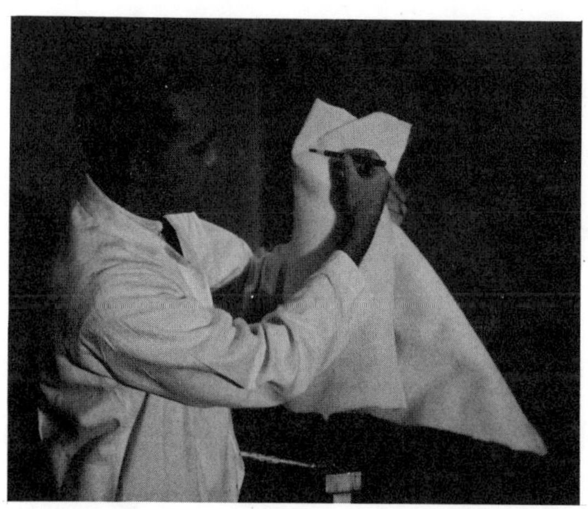

48. Sew the dacron felt along the line made in the previous step to make a bag to fit over the model. Repeat this procedure to make a second dacron felt bag, as two are needed.

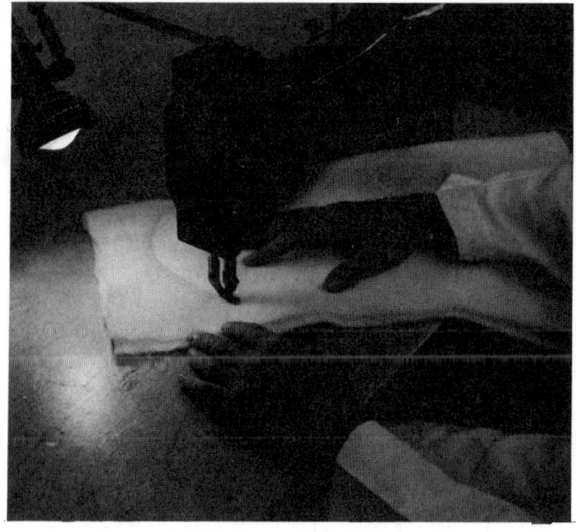

49. Trim the dacron felt bags to the seams, using shears.

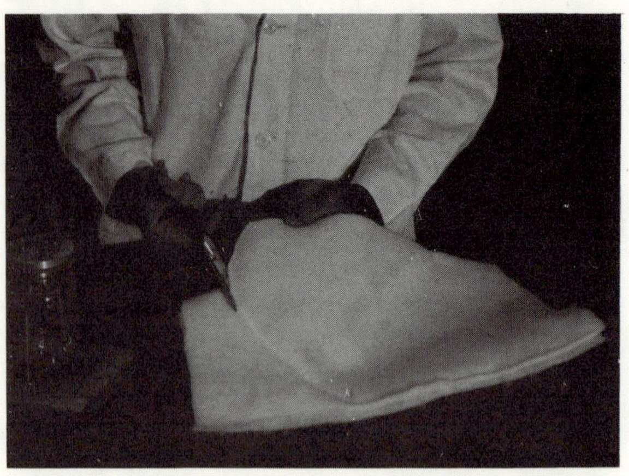

50. Pull the dacron felt bags over the model with the seams over the shoulder and down the medial side of the model.

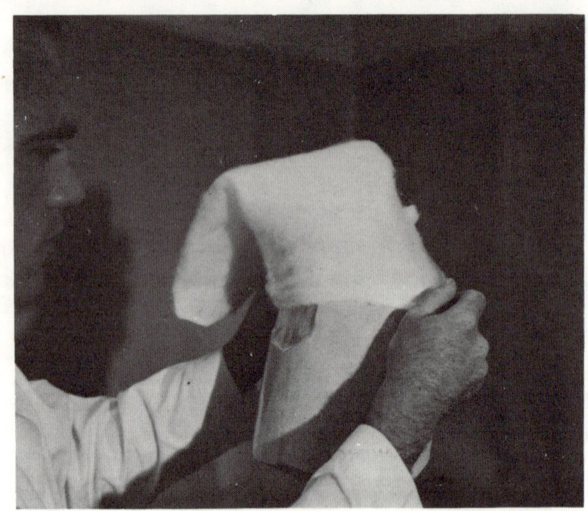

51. With the dacron bags pulled down over the model snugly, tie them around the mandrel with a piece of string.

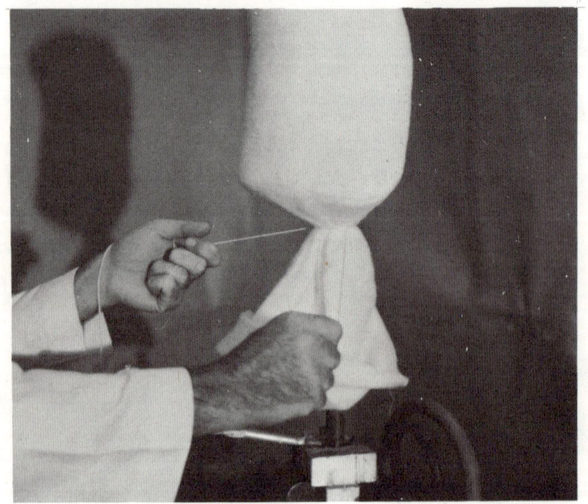

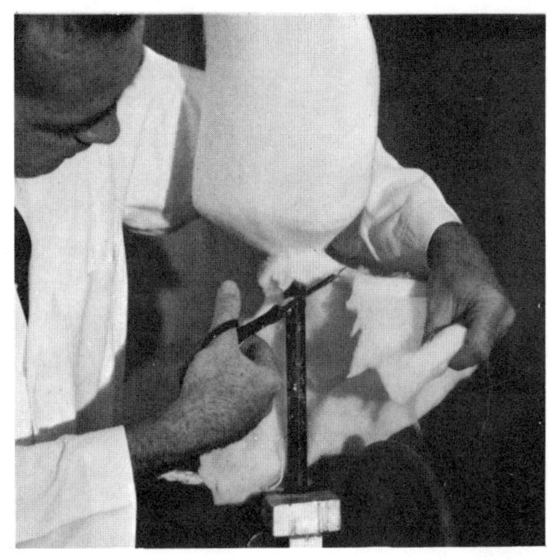

52. Trim the excess dacron felt from around
     the binding at the mandrel. Save the
     scraps of felt for use as filler material
     later.

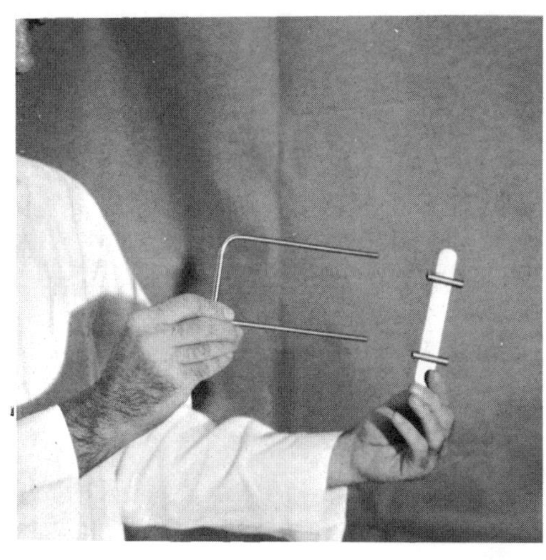

53. The shoulder cap provides a stable foundation
     for various combinations of functional arm
     bracing equipment that, when installed, will
     aid the patient in flexing his shoulder, his
     elbow, or both. It is necessary to provide a
     means for mounting this equipment onto the
     shoulder cap.   This is done by laminating a
     "U-bar mounting bracket" into the plastic of
     the shoulder cap at the time it is made. The
     complete assembly, as shown in the illustration
     consists of a stainless steel plate with two
     hollow tubes welded to it, and a U-bar, the
     ends of which fit into the two hollow tubes.
     The various types of shoulder joints can be
     clamped to the base of the U-bar, which can
     be readily removed from the mounting bracket
     any time it may be necessary.

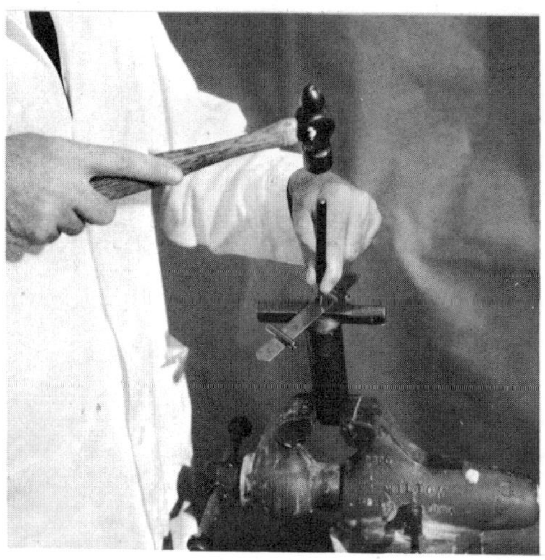

54. Center punch the U-bar mounting bracket
     in four places on the side to which the
     tubes are welded.

55. Drill four Number 18 holes in the U-bar mounting bracket at the points center punched. The purpose of these holes is to allow the plastic to flow through and thus lock the U-bar mounting bracket in place.

56. Shape the U-bar mounting bracket to conform to the contour of the top of the shoulder on the layup. Shape it to a U-shape curve using a bending fork as illustrated.

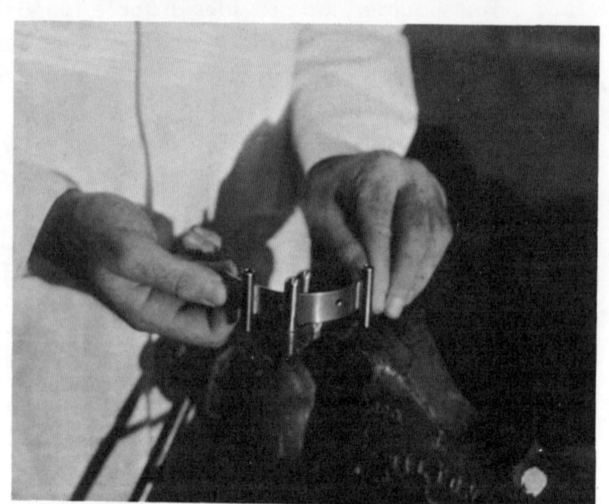

57. Check the contour of the U-bar mounting bracket on the layup as illustrated. It should fit the contour of the top of the shoulder without any gapping.

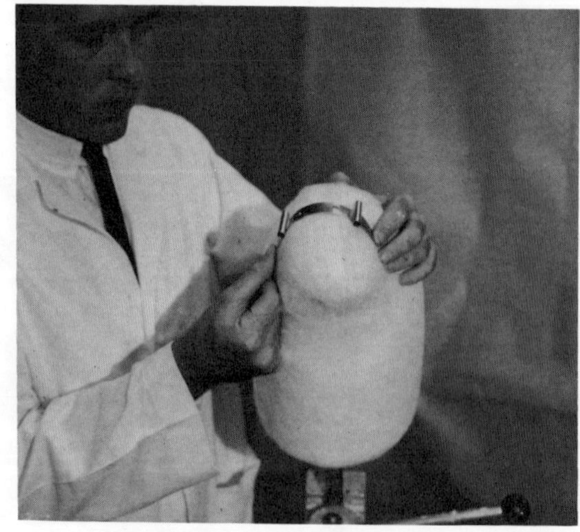

58. Adjust the antero-posterior placement of the U-bar mounting bracket. A straight edge resting on the mounting bracket tubes should be slanted down from anterior to posterior about 5 degrees from a horizontal line. The horizontal line can be established by using a spirit level, and aligning the center line of the model as nearly vertical as possible.

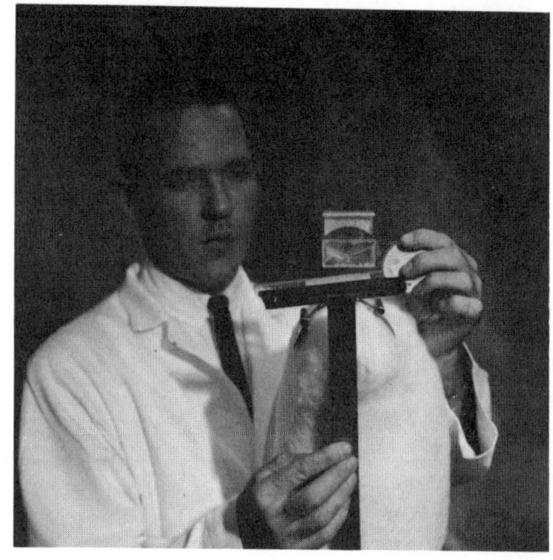

59. Adjust the medio-lateral location of the U-bar mounting bracket. It should be just medial of the acromion, and the top surface of the bracket should be parallel to the top surface of the shoulder.

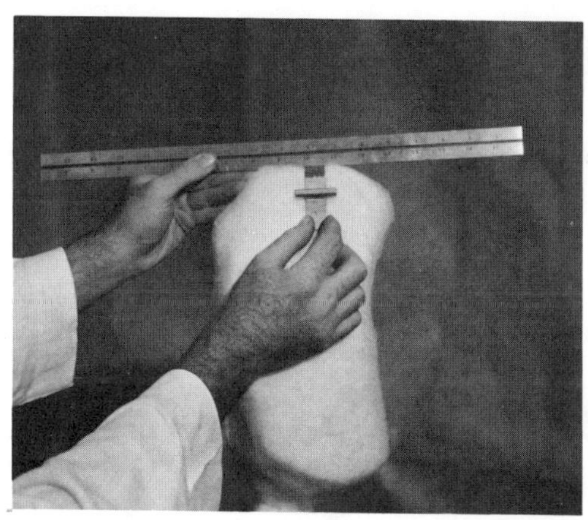

60. Mark the location of the U-bar mounting bracket on the dacron felt by outlining it with a skin pencil.

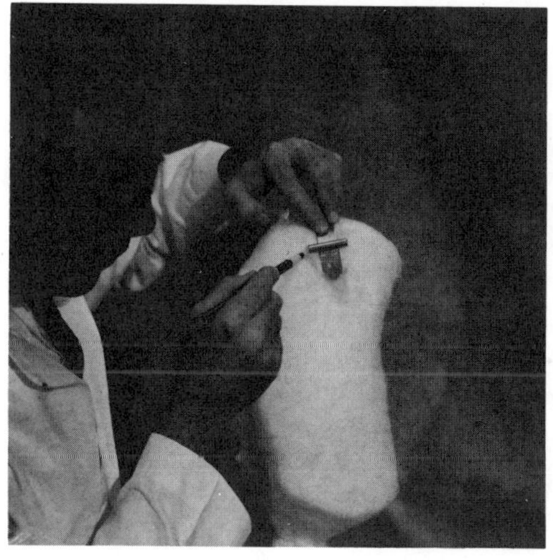

61. Plug the openings in both ends of the tubes on the U-bar mounting bracket with modeling clay to prevent the plastic from working into them and making extra work later trying to clean them out.

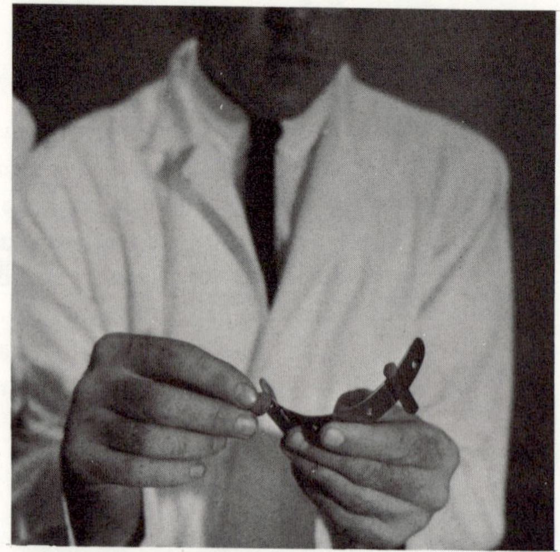

62. Secure the U-bar mounting bracket in place on top of the layup using a piece of cord tied to one of the mounting bracket tubes, then bringing it down under the model, back up the other side, and fastening it to the other tube. Be sure to take a couple of loops around the mandrel with the cord to prevent the bracket from sliding fore and aft.

63. To fair the U-bar mounting bracket into the contours of the shoulder, apply the scraps of dacron felt that were cut off when trimming the excess material in an earlier step. Apply enough of the scraps to build up the contour of the shoulder so no ridges or projections are present.

64. Cut off a piece of 5" nylon stockinette approximately 2-1/2 times the length of the model and pull it halfway down over the model.

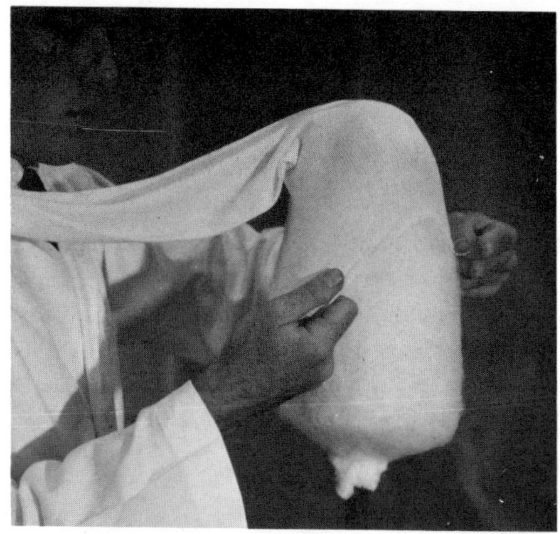

65. Pull the stockinette towards the side of the model that will be trimmed off later and twist it once as illustrated. Be very careful not to disturb the dacron felt build-up around the U-bar mounting bracket.

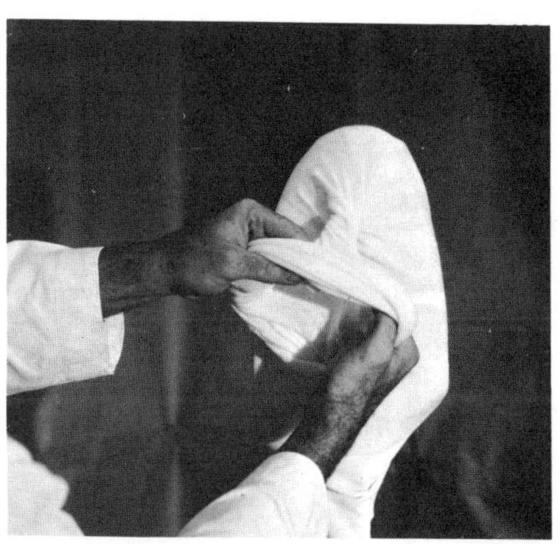

66. After making the twist to close off the first layer of stockinette, double it back and pull the second half of the stockinette down over the model as illustrated. Repeat the last two steps so that a total of four layers of nylon stockinette will be applied over the dacron felt. Smooth out all wrinkles. Tie the four layers of stockinette to the mandrel with a piece of cord and trim off the excess material.

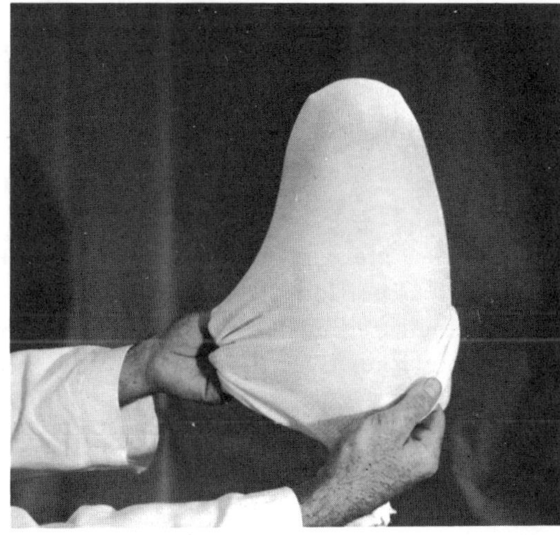

67. Make another P.V.A. bag following the procedure explained earlier. Moisten it in a towel then pull it down over the layup so it conforms to the contours and has no wrinkles. Tie off the top portion of the bag with a bowknot that can be removed quickly.

68. With a sharp pointed instrument, such as an awl or needle, punch a small hole through the inner P.V.A. bag at a point as close to where the mandrel enters the plaster model as possible. The purpose of this hole is to permit the vacuum to get behind the inner bag and hold it closely to the layup while the laminating procedure is taking place.

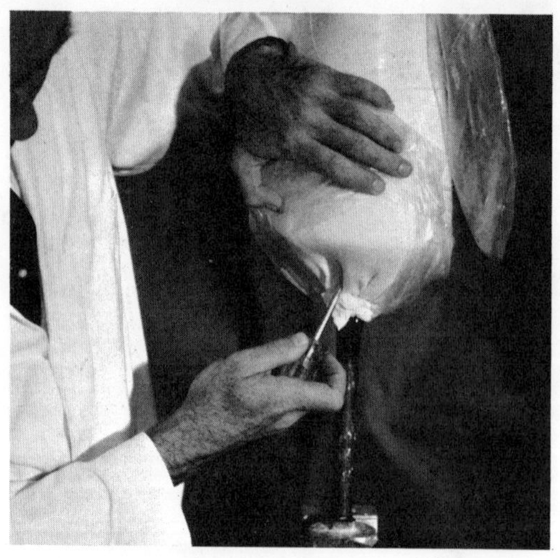

69. With a piece of P.V.A. material tie off the bottom portion of the outer P.V.A. bag, being sure that the tie is below the one previously made for the inner P.V.A. bag.

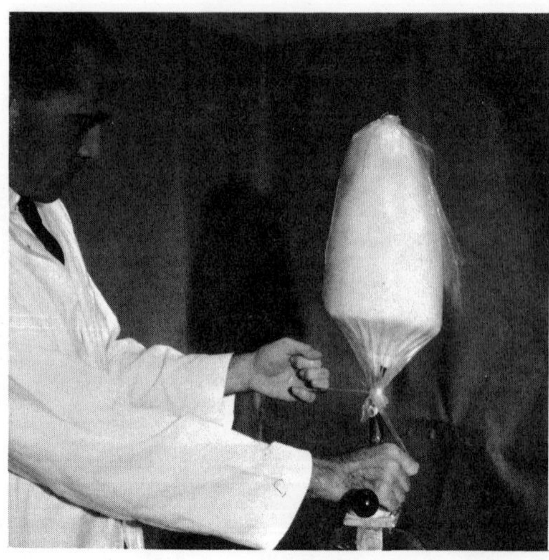

70. Seal the outer P.V.A. bag to the mandrel using plastic tape. Maintain 5 to 10 inches of vacuum on the layup at all times from this point until the resin has been applied and has set up. If a dual vacuum system is being used, apply 6 to 8 inches to the inner P.V.A. bag, 2 to 4 inches to the outer.

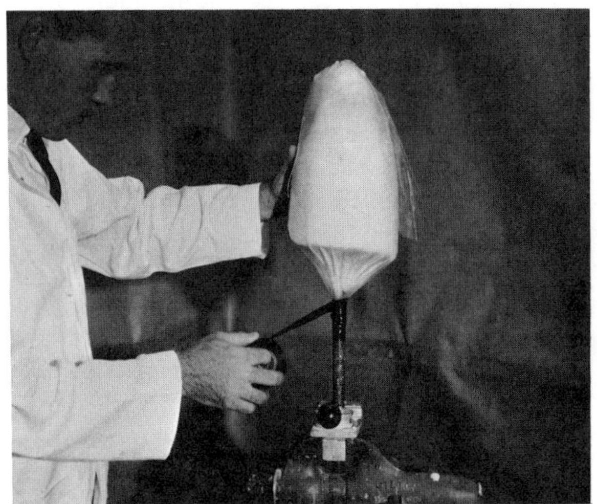

HOW TO LAMINATE THE LAYUP FOR THE SHOULDER CAP

1. Measure out enough resin to laminate the lay-up. Use 4110 Laminac Rigid Cold Cure resin. The amount required will vary according to the number of layers of filler material and the size of the lamination. A standard shoulder cap similar to the one illustrated required 1500 grams of resin. The larger shoulder caps that extend down over the crest of the ilium will require up to 2500 grams of resin. Weigh out the amount required, using a scale. Use only rigid cold cure resin, as blends of rigid and flexible resins are not satisfactory.

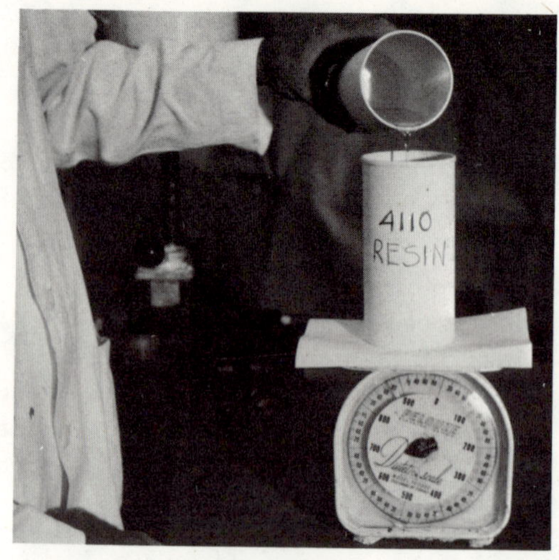

2. Calculate how many grams is two percent of the weight of the resin. In the case of 1500 grams of resin, two percent is 30 grams, for example. Add this weight of catalyst to the resin. Either Luperco ATC or Garox BZP catalyst is satisfactory.

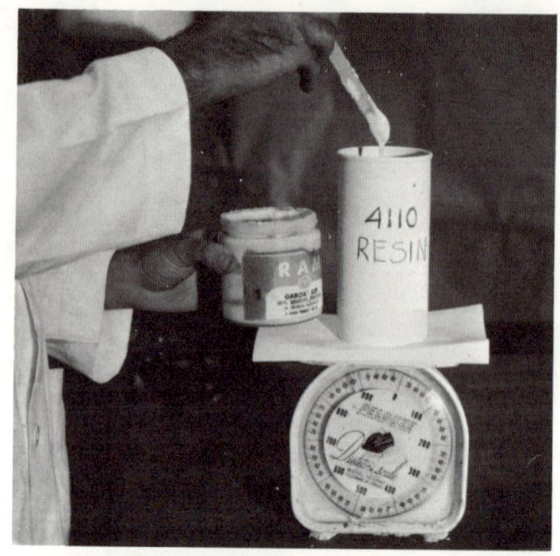

3. Mix the catalyst into the resin with a tongue depressor until it is thoroughly blended. When opaque streaks of catalyst can no longer be seen in the resin the blend is satisfactory.

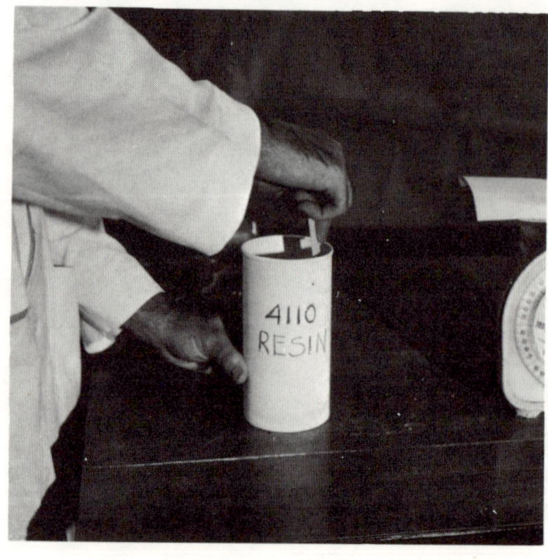

4. Mix pigment into the resin to match the patient's skin color. The amount of pigment required is different for the various colors, and is measured in terms of percent to the total weight of the resin to be colored. For caucasions add 2% of standard color pigment, for negroids add 1/2% of negroid pigment, and mix thoroughly into the resin. Occasionally a pure white color is desired. In this case use 2% white pigment and Lupersol DDM catalyst. The Luperco ATC or Garox BZP catalysts produce an unattractive yellowish white, and so should not be used with white pigment.

5. Cut the bottom out of a small paper cup for use as a funnel in the mouth of the P.V.A. bag to make it easier to pour in the resin.

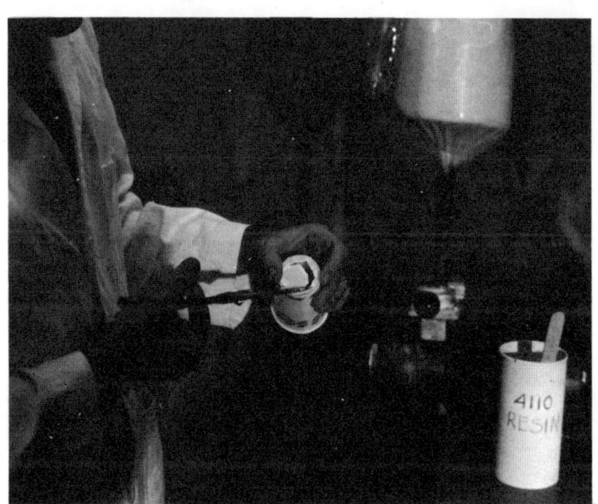

6. Tape the paper cup into the P.V.A. bag with plastic tape so it will not fall out accidently when the resin is being poured.

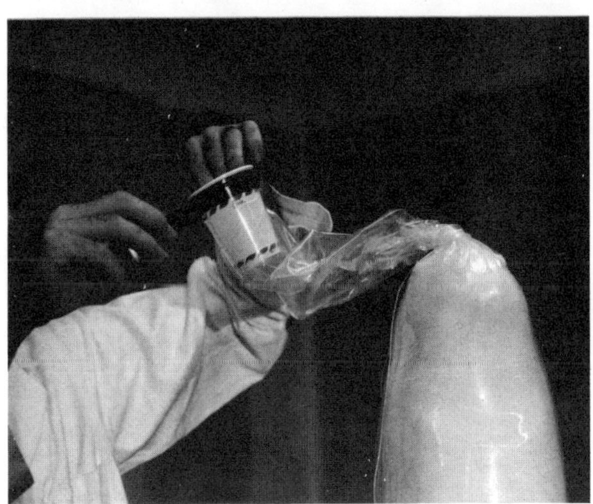

7. **Add** one half cubic centimeter of Naugatuck No. 3 promoter per each hundred grams of resin, and mix it well into the resin. This proportion of promoter gives you approximately twenty minutes of working time at 72° F. room temperature before the resin sets up. Warmer room temperatures speed up the reaction, cooler temperatures slow it down.

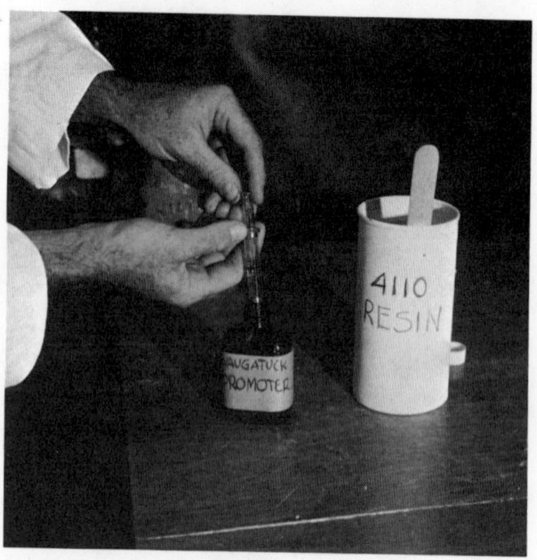

8. Loosen the knot in the cord that was used to tie off the neck of the P.V.A. bag and pour the resin into the bag through the paper cup. If the neck of the bag is not long enough to hold all the resin, hold some of it back and add it after part of the resin is worked into the laminate. Be careful to avoid getting resin on the hands or clothing, as it is somewhat toxic and may cause dermatitis. It is practically impossible to get it out of clothing. If some resin accidentally gets on the skin, wash it off as soon as possible with pure ethyl alcohol.

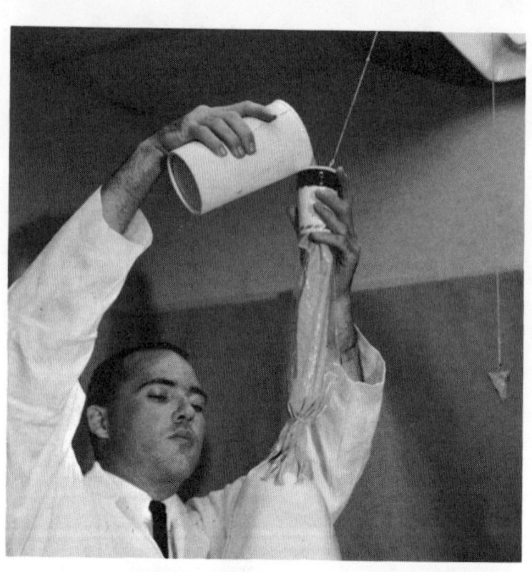

9. Grasp the neck of the P.V.A. bag and twist it in such a way that the resin cannot flow upward, and force it down into the stockinette and dacron felt.

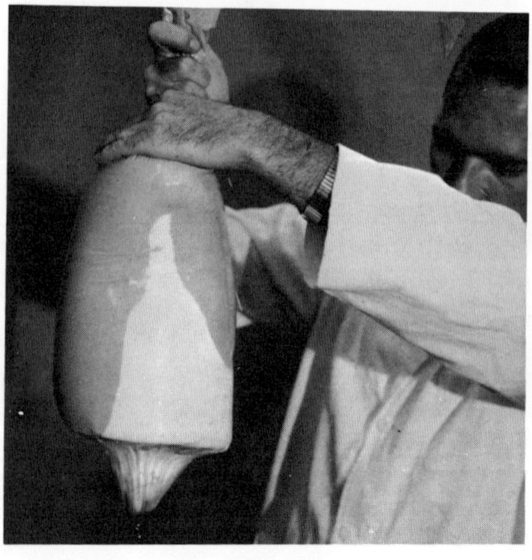

10. Distribute the resin evenly and thoroughly through all the layers of the layup, using a string as illustrated.

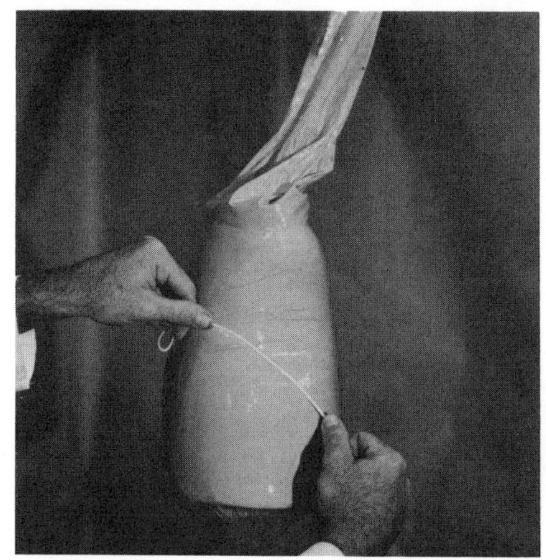

11. To make sure that the resin is thoroughly saturated into undercut and hollow areas, use a dowel or small rolling pin to work the resin thoroughly into the fibers of the layup material. This will prevent the formation of starved areas in the finished laminate.

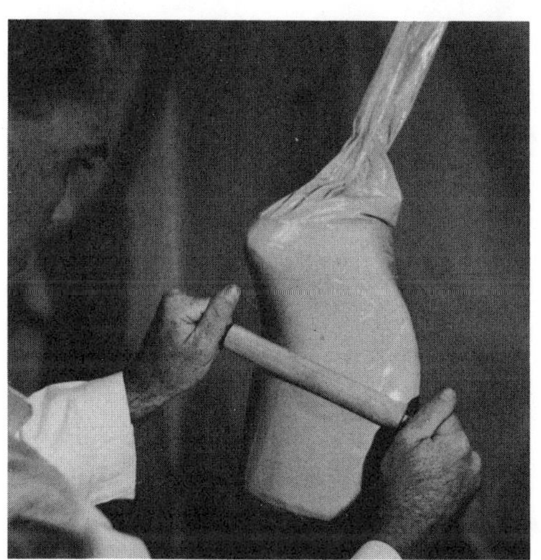

12. Firmly grasp the neck of the P.V.A. bag and pull it toward the medial side of the model that will be trimmed away later. This is to eliminate all wrinkles over the top of the shoulder area.

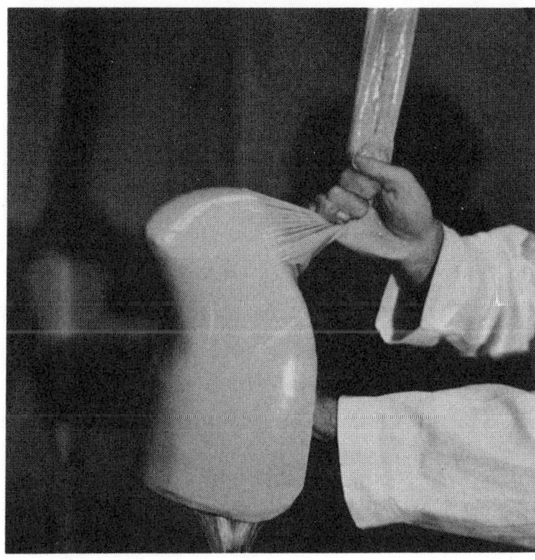

13. Hold the neck of the P.V.A. bag in place while applying plastic electrical tape to the trim line to hold the P.V.A. bag firm-ly in place and to help to contour the laminate.

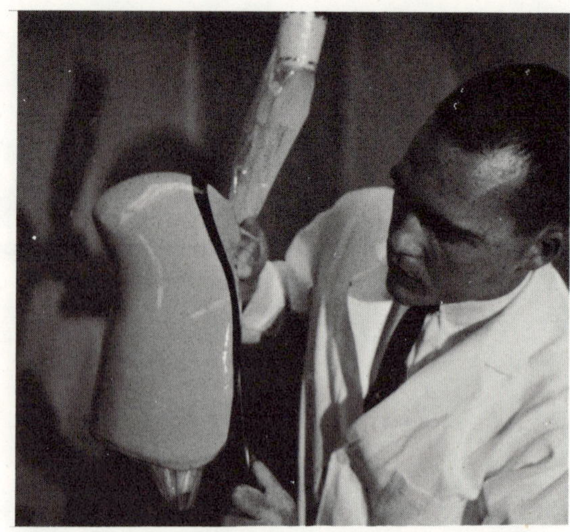

14. Put several turns of tape around the neck of the P.V.A. bag to hold it in the posi-tion to which it has been stretched.

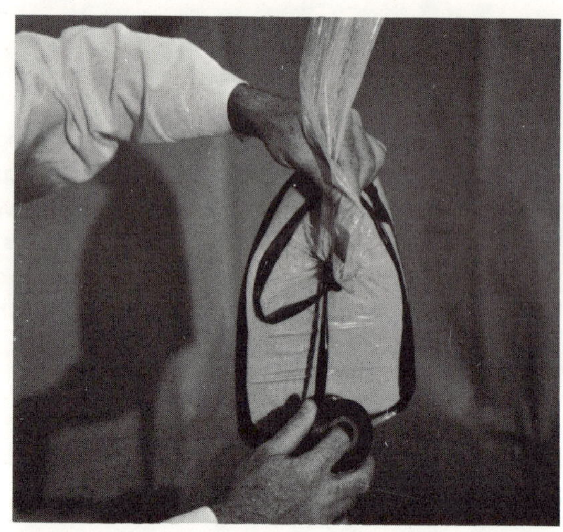

15. Do final stringing of the resin making sure that no bubbles of air are allowed to re-main and that the resin is thoroughly dis-tributed through the laminate. Continue this until the resin sets up. Be sure to check the vacuum occasionally to make sure it has not lowered, as loss of vacuum may cause the layup material to delam-inate or bridge across hollow areas.

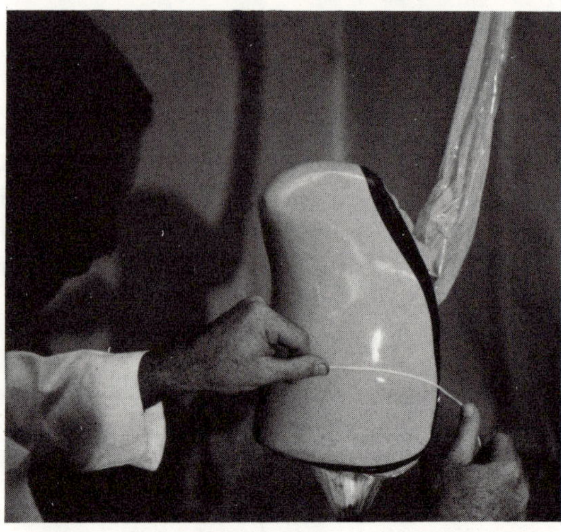

16. Mark the trim area for the arm hole in the shoulder cap with a skin pencil. Plan to remove just enough material for the patient to be able to get his arm through the hole.

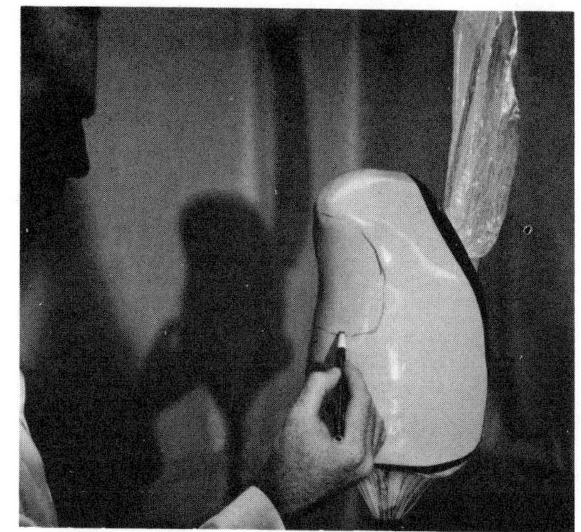

17. With a sharp knife cut out the lamination at the arm hole. This must be done while the plastic is still warm. If it is allowed to set completely, a knife will not cut It and it will be necessary to use an elec- trical cast cutter to do the trimming.

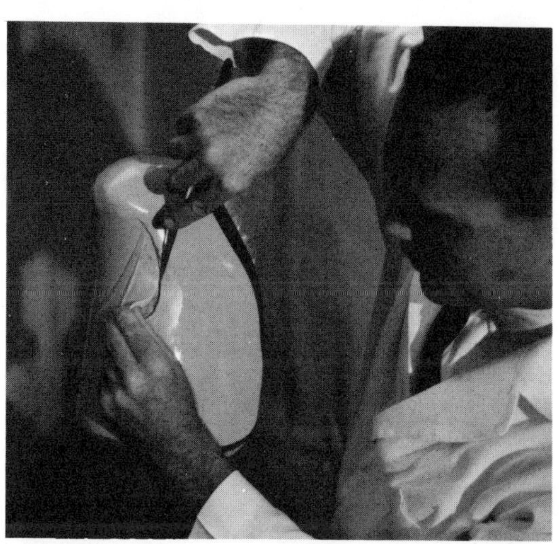

18. While the laminate is still warm trim along the outside edges until the entire shoulder cap can be removed.

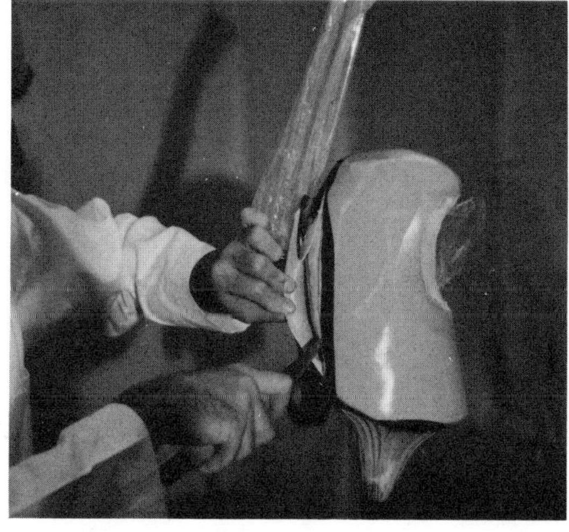

19. With a sharp knife open up the area over the lateral ends of the U-bar mounting bracket tubes enough to allow the entrance of a drill.

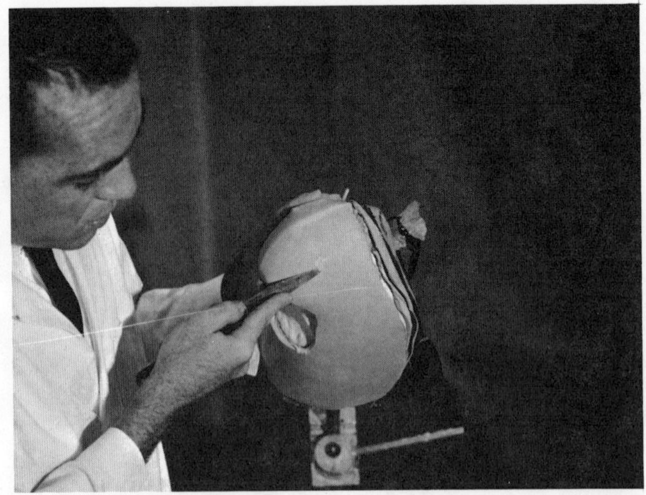

20. Remove the shoulder cap from the model and drill out the U-bar mounting bracket tubes with a 3/16 inch drill.

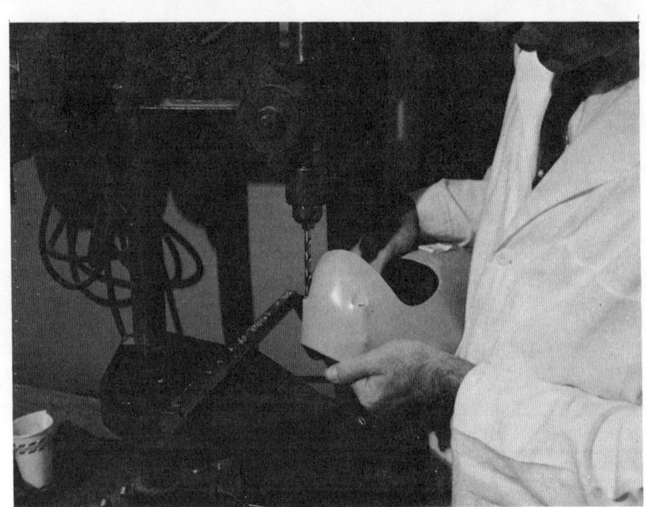

21. Place the shoulder cap on the patient's shoulder, hold it firmly in place with one hand, then mark the arm hole trim lines with a skin pencil. Plan to trim the arm hole so that the upper arm will have adequate clearance without rubbing against the edges of the shoulder cap.

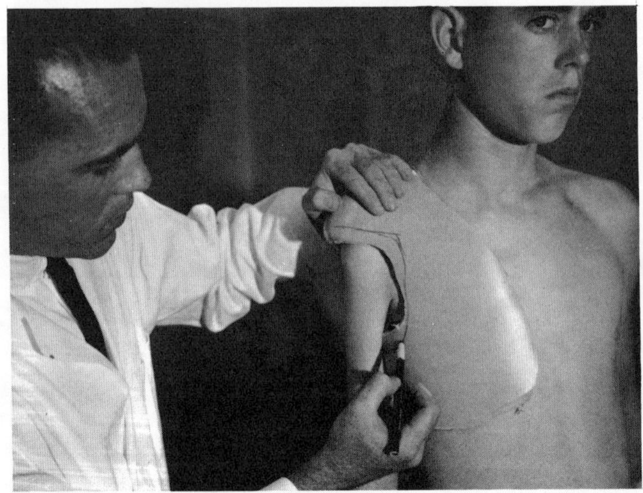

22. Holding the shoulder cap in firm contact with the patient's shoulder, mark the trim line around the neck so that the edges of the cap will not come in contact with any sensitive tissues in that area.

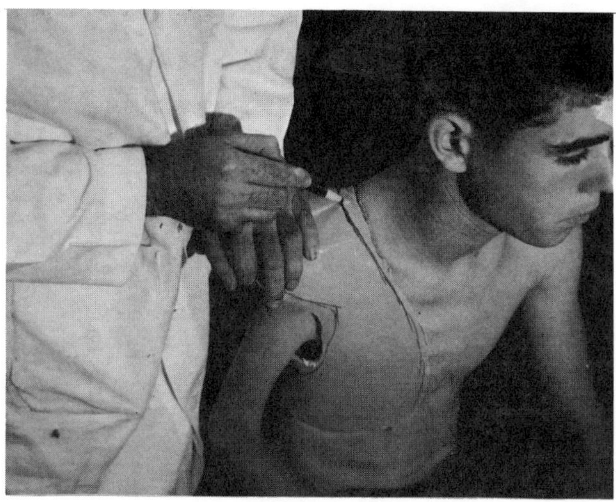

23. In the case of the long shoulder cap with weight bearing on the crest of the ilium, it is necessary to trim and fit the lower portion of the cap so that it fits comfortably and snugly in the pelvic region. Mark the trim line so that motion will not be restricted and the edges of the cap will not cut into sensitive tissues.

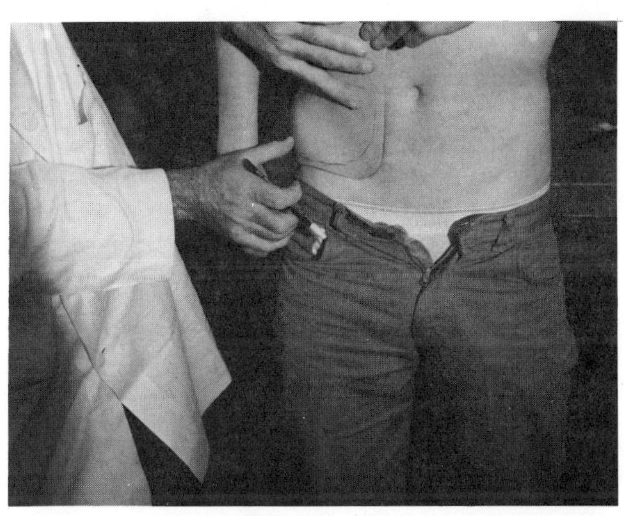

24. Place the shoulder cap back on the model and use an electrical cast cutter (Stryker or equivalent) to trim the shoulder cap to trim lines made in the previous steps.

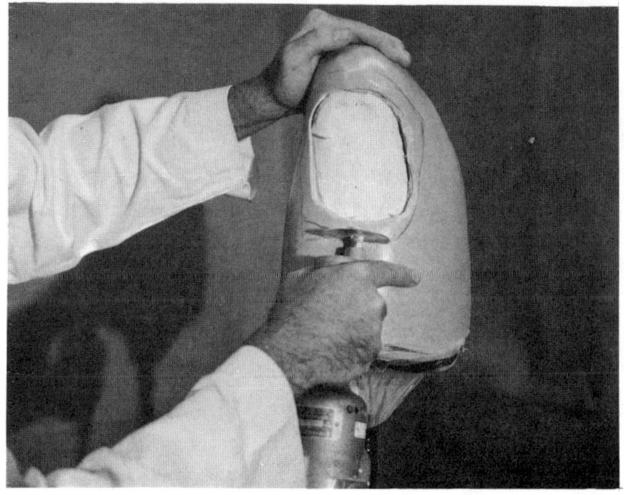

25. Smooth and round all edges of the shoulder cap on a cone sander or similar piece of equipment, or by hand-sanding if power equipment is not available.

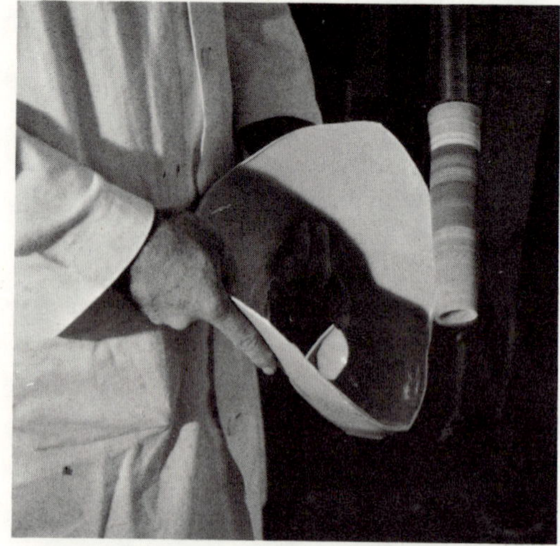

26. Put the shoulder cap in place on the model over the "Kemblo" and leather liner, and mark the trim line on the liner with a skin pencil.

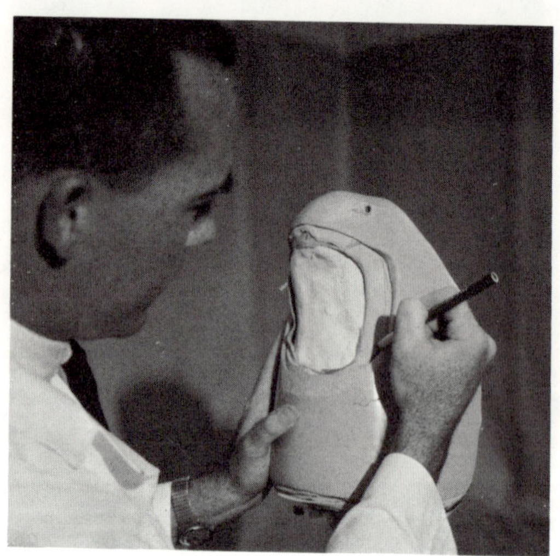

27. Remove the liner and use a pair of shears to trim it to the marks made in the previous step.

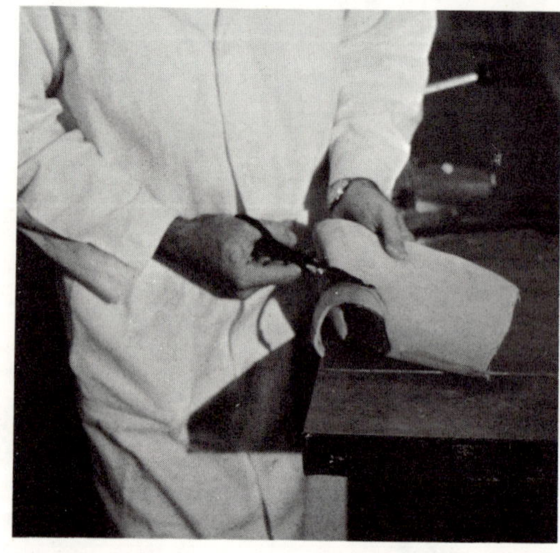

28. With the liner in place in the shoulder cap, try it on the patient and make certain that it neither restricts his motions nor cuts into sensitive areas and fits comfortably even with considerable downward pressure on the shoulder.

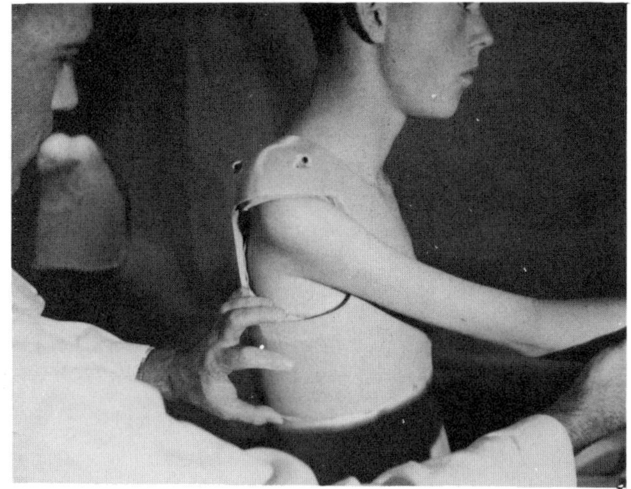

29. Remove the liner from the shoulder cap and skive the edges of the "Kemblo" to a thin edge using a wire wheel brush.

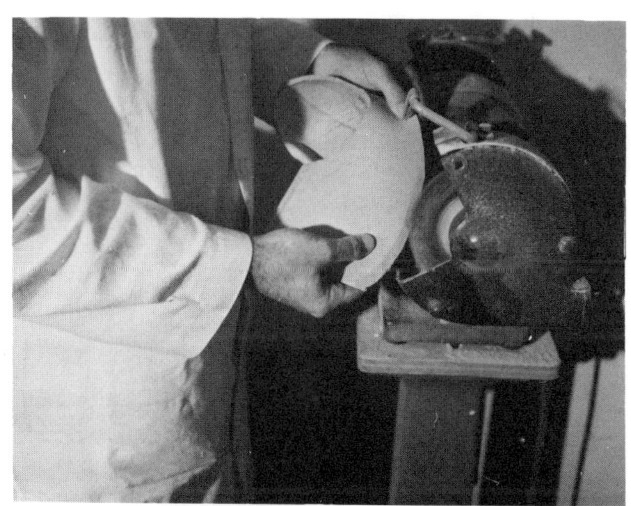

30. Secure the liner into the shoulder cap with rubber cement. Apply a coat of rubber cement to both the liner and the shoulder cap, allow it to dry, then press the two firmly together.

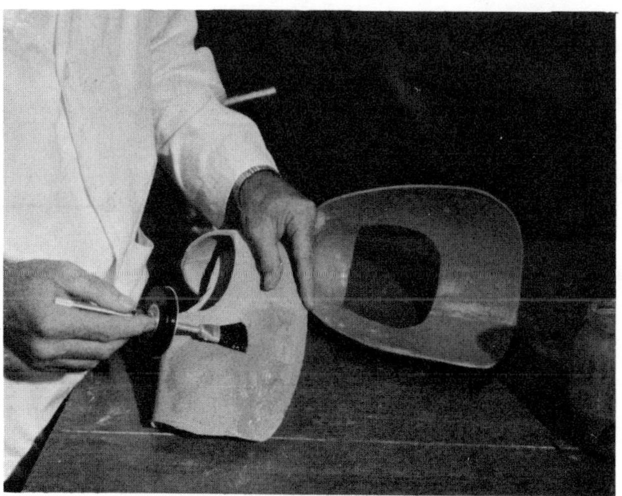

31. Trim the leather with a sharp knife so that it is even with the edges of the shoulder cap all around.

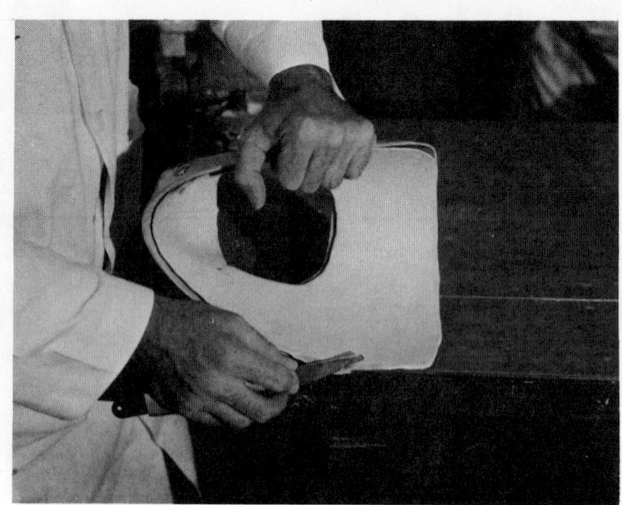

HOW TO MAKE AND ATTACH A CHEST STRAP TO THE SHOULDER CAP

1. Mark the shoulder cap on the anterior aspect opposite the axilla, about two inches in from the edge.

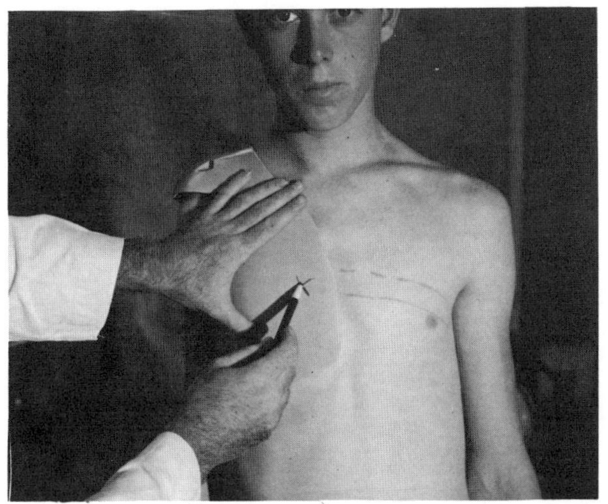

2. Mark the posterior aspect of the shoulder cap opposite the axilla in the same manner as in the previous step.

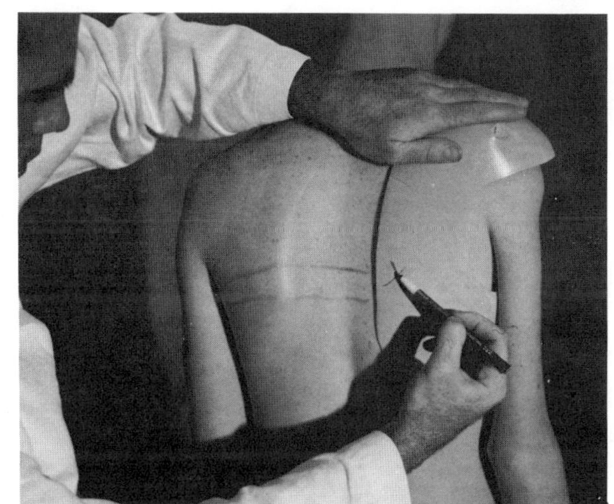

3. Drill Number 18 holes in the shoulder cap at the points marked in the two preceding steps.

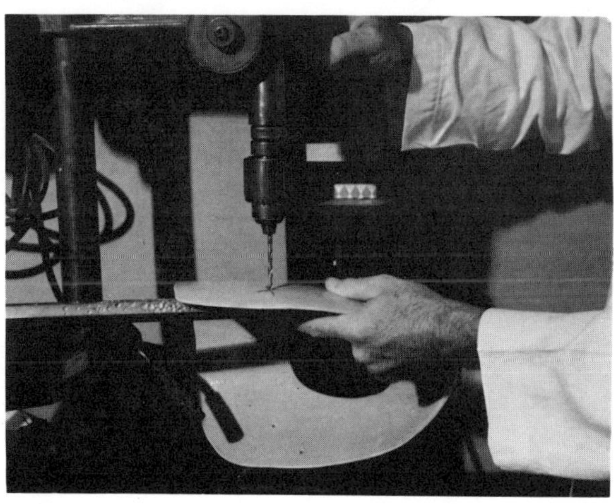

4. Sew a billet of one inch dacron webbing onto a one inch truss hook and rivet it into place on the shoulder cap at the anterior location, using a Number 48-29 "Speedy" rivet.

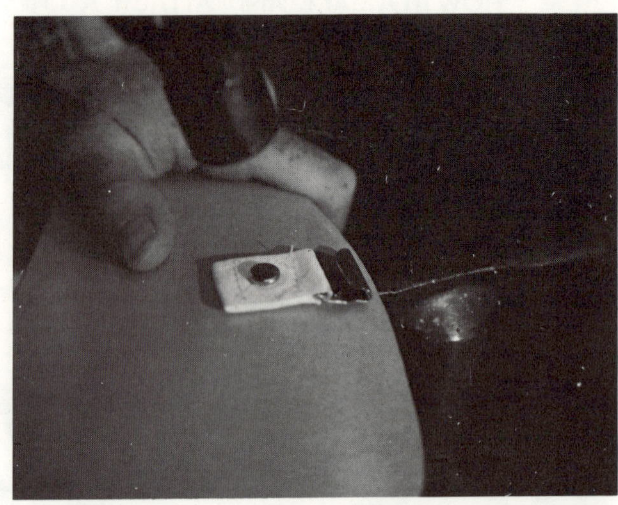

5. Double one end of a piece of one inch dacron webbing about one inch from the end. Punch a 3/16 inch hole into the center of the doubled webbing, and rivet it to the posterior aspect of the shoulder cap in the hole provided for it.   Use the same type of rivet as in the preceding step.

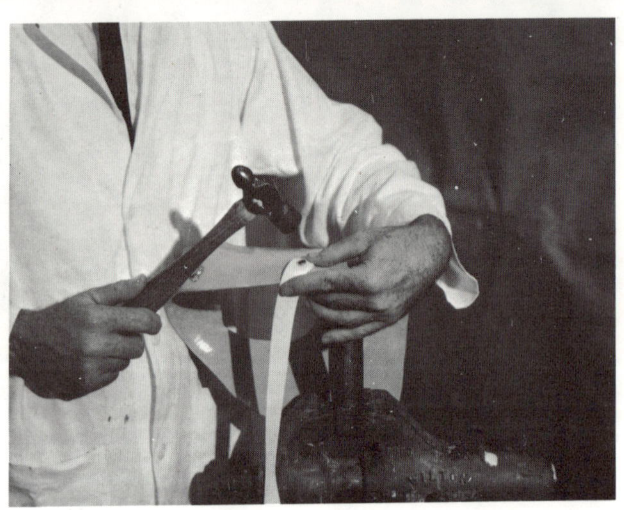

6. Fit the shoulder cap onto the patient, bring the chest strap around his body and fasten it through a one inch two-prong safety buckle of the type that will clip onto the truss hook.

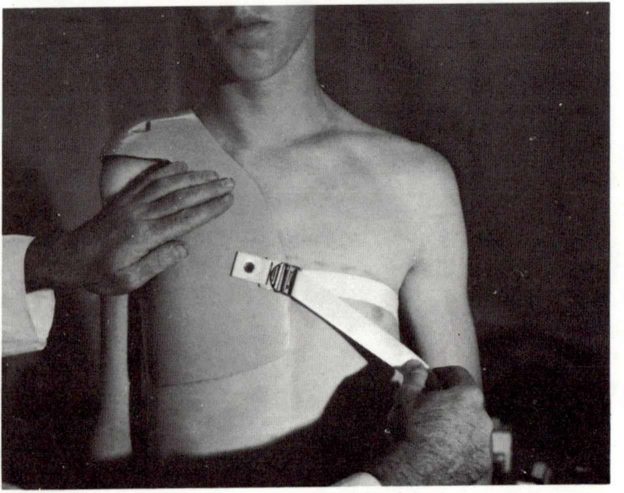

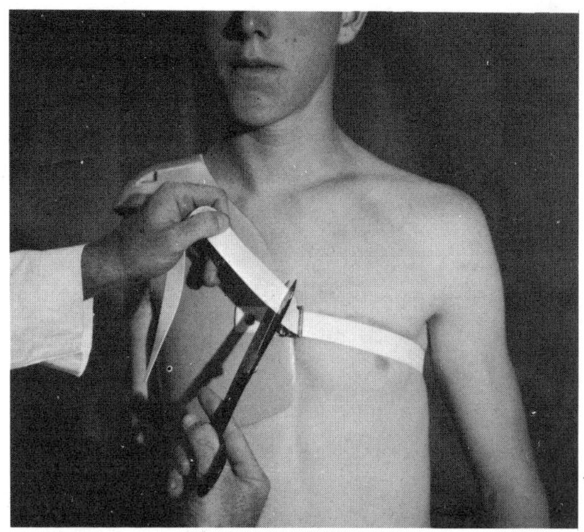

7. Adjust the chest strap to the correct length on the patient and cut off the surplus webbing.

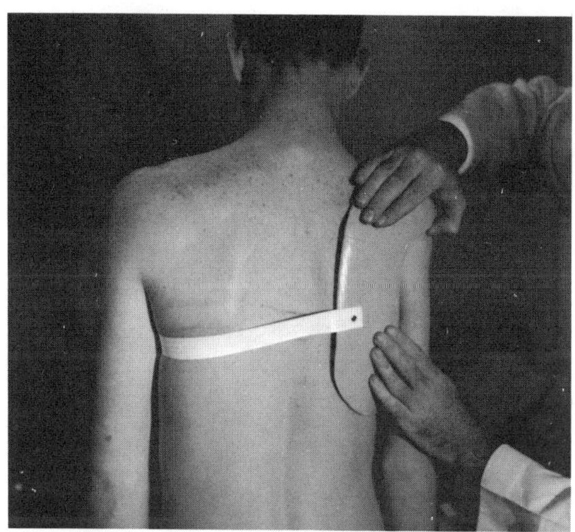

8. Check the fit of the chest strap across the patient's back making sure it does not cut into the flesh, and that it holds the shoulder cap in alignment on the patient's shoulder.

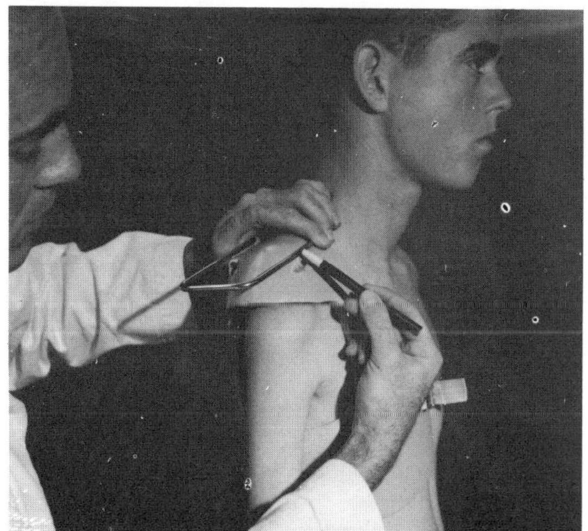

9. Hold the U-bar in place on the shoulder cap above the mounting bracket tubes and adjust it so the end of the U-bar does not extend more than 3/4 of an inch beyond the lateral edge of the shoulder cap. Mark the legs of the U-bar about one inch back from the openings to the tubes, so when the U-bar legs are cut off they will be inserted into them at least one inch.

10. Saw off the legs of the U-bar at the points marked. Remove the burrs from the cut ends.

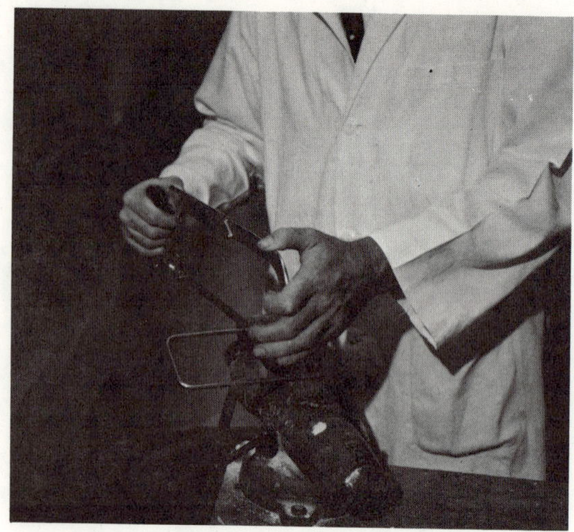

11. Insert the U-bar into the U-bar mounting bracket. It should fit firmly and snugly and not work out easily.

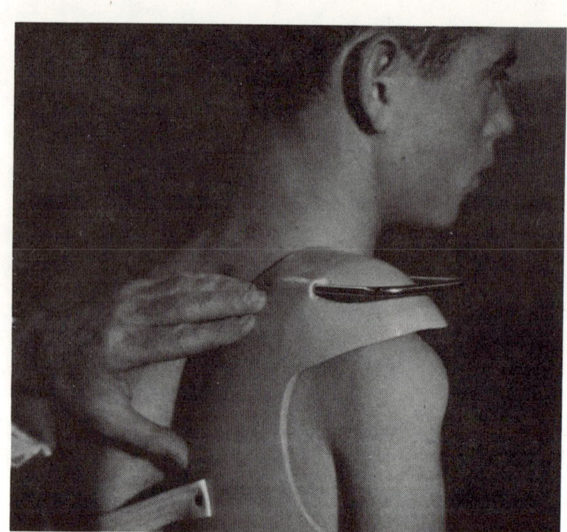

HOW TO FIT A SHOULDER FLEXION ASSIST ON THE SHOULDER CAP

Introduction

The shoulder flexion assist fitted to a shoulder cap has been found helpful in those cases where the patient has good shoulder extensors but his flexors are too weak to provide useful function. The device consists of a non-locking or "free" joint fastened by a clamp to the U-bar on the shoulder cap. An extension strap attached to the lower strap of the joint extends about two-thirds down the patient's arm. A plastic half-cuff is riveted to the lower strap so it can exert pressure on the posterior aspect of the arm. Rubber bands are looped over band posts on the clamp and on the strap so they will be stretched when the shoulder is extended. When the rubber bands contract, they provide energy to assist the patient to flex his shoulder.

The shoulder flexion assist is the simplest of the functional arm braces, and has been well-accepted by a number of patients with good elbow function and either normal grasp or prehension assistance provided by a hand splint. The shoulder flexion assist enables these patients to place the hand to the mouth, and other useful positions, which is appreciated even by many unilaterally involved individuals.

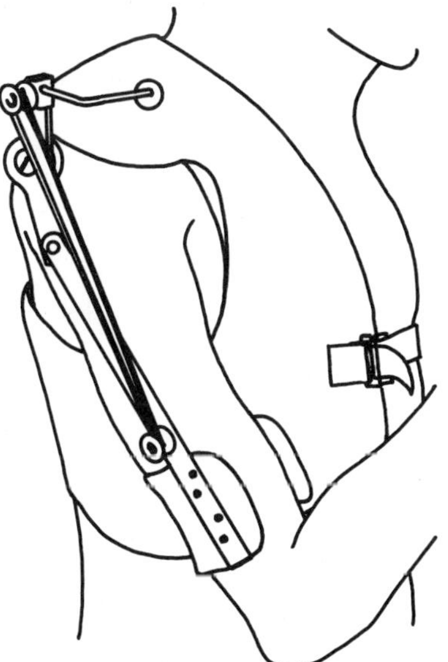

Figure 104. Shoulder Flexion Assist

<u>Parts Used</u>

The parts used in fitting the shoulder flexion assist are illustrated in the photograph at right. From left to right, they are the FAB-13 plastic cuff with attaching rivets; the FAB-24 extension strap and attaching screws; the FAB-19 free hinge with stop pin and band post. A third band post is provided to be fastened to the extension strap just above the plastic cuff.

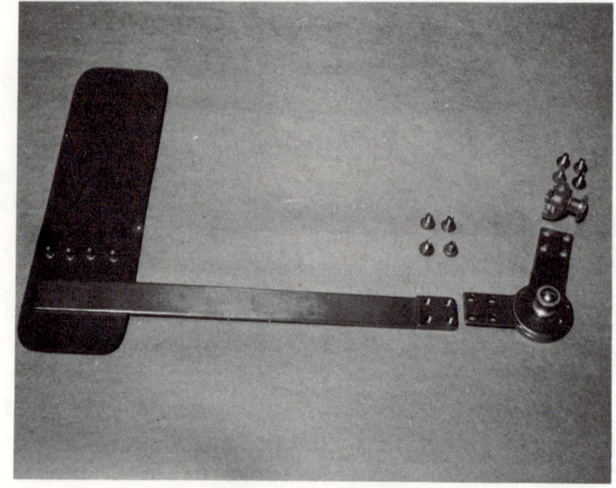

The free hinge is made of stainless steel, with an oilite bearing and teflon washers to keep friction to a minimum.

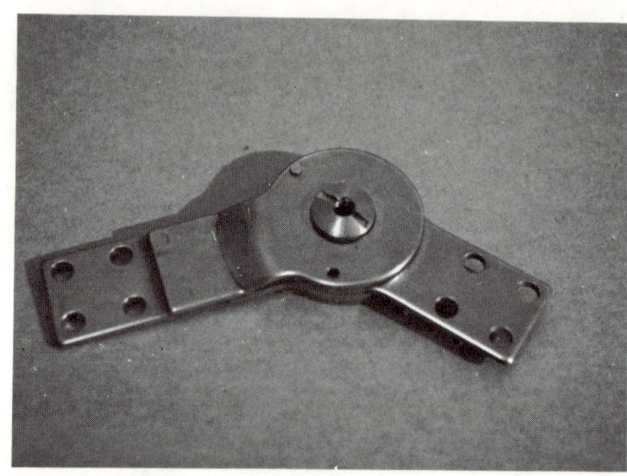

The band post is turned out of aluminum. The thick flange spaces the post out 1/8 inch to avoid the rubber bands interfering with the other parts of the brace. The rubber bands loop around the posts, making the older type hooks unnecessary.

The clamp is aluminum, with four 8-32 threaded holes for the clamp screws. The clamp is milled to fit 1/4 inch rod in one slot, 3/16 inch in the other.

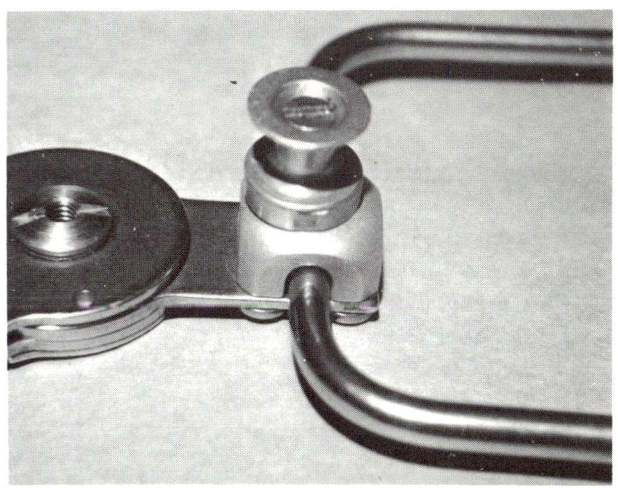

The hinge is assembled on the U-bar with the clamp and band post as illustrated.

Procedure

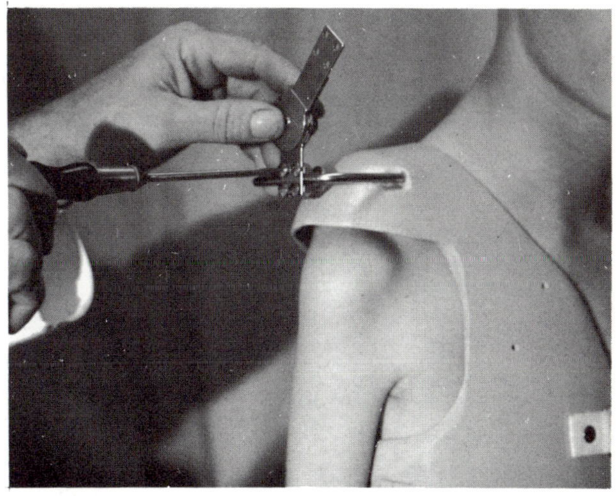

1. Assemble the free hinge onto the U-bar.

2. Assemble the band posts on the hinge and clamp.

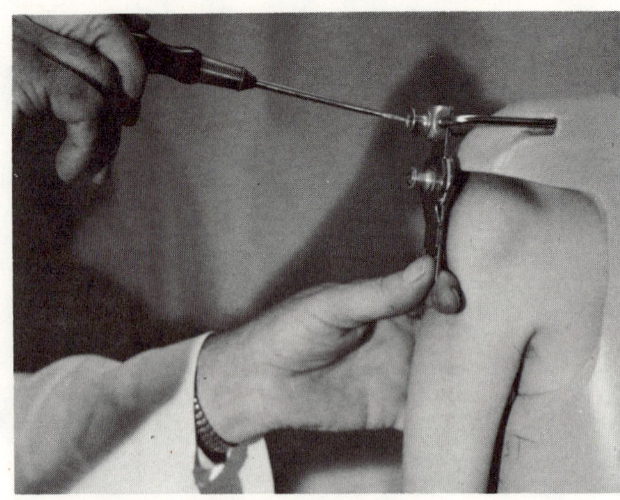

3. Shape the plastic cuff. Heat the Royalite plastic in an oven until it is pliable, then shape it into a half-cuff of correct diameter to fit the posterior portion of the patient's arm just above the fold at the elbow. A wooden form facilitates shaping the cuff quickly and accurately.

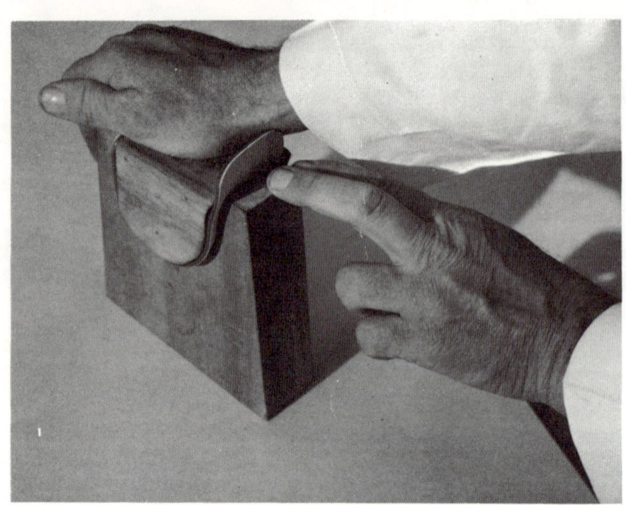

4. With the cuff in place, mark the extension strap even with the lower edge of the cuff.

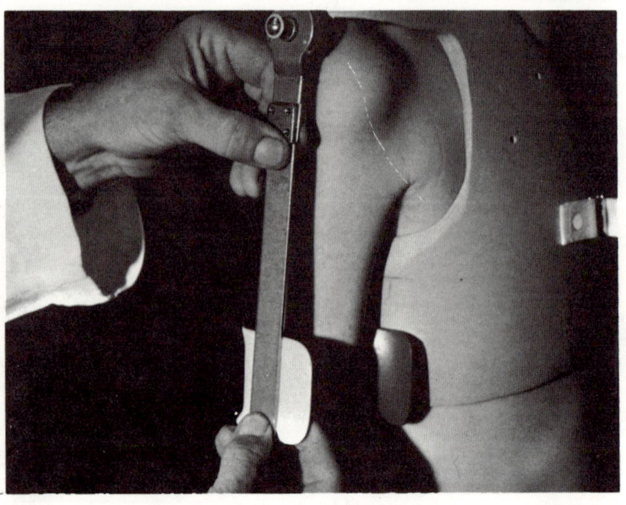

5. Drill four Number 30 holes through the strap and cuff while they are clamped together, then rivet them with 1/8 inch aluminum rivets.

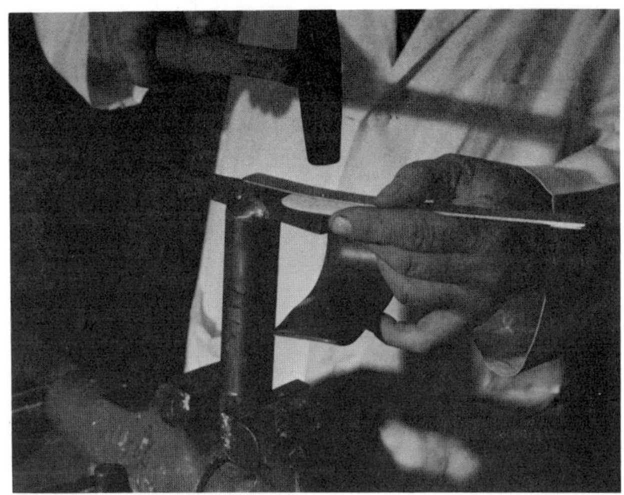

6. Saw off the excess length of strap below the cuff, then smooth the cut edge and round the corners.

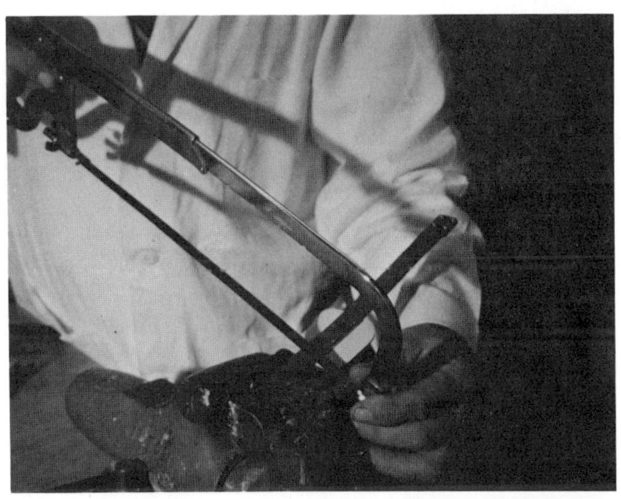

7. Trim the excess material from the cuff if necessary.

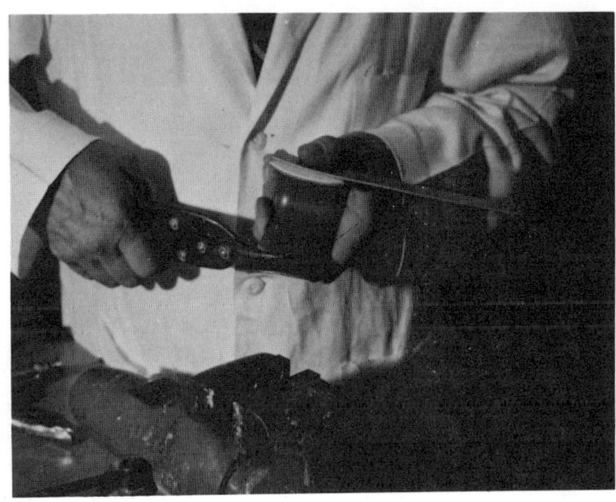

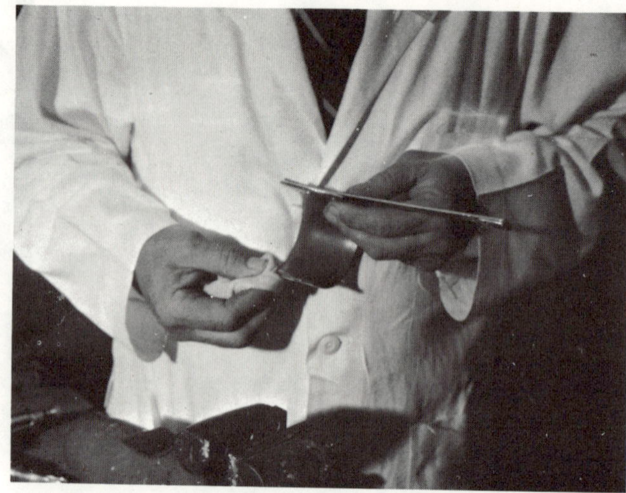

8. Polish the edges of the cuff with a rag and a little solvent.

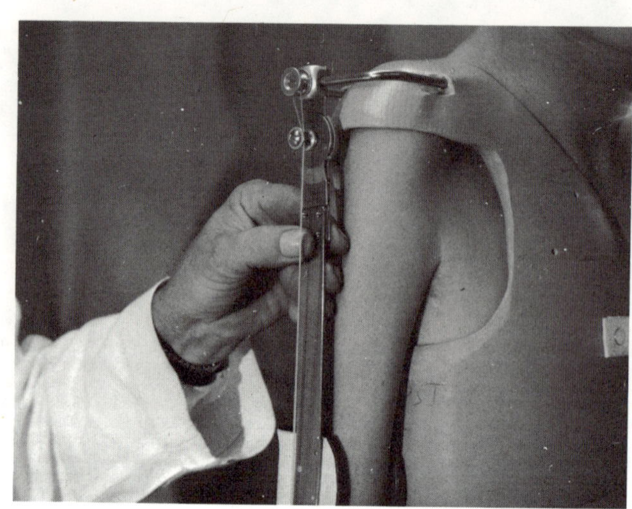

9. Hold the strap in position on the hinge, then stretch a rubber band from the top band post and determine the location of the distal band post. It must be placed back far enough so the rubber bands bend over the hinge band post when the arm is hanging relaxed. In this position the pull of the rubber bands is all exerted lengthwise on the strap, and the patient does not have to tire his muscles keeping them from flexing his shoulder when he does not want it flexed. Mark the location of the distal band post.

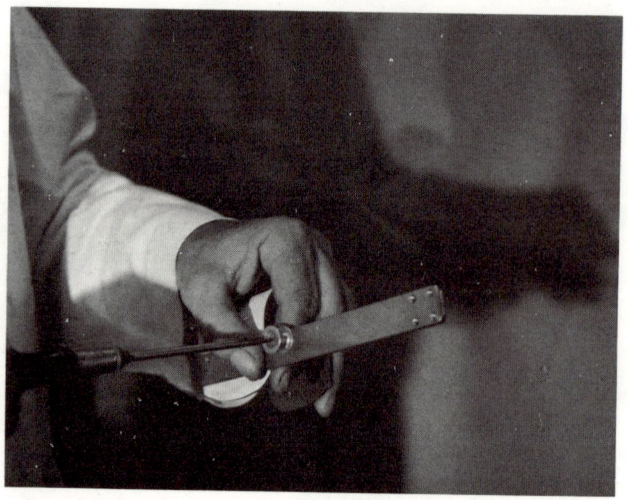

10. Drill a Number 29 hole and tap for an 8-32 screw in the strap at the point marked in the preceding step. Assemble the band post to the strap.

11. Assemble the strap and cuff to the hinge.

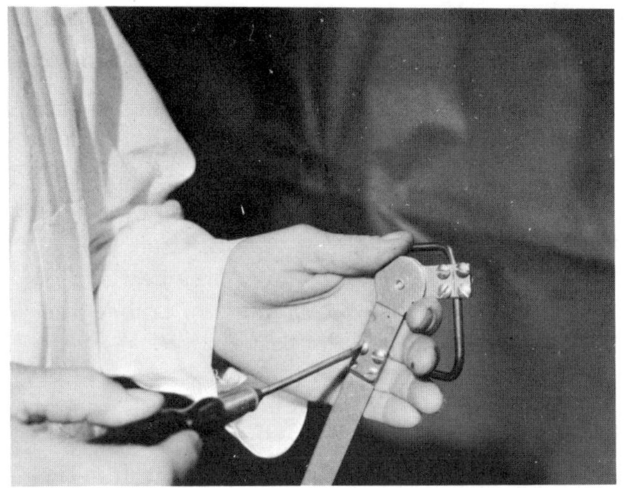

12. Loop Number 32 rubber bands onto the band posts, starting at the top post, bending around the distal post, then back up, looping the end of the band around the top post.

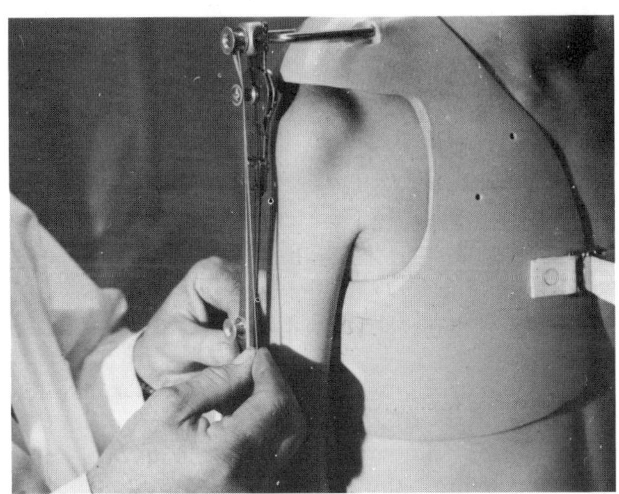

13. Continue to add rubber bands until the patient has a functional balance between his ability to extend his shoulder against the tension of the rubber bands and the amount of assist he needs to flex his shoulder. Use just enough rubber bands to obtain maximum possible function.

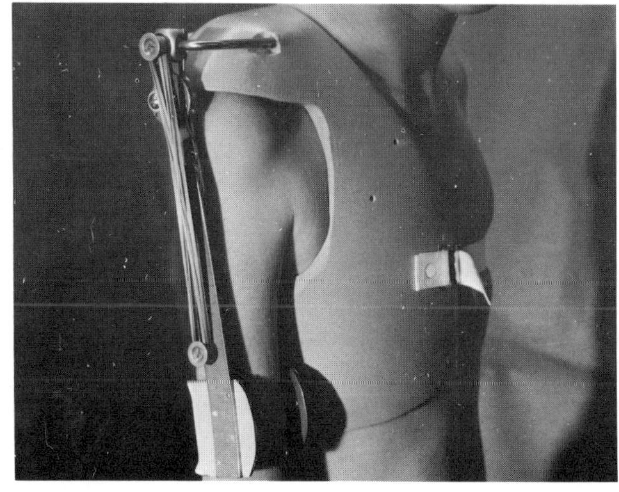

HOW TO FIT A LOCKING JOINT FUNCTIONAL ARM BRACE WITH ELASTIC FLEXION ASSISTS

Introduction

Ratchet shoulder and locking elbow joints are the basic elements of a versatile functional arm brace that can be adapted to provide function for patients with various combinations and degrees of shoulder and elbow joint dysfunction.

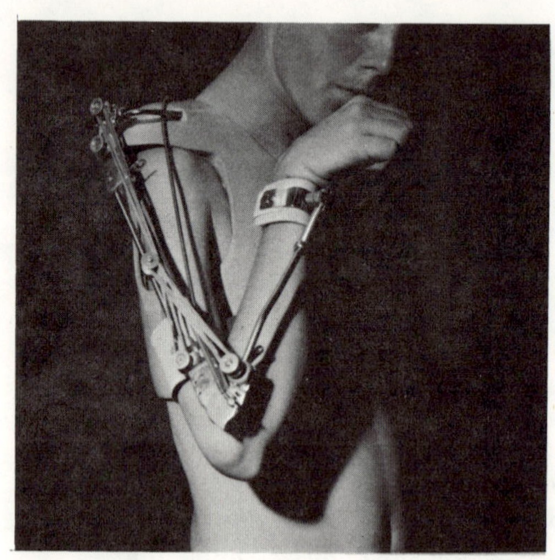

This device is called the "Locking Joint Functional Arm Brace". Elastic flexion assists on the locking joint functional arm brace furnish a relatively simple solution to the problem of providing shoulder and elbow flexion in those cases where the extensors are strong enough to simultaneously extend the joints and stretch the rubber bands. In those cases where the extensors are too weak to stretch rubber bands, outside power is necessary. The application of the artificial muscle, powered by carbon dioxide gas to furnish outside power, will be explained in a later section. A suitable hand splint may be used in combination with the arm brace to provide grasp if hand function is impaired.

Locks are provided on both shoulder and elbow joints to relieve the patient's weakened muscles of the strain of holding the joints in the flexed position. Even with the elastic assists it is very difficult to carry loads like a coat or a cafeteria tray without the locks to provide stability. Used with a plastic shoulder cap, the locking joint functional arm brace provides the arm paralytic with greater stability and more precise function and control than he has been able to obtain in the past.

Parts Used

Before starting to learn how to fit the locking joint functional arm brace, it is first necessary to become familiar with the names and characteristics of the parts used.

Figure 105 on page 285 illustrates the parts that make up the typical locking joint functional arm brace. To avoid confusion the part numbers and nomenclature used by the parts manufacturer, A. J. Hosmer Corp., are used consistently. Each part is named and described, starting with the U–Bar Mounting Bracket and proceeding distally to the cuff strap. The brace illustrated in the drawing is for a left arm, and when ordering, it is customary to specify either left or right. However, the same parts can be assembled either way, as will be explained later.

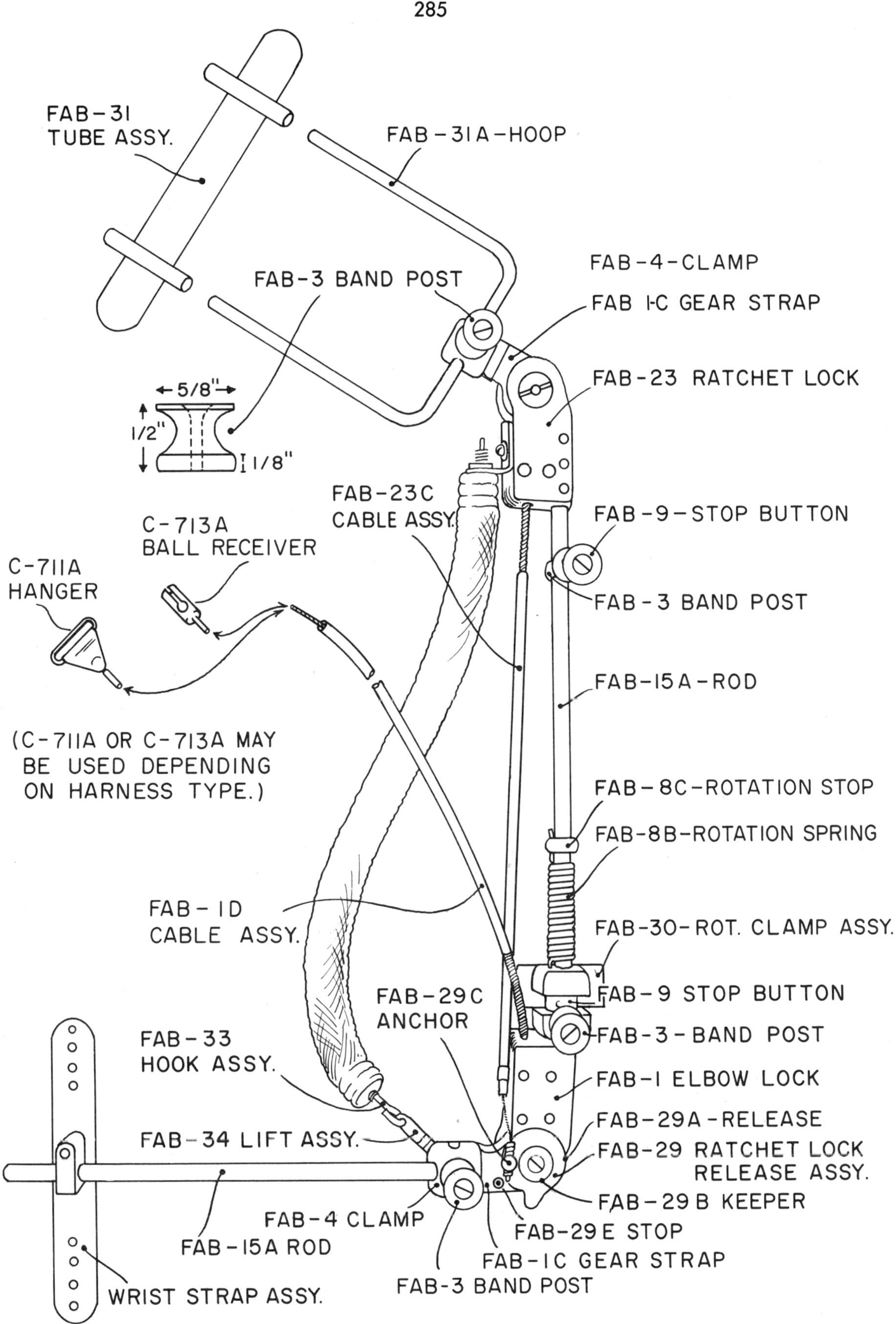

Figure 105. Locking Joint Functional Arm Brace Parts

FAB-31   U-Bar Mounting Bracket, a flat piece of 3/4 x 5-1/2 inch stainless steel with two pieces of 1-1/2 inch stainless steel tubing silver soldered to it at right angles to its long axis on 3 inch centers. The inside dimension of the tubing is 3/16 inch. It will be recalled that the U-bar mounting bracket was shaped to the contour of the plaster model of the shoulder, and laminated into the shoulder cap in an earlier procedure. The parallel metal tubes on the bracket open flush with the surface of the shoulder cap, and provide a convenient means for "plugging in" the U-bar, onto which the rest of the brace is mounted. The friction of the legs of the U-bar in the tubes is all that is needed to keep them in place in ordinary use.

FAB-31A   U-Bar, a horse-shoe shaped piece of 3/16 inch stainless steel rod. The legs are 5 inches long and spaced 3 inches apart to match the spacing of the tubes on the bracket. The legs are shortened to fit when the U-bar is assembled.

FAB-1C   Gear Strap, a lever which pivots in the FAB-23 Ratchet Lock and serves to attach it to the U-bar.

FAB-4   Clamp, a square piece of aluminum held to the gear strap with four screws. When the screws are tightened, the clamp grips the U-bar between itself and the gear strap.

FAB-3   Proximal Shoulder Band Post attached to the center of the clamp, a spool-shaped aluminum fitting that serves as the upper anchor for the rubber bands that provide the shoulder flexion assist. (See Figure 106.) All band posts are FAB-3's, as the parts are the same although they are used in various locations on the brace. FAB-23, the ratchet lock, includes a bracket provided for mounting the proximal end of the artificial muscle when the latter is used on the brace. It may be discarded when elastic flexion assist only is used.

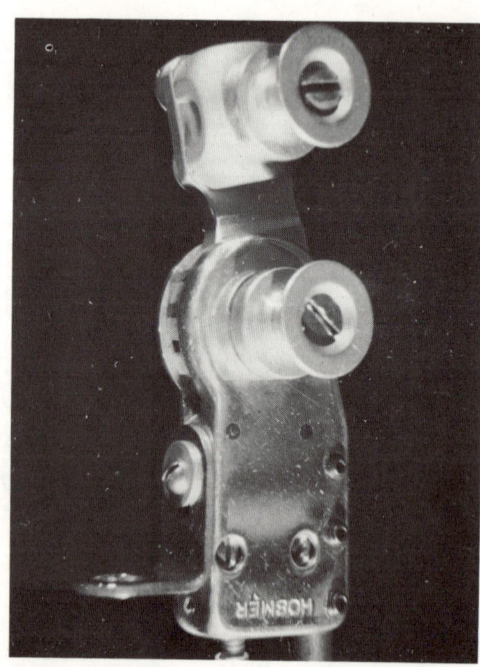

Figure 106. Ratchet Lock with Band Posts

FAB-23   Ratchet Lock, serves as the shoulder joint for the brace. (See Figure 106.) It is so designed that the ratchet action of the lock allows the joint to progressively flex through five stops, a range of 110 degrees, which is adequate for shoulder flexion. Figure 107 illustrates the ratchet mechanism in the joint. The body of the joint, containing the spring-loaded pawl, revolves counter-clockwise around the gear strap, as the pawl clicks into each stop in succession. The right shoulder on the pawl in the illustration is at an angle on one side, allowing it to ratchet counter-clockwise past the teeth on the gear. The other shoulder is parallel to the teeth, so the pawl locks the gear against clockwise movement. To unlock the joint, the pawl is pulled out of the stop by the control cable, allowing the shoulder to extend from the action of gravity on the arm. The control cable is linked to the FAB-29 Ratchet Lock Release Assembly on the FAB-1 Elbow Lock so that when the elbow is extended the control cable is pulled, and the shoulder ratchet lock is released, allowing the shoulder to extend.

Figure 107.   Ratchet Lock

FAB-9   Stop Button, with Distal Shoulder Band Post, is clamped to the FAB-15A Rod (Figure 108). The rod, of 1/4 inch stainless steel, connects the ratchet lock to the rotation clamp assembly, which in turn is connected to the elbow lock. The stop button and distal shoulder band post assembly is fastened to the rod just below the ratchet lock. Rubber bands are fastened between the proximal and distal shoulder band posts to provide shoulder flexion assist action.

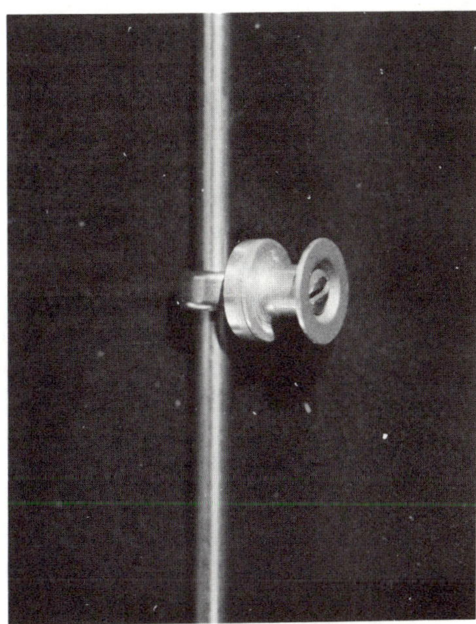

Figure 108. Stop Button with Shoulder Distal Band Post

FAB-8A   Spring Anchor Ferrule, and FAB-8B, Assist Spring, (Figure 109) fit onto the FAB-15A Rod just above the FAB-30 Rotation Clamp Assembly. The purpose of the assist spring in this location is to act as a humeral rotation assist. The assist springs are wound left or right hand depending on whether external or internal rotation assist is wanted. In the illustration, Figure 109, the spring on the left is wound right hand, the one on the right is wound left hand. A left-hand wound spring provides external humeral rotation assist on the left arm or internal rotation assist on the right arm. The action of a right-hand wound spring is the opposite.

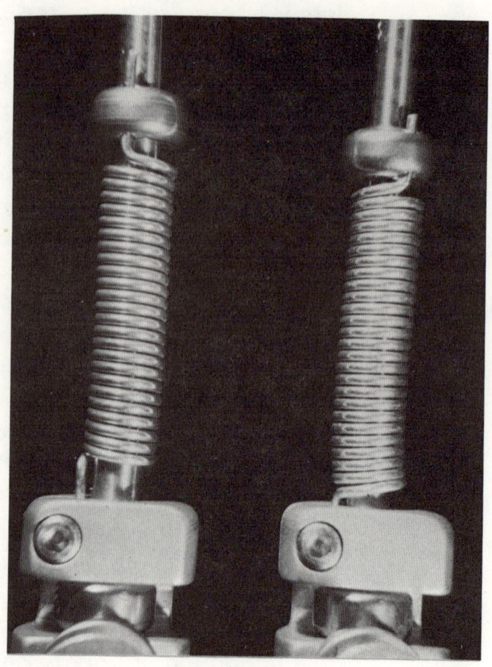

Figure 109. Spring Anchor Ferrule and Assist Spring

FAB-30   Rotation Clamp Assembly, is a "U"-shaped aluminum casting that is fastened to the distal end of the FAB-15A Rod (Figure 110). At first glance the rod appears to go completely through the rotation clamp assembly, but this is not the case, as it terminates in the slot of the "U" where it is retained by an FAB-9 Stop Button. The arm of the "U" through which it passes is slotted and provided with an adjustment screw which makes it possible to lock the rotation clamp on the rod, allow it to rotate freely, or provide various degrees of friction. When a humeral rotation assist spring is used, the adjustment is set to allow free rotation. The proximal elbow band post is fastened to the rotation clamp assembly, and the clamp assembly itself is attached to the elbow lock by a short

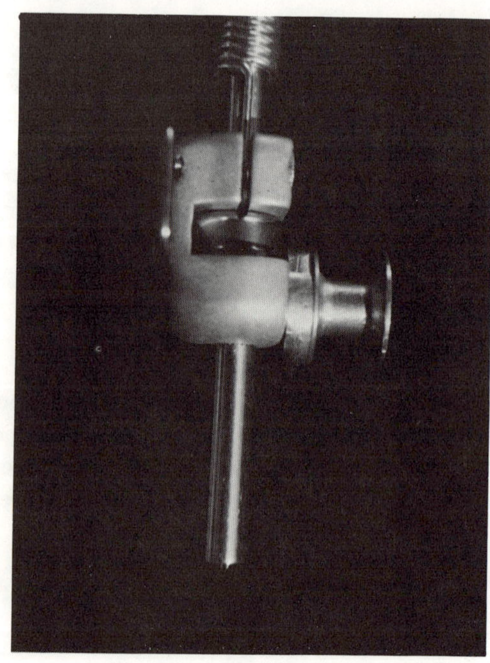

Figure 110. Rotation Clamp Assembly

rod that is a splined press fit into the clamp casting (Figure 111). Note that one side of this short splined rod is milled flat on the end opposite the splines. This end of the rod inserts into the FAB-1 Elbow Lock and is held in place by three set screws that bear on this flat surface. The orientation of this short rod when it is pressed into the rotation clamp assembly determines the position of the latter in relation to the elbow lock, and since there is a 180 degree difference in the location of the short rod in the rotation clamp as between left and right hand units, making this change is the chief problem in adapting a left hand unit to a right, and vice versa, as will be discussed later.

The FAB-13 Humeral Cuff (Figure 112) is also attached to the clamp casting, either directly or with the small stainless steel reinforcing plate provided. The FAB-13 humeral cuff is supplied by the factory as an unformed piece of Royalite plastic 8 x 2-1/4 x 1/8 inches in size. (See Figure 112.) When heated in an oven the material can be easily shaped to fit the patient's arm. When it cools it will retain its new shape. It can be reheated and reshaped without damaging it in any way.

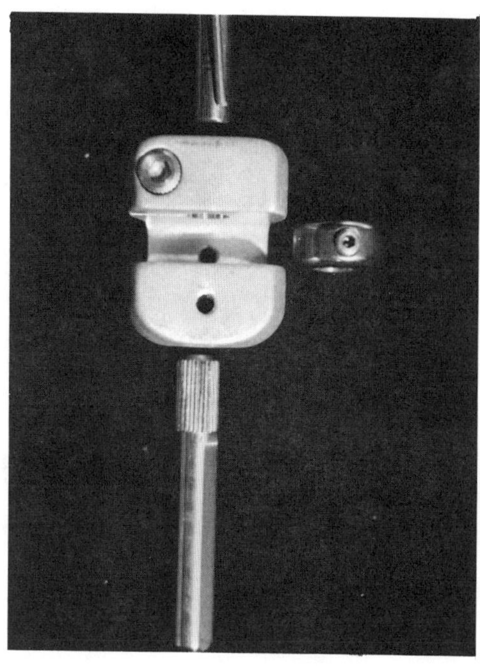

Figure 111.

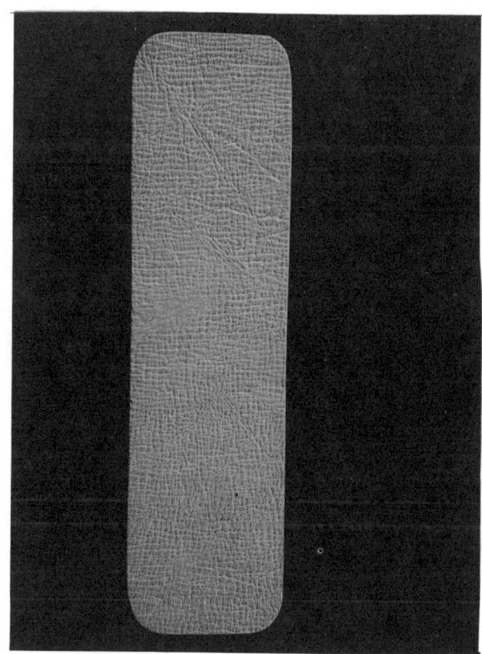

Figure 112.

FAB-1   Elbow Lock (Figure 113) is a locking elbow joint with a locking mechanism that may be operated as an alternator (pull-release to lock, pull-release again to unlock), or as a spring lock (pull and hold to unlock, release and relax to lock). When the unit is used with elastic flexion assists the alternator is satisfactory, but when used with outside power it must be changed over to spring lock operation to prevent the artificial muscle from exerting its power against a locked joint, which may cause the muscle to blow out. The changeover is simple and will be explained in the description of the lock mechanism. The FAB-29 Ratchet Lock Release Assembly, a stainless steel disc with a cam cut into its edge, is fastened to the joint center bolt with a screw and FAB-29B nylon keeper that serves as a bearing. An FAB-29C Anchor is riveted with a free fit into a countersunk hole in the disc so it can swivel as the latter rotates. The anchor is threaded for a bushing through which the control cable from the shoulder ratchet lock is passed. A small brass ring is soldered to the end of the cable so the bushing will pull the cable when the ratchet lock release rotates. This movement is caused by the FAB-29E Stop on the gear strap contacting the cam on the lock release assembly when the elbow is extended. The FAB-29E stop is a split pin driven into a hole in the FAB-1C gear strap of the elbow lock, so when the elbow is extended the stop drives the lock release counter-clockwise, the anchor and bushing pull the cable, and the shoulder ratchet is unlocked. The timing of the shoulder ratchet unlocking is adjusted by screwing the bushing up or down in the anchor. The FAB-34 Lift Assembly, the lever fastened to the end of the gear strap, is used as the distal attachment point for the artificial muscle when it is installed on the brace. The FAB-3 Distal Elbow band post and FAB-4 clamp are fastened to the gear strap. An FAB-15A rod extends from the clamp to the wrist unit or hand splint. Rubber bands are looped between the elbow proximal and distal band posts to provide an elbow flexion assist.

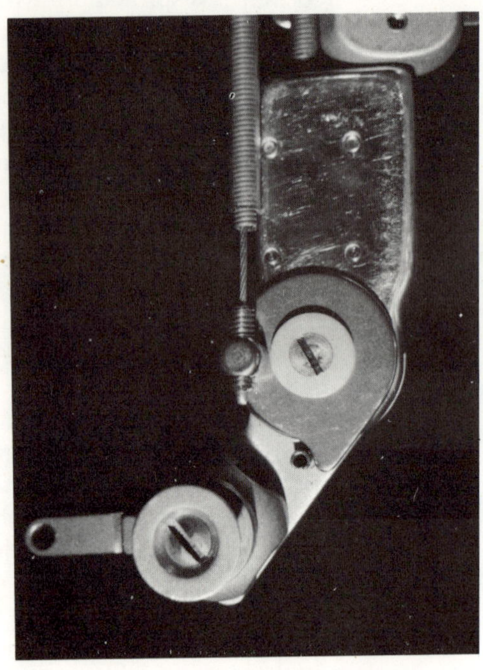

Figure 113. Elbow Lock

The elbow lock comes from the factory with an FAB-1D cable assembly to operate its lock mechanism. A dacron harness is worn by the patient to enable him to operate the cable to lock and unlock the elbow (Figure 114). The locking mechanism when locked is illustrated in Figure 115. The locking bar (A) is forced by spring pressure into the first tooth space in the gear sector, and the driver (B) and keeper (C) are in locked position. Pulling the control cable causes the mechanism to unlock as in Figure 116. The locking bar (A) is clear of the tooth space in the gear sector, and is held in the position by the action of the keeper (C) on the driver (B). The cable can be relaxed and the lock will remain unlocked, until the cable is again pulled to re-lock it. This action forces the keeper off the shoulder of the driver, and the springs force the driver and locking bar into one of the tooth spaces, re-locking the joint, when the cable is released. This kind of lock is called an alternator, as it requires that the control cable be alternately pulled and released to lock it again.

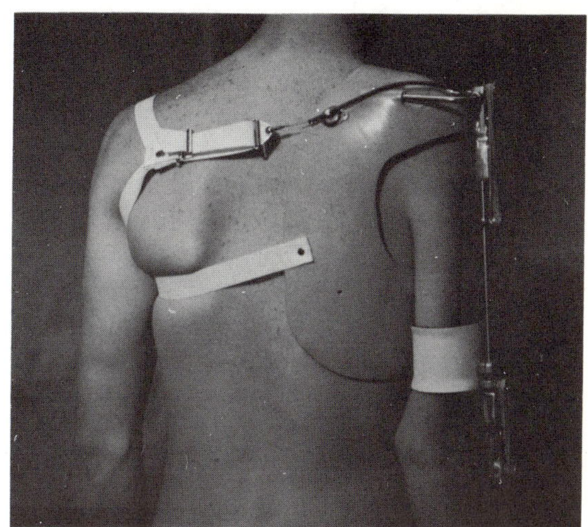

Figure 114.

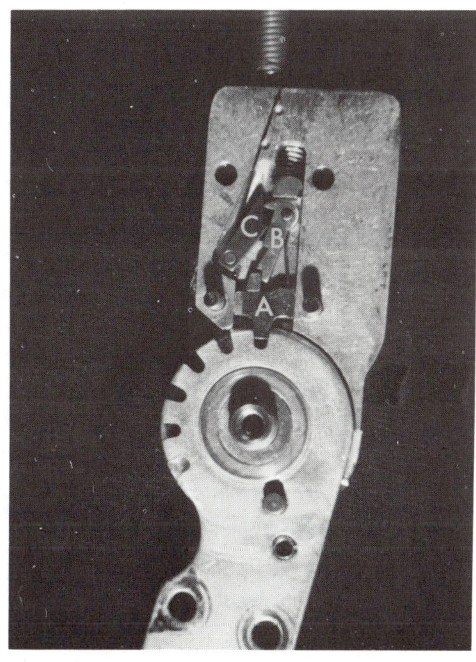

Figure 115.

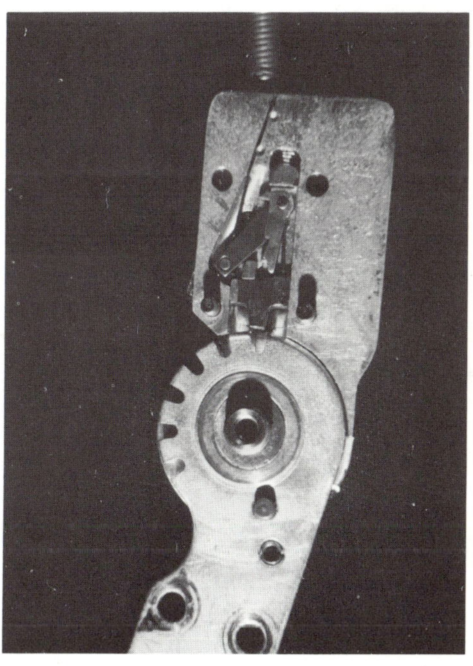

Figure 116.

To change the FAB-1 elbow lock to a simple
spring lock (pull-hold to unlock, release-relax
to lock) it is necessary to remove the two case
screws, the hinge screw and the cover from the
elbow lock, then remove the keeper ("C" in
Figures 115 and 116). The lock now appears as
in Figure 117, with the driver (B) forcing the
locking bar (A) into the first tooth space in
the gear sector. When the control cable is
pulled, as shown in Figure 118, the locking bar
is withdrawn from the tooth space in the gear
sector, leaving it free to rotate. However, in
the absence of the keeper, the locking bar will
enter any adjacent tooth space when the cable
is released and thus again lock the joint. As
mentioned earlier, the locking elbow must be
converted to a simple spring lock when the artificial muscle is used for elbow flexion, as the
muscle is controlled by a valve operated by the
same harness that operates the elbow lock.
Every time the patient uses the harness to open
the valve that feeds carbon dioxide gas to the
artificial muscle, the same movement that opens
the valve must unlock the elbow. If the elbow
is locked when the valve opens, the gas pressure
builds up in the artificial muscle, causing it to
suffer serious damage. With an alternator lock
on the elbow this would happen every other time
the patient used his artificial muscle, making it
necessary to change the lock to the simple spring
type. When this is done, each time the cable is
pulled and the valve is opened, the elbow is unlocked. When the cable is released and the
valve is closed, the elbow locks again, which
is the desirable action when using the artificial
muscle on a functional arm brace.

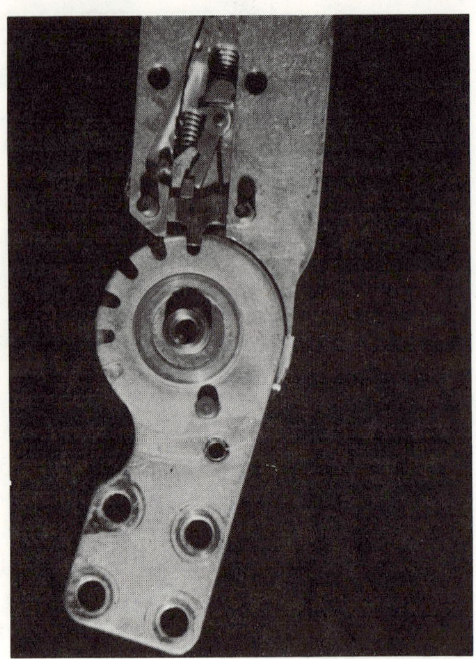

Figure 117.

Figure 118.

FAB-12   Cuff Strap (Figure 119) is a stainless steel strip that is shaped to fit the patient's wrist and lined with flexible plastic leather, or felt. It is fastened to an FAB-7 Friction Wrist Control with a screw and bushing so it can swivel freely. The friction wrist control may be so adjusted that it is locked onto the rod, or it may be adjusted to permit passive friction control of wrist supination and pronation. Active assistance to either supination or pronation may be provided by the same type of assist springs used for humeral rotation assist (Figure 120). To accomplish this, the friction wrist control is adjusted to rotate freely on the rod, and is prevented from sliding off by an FAB-8C Rotation Stop Button, which clamps onto the rod with a set screw and limits the range of supination or pronation. One end of the FAB-8B Assist Spring fits into a hole in the FAB-8A Spring Anchor Ferrule which clamps firmly to the rod with a set screw. In operation, the cuff strap is attached to the patient's wrist and to the friction wrist control which is free to swivel on the rod within the limitations imposed on it by the assist spring and the rotation stop. If the mechanism illustrated in Figure 120 was part of a brace on a patient's right arm, it would serve as a pronation

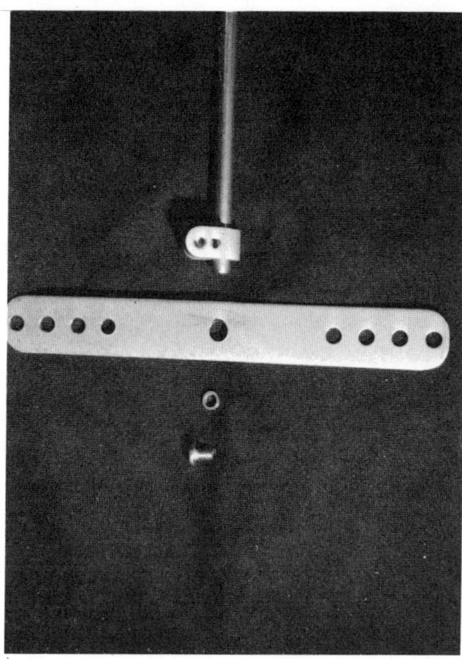

Figure 119. Cuff Strap

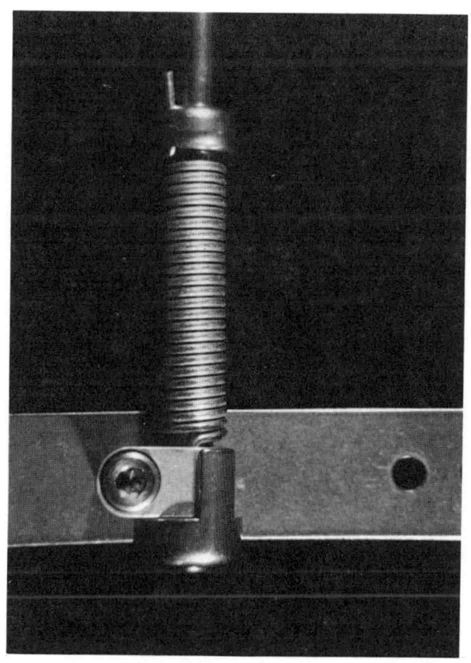

Figure 120.

assist, because supinating the wrist would wind up the spring, storing energy in it. When released, the energy would assist the patient to pronate his wrist. On the left arm, the action would be reversed, and the device would be a supination assist. Figure 121 illustrates the same mechanism with the spring wound right hand instead of left to provide supination assist on the right arm, pronation assist on the left.

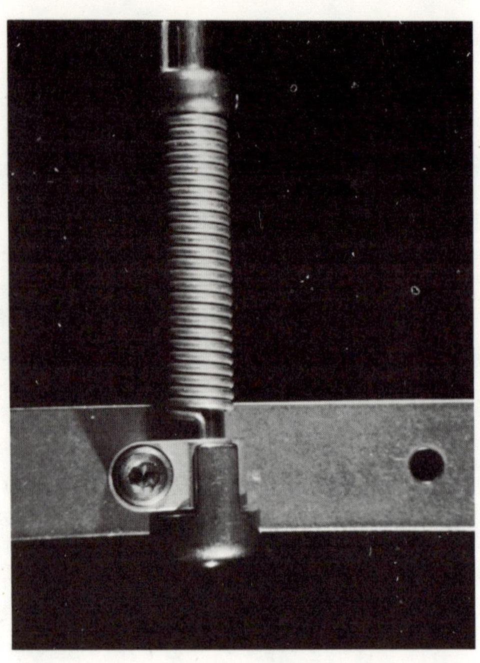

Figure 121.

FAB-1D   Cable Assembly operates the elbow lock. It is long enough to extend over the shoulder to the harness, and is made up of a stainless steel cable operating in a housing of the same material, covered with a plastic sheath to protect the patient's clothing. The cable is cut to the correct length for the patient, and a C-714A Hanger (Figure 122) is swaged onto it so it can be fastened to the harness. (A C-173 A Ball Receiver is substituted for the C-711A Hanger to connect to the control valve when the artificial muscle is used.)

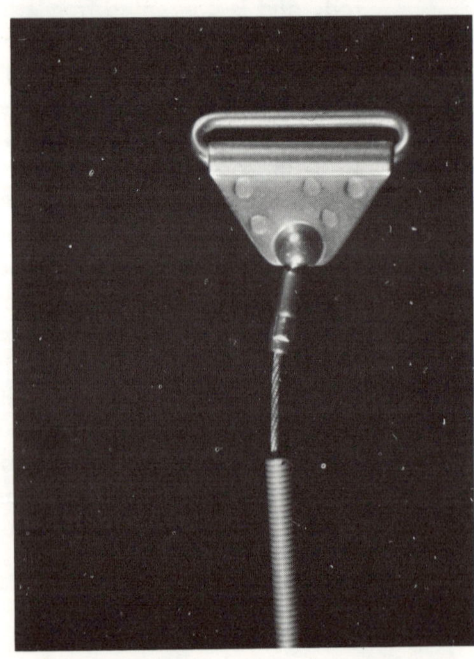

Figure 122. Cable Assembly

Changing Parts to Adapt Right Hand Brace to Left, and Vice Versa

The clamp and both band posts are removed and mounted on the opposite side of the FAB-23 ratchet lock (Figure 123). In the illustrations, the assembly shown to the left is for the left arm, the one on the right is for a right arm. The rotation clamp assembly must be reversed (Figure 124), but simply rotating the rod in the hole in the elbow lock case will not work, as it has a flat milled on it to take the three set screws. The short splined rod must be pressed out and reinstalled in a position exactly 180 degrees from its original position. This is difficult to do without a press and may result in breakage of the rotation clamp casting. If proper equipment is not available it is best to substitute a correctly oriented factory-assembled unit.

The FAB-29 ratchet lock release assembly is sent from the factory with the anchor unassembled, to make it easy to put the unit together for either a right or left brace. To avoid confusion, it is best to try the lock release assembly on the elbow joint before attaching the anchor. Make certain that the disc is on the lock so the flat of the cam on its edge contacts the FAB-29E stop as shown in the illustration. If the stop is on the wrong side of the gear strap, press or drive it through to the other side. The anchor hole must be countersunk on the side away from the anchor, then the anchor pin is inserted and riveted into the countersunk space. Rivet it just enough to secure it without binding the swivel action.

The elbow band posts, clamp, and lift assembly must be assembled on the same side of the elbow lock as the lock release assembly.

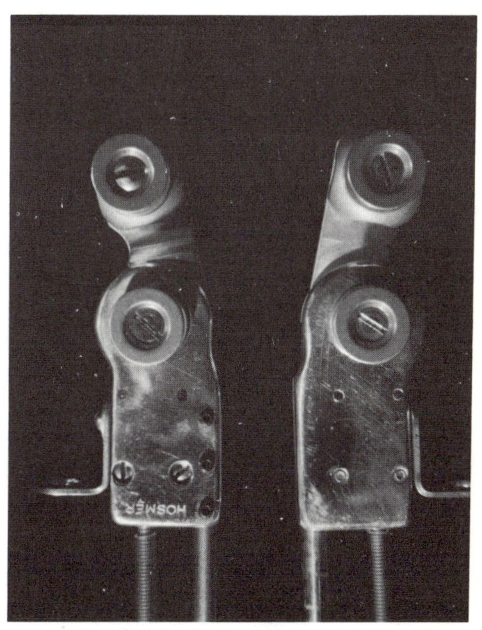

Figure 123.

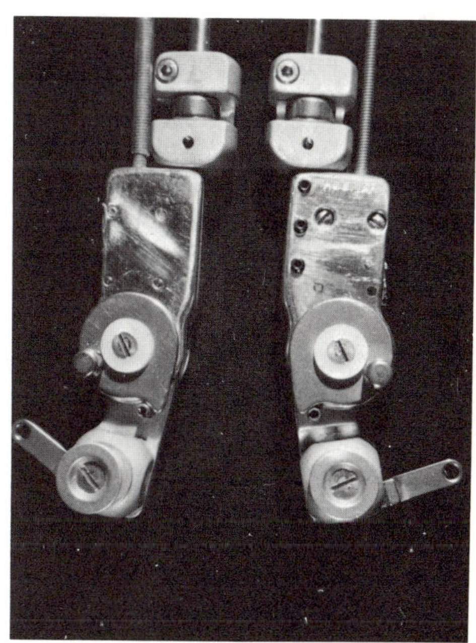

Figure 124.

Fitting Procedure

1. The Locking Joint Functional Arm Brace comes from the factory as a kit with the necessary parts for fitting the brace with elastic flexion assists, alone or in combination with an artificial muscle. Assemble the ratchet lock onto the U-bar. The clamp must be inside, or on the side next to the patient's arm, the band post in the upper posterior screw hole. The clamp is assembled with the small groove on the U-bar if the brace is to be clamped tight, with the large groove if it is to be free to abduct. Stop buttons may be used on either side of the clamp to prevent the latter from sliding on the U-bar in those cases where the clamp is to be set for passive humeral abduction control.

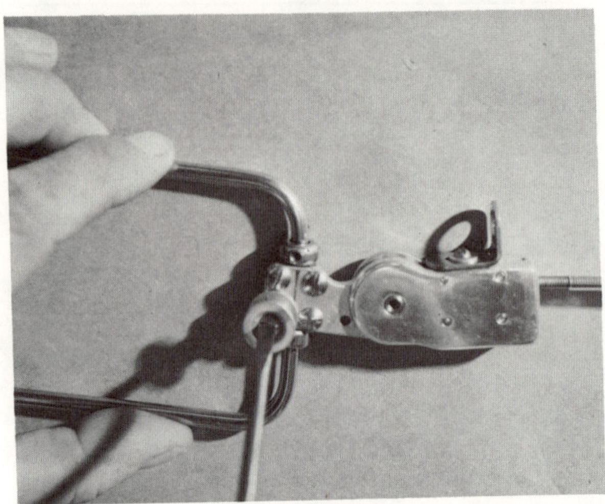

2. Put the shoulder cap on the patient and insert the U-bar into the U-bar mounting bracket. Use a screwdriver and 1/16 inch Allen wrench to remove the stop button and band post, rotation stop, and the two assist springs from the rod in the elbow lock. (The factory supplies one each of left and right assist springs with each kit.) Place the elbow lock and rod assembly against the patient's arm, with the joint center of the elbow lock adjacent to the patient's elbow center. Mark the rod one inch above the lower edge of the ratchet lock case, and cut it off at that point with a hacksaw. Smooth and round off the cut edge of the rod.

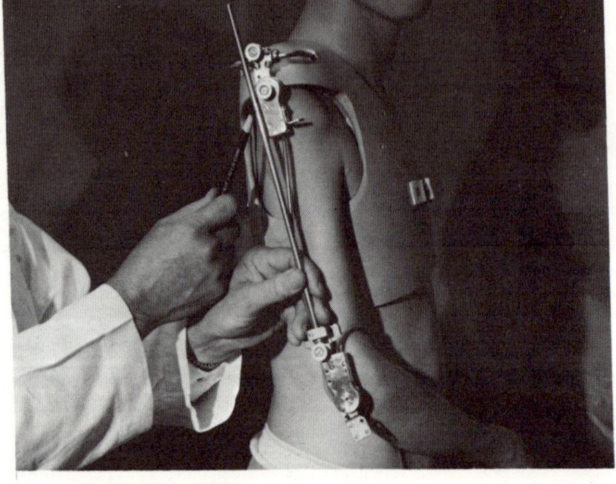

3. The FAB-15A rod is now the correct length for the patient's arm, but the three set screws in the ratchet lock cannot grip it securely without a flat on the side of the rod. Measure 1-1/4 inches back from the end of the rod and mark it at that point.

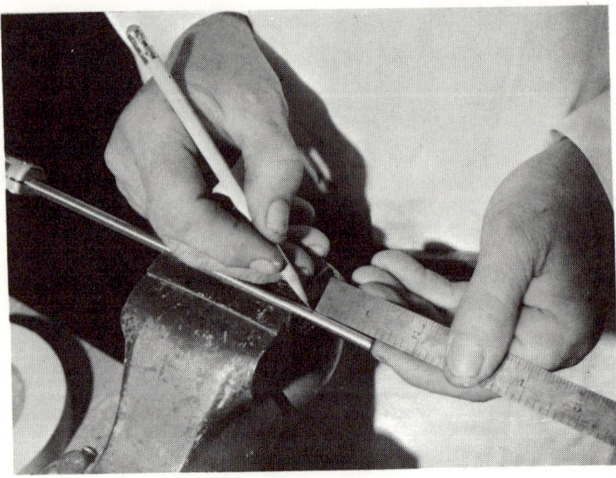

4. With a smooth file, cut a 1-1/4 inch flat on the rod, about 1/32 inch deep.

5. Assemble the rod and elbow lock into the hole in the ratchet lock; use a 1/16 inch Allen wrench to tighten the three set screws in the ratchet lock case so the rod will not come loose.

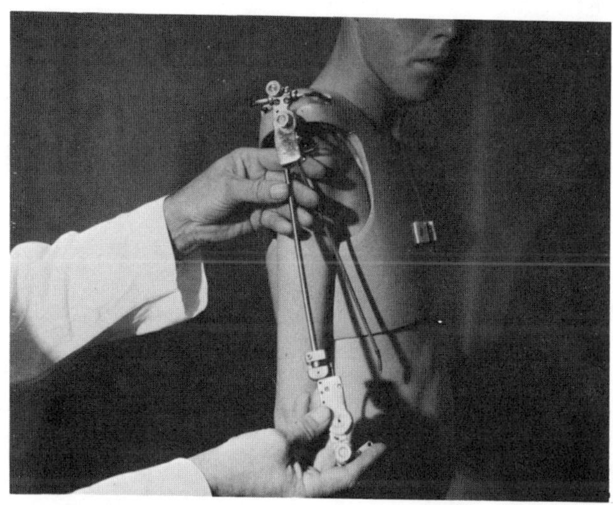

6. With the elbow lock fully extended and the cam of the ratchet lock release contacting the stop, stretch the FAB-23C ratchet lock cable alongside the ratchet lock release. Mark the cable slightly below the anchor hole in the ratchet lock release.

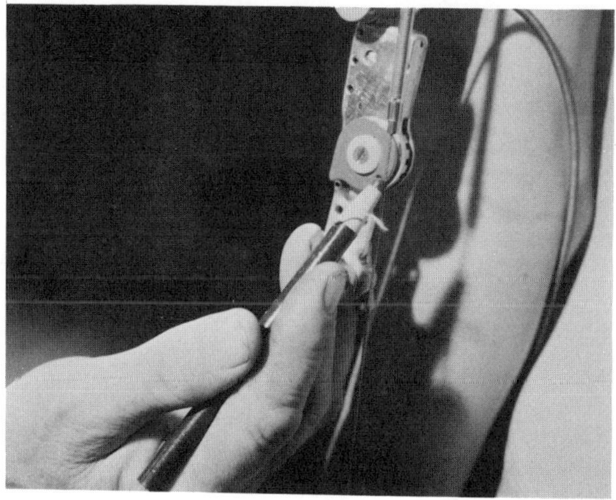

7. Use a sharp pair of diagonal cutting pliers to cut the cable at the mark; avoid fraying the strands of the cable.

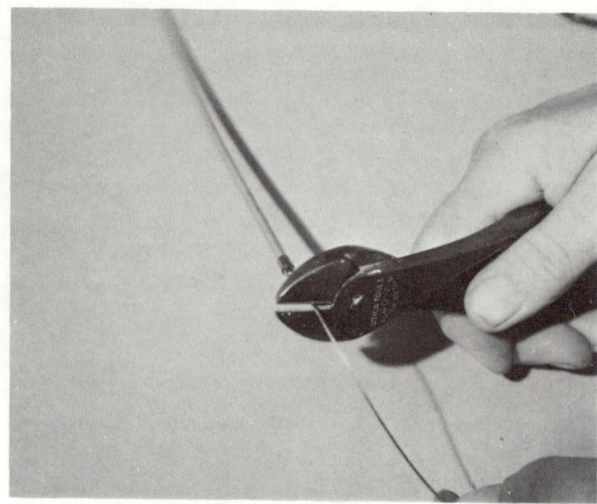

8. Slip the threaded anchor bushing over the cable end.

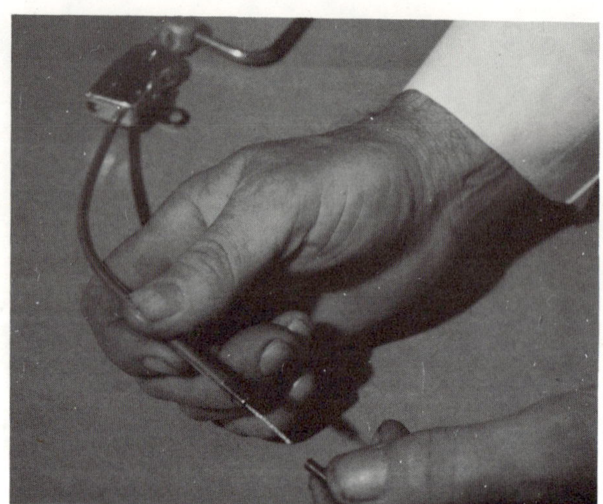

9. Slip the brass collar over the cable end.

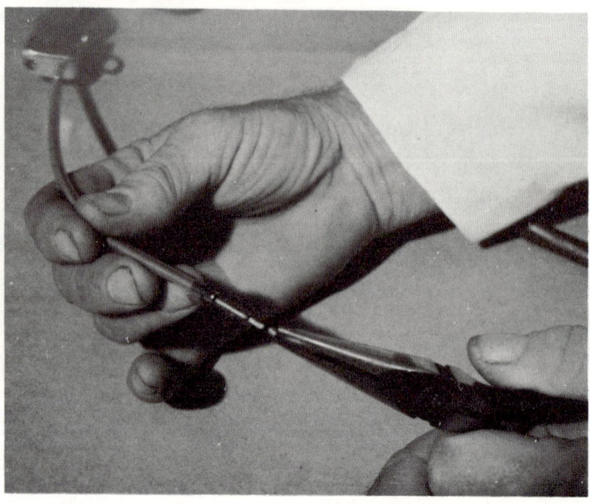

10. Solder the brass collar to the end of the cable. "Lloyd's Stainless Steel Soldering Flux" and soft solder have been found satisfactory, although any good stainless steel flux will work. Do not use too much solder, as the solder and bushing on the end of the cable must be large enough to be unable to slip through the bushing, but must be small enough to go through the threaded hole in the anchor. This is important, as there is no other way to uncouple the cable if the brace must be disassembled later for adjustment or parts replacement.

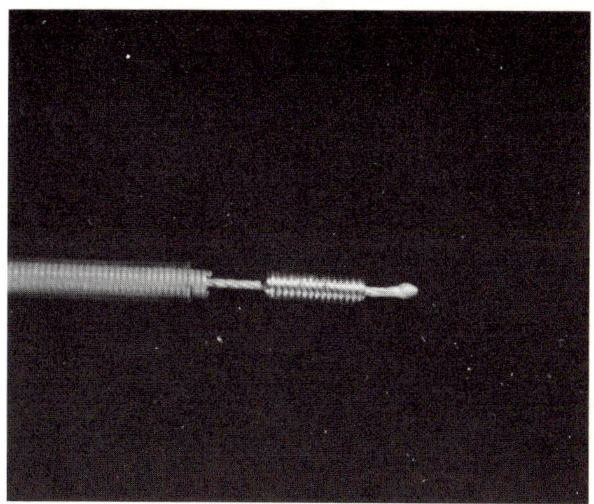

11. Remove the ratchet lock release assembly. It is advisable to first mark the top side of the disc, as it is easy to install this part backward.

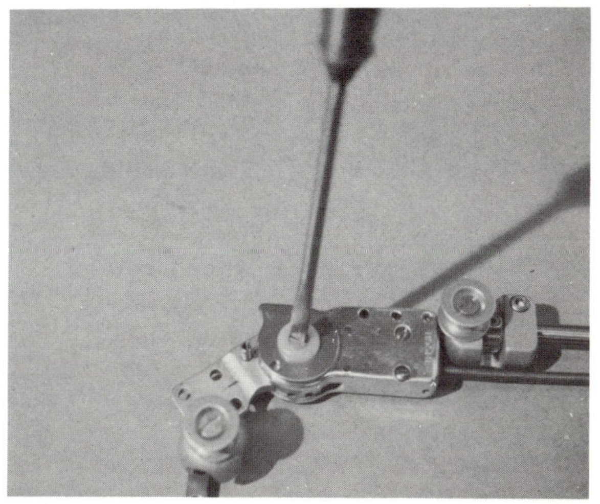

12. Countersink the anchor hole in the ratchet lock release enough to allow the anchor pin to be expanded to retain it in place without binding, but use care to avoid enlarging the hole. Be sure to countersink the hole on the side of the ratchet lock release opposite the side marked "top."

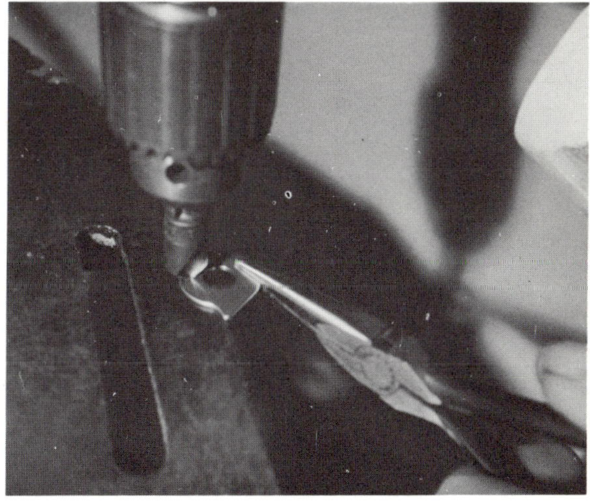

13. Insert the anchor pin into the hole, making sure the anchor is on the side of the ratchet lock release marked "top", then carefully peen the anchor pin enough to hold in the countersunk part of the hole, but not enough to bind.

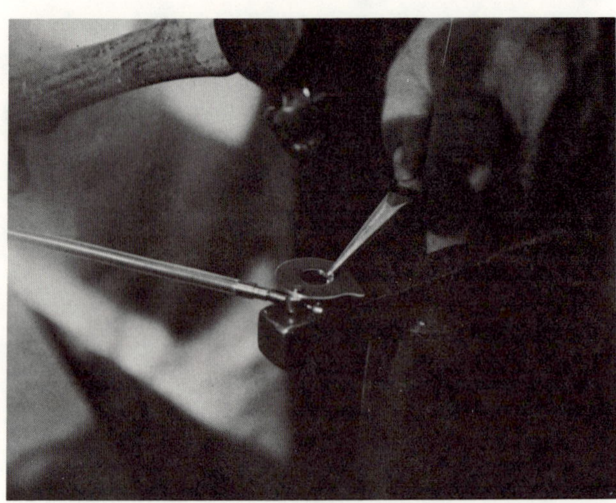

14. Replace the ratchet lock release assembly on the elbow lock, and thread the anchor bushing and cable into the anchor. With long nose pliers adjust the anchor bushing so the ratchet lock release is rotated by the stop contacting the lock release cam, when the elbow lock is fully extended. This adjustment must be set so it does not prevent the elbow lock from extending fully, yet it must completely unlock the ratchet lock when the elbow lock is in full extension.

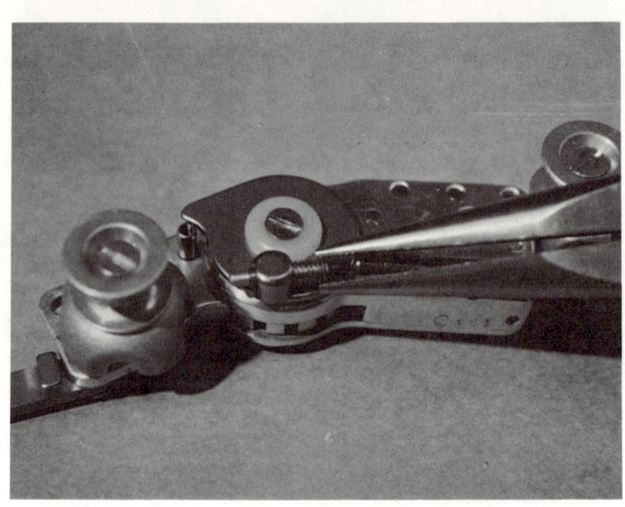

15. Heat the Royalite humeral cuff in an oven and shape it to fit the patient's upper arm, using a form or doing it by hand.

16. Hold the cuff in place on the patient's arm, with its lower edge slightly above the level of the upper surface of the elbow lock case, then mark the cuff at a point opposite the center of the slot in the rotation clamp assembly.

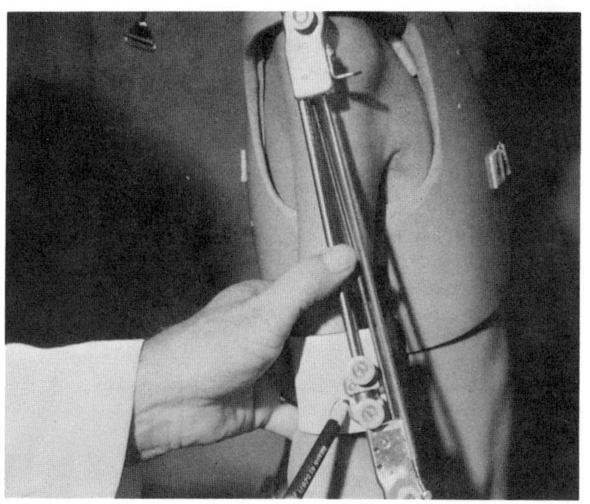

17. Draw a line on the cuff parallel to and one inch from the edge, then transfer the mark made in the preceding step to this line. Set a pair of dividers with the points 21/32 inch apart, set them on the line so the mark made earlier is centered between them, and make a mark on the line at each point.

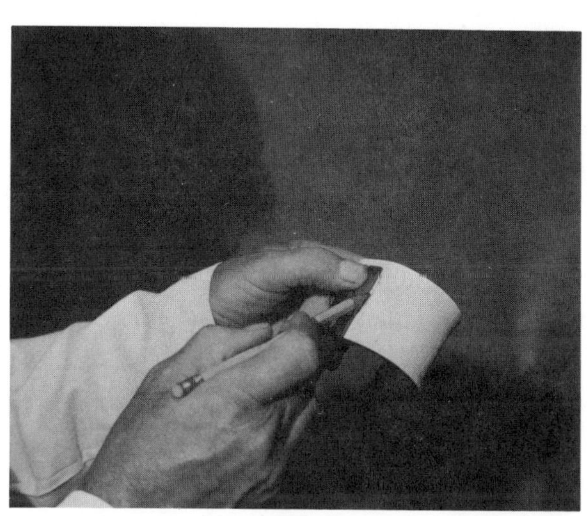

18. Drill No. 28 holes in the cuff at each of the two marks and fasten the cuff to the rotation clamp assembly with two 6-32 5/16 inch truss head screws. In most cases the two screws directly through the Royalite plastic will give good service. If a patient subjects the cuff to unusual strains it may be necessary to reinforce it with a piece of sheet stainless steel.

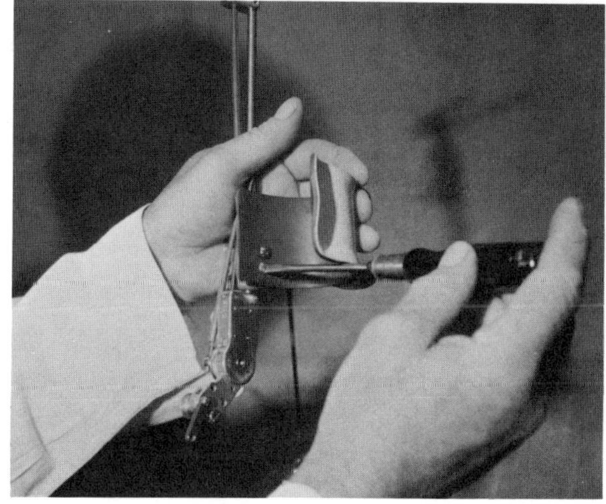

19. Check the cuff on the patient and make certain it fits comfortably. Adjustments in shape of the cuff can be made by heating it. Flaring and rounding the edges helps to prevent irritation and painful pressures on the patient's arm. Some patients are so sensitive that the cuff has to be lined with felt for them to tolerate it.

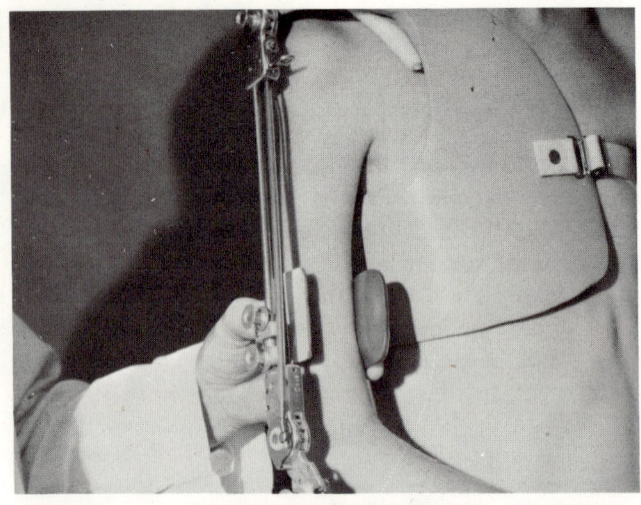

20. Insert the forearm rod in place in the elbow lock clamp, and judge where it must be bent to conform to the contours of the patient's forearm. Mark the rod for the first bend, usually about four to six inches from the clamp. Estimate about how much it must be bent to follow the curve of the arm.

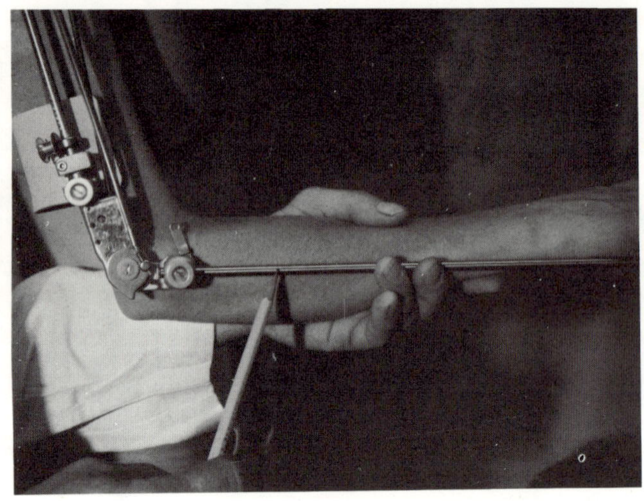

21. Use a bending fork to bend the rod.

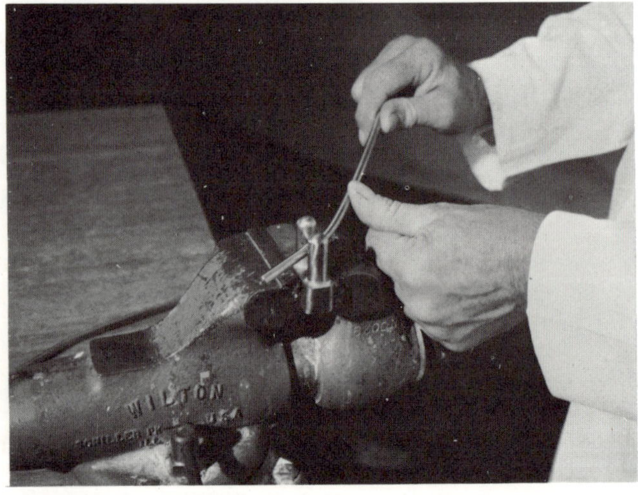

22. Try the rod on the patient and continue bending it until the first portion
is a good fit, then mark it for the reverse bend that will bring the distal
end of the rod under and parallel to
the ulnar aspect of the wrist.

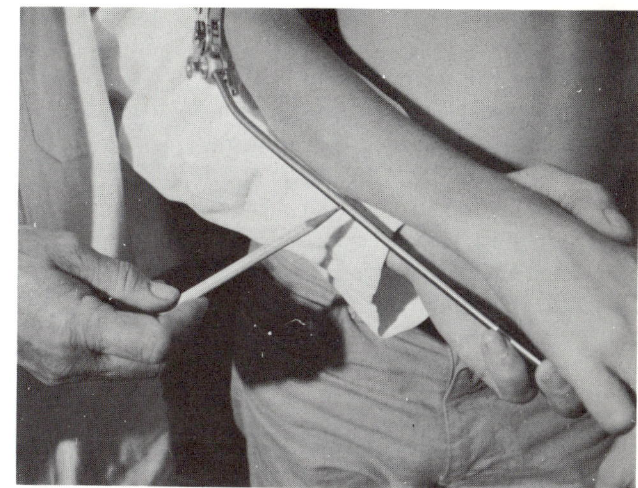

23. Put the reverse bend into the rod,
try it on the patient, and re-bend
it until it fits. The wrist end of
the rod should terminate under the
ulnar side of the wrist when the
patient's thumb is up, as support
in this manner enables the patient
to use gravity to improve control
of forearm rotation.

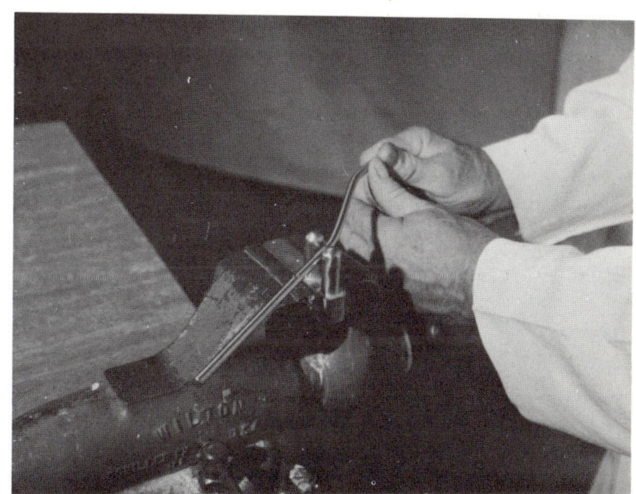

24. Select the correct assist spring for
the forearm rotation assist. In the
illustration, the upper spring is
wound right, the lower one is
wound left. For a right arm brace
use the left-wound spring for pronation assist, the right-wound
spring for supination assist; for a
left arm brace use the left-wound
spring for supination assist, the
right-wound spring spring for pronation assist.

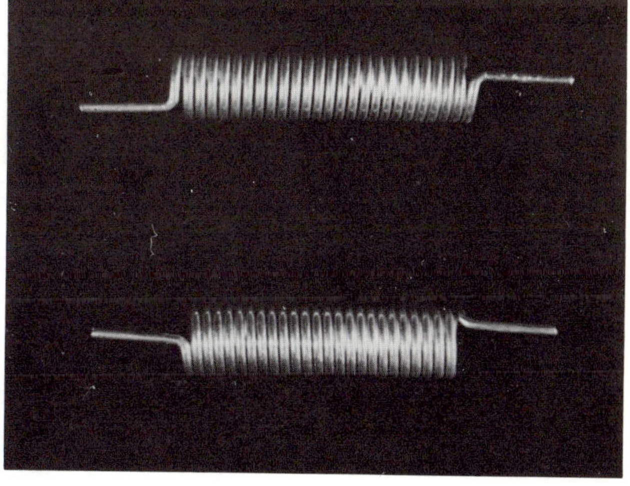

25. Slide the spring anchor ferrule, the assist spring, the friction wrist control, and the rotation stop button onto the wrist end of the rod in the order named. Fit the ends of the spring into the holes in the ferrule and stop button, then tighten the set screws in both with a 1/16 inch Allen wrench. The friction wrist control must turn freely on the rod.

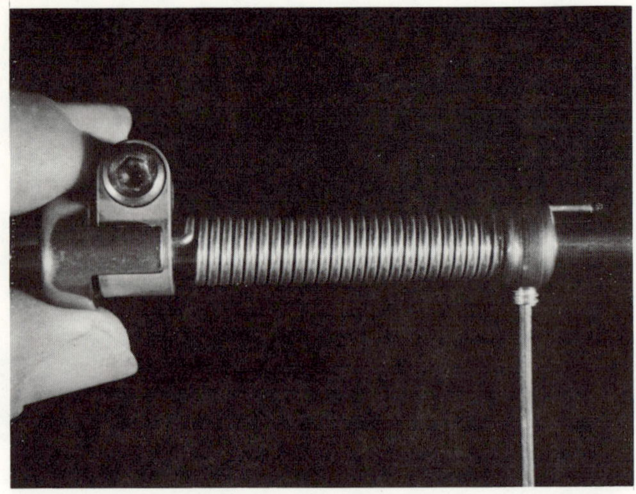

26. Lay out the thickness of the patient's wrist on the cuff strap, an equal amount on each side of the center hole.

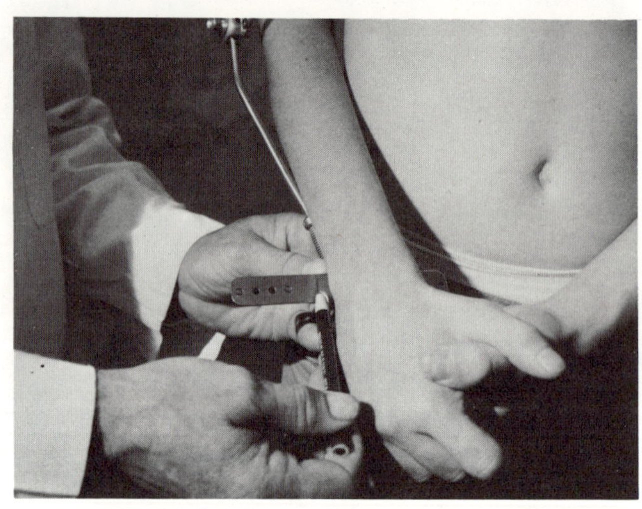

27. Bend the cuff strap on the marks into a "U" shape to fit the contours of the patient's wrist.

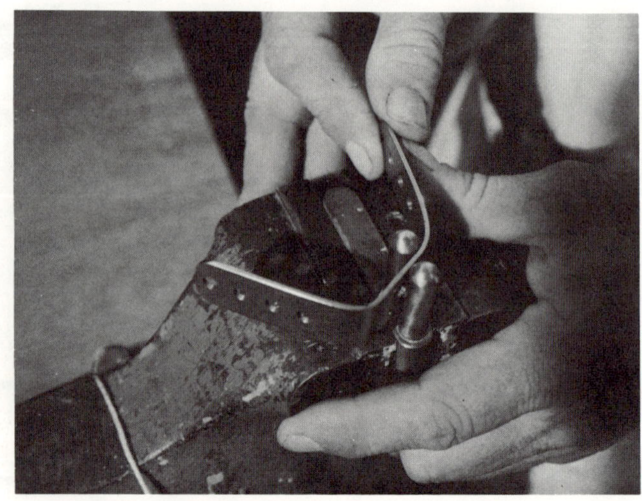

28. Trim the ends of the cuff strap so they do not extend past the radial aspect of the patient's wrist. Round and smooth the cut edges.

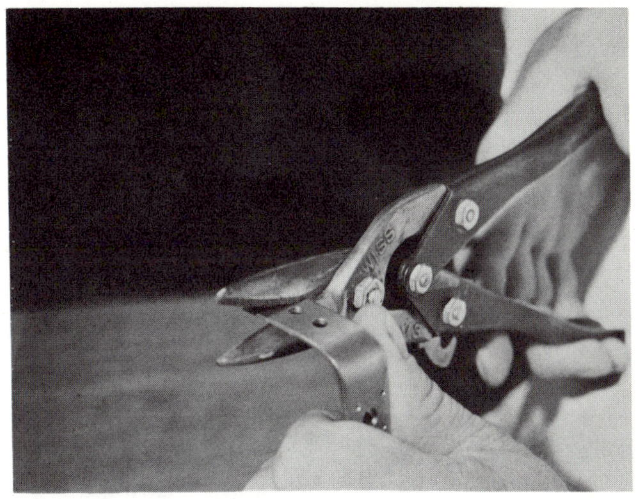

29. Lay out a liner for the cuff on flexible plastic, leather, or felt. Flexible plastic is easier to clean and does not absorb moisture, but the softer leather or felt should be used if the patient's skin is unusually tender or sensitive. Cut the material 2 inches wide and long enough to extend 1/2 inch beyond the ends of the cuff.

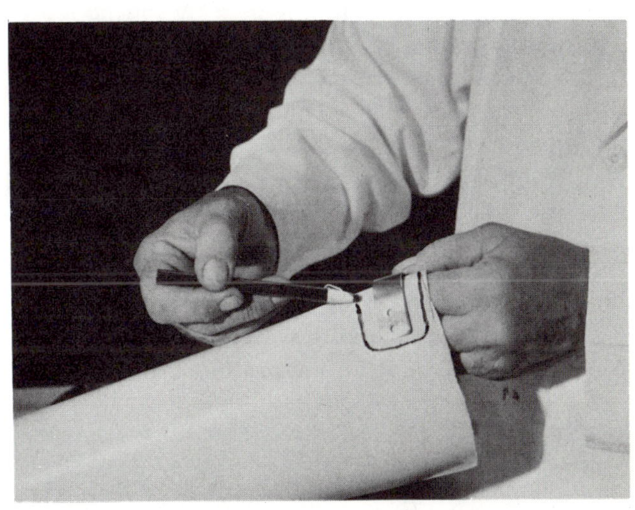

30. Fasten the wrist cuff to the friction wrist control with the screw and bushing supplied. The cuff must swivel freely with the screw tightened securely.

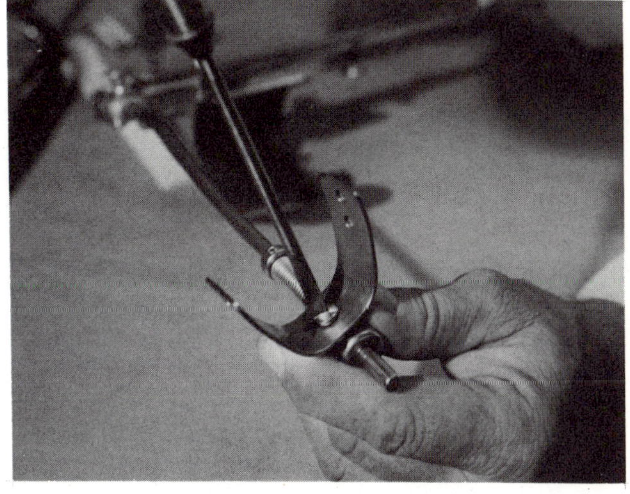

31. Cut two pieces of matching Velcro for a
closure, rivet the cuff liner and Velcro
inside the cuff with Speedy rivets.

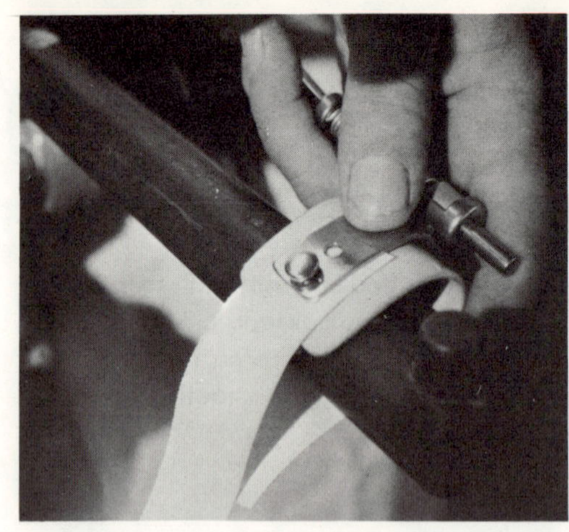

32. With the brace on the patient, position the
cuff slightly proximal to the wrist joint. Adjust the forearm rotation assist by setting the
tension of the assist spring. If the patient
can pronate, but not supinate, adjust the
anchor ferrule so he has to wind up the
spring when he pronates. Now when he
wants to supinate his wrist the spring will
release energy as it unwinds and will help
him make the motion. If the patient can
supinate but not pronate, the procedure is
the same, except the anchor ferrule is
turned in the opposite direction to tighten
the spring. Cut off the rod flush with the
rotation stop, smooth and round the cut
end of the rod.

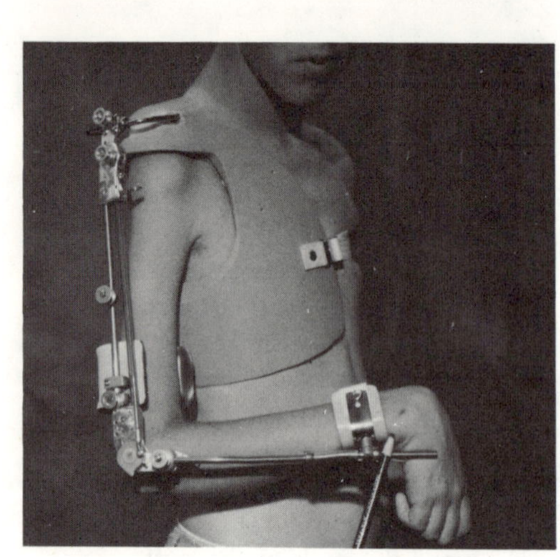

33. Sew a shoulder loop of 1 inch dacron webbing
to fit comfortably under the patient's axilla
and over his shoulder. Sew a 1 inch buckle
with a billet to the point where the loop crosses, and extend a strap from this point so it
will thread through the hanger on the elbow
lock control cable and back to the buckle.
Bring the cable over the patient's shoulder,
thread a C-709 retainer onto the housing, fasten a C-708 base plate to the shoulder cap as
illustrated, insert the retainer into the base
plate, check the slack in the housing. Adjust
the retainer on the housing until all slack is
removed, then cut off the excess housing and
cable. Swage a C-711 hanger onto the end
of the cable, and buckle the shoulder loop
through it. Have the patient try locking and
unlocking the elbow lock, adjust the shoulder
loop for easiest operation without discomfort.

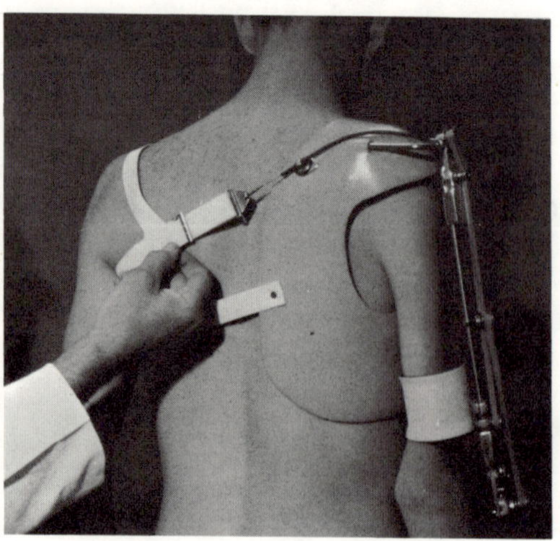

34. Apply No. 32 rubber bands to the shoulder and elbow band posts by looping the end of a band over one of the band posts, bending it around the other and returning to loop over the one from which it started. Add rubber bands until the patient cannot extend his shoulder and elbow, then remove just enough so he can do so easily. His arm should hang in a natural manner, and require no straining of his muscles to keep it down.

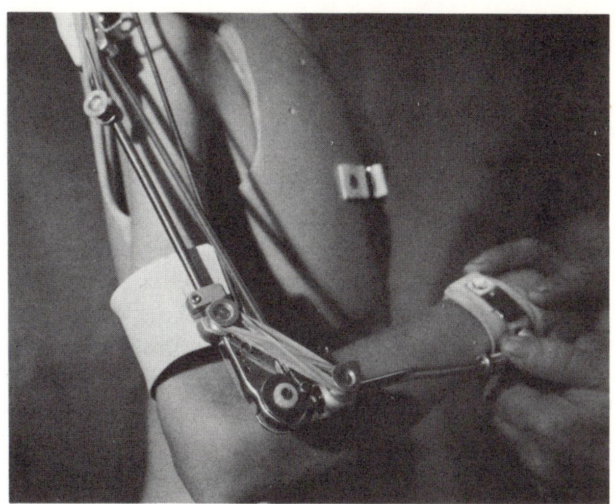

35. Help the patient learn to manipulate the ratchet and elbow locks. Direct him to pick up an object with his hand, elbow lock unlocked, then flex his elbow and lock the elbow lock by scapular abduction. Then direct him to flex his shoulder to bring the object to his mouth. Next he should unlock and extend his elbow, and at full extension it should unlock the shoulder ratchet lock, allowing the shoulder to extend. If possible, arrangements should be made for the patient to have training in the use of his brace by a physical or occupational therapist.

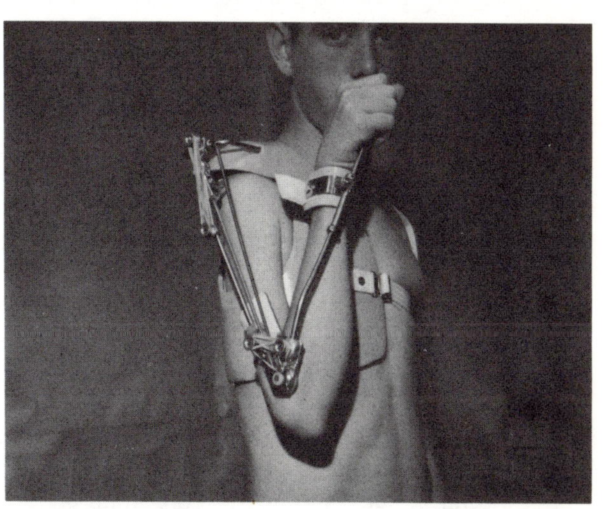

36. As the patient practices using his brace, observe how well he is able to control humeral rotation. If he has no active control of humeral rotation, it is best to apply considerable stabilizing friction by tightening the FAB-30 rotation clamp with a 1/8 inch Allen wrench. If he has power to rotate his humerus internally but not externally, or vice versa, the addition of an assist spring will be helpful (see illustration). Use a right wound FAB-8B assist spring with a spring anchor ferrule for external rotation assist on a right arm, left wound for internal rotation assist, the opposite for a left arm. The distal end of the spring fits into the slot in the rotation clamp, which must be left free on the rod.

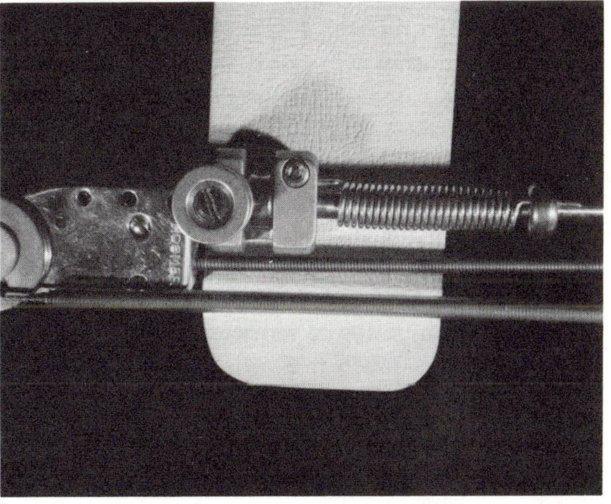

HOW TO INSTALL AN ARTIFICIAL MUSCLE ON A LOCKING JOINT
FUNCTIONAL ARM BRACE

Introduction

The elastic flexion assist is useful for the patient who has some strength in his extensors, particularly in the elbow. If he has neither flexors nor extensors with enough strength to provide an acceptable minimum of function it is necessary to provide a source of outside power to supplement any muscular strength the patient may have. Compressed carbon dioxide gas has been found to be a convenient and inexpensive source of such power, and by using the artificial muscle, this source of power can be applied to the locking joint functional arm brace to provide positive elbow flexion.

Parts Used

A review of the parts needed to install an artificial muscle on a locking joint functional arm brace follows. The numbers and nomenclature of the parts is in accordance with those established by the manufacturer, Orthopaedic Supplies Co., Incorporated.

1502 CO_2 Tank, Figure 125, stores the carbon dioxide used to power the artificial muscle. The handle operates the shut-off valve, which is opened by turning counter-clockwise. The tank is made of steel, and the gas is normally contained under a working pressure of 850 pounds per square inch at 70 degrees Fahrenheit. To assure safety from excessive pressure, a pressure blowout disc is provided, which ruptures and releases the gas if it reaches a pressure of 1800 pounds per square inch. The tanks are available in

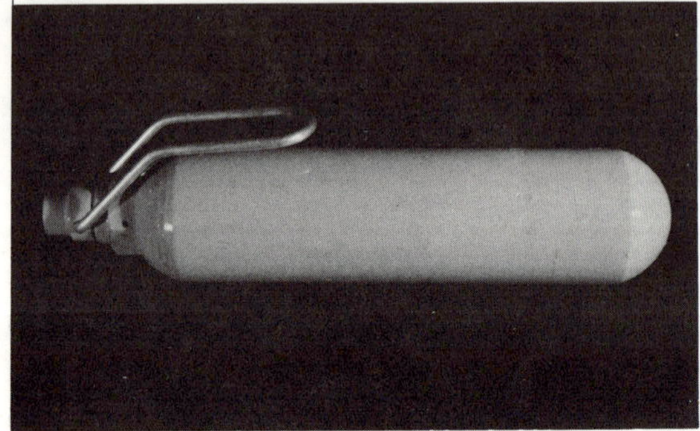

Figure 125. CO_2 Tank

two sizes; the 1502, shown in Figure 125, has a capacity of 1 pound, 1 ounce of CO_2, and is 7 inches long and 2-1/2 inches in diameter. The 1502 tank is small enough to be carried by ambulatory patients, either in a pocket or leather pounch slung around the waist or other convenient means. The 1501 tank has a capacity of 2 pounds, 13 ounces, and is 13 inches long and 1-7/8 inches in diameter. This larger tank is designed for wheel chair patients, and can be attached to the chair in a convenient location by clips or brackets. The frequency with which the tanks must be refilled depends on how much the patient uses his artificial muscle. The average is two to four days for the small tank, proportionately longer for the larger tank.

1601 CO_2 Pressure Regulator (Figure 126) screws into the shut-off valve in either size CO_2 tank. With the regulator valve turned out (counter-clockwise) the pressure output is zero, when turned all the way in the pressure is a maximum of 90 pounds per square inch. The 1/4-28 fitting on the side of the regulator is for connecting the plastic tubing that feeds the gas pressure to the control valve.

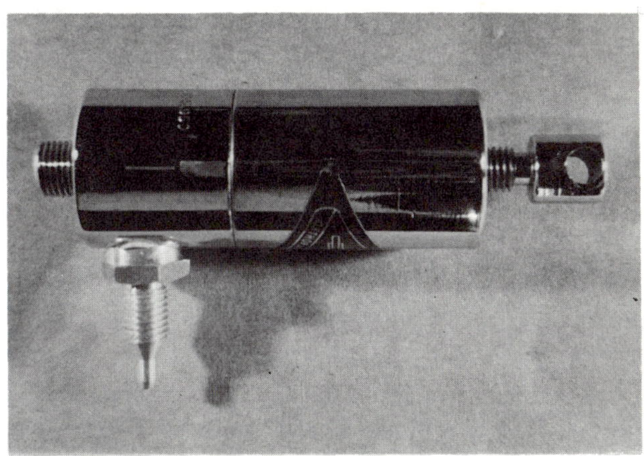

Figure 126. CO_2 Pressure Regulator

1901 Connecting Tube (Figure 127) is clear polyethylene plastic tubing, 3/16 inch outside diameter, 1/16 inch inside diameter. The aluminum bushings are 7/16 inch long and 5/16 inch in diameter, and have a 1/4-28 inside thread. The tapered shoulder on the male fittings flares the plastic tubing when it is heated with a match and forced onto the fitting over the nipple. When the bushing is tightened it seals the flare to the fitting, making a very simple leak proof joint. No tools are needed, as the joints hold pressure when the bushings are turned finger tight.

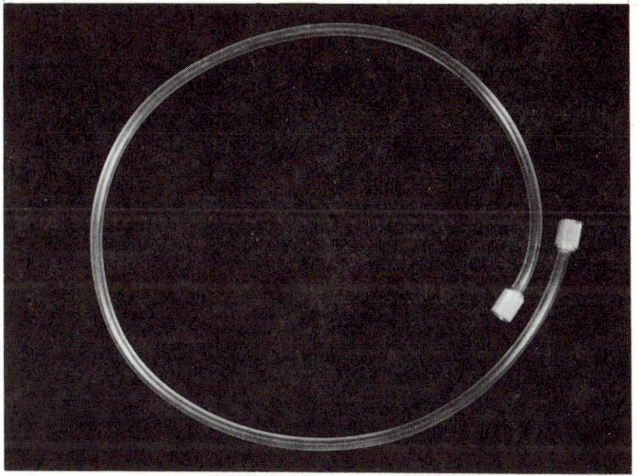

Figure 127. Connecting Tube

1702 Slide Valve (Figure 128) enable the patient to control the flow of gas from the tank to the artificial muscle, to maintain pressure in the muscle as long as it is needed, and to exhaust the gas from the artificial muscle to complete the cycle. The valve is inserted between the control strap and control cable of a shoulder type harness, so the patient can operate the valve by scapular abduction. The control motion pulls the flexible cable in the control

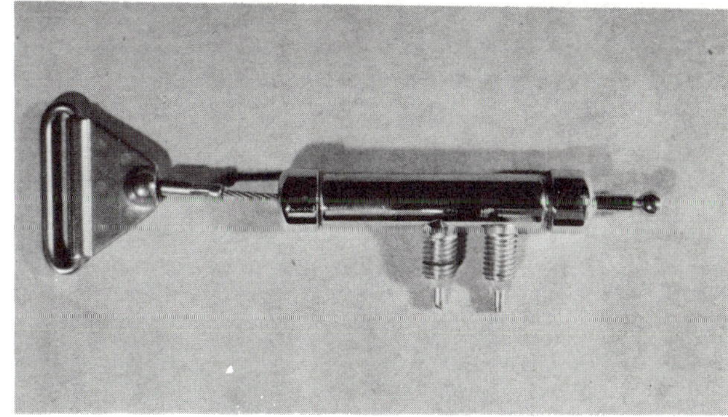

Figure 128. Slide Valve

valve, and the cable is in turn connected to a spring-loaded ported piston in the valve cylinder. The total excursion of the piston in the cylinder is 1/4 inch, and 1-1/2 pounds of force are required to compress the spring to operate the valve. Pulling the valve out the full 1/4 inch stroke permits gas to flow from the tank to the artificial muscle, causing it to operate. This is the "fill" position. When allowed to completely retract the valve is in the "hold" position, and gas cannot flow into or out of the artificial muscle. Pulling the valve out 1/8 inch, or half stroke, allows the gas in the muscle to flow out; this is the "exhaust" position. Since the control cable from the valve operates the elbow lock, the valve spring pressure must be strong enough so that when the patient pulls on the valve, enough force will be transmitted down the cable to unlock the elbow before the valve opens and activates the artificial muscle. If the valve is opened with the elbow locked, the artificial muscle may be damaged. The valves come with 3/4 pound springs for use with the flexor hinge splint, so it is important to be sure the valve used with the locking joint functional arm brace has the 1-1/2 pound spring if it is to operate properly. In ordering these valves it is important to state which spring is required.

Be sure to check the elbow lock and make certain it has been changed over from alternator to spring lock operation.

1702 S   Slide Valve (Figure 129) is the same as the 1702, except that a webbing loop is substituted for the ball stud used on the latter. This may be needed where it is desirable to use webbing on both ends of the valve, as for example, in cases where the control cable would otherwise contact the patient's back. It may also be used in special chest strap harness control systems and in others where shoulder elevation is the only available control motion. This valve comes with either 3/4 or 1-1/2 pound valve spring.

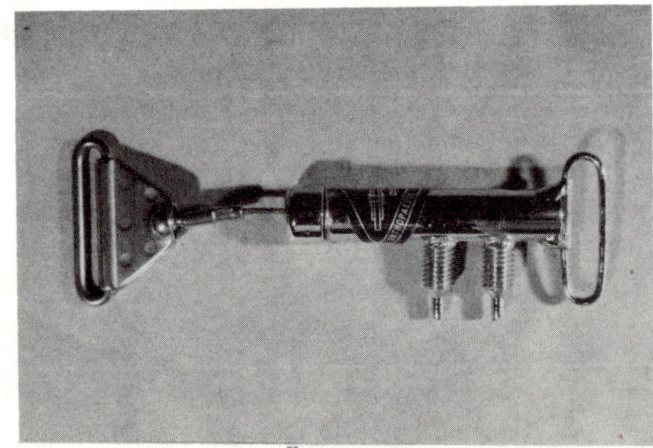

Figure 129. Slide Valve

C-713 Ball Receiver (Figure 130) is swaged
C-714 onto the end of the 1/16 inch diameter control cable that operates the
elbow lock, and connects it to the
3/16 inch ball on the end of the
1702 slide valve. This part is made
in two types, the C-713 (Figure 130,
right), which has a safety latch to
prevent the ball from accidentally
becoming dislodged, and the C-714
(Figure 130, left), which does not
have this latch. These parts are
made by the A. J. Hosmer Corp.

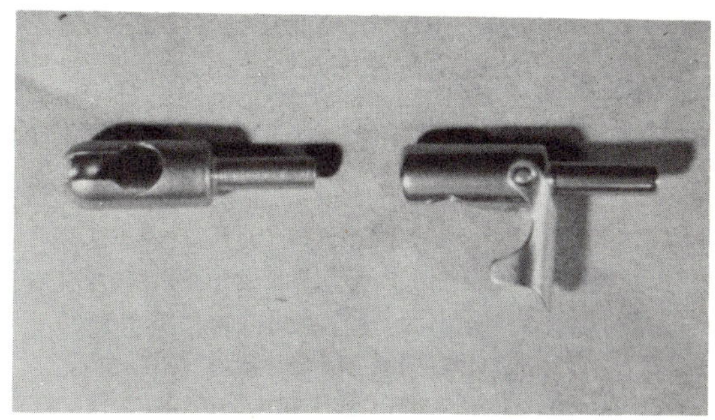

Figure 130. Ball Receiver

1704 Flow Control (Figure 131) is inserted
in the tubing between the control
valve and the artificial muscle to vary
the rate of flow of the gas to the muscle. Turning the small adjusting
screw located on the side of the control clockwise reduced the speed with
which the gas gets to the muscle;
turning it in the opposite direction
increases it.   The artificial muscle
contracts instantly when the control
valve admits full pressure of the gas
to it. The effect which this has on
the arm brace is to snap the forearm
up very quickly, causing the patient

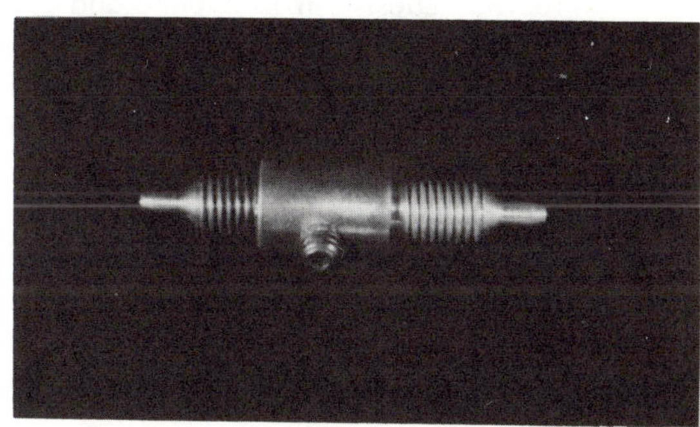

Figure 131. Flow Control

to lose close control of the operation he is trying to accomplish, and sometimes even
making it impossible for him to accomplish it. By trying different adjustments on the
flow control, it is possible to slow down the action of the muscle to a speed that is
about right for the activities the patient wants to do and for his ability to coordinate.

312

1801
1802 Artificial Muscle (Figure 132)
is made of a latex rubber tube
enclosed in a nylon helical
weave. A metal ferrule seals
one end of the rubber and nylon
combination, and is drilled and
tapped to accept the FAB-23
hook assembly. The other end
has a similar ferrule, except
that it has the same type 1/4-28
threaded nipple used on the
other tubing fittings, and serves
as the inlet for the CO_2 gas.
A 1/4-28 nut applied to the
nipple is used to mount the proximal end of the muscle in the
bracket on the ratchet lock.
When the carbon dioxide gas is

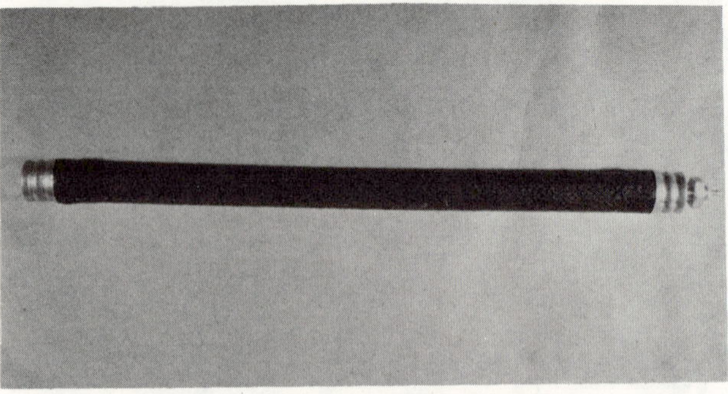

Figure 132. Artificial Muscle

admitted to the muscle, the rubber tube expands against the helical weave of the nylon
tubing, causing it to expand and at the same time, shorten. The artificial muscle shortens
approximately 30% of its length when fully expanded, thus providing the force and excursion needed to drive the brace. For improved performance on the locking joint functional arm brace, a looser weave nylon tubing coated with latex has been developed.
These can be distinguished from the standard muscle by the coating of black latex on the
outer surface of the tubing. Because of the looseness of the nylon weave, the latex coating must be provided to prevent the inner tube from blowing out through the spaces between
the threads.

FAB-33 Hook Assembly (Figure 133) is
a stainless steel hook provided
to connect the distal end of the
artificial muscle to the lift
lever assembly. It is made
longer than is usually necessary,
and a hollow threaded bushing is
provided to be silver soldered to
the end of the hook when it has
been cut to size. The threaded
bushing fits the tapped hole in
the distal end of the artificial
muscle.

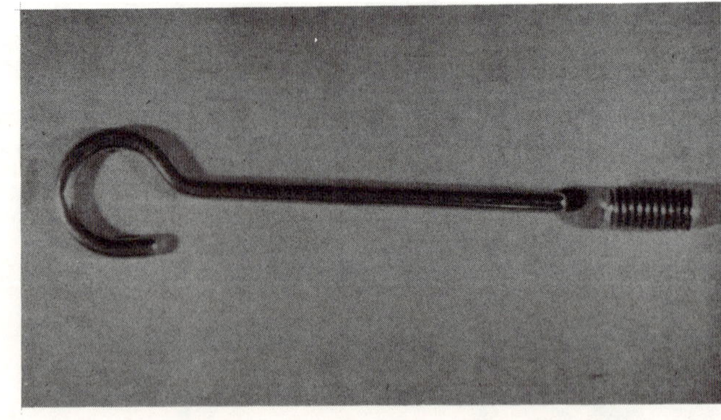

Figure 133. Hook Assembly

Fitting Procedure

1. Measure the distance between the eye on the lift assembly on the elbow lock, and the artificial muscle mounting bracket on the ratchet lock. Be sure the elbow lock is fully extended and the lift assembly is in position with its stop firmly against the gear strap. Check the elbow lock to make sure it has been changed over to spring lock (pull-hold to unlock, release-relax to lock), instead of alternator lock (pull-release to unlock, pull-release to lock). If this has not been done, make the necessary changes as explained on page 292. Select an artificial muscle one inch shorter in length than the measurement made.

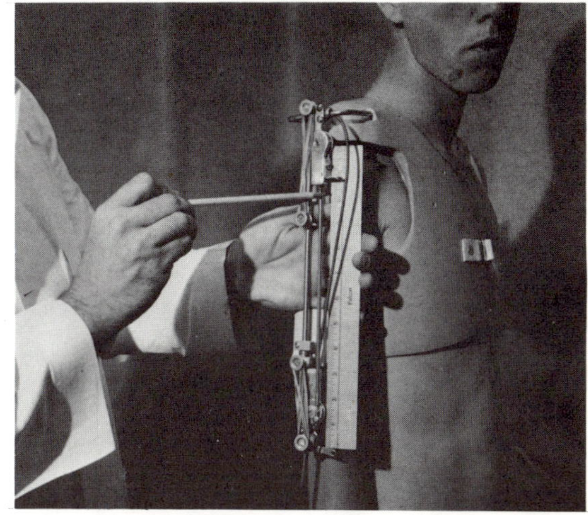

2. Insert the nipple end of the muscle into the hole in the mounting bracket on the ratchet lock, and secure it in place with the 1/4-28 nut provided.

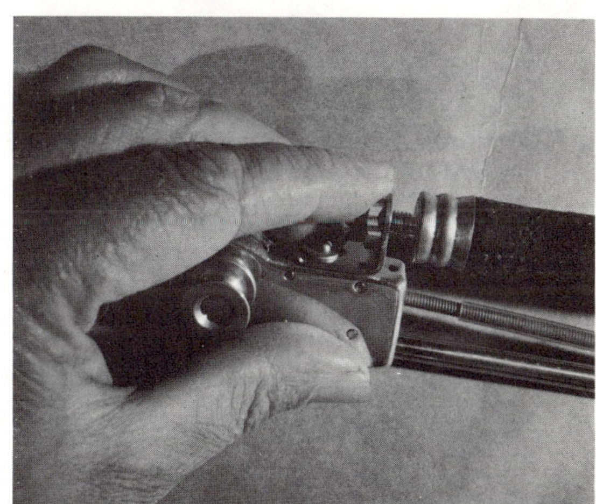

3. Insert the FAB-23 hook into the FAB-34 lift assembly and mark the hook at a point opposite the end of the muscle.

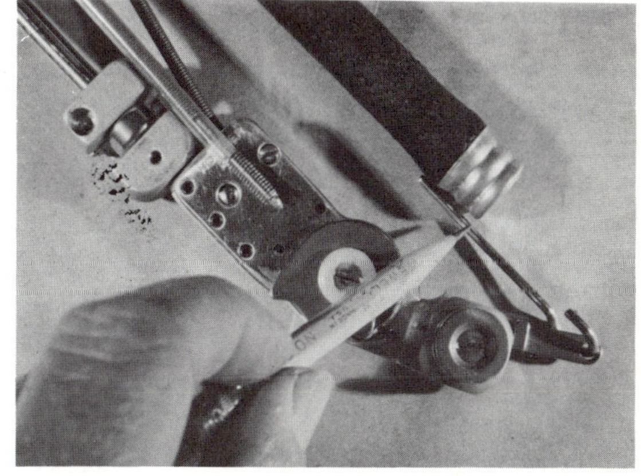

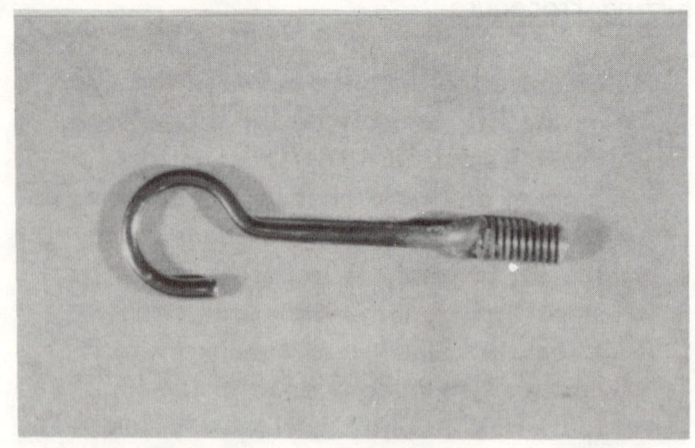

4. Silver solder the bushing to the hook at the mark made in the previous step and cut off any of the shaft of the hook that protrudes through the bushing.

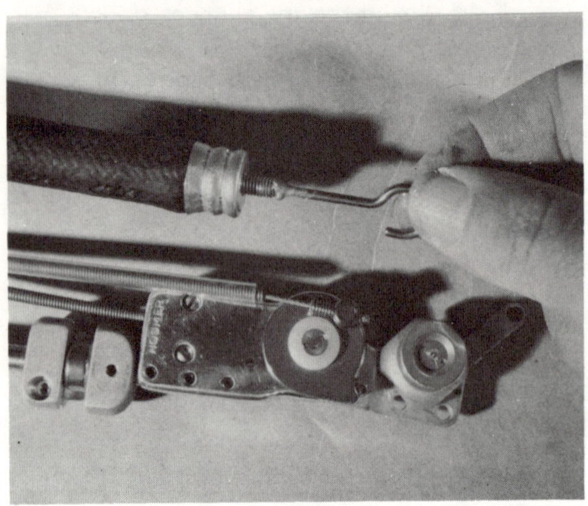

5. Insert the hook into the distal opening in the artificial muscle.

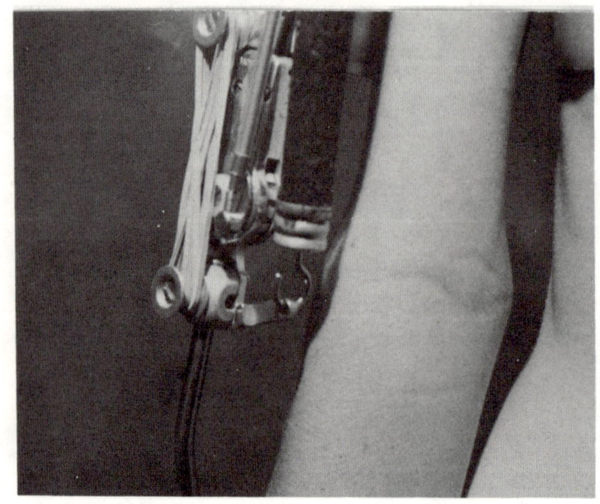

6. Fasten the hook into the eye in the lift lever and extend the elbow. The muscle should be tight when the elbow is fully extended, but not so tight as to prevent full extension. Adjust by screwing the hook in or out as necessary.

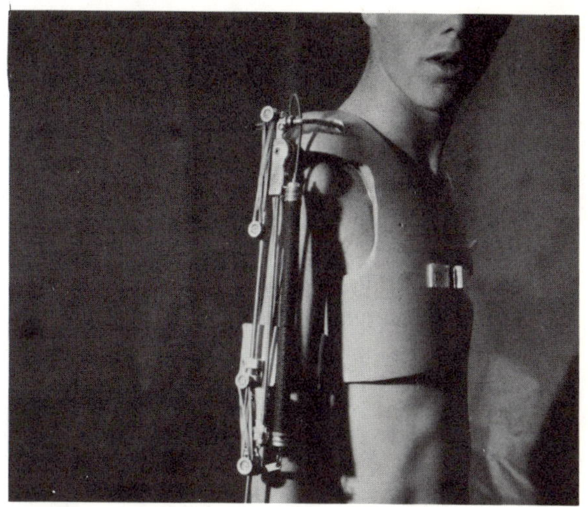

7. When correctly installed and adjusted, the artificial muscle should be straight and tight, but should not restrict elbow extension.

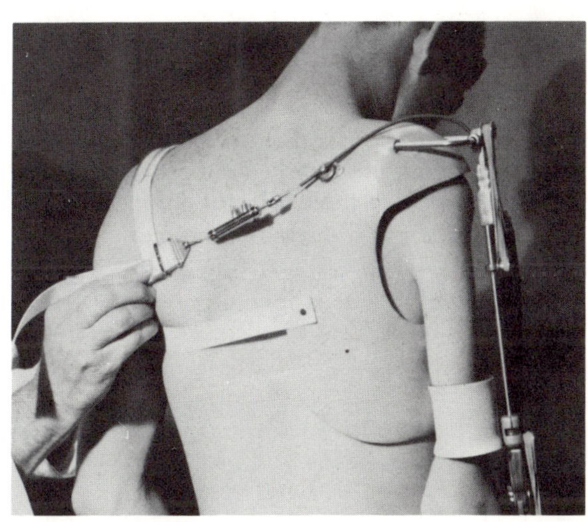

8. Fit a one inch dacron webbing shoulder loop harness onto the patient's sound side, bringing a strap and buckle out at the junction of the loop.   Bring the elbow lock control cable over the shoulder, fasten it to the shoulder cap with a base plate and retainer, then cut off the cable and housing, and swage a C-713 or C-714 ball retainer onto the end of the cable. Fasten a 1702 slide valve to the harness and ball joint, and adjust to take out all the slack, but leave it loose enough for comfort and to avoid accidental operation of the valve.

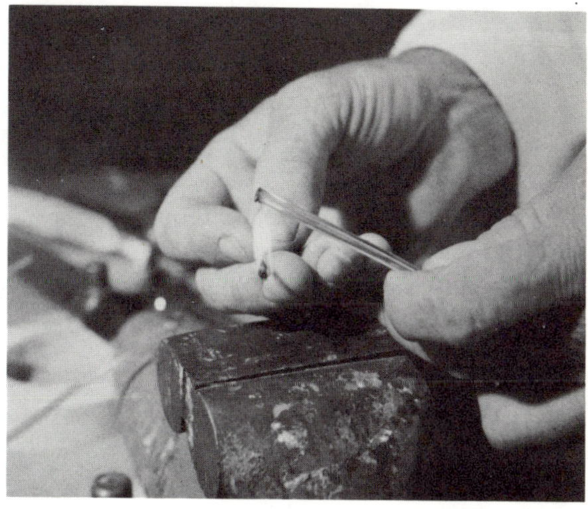

9. Slip one of the tubing bushings over the end of a length of connecting tubing, then soften the end of the tubing with the flame of a match. Do not burn the plastic, just heat it enough to soften it.

10. While the end of the plastic tubing is still soft, force it down over the nipple and shoulder on the artificial muscle fitting. Tighten the bushing onto the fitting, using the fingers.

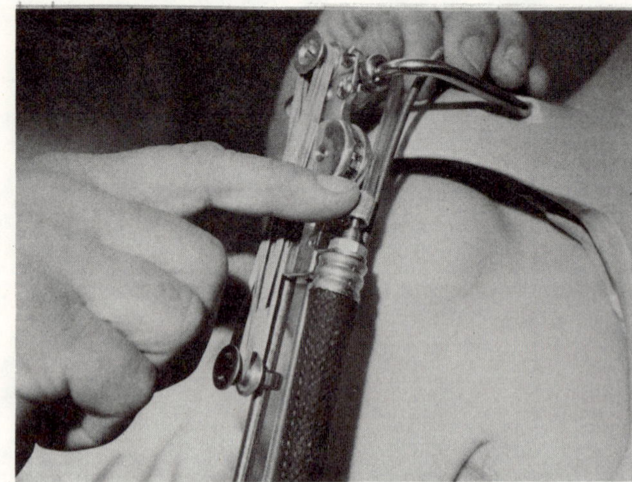

11. Heat the other end of the tubing connected to the artificial muscle, and use the same technique to connect it to the fitting on the slide valve that is located nearest the control cable.

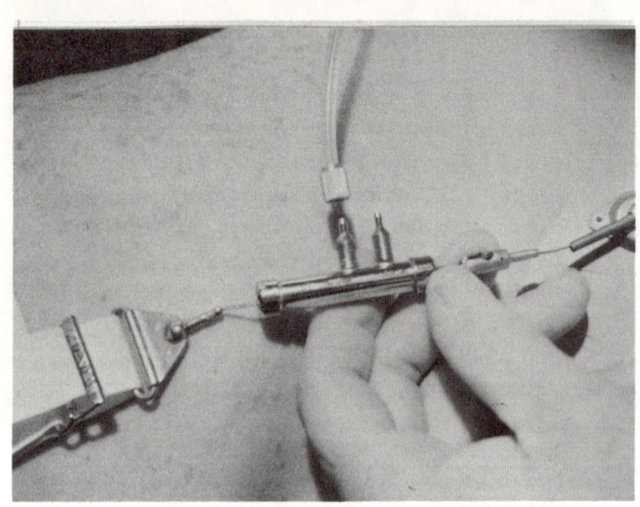

12. Screw the CO_2 pressure regulator into the valve in the CO_2 tank.

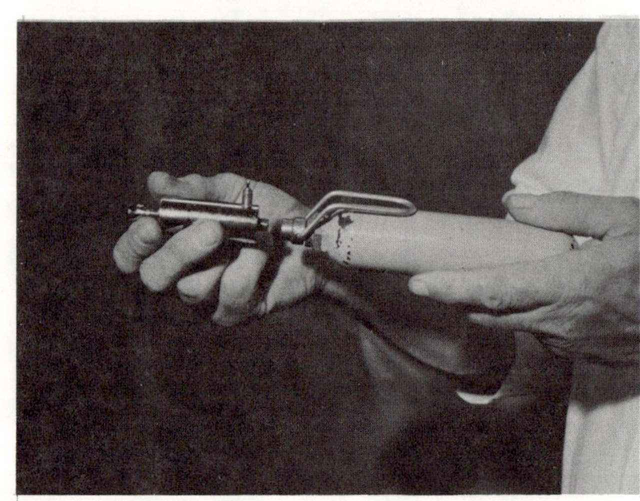

13. Fasten another piece of connecting tubing to the outlet on the CO_2 pressure regulator, using the same technique as before.

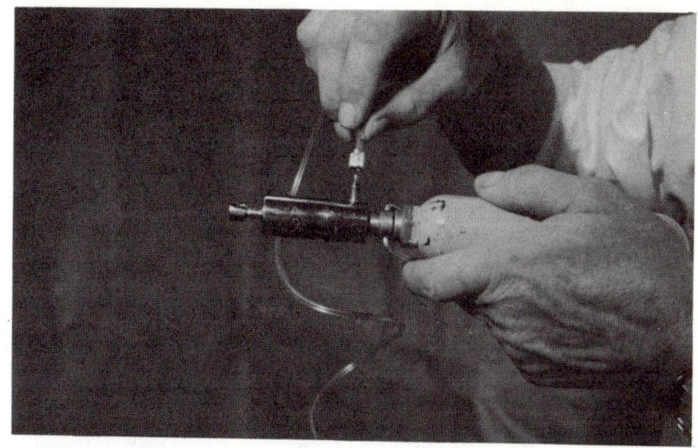

14. Connect the other end of this piece of tubing to the remaining fitting on the slide valve.

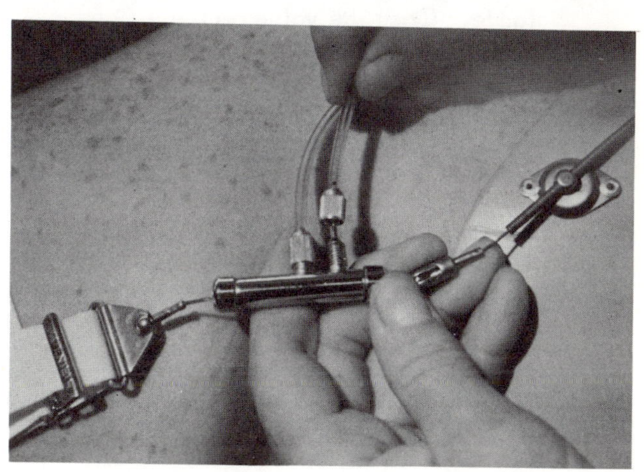

15. Put the CO_2 tank in the patient's pocket or other temporary place. The CO_2 hook-up should look like that in the illustration at this point.

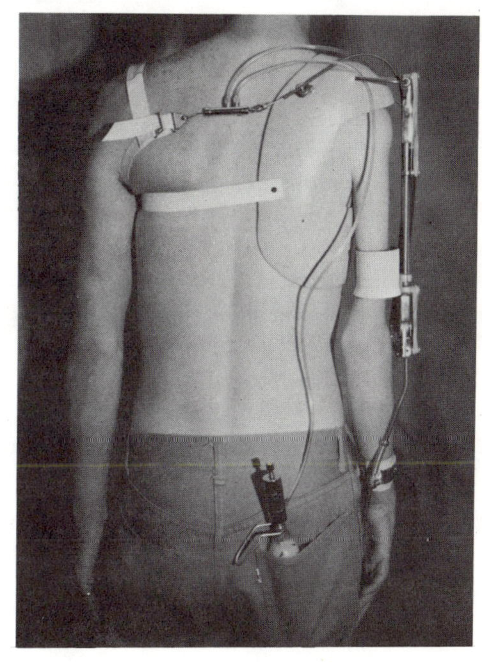

16. Check all CO_2 connections, artificial
muscle length adjustment, functioning of
ratchet and elbow locks, and tightness
of all set screws and fastenings on all
parts of the brace. Check all rubber
band assists at shoulder and elbow for
balance. Open the valve on the CO_2
tank half a turn.

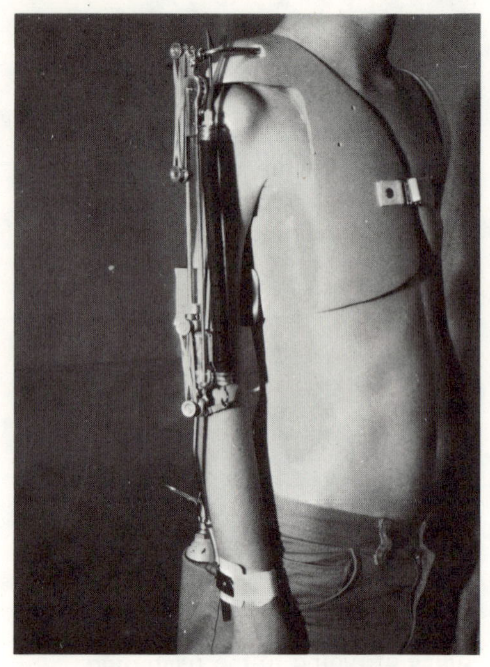

17. Teach the patient to abduct his scapulae
to operate the slide valve.   The artificial muscle should expand and shorten,
flexing the elbow.   If the elbow flexes
too rapidly, it is necessary to insert a
flow control into the tubing between
the slide valve and the artificial muscle.
The valve in the flow control is turned
clockwise to slow the action of the artificial muscle.

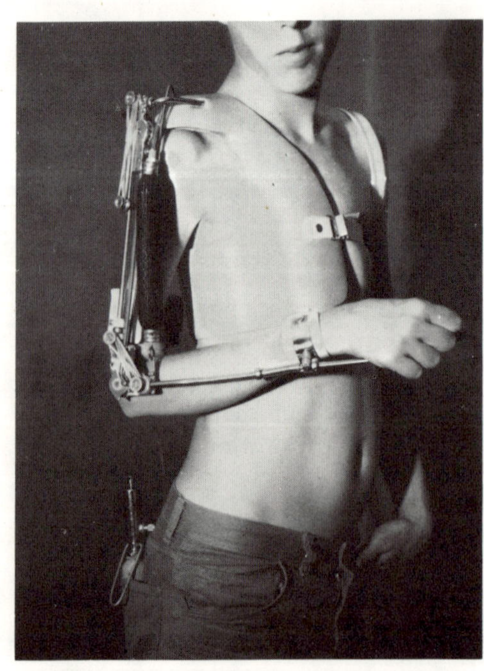

18. Direct the patient to then flex his shoulder. If his shoulder flexors are too weak, bending the trunk forward will allow the ratchet lock to flex. Where only a small amount of help is needed, a little "body english" will swing the arm and flex the ratchet lock. Every click the ratchet gains is held until elbow extension unlocks it.

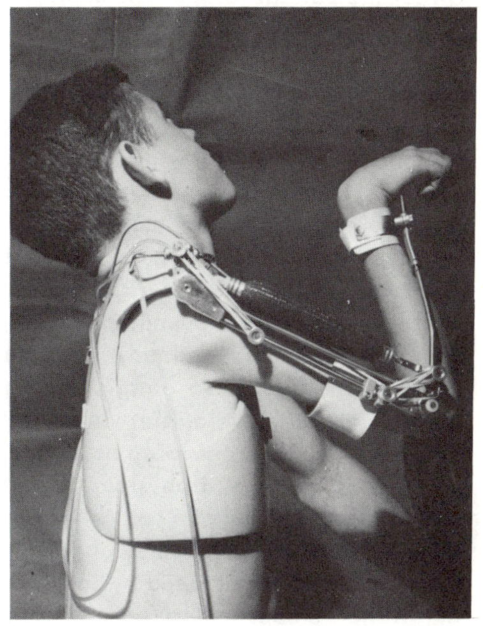

19. Enough shoulder flexion can usually be obtained to get the hand to the face and head. After some practice and instruction, most patients can accomplish this maneuver with a minimum of awkwardness.

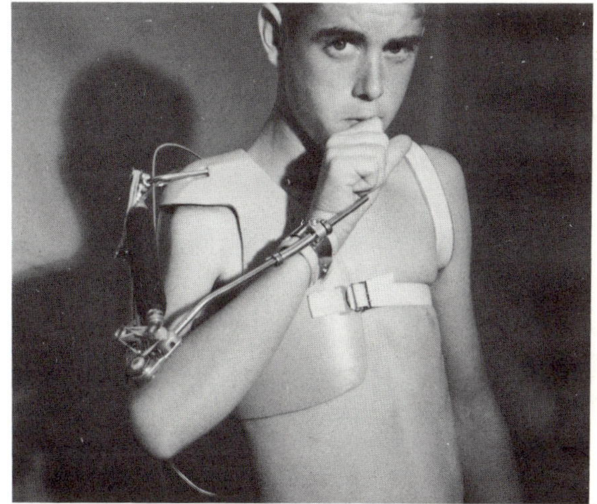

20. Direct the patient to open his control valve to the half-way point to exhaust the gas from the artificial muscle. The elbow should unlock and extend, and as it reaches full extension, the ratchet lock at the shoulder should unlock, allowing the shoulder to extend. The adjustment of the harness so the elbow will unlock at the 1/8 inch open "exhaust" position is delicate, and often requires patience to get it right. The valve must have the 1-1/2 pound spring if good operation is to be obtained. The patient must be encouraged to practice using his brace so he can develop skill in its control and use. If at all possible, he should receive training from a physical or occupational therapist in both controls and use training. Almost all failures of patients to use their braces are caused by lack of training in their proper use. Very few orthotists have the time, professional background, or equipment to properly train patients to use functional arm braces. Cooperation with the therapist in making adjustments and alterations to improve function will go far toward insuring a good training job and a satisfied brace wearer.

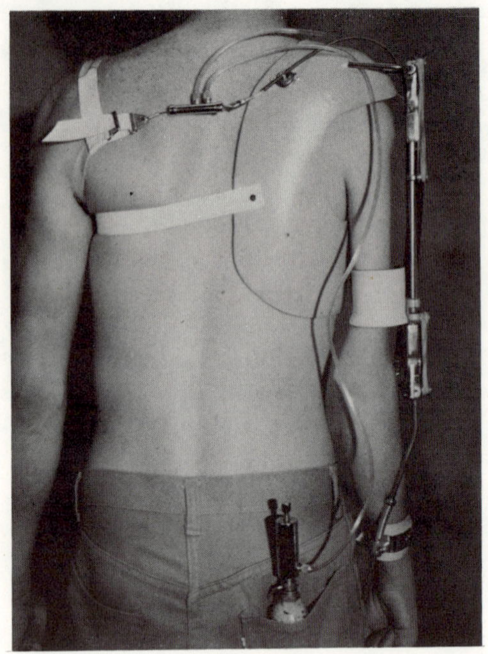

HOW TO MAKE AN ABDUCTION OUTRIGGER

Introduction

The abduction outrigger is a shoulder cap attachment that is used when the
muscles of the shoulder girdle are so weakened that subluxation of the shoulder joint
occurs. It helps to relieve the weak muscles
of the strain of supporting the weight of the
arm.

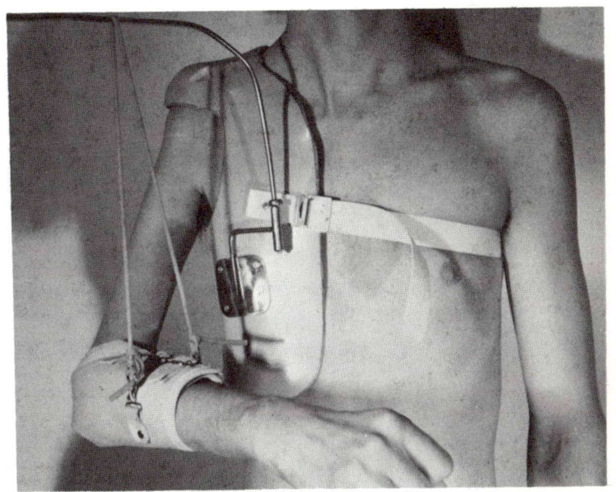

The abduction outrigger consists of
a bracket riveted to the cap, into which a
piece of 3/16 inch stainless steel wire is
inserted.   The wire extends vertically to
the approximate height of the patient's
shoulder, at which point a right angle bend
extends it in a horizontal direction for 12
to 14 inches. The patient's forearm is suspended from the outrigger by a length of
rubber surgical tubing fastened to a leather
forearm cuff.

With this attachment, the patient is able to reach out to his side and then bring his hand
back to his mouth.   The rubber tubing and outrigger relieve the weakened muscles of the work of
abducting the shoulder and flexing the elbow. The level at which the hand is held can be varied
by changing the length of the rubber suspension.

If the patient's shoulder extensors and adductors have some strength, he can use them to
bring his hand down, stretching the rubber tubing, which will then give him help in flexing and
abducting the shoulder.

1. A Hosmer C-710C Cross Bar Anchor Plate
   is used as a mounting plate for the outrigger. A 1 inch piece of 3/16 inch I.D.
   stainless steel tubing is silver soldered to
   the end of a 4 inch length of 3/16 inch
   stainless steel rod. The latter is bent at
   right angles in the middle and is silver
   soldered to the anchor plate as illustrated.
   The horizontal part of the rod may be bent
   to adjust the angle of the outrigger. If
   adjustment is not needed, the 3/16 inch
   I.D. tubing may be silver soldered directly to the anchor plate, with the top end
   extending 1/4 inch above the edge of the
   plate. The plate is fastened to the shoulder cap with speedy rivets so the tubing
   socket is level with the axilla and approximately on the midline of the shoulder cap.

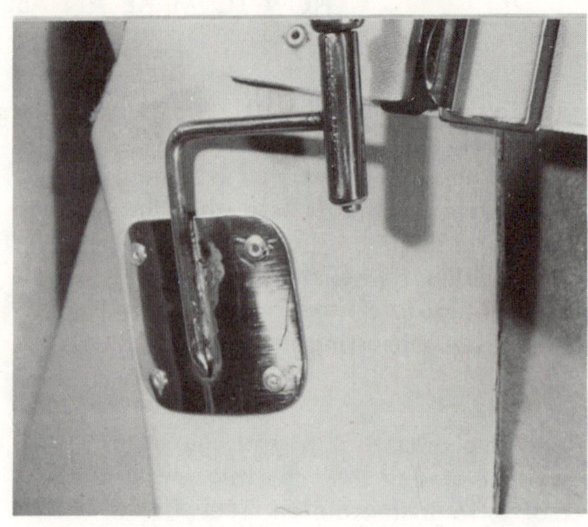

2. A 3/16 x 20 inch stainless steel rod is
   inserted into the socket, with a Hosmer
   FAB-C-9 3/16 inch stop button clamped 2 inches from the end to prevent it
   from entering too far. The rod is bent
   at right angles so the horizontal part is
   level with the shoulder joint.

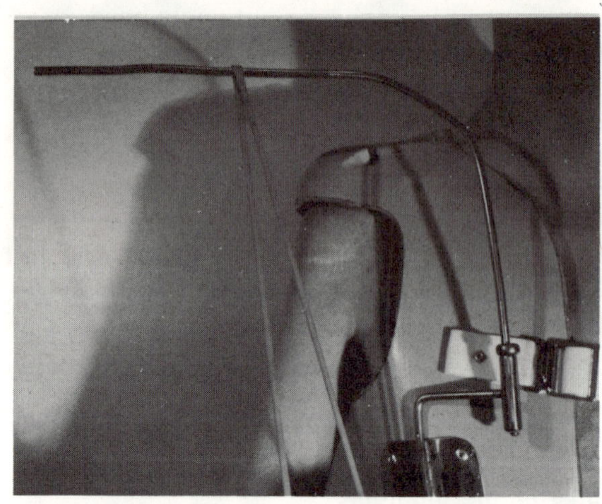

3. Fit a forearm cuff (available from
   Hosmer) to the patient's forearm.
   Cut off a piece of 1/8 inch I.D.
   1/32 inch wall rubber surgical
   tubing long enough to go from one
   attachment loop on the cuff, over
   the outrigger, and down to the
   other attachment loop.

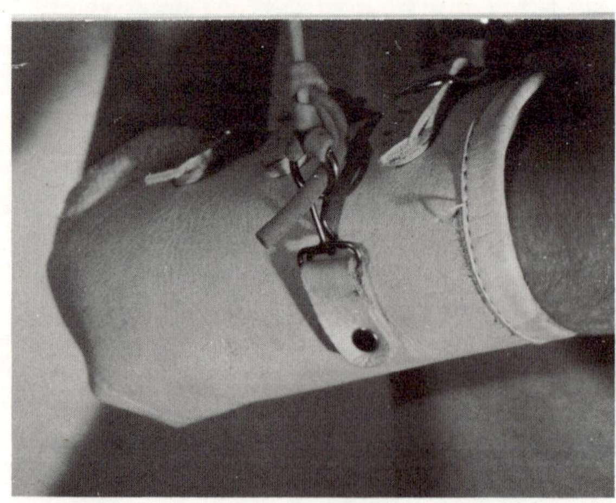

BILATERAL UPPER EXTREMITIES ORTHOTICS PATIENTS

The bilateral patient is far more difficult to fit with splints and arm braces than the unilateral, but because of the severity of his handicap, he is more highly motivated to use his equipment. Some of the most rewarding cases with which the orthotist may have the opportunity to work are bilaterals who are completely dependent on others for feeding, dressing, toilet functions, and all the other activities of daily living taken for granted by normal people. With their splints and braces, many of these patients are able to perform these activities for themselves so that their lives are completely changed.

The splints and braces used on bilaterals are much the same as for unilaterals. The difficulty is deciding whether to fit the patient with devices for both sides or for one side only. This decision must be made on the basis of the evaluation information for each individual case. In Figure 134-A a bilateral patient with poor elbow and shoulder strength is fitted with two functional arm braces, the one on the left driven by an artificial muscle, the one on the right by rubber band assists. In this case a bilateral fitting was selected because of the degree of useful residual function in both hands. The problem was to place the grasp where it would be useful, and the arm braces helped to do this.

In other bilateral cases it has been found more effective to fit one side only. In Figure 134-B we see a young woman with flail arms and hands fitted with an artificial muscle driven functional arm brace on the right side only. She was unable to operate the artificial muscle by any other means than chin pressure on the valves located on the upper chest area, and it was found that she was unable to control more than one brace in this manner. However, the one brace was invaluable to her.

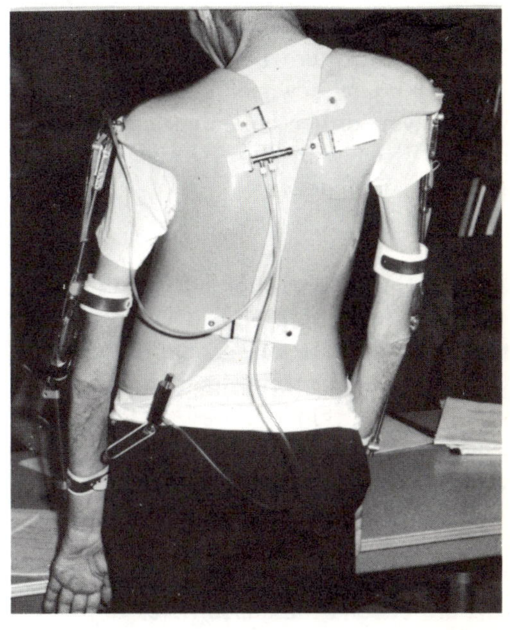

Figure 134-A. Bilateral Functional Arm Brace Patient

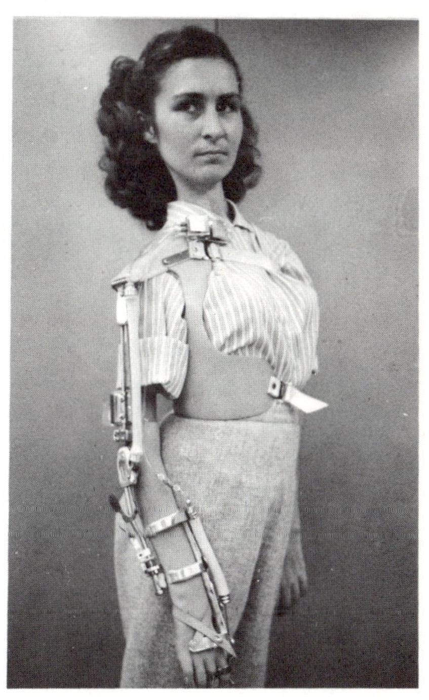

Figure 134-B.

Extremely complex bilateral fittings have been done experimentally at Rancho Los Amigos Hospital. An example is shown in Figure 134-C. The patient is confined to a wheel chair equipped with power-driven ball-bearing feeders, and is fitted with flexor hinge splints similarly equipped.

The only control motions the patient had available were flexion of the toes, so a console of six control valves was set up on a special footboard for the wheel chair, (Figure 134-D). By pressing the various valves, the patient was able to control both hands and arms.

Ultra complex fittings such as this are beyond the capabilities of the average orthotist. Patients needing this kind of help should be sent to an institution comparable to Rancho Los Amigos Hospital where the services of engineers, electronic technicians, and experimental machinists are available for designing and making the special equipment needed.

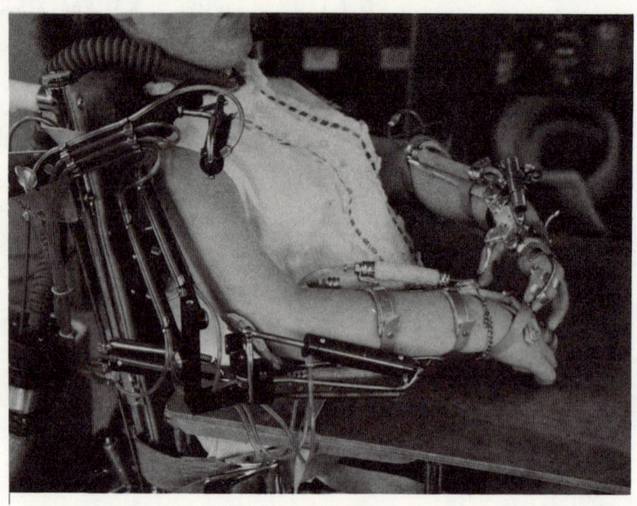

Figure 134-C.

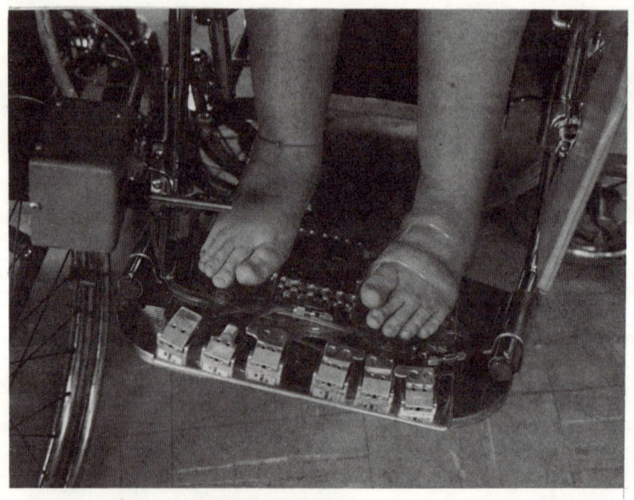

Figure 134-D.

CHAPTER IV. FEEDERS

HOW TO INSTALL A BALL BEARING SUPPORTIVE TYPE FEEDER

<u>Introduction</u>

The patient who has lost the use of the muscles that operate the shoulder and elbow is unable to reach for objects even though his grasp may be strong enough to hold them if they are placed in his hand.   In addition, he is unable to bring the hand to the mouth or other parts of the head.

Feeders are mechanical devices designed to support the hand and forearm in balance on a pivot and swivel arms or suspension straps so that very slight movements of the body or shoulders will produce useful movement of the arms, enabling the patient to reach for objects and transport them to his mouth.   The name "feeders" implies that these devices are primarily for enabling the patient to feed himself, but in addition to this function various special assistive devices make it possible to comb the hair, shave, brush the teeth, apply lipstick, handle a cigarette, and other like activities.

There are two basic types of feeders, suspension and supportive. The suspension feeder supports the arm on an overhead bar, usually attached to the wheel chair, with an adjustable strap and suspension bracket to connect the overhead bar and the metal arm trough.

The supportive feeder is provided with the same metal arm trough used in the suspension type, but in this case it is supported on swivel arms attached by a bracket to the wheel chair. Supportive feeders may be used on tables, attached to a body jacket worn by the patient, or placed on a music stand type tripod for use at a bench, or counter.

There is little fabrication work involved in feeder installation, as the various parts may be purchased for much less than it would cost to make them.

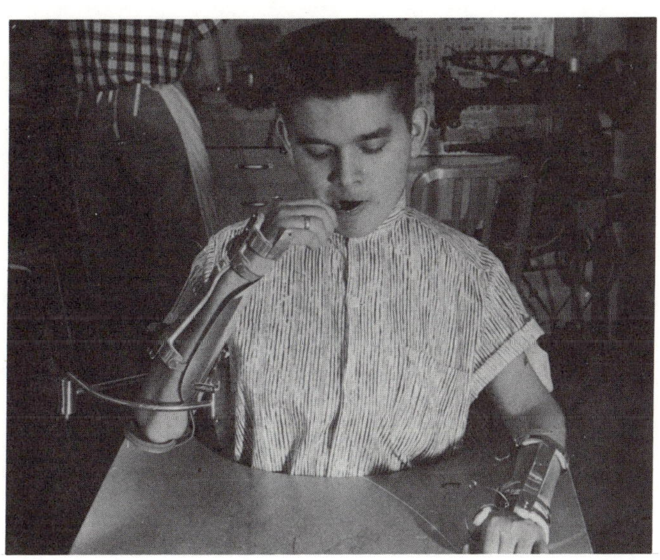

The ball bearing type supportive feeder has proven to be very effective because of the very low friction of the ball bearing swivels. When properly adjusted and balanced, the slightest force will cause the arm to move.

It is not unusual to find a patient who has sufficient muscle power to move his arm in one or more directions, but lacks power to move it in others. For example, he may be able to externally rotate his humerus, causing the forearm to move away from his body, but be unable to internally rotate it to bring the forearm back toward his body. In such cases it is often possible to increase function by so adjusting the swivels and lever arms that gravity substitutes for the weak muscles in moving the arm to the desired position.

In the following steps the procedure for installing and adjusting the ball bearing type feeder will be described in detail.

It is less expensive to purchase kits of parts from which to assemble ball bearing feeders than to attempt to fabricate them from raw materials. Three kits are available:

No. 1 Adult size for upright wheel chairs, may be had with standard or supinating rocker arm assemblies, and with or without up and down stops.

No. 2 Adult size, same as #1 except for reclining wheel chairs only.

No. 3 Child size, same style as #1. (Not made for reclining wheel chairs.)

1. Measure distance from ulnar styloid to olecranon on patient's arm.

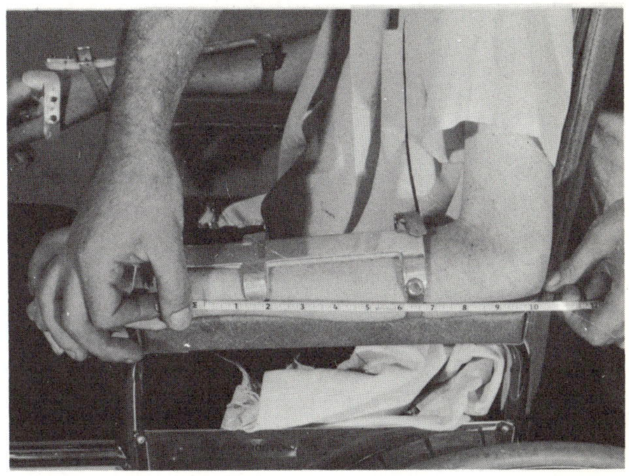

2. Subtract 2-1/2 inches from the dimension taken in "1", lay out this amount from shoulder to tip of a stock arm trough, cut off the excess metal, being sure to retain the same curvature. Smooth the edges where cut.

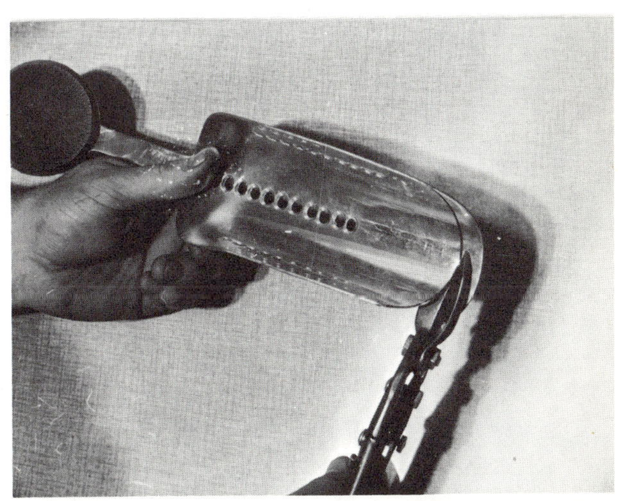

3. Using hardwood form block with a rounded face, flare the tip of the feeder trough. This is to prevent the patient's arm from contacting the edge of the metal.

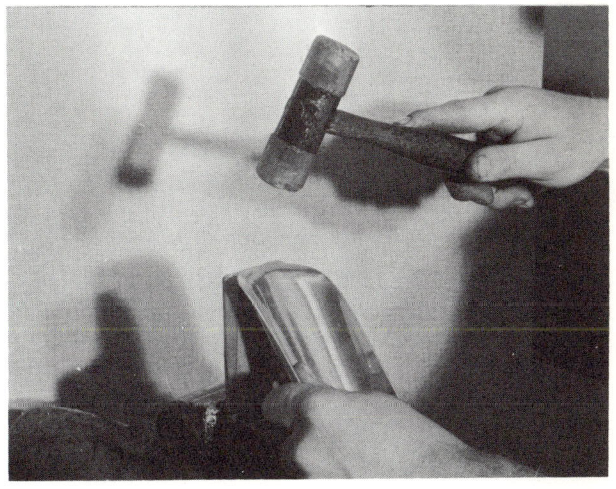

4. Flare the medial shoulder of the trough in the same way.

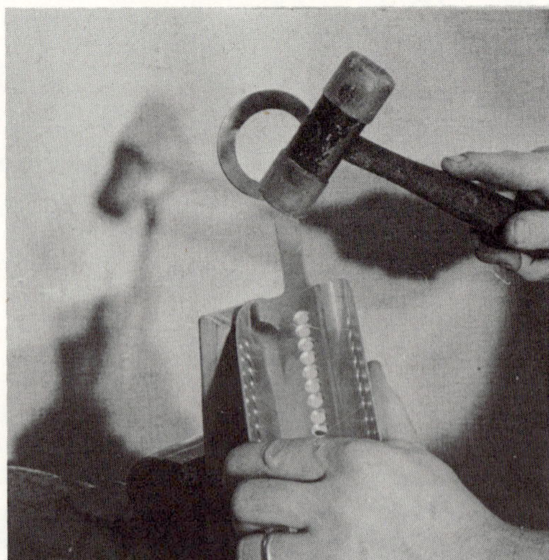

5. The feeder trough should be flared at tip and shoulders as shown in the illustration.

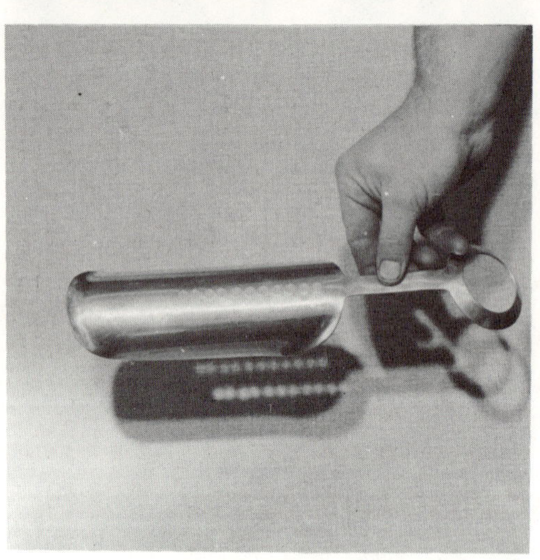

6. Bring the dial into proper position by first bending the dial arm down about 60 degrees.

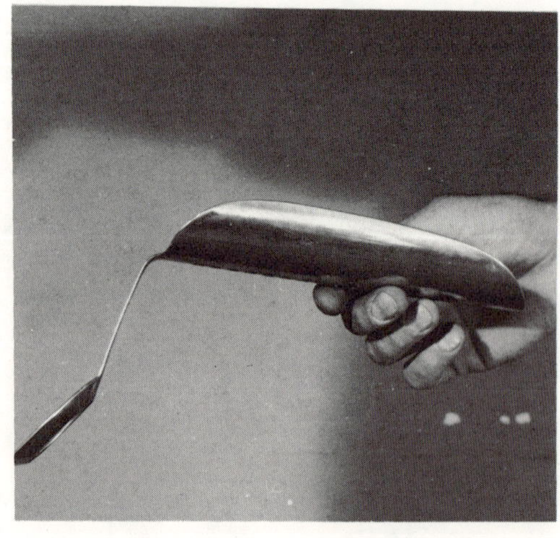

7. Shape the dial arm into a curve so the dial is placed at right angles to and about 2 inches from the shoulders of the trough.

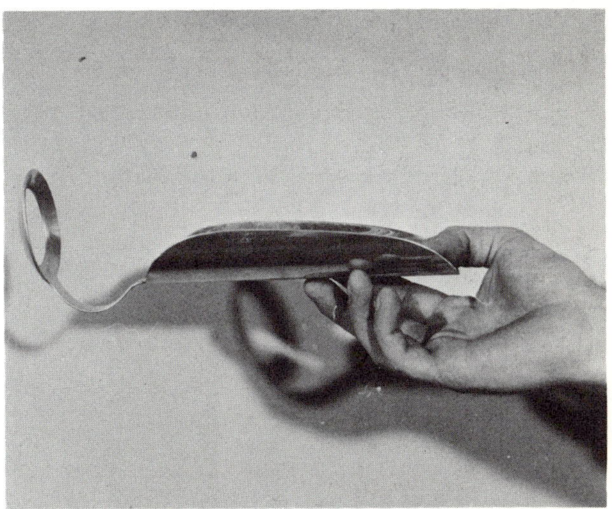

8. The dial should be aligned so it is concentric with the curve of the trough, as illustrated.

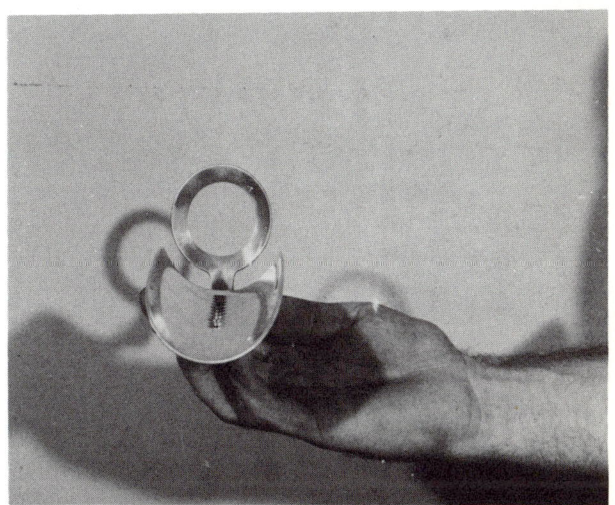

9. Install rocker arm assembly.

   a. A number of holes are provided so the rocker arm can be placed in several positions to achieve balance.

   b. The back screw in the fifth hole from the shoulders is a good starting point.

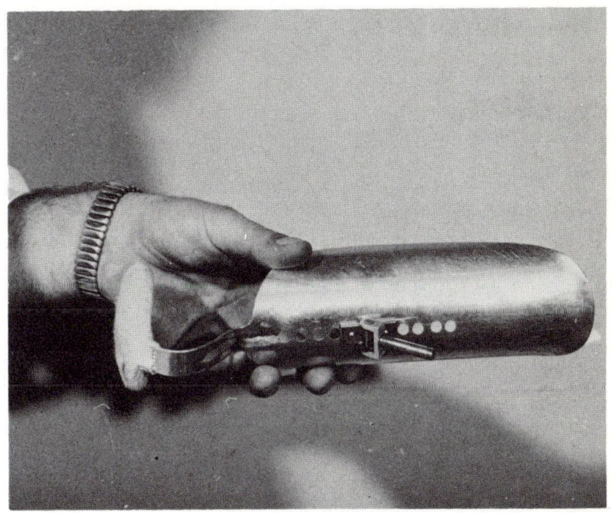

10. Install ball bearing bracket assembly.

   a. A left bracket is illustrated; note
      that the bearing housing is held
      in a vertical position, even though
      the chair back is at an angle.

   b. Leave the bracket slightly loose
      for adjustments.

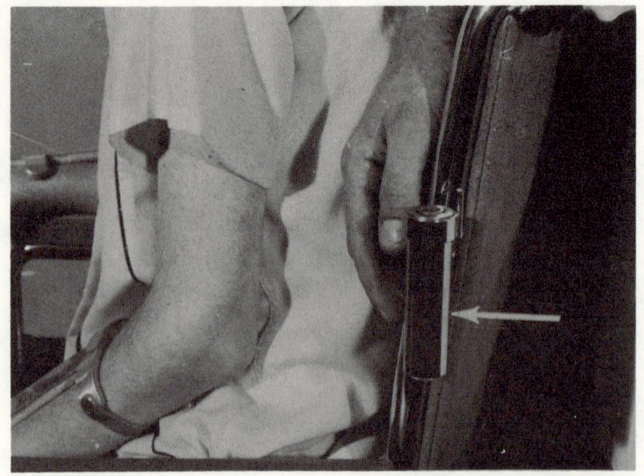

11. File the shoulders on the swivel arms.

   a. The part of the swivel arm that
      goes into the bearing will bind
      unless it is filed down on the
      upper part of the inner surface,
      as illustrated.

   b. File just enough so the arm can
      be easily removed and installed.

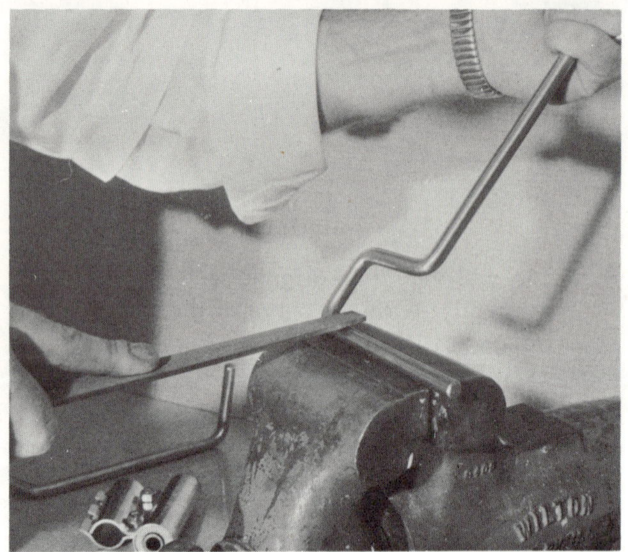

12. When properly filed, the shoulder of
    the swivel arm will look like the right
    hand example in the illustration; the
    left hand example has not been filed.

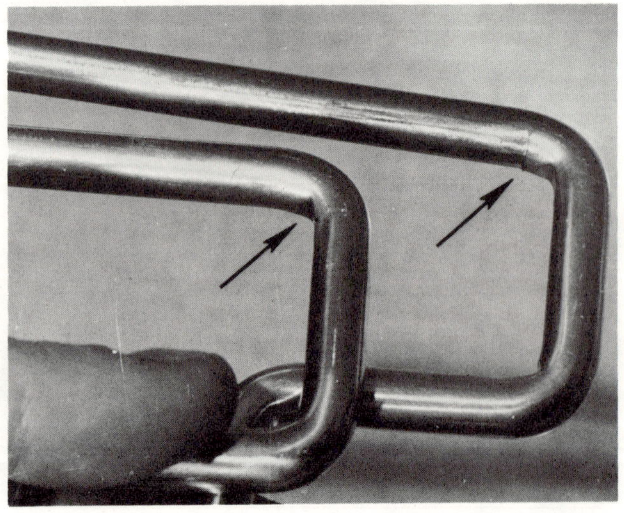

13. Adjust the height of the bracket on the wheel chair back. In the illustration the bracket is much too high, forcing the patient to elevate his shoulders in an uncomfortable position.

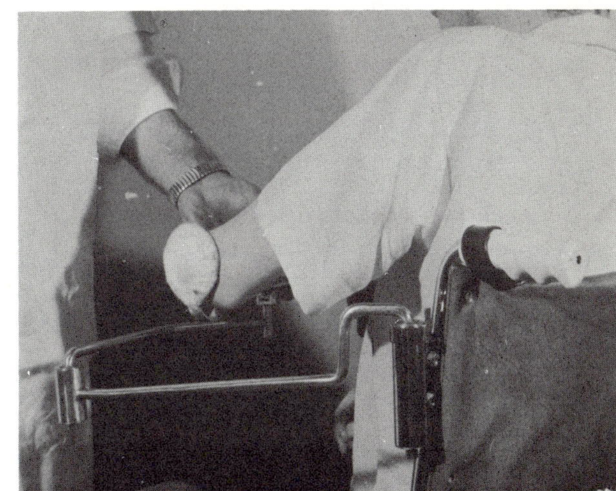

14. In this illustration the bracket is too low, so the patient is unable to reach his mouth, even with help.

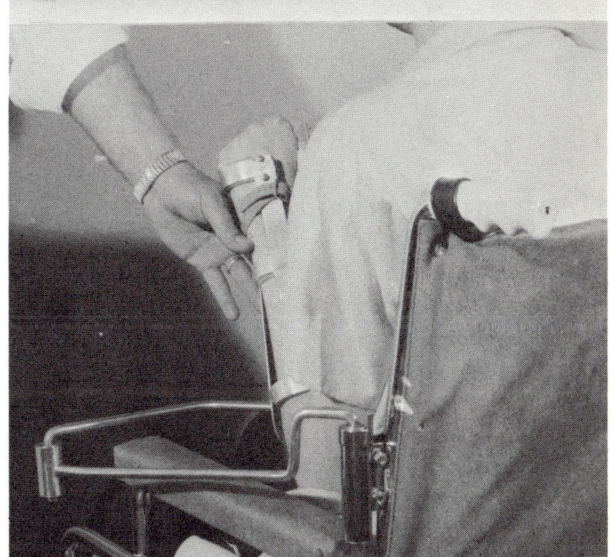

15. Here the bracket is adjusted to about the correct height; the patient can reach his mouth and his shoulders are not elevated above normal.

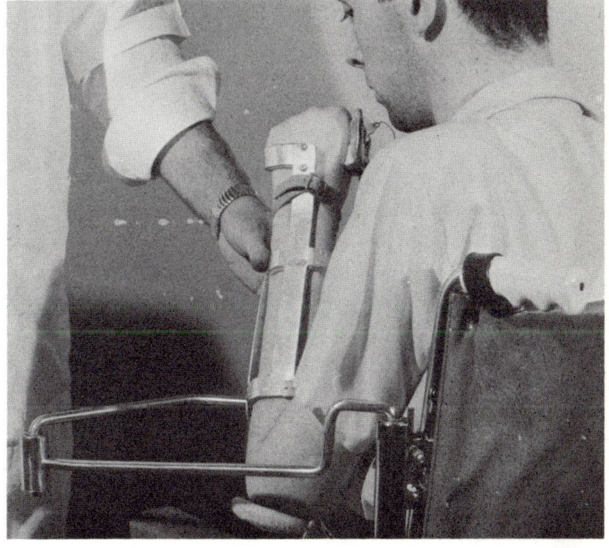

16. Check the bracket with a bubble level to see if it is at right angles to the floor. Small corrections can be made by fore and aft movements of the bracket.

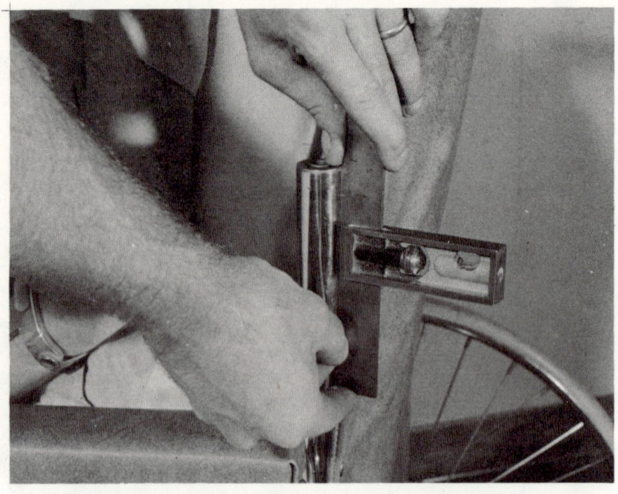

17. If fore and aft movement is not sufficient to adjust the bracket to a vertical position, make sure the clamp screws are loose, then bend the bracket slightly until it is square. If this is done with the screws tight, the bracket will be broken.

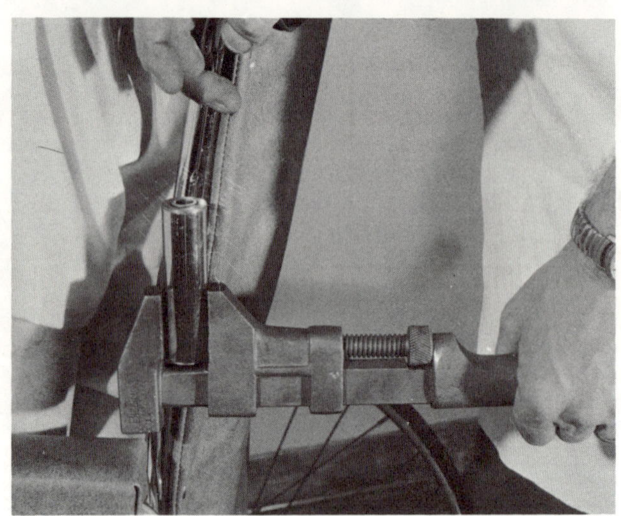

18. Place the swivel arm in place in the bracket. If the bracket is vertical to the floor, the swivel arm will not swing by itself but will remain wherever it is placed. This is the neutral point from which all adjustments should start.

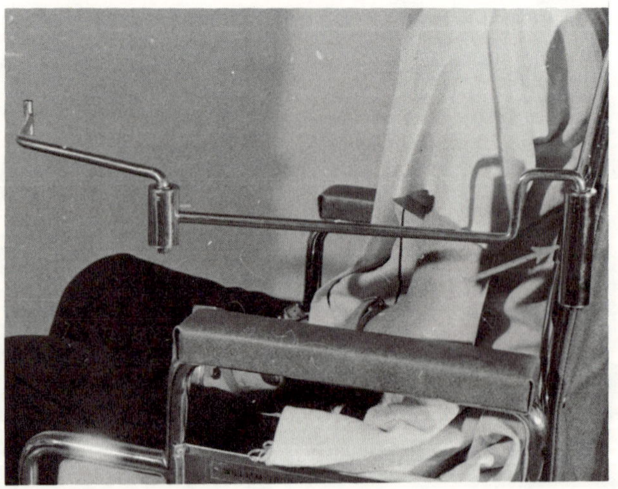

19. Tighten the clamp screws securely using a 7/16 inch socket wrench.

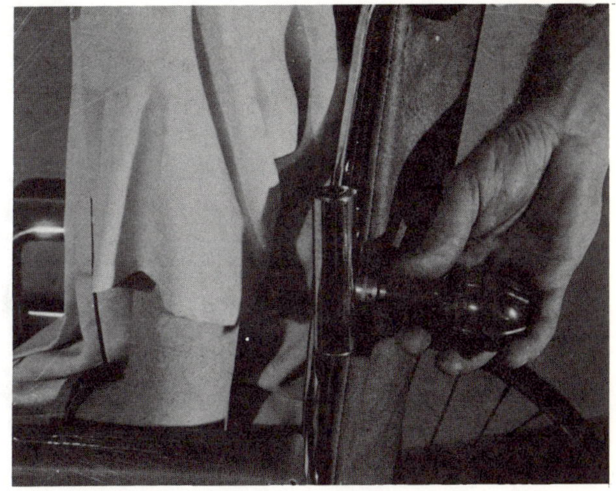

20. To balance the feeder trough, first place the patient's arm in the trough and tilt it so his hand is near his mouth.

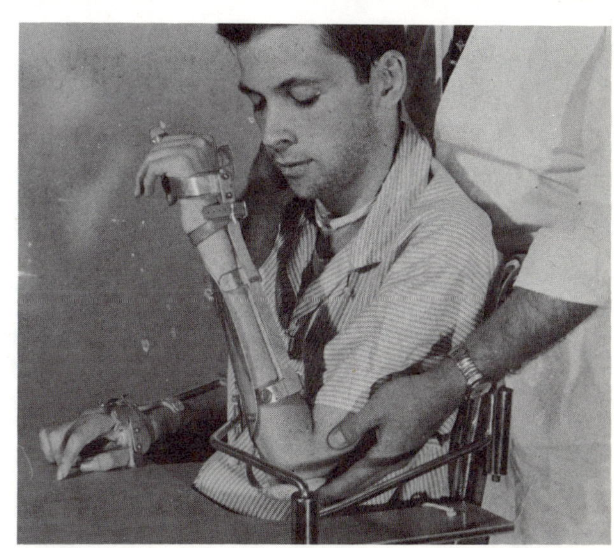

21. With his arm in this position, direct the patient to reach down to pick up something offered to him at a distance that would only require tilting of the trough on the rocker arm. In the illustration, the patient is unable to reach down and take the screwdriver, indicating that there is too much weight in the elbow end of the trough and not enough in the hand end.

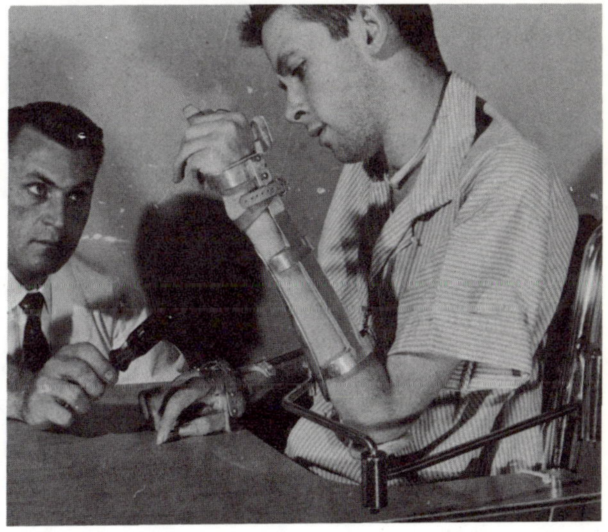

22. In this case, the illustration shows the position of the rocker arm on the trough, the back screw being seven holes from the shoulders.

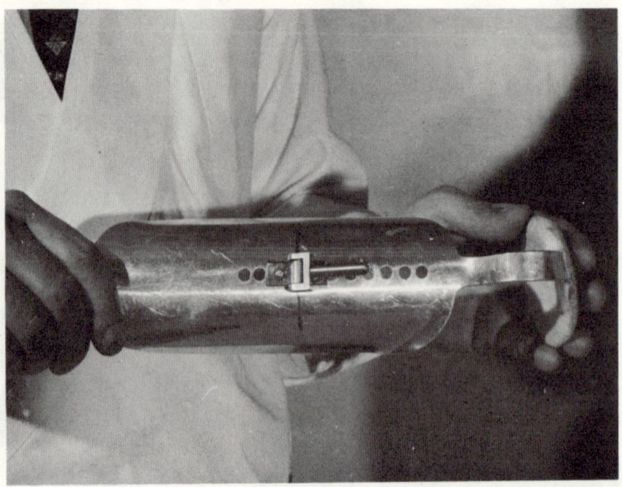

23. To place more weight toward the hand end of the trough, the rocker arm is moved two holes toward the shoulders.

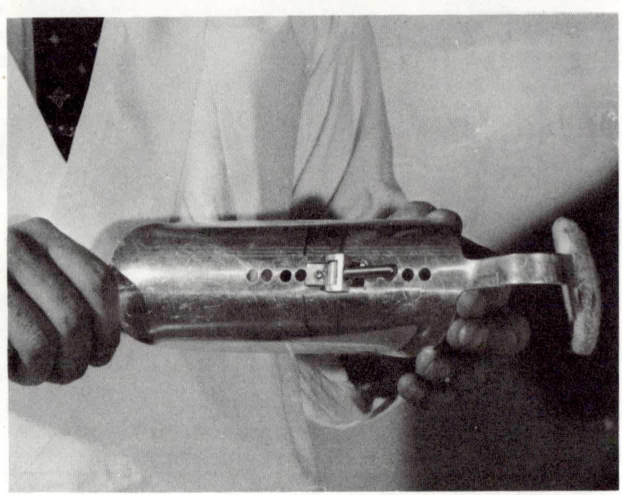

24. This relocation of the rocker arm slightly toward the elbow makes the hand end of the trough heavier, and the patient is able to reach down.

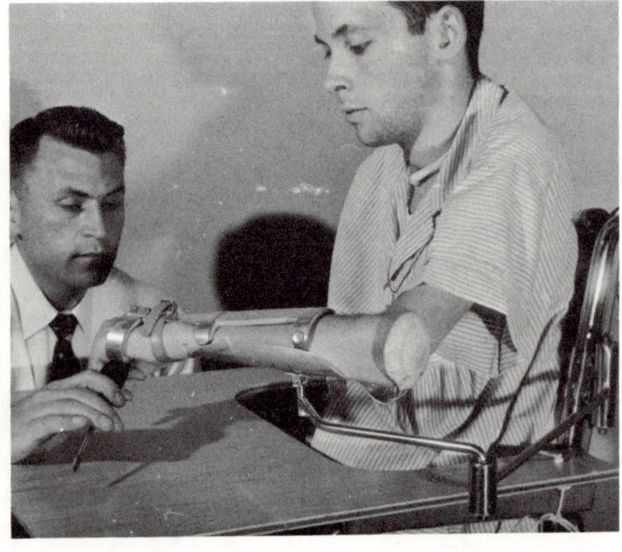

25. Direct the patient to lift the object to his mouth. In this case, the patient was able to do so, so the balance was so nearly perfect that he could both reach for objects and bring them to his mouth.

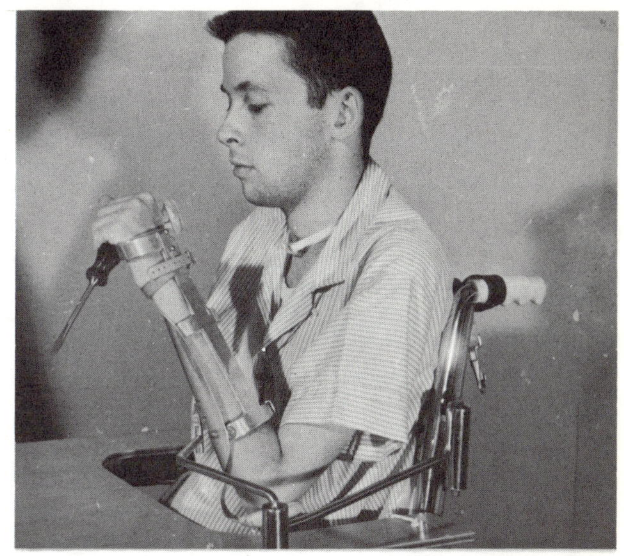

26. Let's assume the patient's trough is hand-heavy, and when we direct him to reach up for the screwdriver, he is unable to do so.

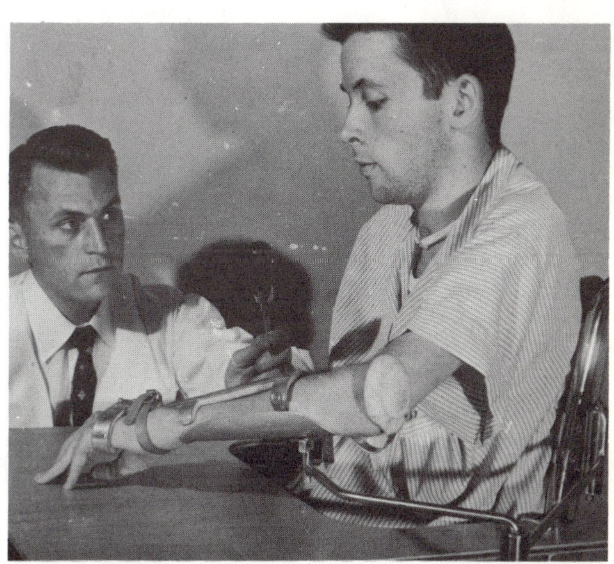

27. We find the rocker arm located on the trough as illustrated.

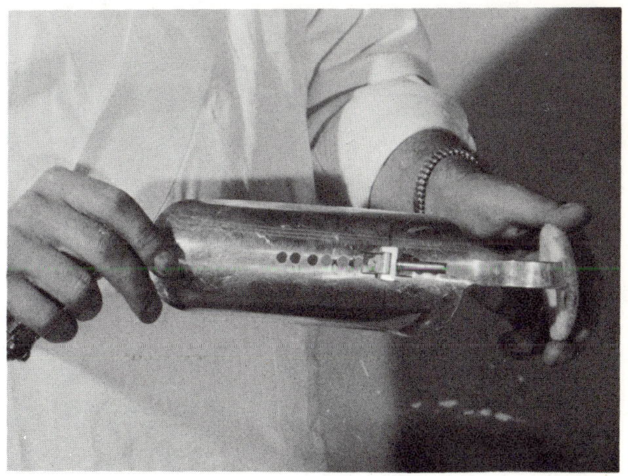

28. The rocker arm is moved two holes toward the tip of the trough to bring more weight toward that end.

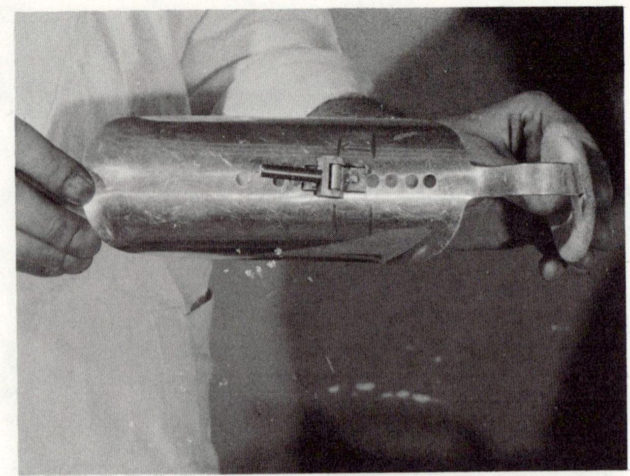

29. The patient tries again, this time with a paper cup.

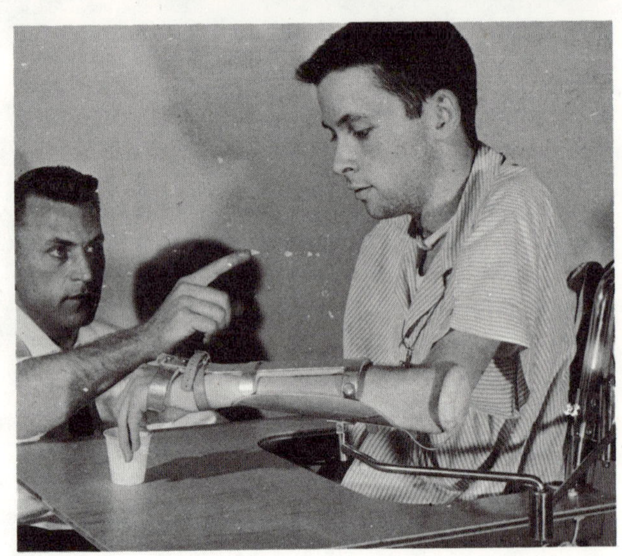

30. He is able to bring it to his mouth.

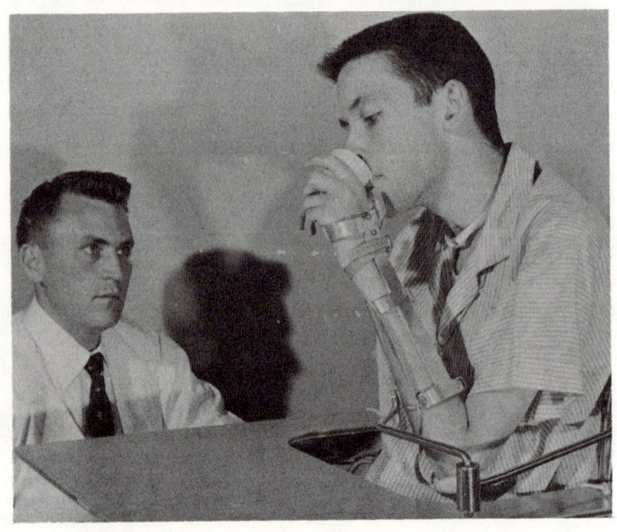

31. He is able to place it back on
the table.

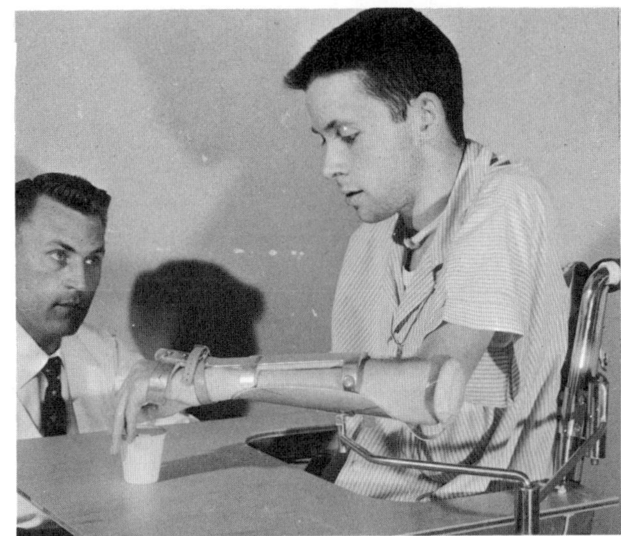

32. He can tilt his arms up.

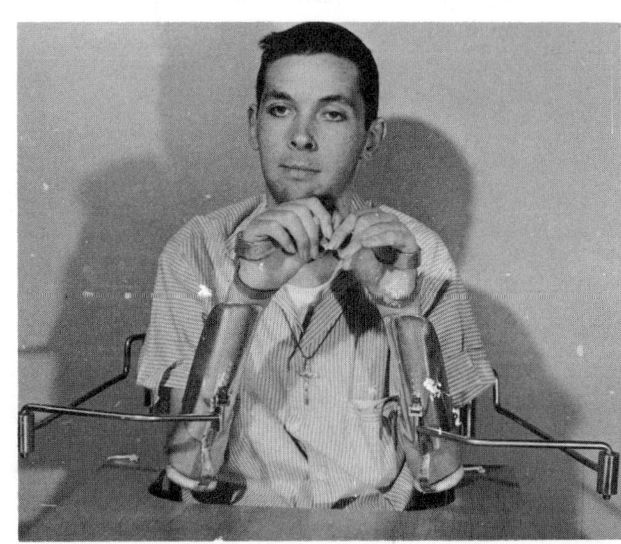

33. He can tilt them down.

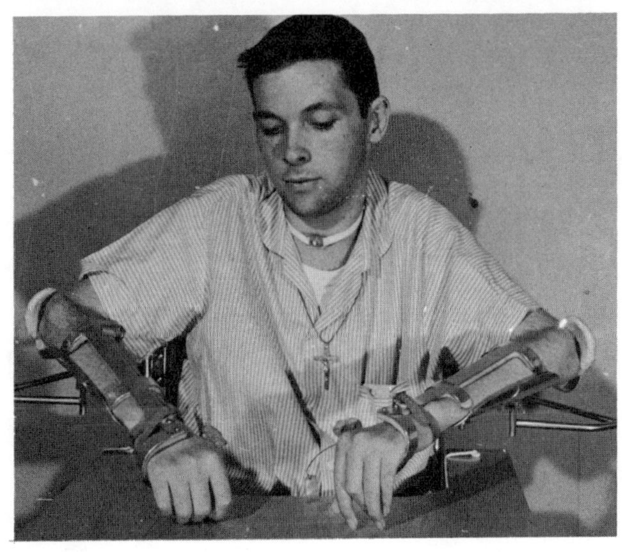

34. Adjust proximal swivel arm for gravity assistance.

    a. If the patient is able to move his arm laterally away from his body, but cannot bring it back again, move the bracket forward on the wheel chair bar.

    b. If the patient is able to move his arm toward his body, but cannot move it away, rotate the bracket back on the wheel chair bar.

    c. The effect of moving the bracket is to tilt the ball bearing unit slightly one way or the other, which in turn makes it necessary for the patient to push a little harder "uphill" one way but enables him to "coast" the other.

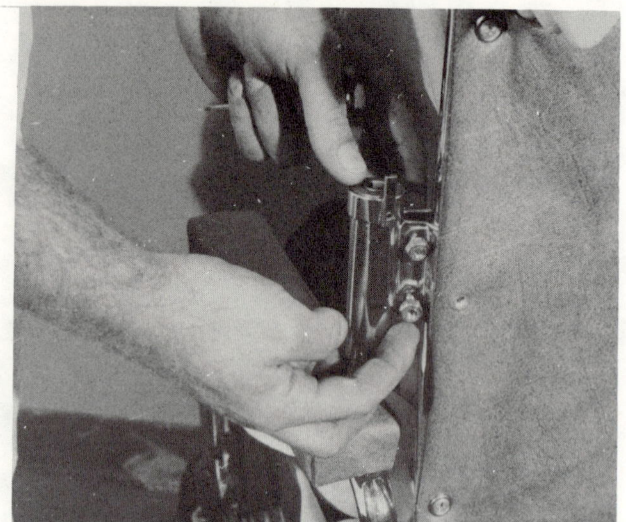

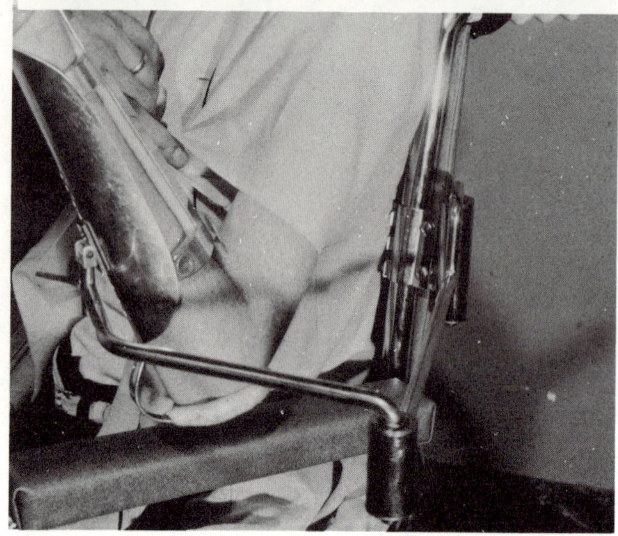

35. Adjust distal swivel arm for gravity assistance. If the patient cannot reach out in front of himself, but is able to pull his arm back if it is extended for him, he needs gravity assistance for reaching.

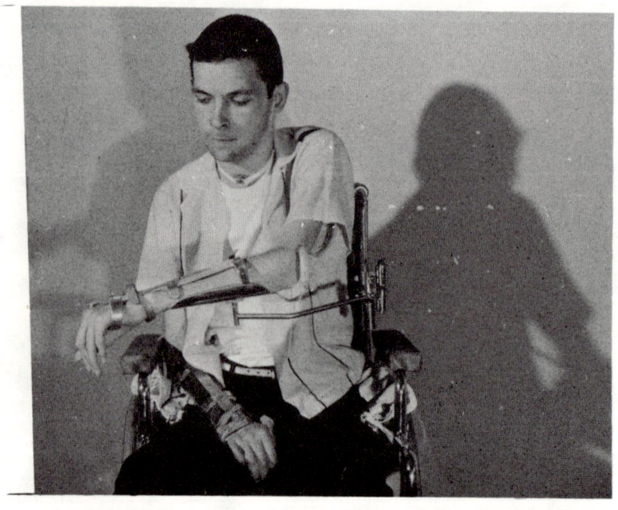

36. Make a slight bend in the swivel arm so the upper end of the bearing is tilted outward as shown in the illustration. This will give gravity assistance in reaching.

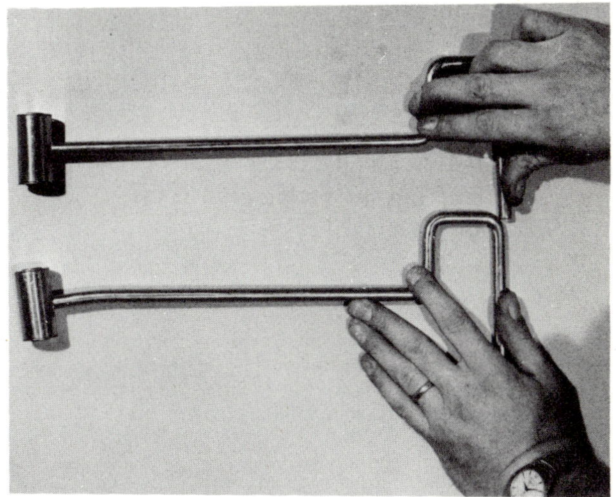

37. If the patient is able to reach, but cannot draw his arm back again, make the bend in the opposite direction. This will give gravity assistance in drawing the arm back after reaching.

A very careful adjustment is necessary to arrive at a balance between the force of gravity and the force the patient is able to exert.

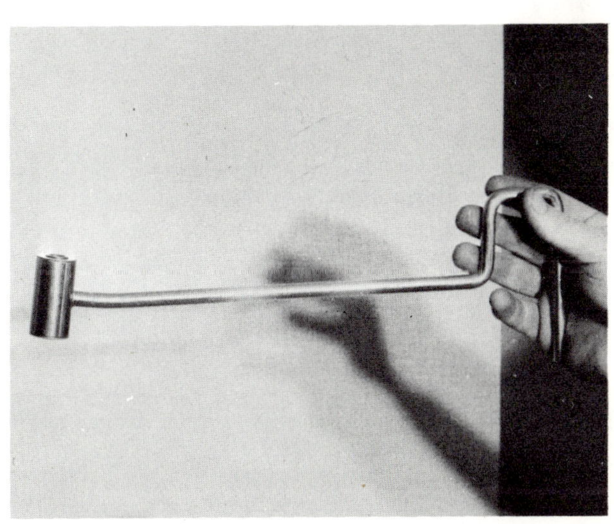

38. Install stops if needed.

a. In some cases it is possible for the patient to get his arm into a position from which he cannot extricate himself. To prevent this, stops are installed at the bearings.

b. The stops can be adjusted so the swivel arms are limited to certain positions. Tighten the set screws securely after the adjustments are made.

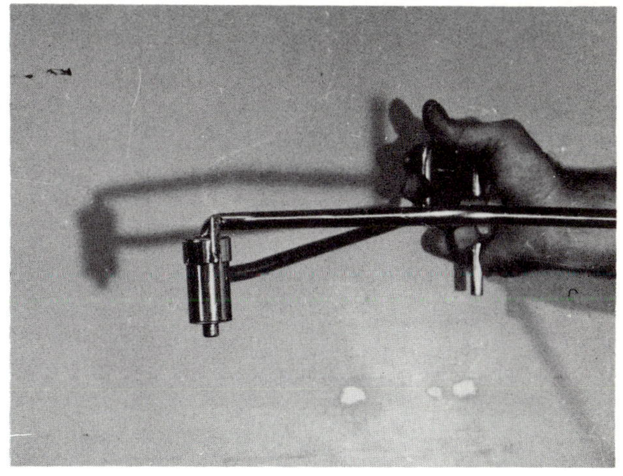

39.   Pad the dial. First, cut a circular piece of 1/4 inch foam plastic the size of the dial.

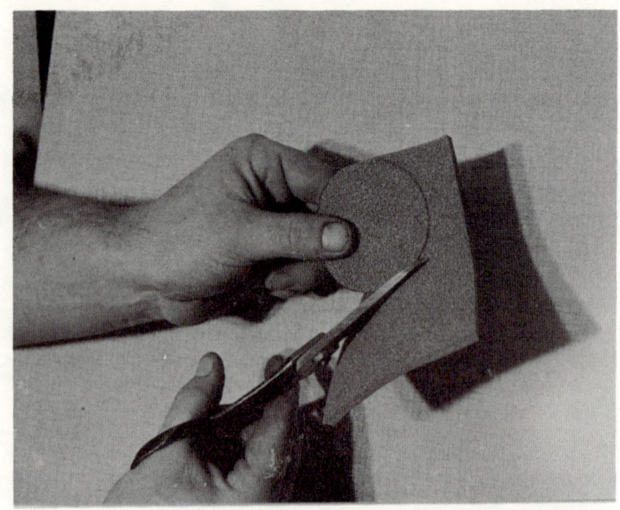

40.   Using Barge cement, cement the pad in place, cover with moleskin.

41.   Install supinator if needed. In some cases the plain rocker action of the rocker arm is not satisfactory. The boy in the illustration is using a feeder with a plain rocker arm.

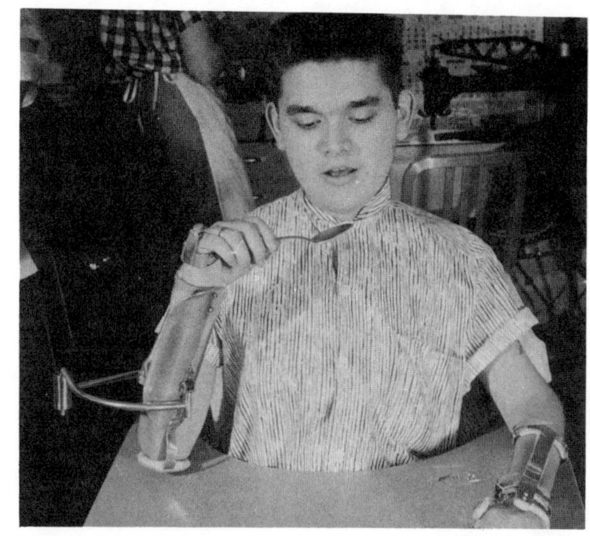

42. Note that the boy cannot get the spoon to his mouth.

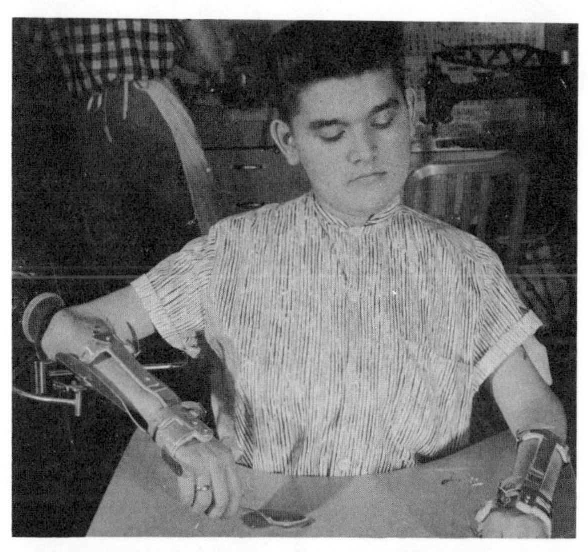

43. The same boy using a feeder equipped with a supinator.

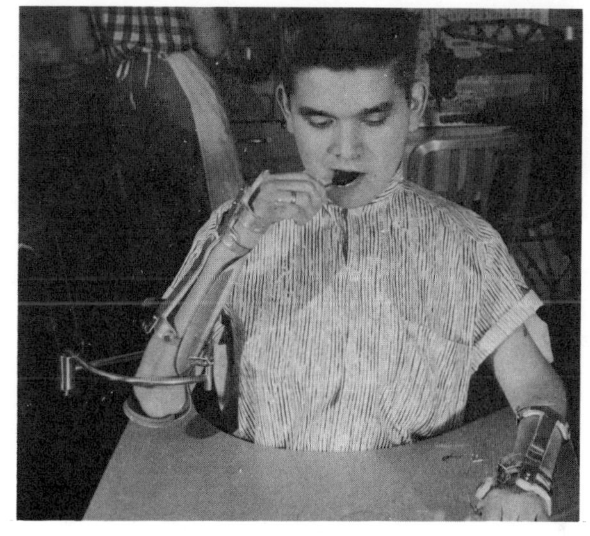

44. Note that the forearm is supinated slightly, bringing the spoon to the boy's mouth.

45. The mechanism of the supinator
is simple. In the illustration
the trough is removed so the parts
can be seen, but if still in place
it would be in the down position.
Note the angle of the plate to
which the trough would be attached.

46. In this illustration the trough would
be in the up position. Note the
changed angle of the mounting plate.

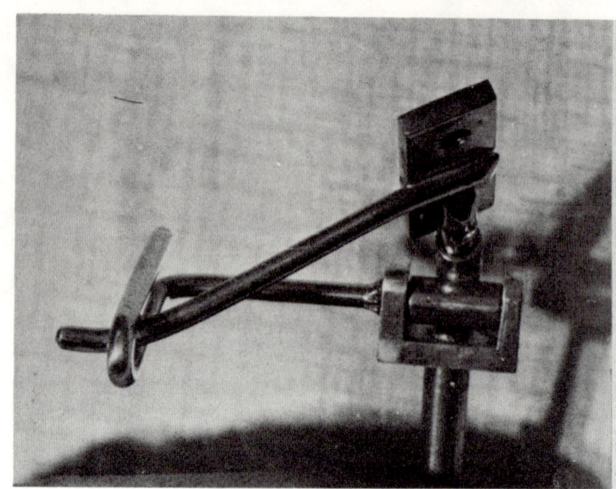

HOW TO ASSEMBLE AND ADJUST A SUSPENSION TYPE FEEDER

Introduction

The suspension type feeder is easily assembled of stock parts, and is usually installed on the patient's wheel chair. Quite often this type feeder is the first equipment of this kind given to the patient as he begins to show progress. Later, if improvement continues, the ball-bearing supportive feeder may be substituted for the more cumbersome suspension type.

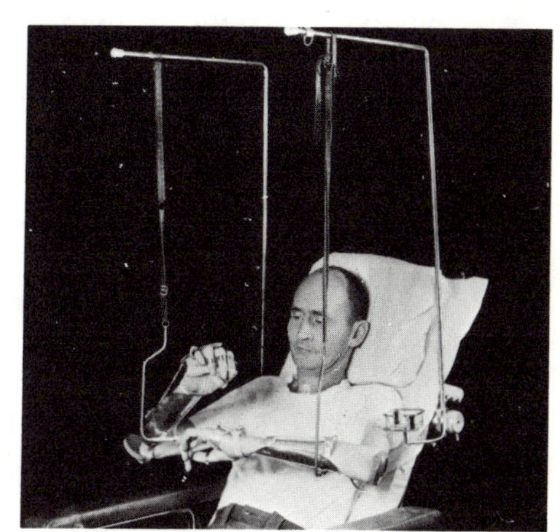

The suspension feeder consists of an adjustable clamp fastened to the back bar of the wheel chair, an "L"-shaped overhead rod that is held in the clamp, a suspension spring and strap, and a suspension bracket to hold the arm trough. All of these parts may be purchased at lower cost than they could be made in the shop, so the work consists of assembling and adjusting them.

As with the ball-bearing feeders, it is less expensive to purchase kits of parts for assembling suspension feeders than to attempt to fabricate them from raw materials. The following kits are available.

No. 1   Suspension feeder for upright wheel chairs.

No. 2   Suspension feeder for reclining wheel chairs.

No. 3   Supinator assembly for feeder.

1. Fasten the adjustable clamp to the back bar of the wheel chair. This clamp can be adjusted to several positions so it will always hold the rod vertically when the wheel chair back is adjusted to different angles.

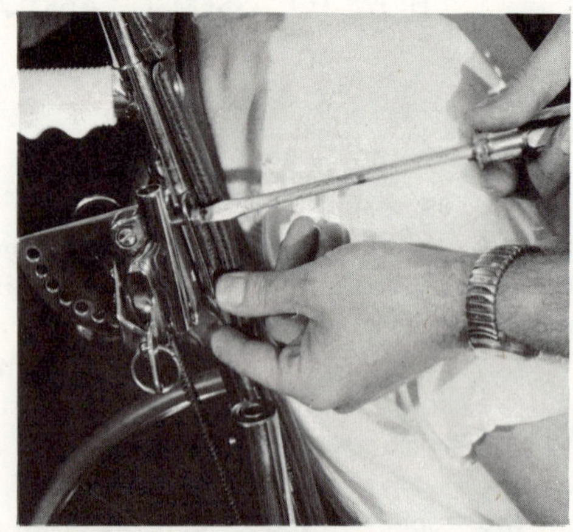

2. Install the rod in the clamp.

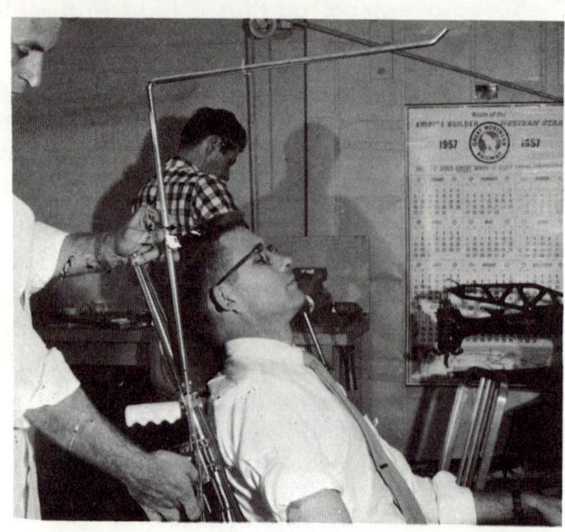

3. Assemble the arm trough, suspension bracket, and suspension strap and spring onto the horizontal part of the rod. The arm trough is trimmed, flared, and balanced exactly the same as for a ball-bearing supportive feeder.

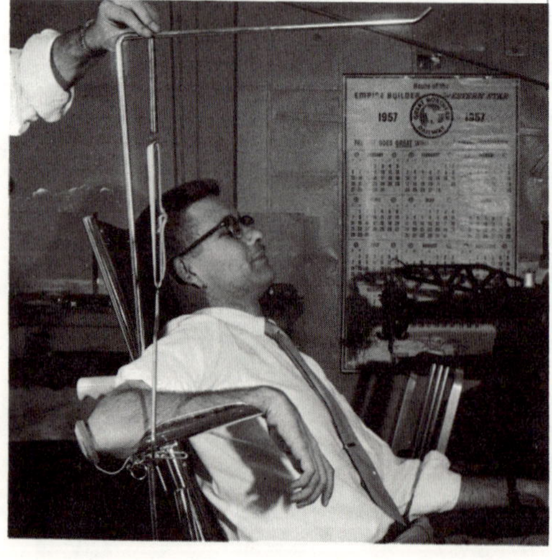

4.   Adjust the apparatus for maximum
     function for the patient.

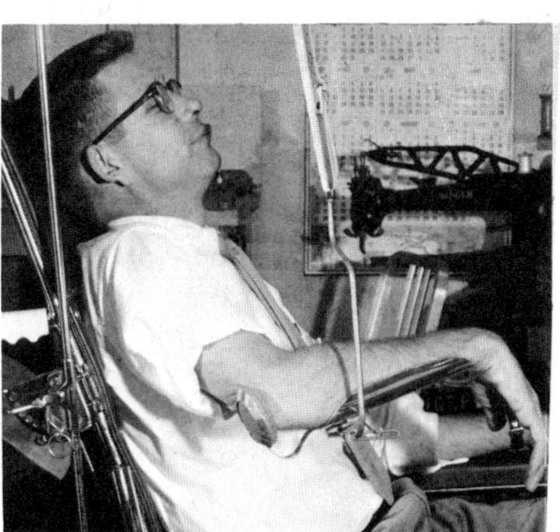

HOW TO MAKE A T-BAR FOR A FEEDER

<u>Introduction</u>

Many patients do not have enough strength to use their special assistive devices with their hands. If these patients are equipped with feeders, the addition of a T-bar to the feeder makes it possible to attach many of the special assistive devices to it. In many cases this makes it possible for the patient to comb his hair, brush his teeth, and do other things of this nature by manipulating his feeder.

1. Install two attachment studs on the bottom of the feeder, about 1/4 inch back from the front edge. Follow the same technique as when installing attachment studs on hand splints.

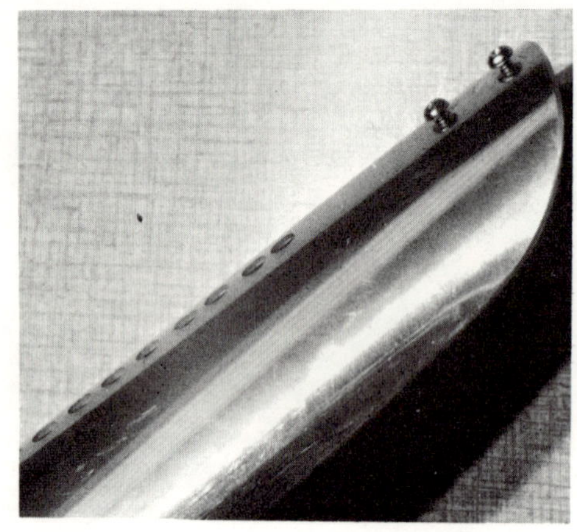

2. Make an attachment bar about 6 inches long to fit the holding screws. Again follow the same technique as described for attachment bars for hand splints.

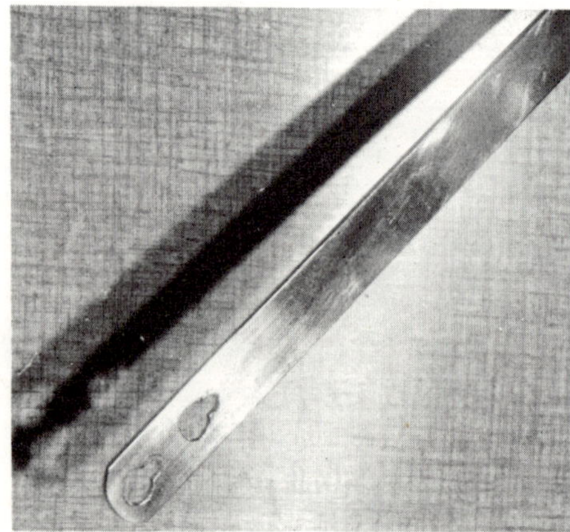

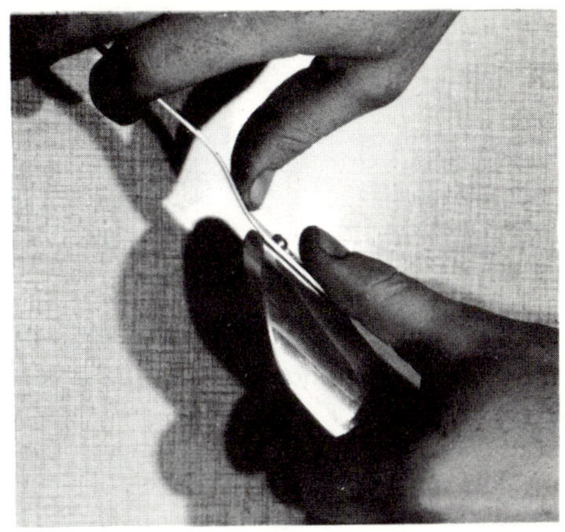

3. Bend the attachment bar in two places to allow clearance for the heel of the palm, as shown in the illustration. The amount of offset is about 3/8 inch.

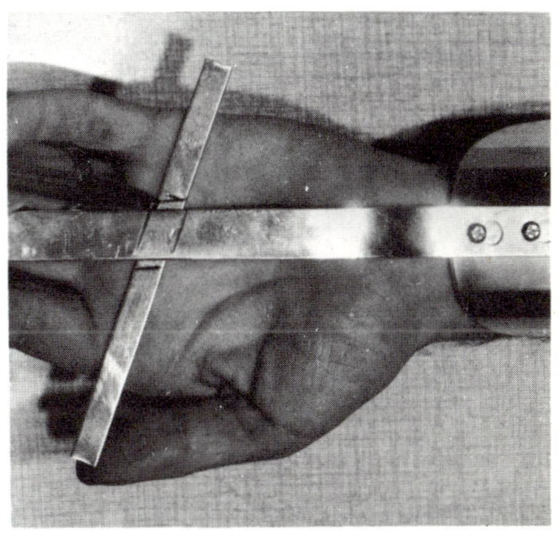

4. Make a crosspiece 3/8 inch wide and equal in length to the width of the hand at the metacarpophalangeal joints plus 1-1/2 inches. With the patient's hand in position place the crosspiece so it lies along the metacarpophalangeal pads, then mark parts as shown. Round corners and smooth edges.

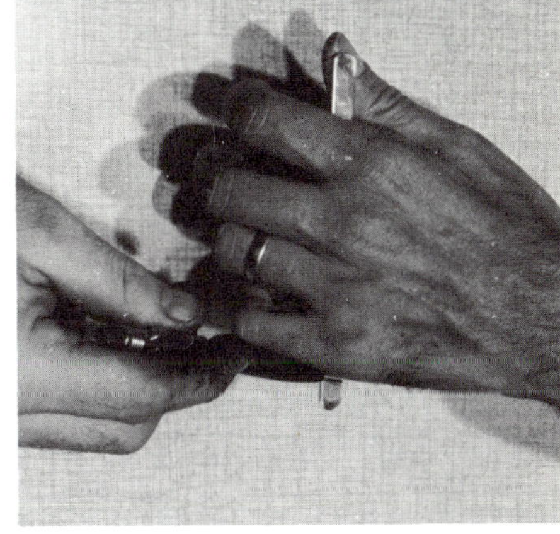

5. Drill #40 holes through attachment bar and crosspiece, then rivet them together. With the patient's hand in position, mark the location of the ulnar border of the hand on the crosspiece.

6. Bend the ulnar end of the crosspiece at
the mark to form an extension at right
angles to the crosspiece. This is to
prevent the hand from sliding off.

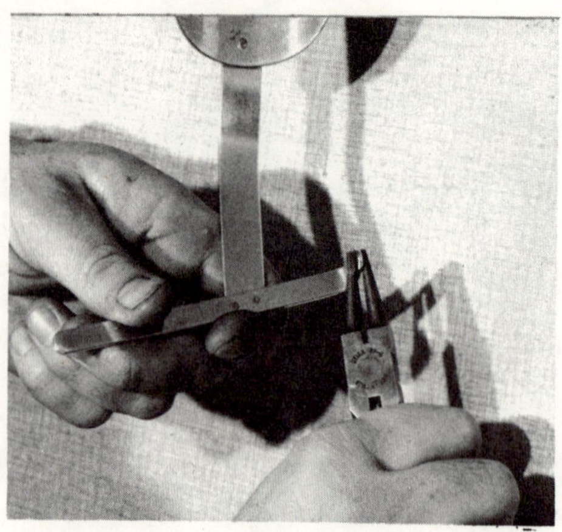

CHAPTER V. SPECIAL ASSISTIVE DEVICES

Introduction

After a patient has recovered enough strength to perform a few functional acts using splints and feeders he may broaden the range of his activities by using a variety of special assistive devices.

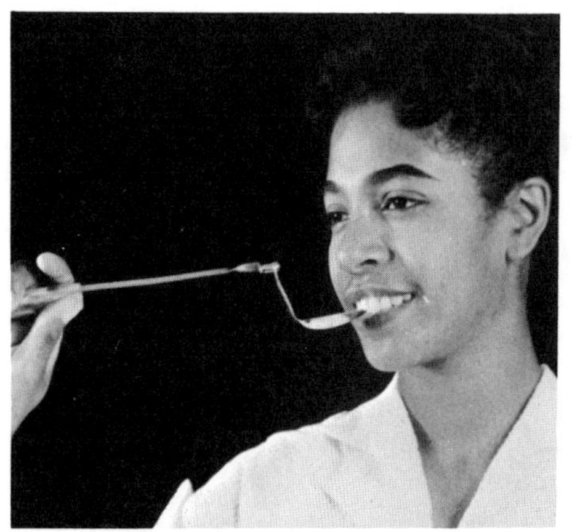

Over the years, experience has shown which devices are most useful to the patient, and only those of proven value will be explained in this section.

Since the ability to feed himself is very important to the patient, special swivel spoons, dishes, and other eating devices are much in demand. Devices that make it possible for the patient to care for his person without help, such as combing the hair, brushing the teeth, shaving, and applying lipstick are all of great importance in encouraging self-reliance and independence.

In this section detailed instructions are given for making the basic special assistive devices; however, once the principles involved are understood many others will occur to the person of imagination and ingenuity.

HOW TO MAKE A BUTTONHOOK

Introduction

The conventional b ttonhook used to
fasten button-type shoes is not the kind of
buttonhook that is useful for the person whose
fingers are too weak to button shirts, dresses,
and other garments. A special device with a
diamond-shaped wire "hook" makes it possible
for such an individual to button a freshly starch-
ed shirt with comparative ease, assuming he has
enough grasp to hold the device and arm motion to pull on it.

Buttonhooks of this type are also very
useful for amputees, particularly bilateral
amputees.

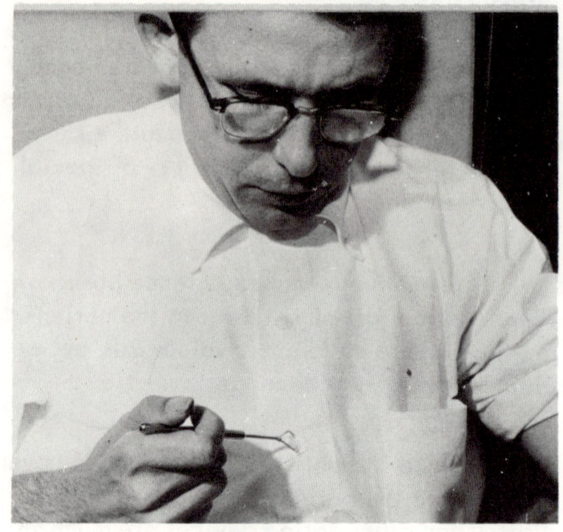

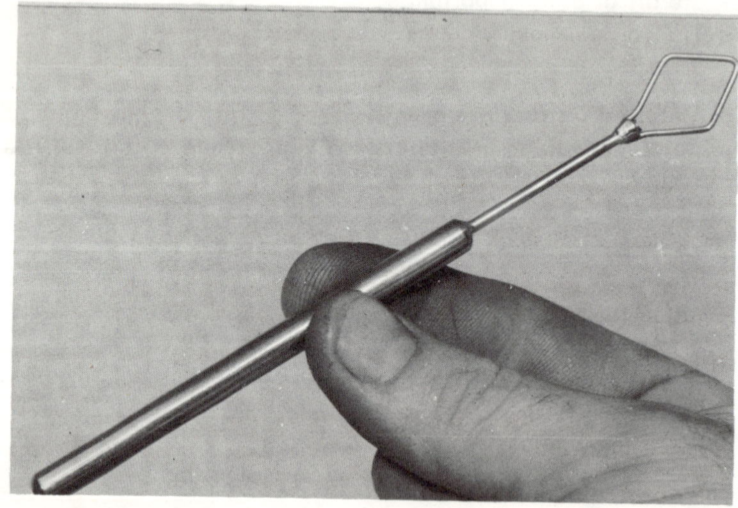

1. Make hook.

    a. Cut a 4 inch piece of .047 inch
        stainless steel piano wire, and
        bend it into the shape of a dia-
        mond, each side about 1/2 inch
        long.

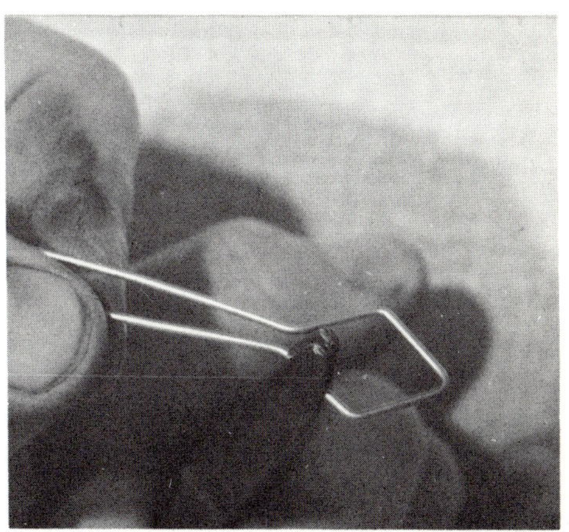

    b. Cut off the stem to a length of
        about 1/4 inch.

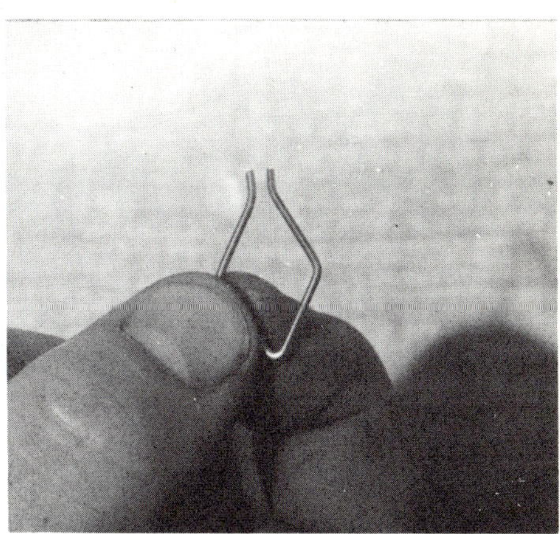

2. Cut a piece of 3/32 inch stainless
    steel wire 1-1/2 inches long, and
    silver solder the hook to one end of
    it. This should be done with the
    parts held in the vise as shown; the
    metal of the vise serves as a chill to
    prevent the heat from annealing the
    hook.

3. Clean and polish the hook and extension.

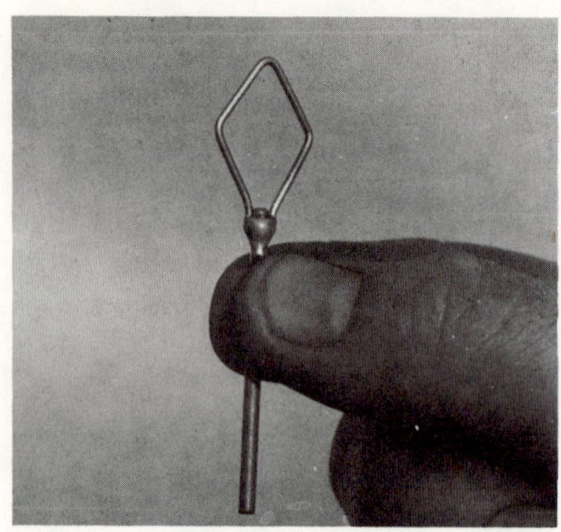

4. Cut a piece of 1/4 inch stainless steel rod 3-1/2 inches long for the handle. Round both ends.

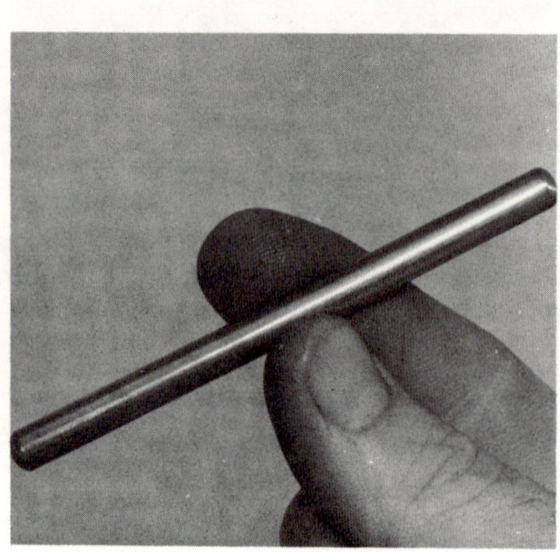

5. Center punch and drill one end of the handle about 1/4 inch deep with a #40 drill.

6. Silver solder the hook extension
into the handle.

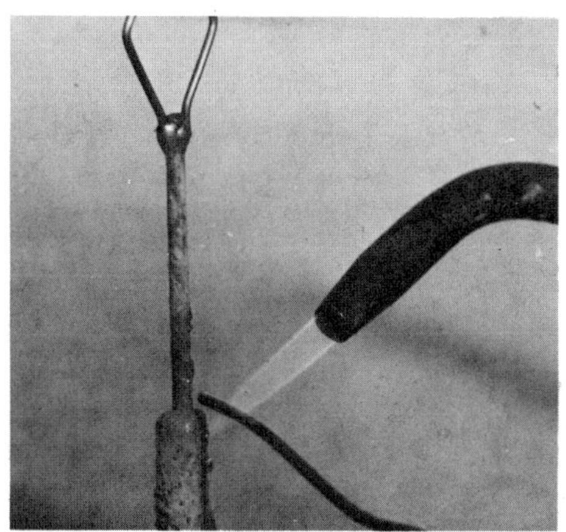

7. Clean and polish.

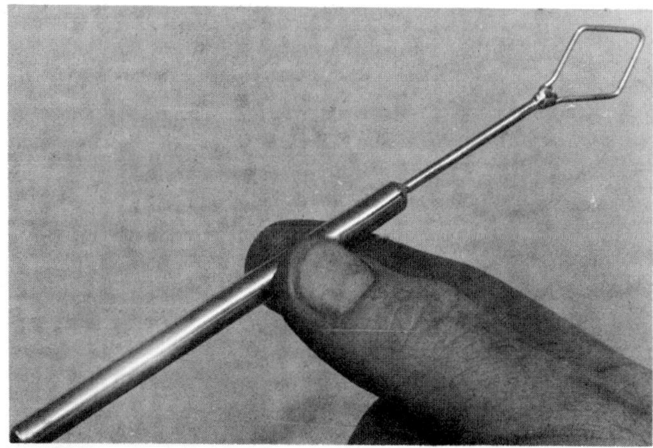

HOW TO MAKE A TOOTHBRUSH HOLDER

Introduction

It means very much to the patient to be able to care for his own personal needs, such as combing his hair and brushing his teeth. In many cases the hands are too weak to hold an ordinary toothbrush tightly enough to use it. However, with a toothbrush holder, extension and handle or clip for attaching to a "T"-bar, many patients can brush their teeth with little difficulty.

In the instructions that follow the toothbrush holder is made with a clip to attach to a "T"-bar so it can be used with a feeder. However, it would be quite simple to solder a standard 1 inch aluminum handle to the extension for patients with enough grasp to use the brush in that manner.

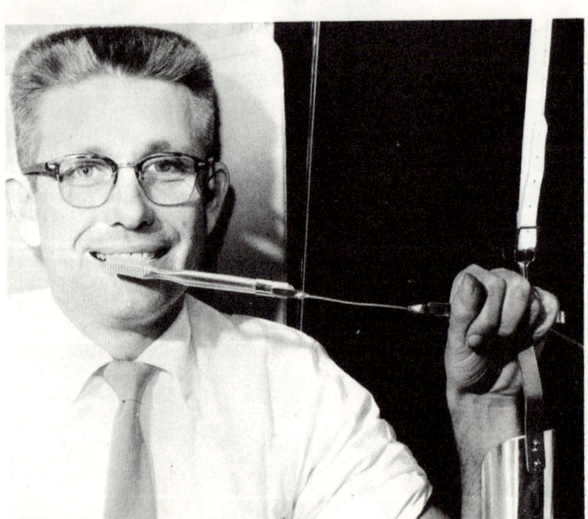

1. Make the toothbrush clip.

    a. Measure the circumference of the
        toothbrush handle at the end. Be
        sure to wrap clean paper around
        the bristles of the brush to pre-
        vent soiling.

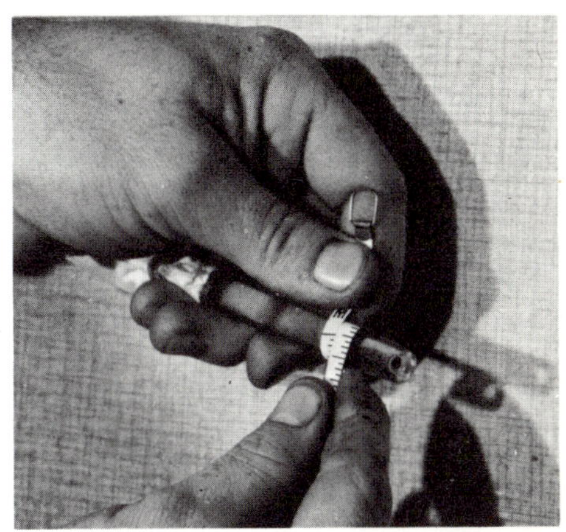

    b. Cut a piece of .025 inch stain-
        less steel 1 inch long and width
        equal to the circumference of the
        toothbrush handle.   Round the
        corners and smooth the edges.

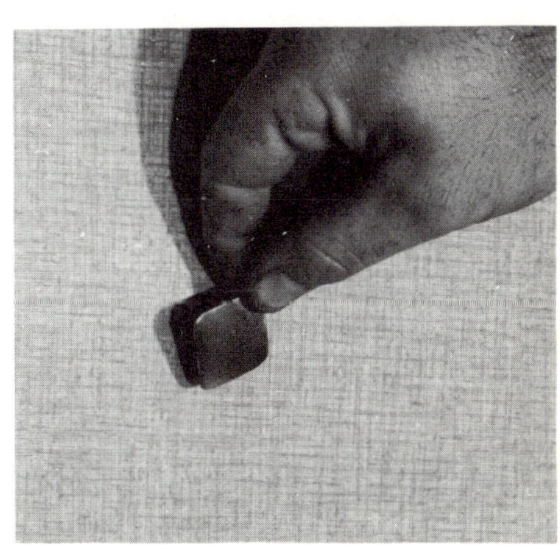

    c. Grip the piece of stainless steel
        against the toothbrush handle in
        the vise, and carefully form the
        clip using a ball pein hammer.
        Be careful to avoid nicking or
        cracking the handle of the brush.

d. The finished clip should look like the one in the illustration and should fit the toothbrush handle tightly enough so it will not come off in use.

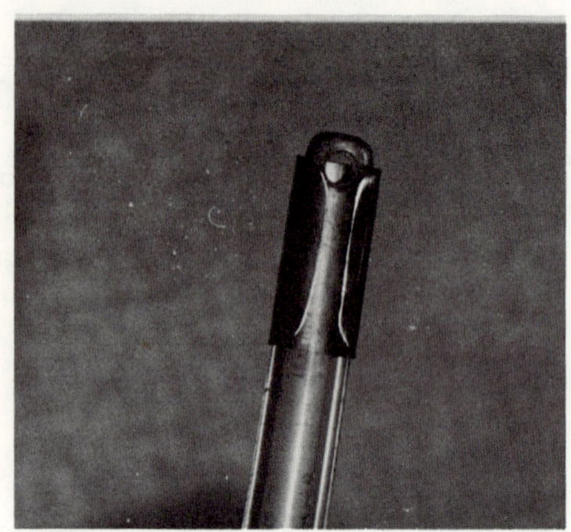

2. Make the "T"-bar clip.

a. Cut a piece of .025 inch stainless steel 1 by 3/4 inches, round the corners and smooth edges.

b. Using a piece of .064 by 3/8 inch steel for a mandrel, form the "T"-bar clip.

c. The completed clip should look
like the illustration. Adjust it
for a snug fit on the "T"-bar.

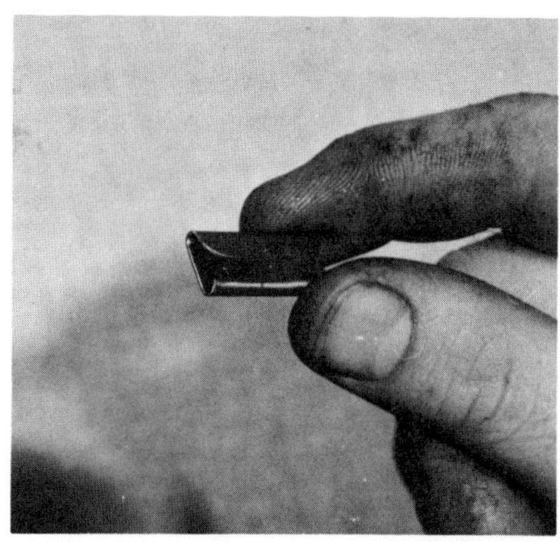

3. Cut a piece of .064 by 1/4 inch stainless steel long enough to reach the
patient's mouth from the "T"-bar,
then silver solder the clips to the ends
of this extension. Smooth and polish.

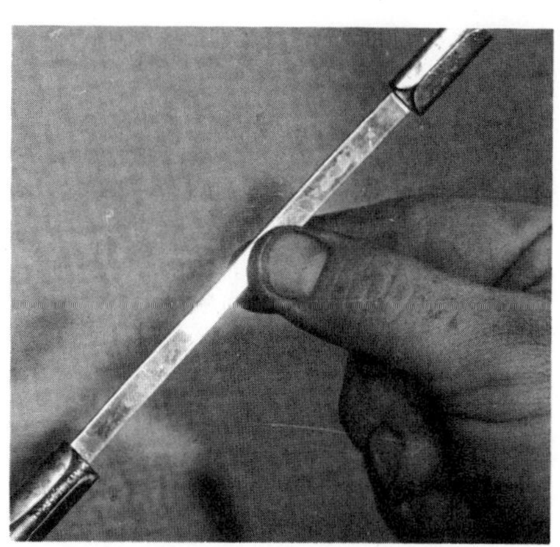

4. Put a 90 degree twist in each end of
the extension at the clip.

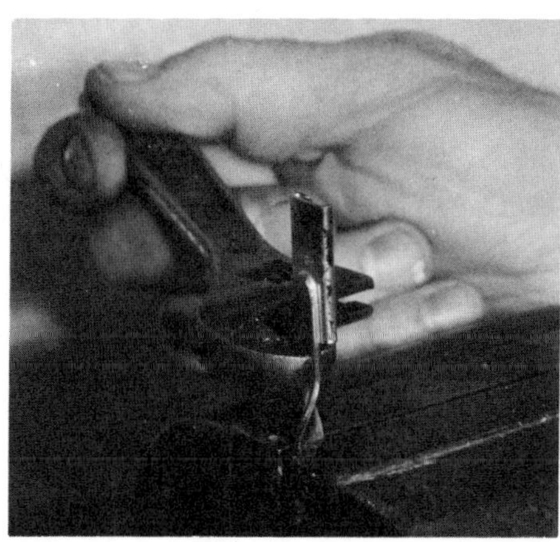

5. Assemble the toothbrush holder to the "T"-bar and make any small adjustments necessary. It is often necessary to bend the extension one way or another so the patient can get the brush into his mouth at the right angle. The clips should be adjusted so they will not come off accidentally.

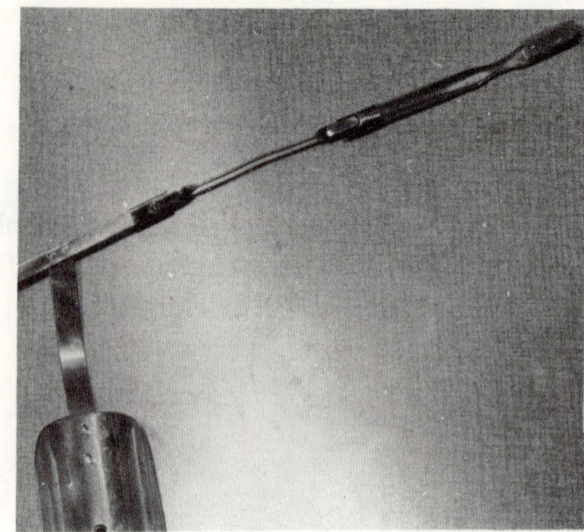

HOW TO MAKE A PAGE-TURNER FOR USE ON A T-BAR

<u>Introduction</u>

Reading helps the person who is confined to a wheel chair pass the time profitably, but turning the pages of a book or magazine can be very difficult if not impossible for the person with paralyzed hand or arm muscles.

If the patient uses a feeder with a "T"-bar a simple page turner can be made that will prove very useful

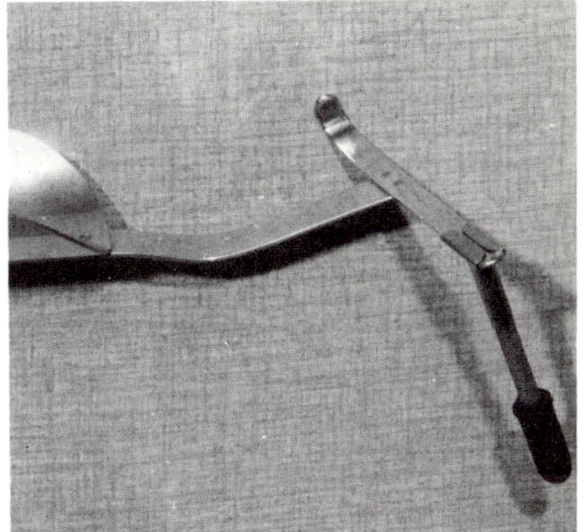

1. Cut a piece of .037 inch stainless steel 1 x 3/4 inch to make the socket; use a piece of 1/4 inch steel rod for a mandrel and form the piece into a cylinder 1 inch long.

2. Round the corners and smooth the edges.

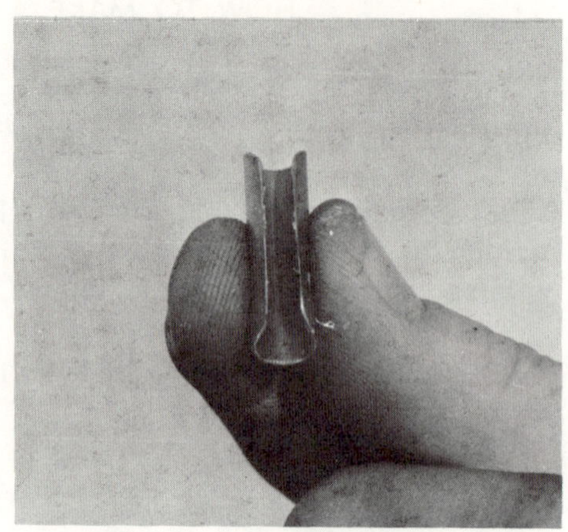

3. Make a standard "T"-bar clip, then silver solder the socket to the clip in the position shown in the illustration. Clean and polish.

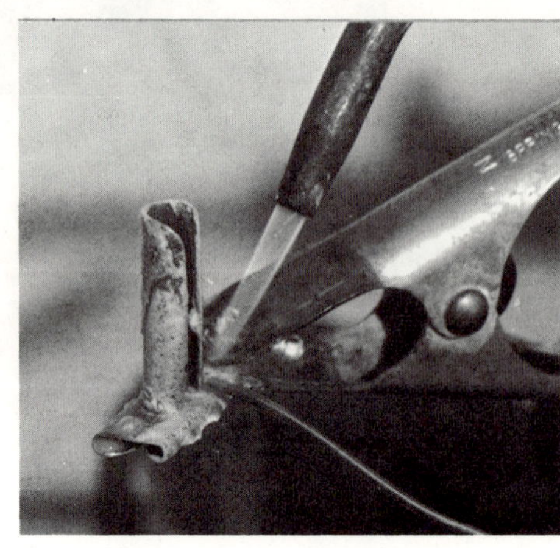

4. Cut a piece of 1/4 inch wooden dowel, 3 inches long is usually about right, insert it into the socket; slip a medicine dropper bulb onto the end of the dowel.

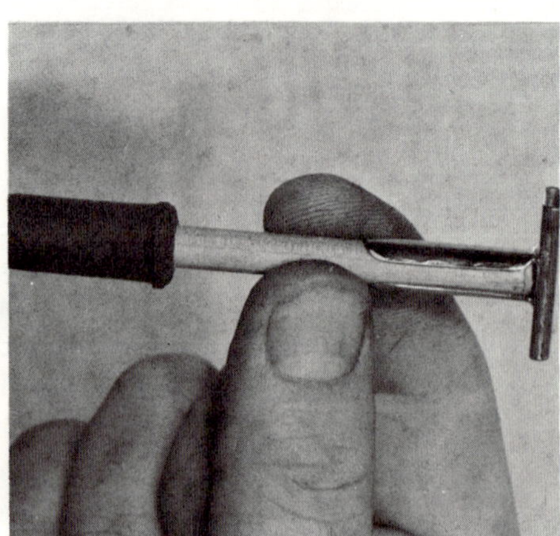

5. Assemble the page turner onto the "T"-bar, and make adjustments as needed so the patient can use it.

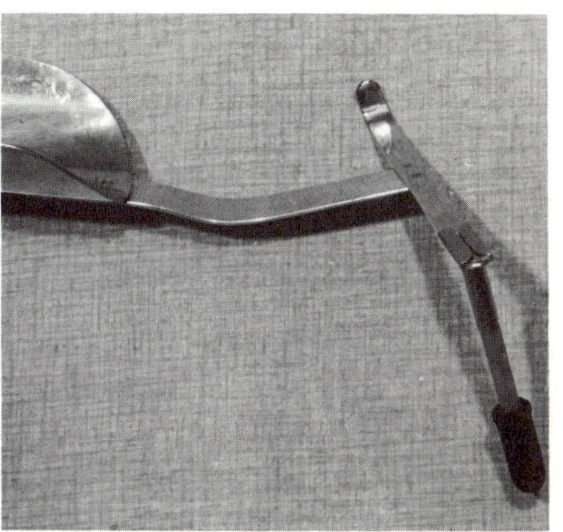

HOW TO MAKE A SWIVEL SPOON

Introduction

The patient who is equipped with a feeder often does much better in getting the food to his mouth without spilling it if he uses a swivel spoon rather than a standard type.

The swivel spoon is a standard spoon which has been altered so the bowl of the spoon is suspended from the end of the handle in such a way that the bowl will not tilt sideways when the handle is turned.   After a little practice, many patients are able to feed themselves with a swivel spoon.

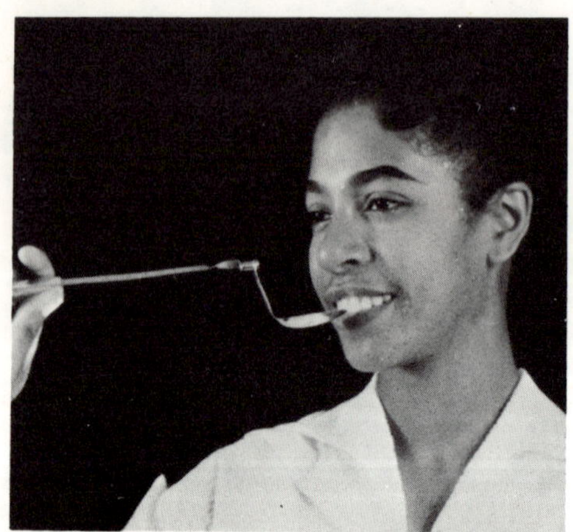

1. Make the swivel bushing and install it on the spoon handle.

   a. Measure off 5/16 inch of stainless steel tubing, size 1/4 inch outside diameter, .049 inch wall thickness, and mark it.

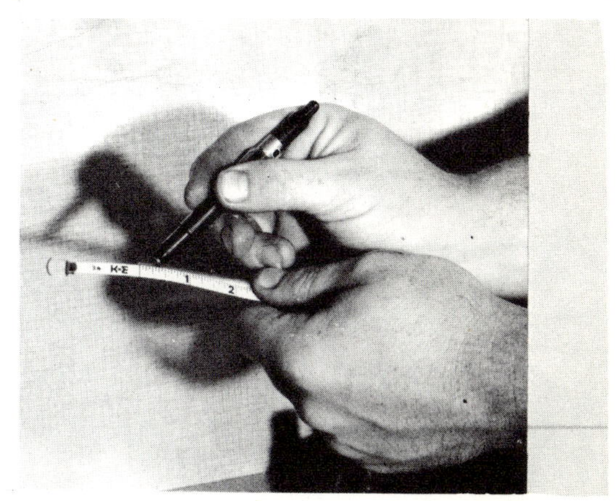

   b. Cut off the bushing at the mark.

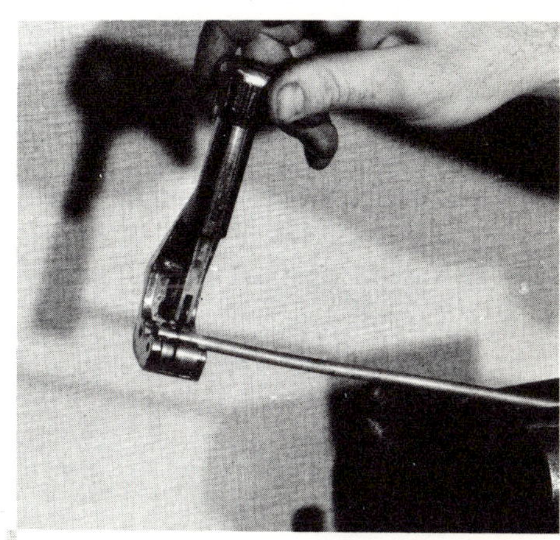

   c. Place the bushing on the handle of a stainless steel spoon, about 1-1/2 inches from the bowl, then silver solder it in place.

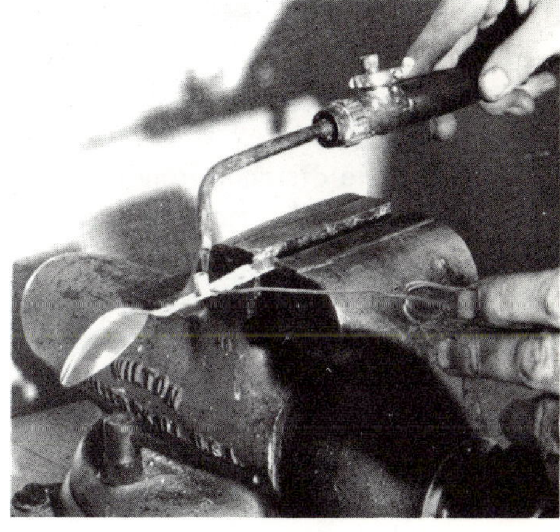

d. Cut off the part of the
    handle extending beyond
    the bushing.

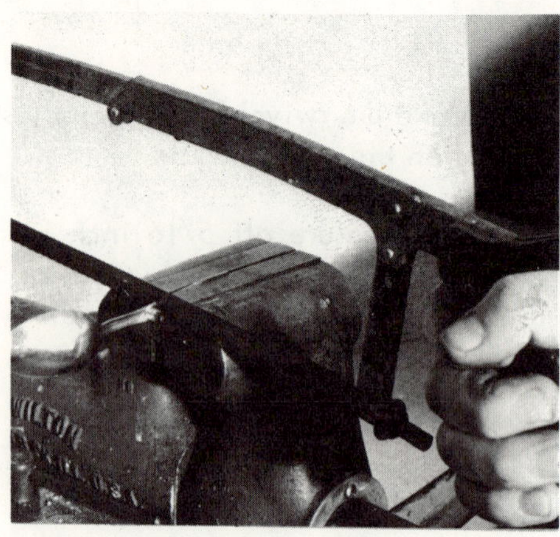

2. Drill a #19 hole through the
    bushing and handle of the spoon.
    Round all corners, smooth edges,
    and polish.

3. Make handle extension.

    a. Cut a piece of 1/4 inch by .064
        inch stainless steel. The length
        is the distance from the patient's
        hand to his mouth when he has
        his hand as close to his mouth as
        he can get it.

    b. Round the corners and smooth
        the edges.

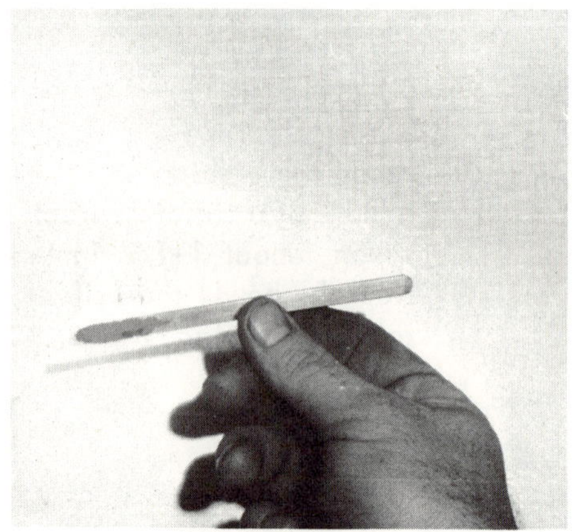

4. Make the handle.

   a. Select a piece of 1 inch 24ST aluminum tubing with .049 inch wall thickness, and mark off 3-1/2 inches.

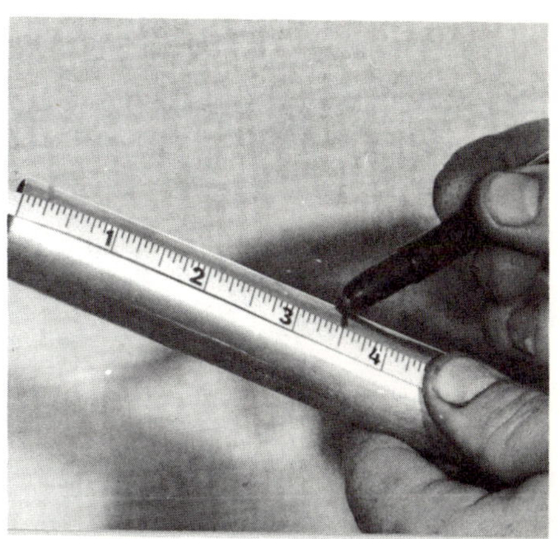

   b. Cut off the piece of tubing at the mark.

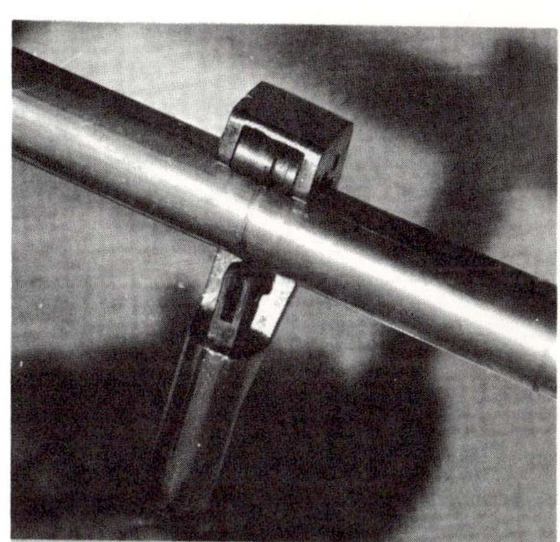

   c. Solder the extension to the handle, using Alumaweld flux and solder.

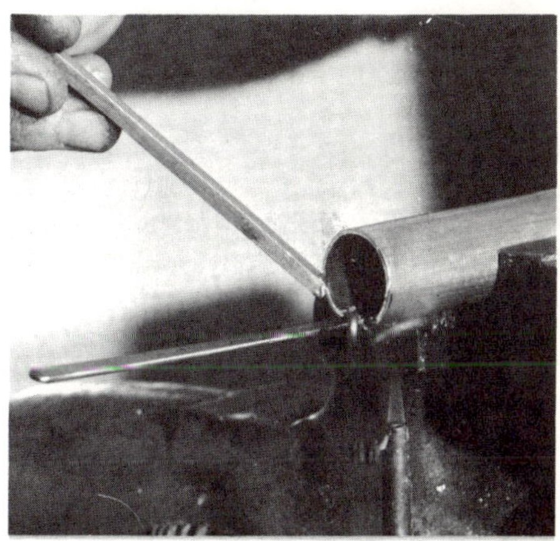

5. Drill a #29 hole in the end of the handle extension.

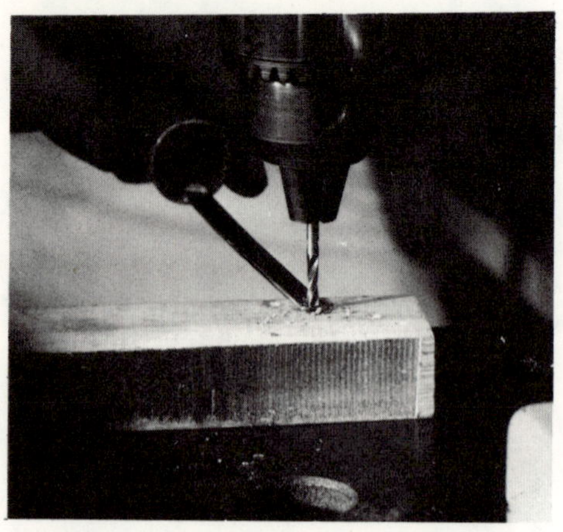

6. Tap the #29 hole with an 8-32 tap.

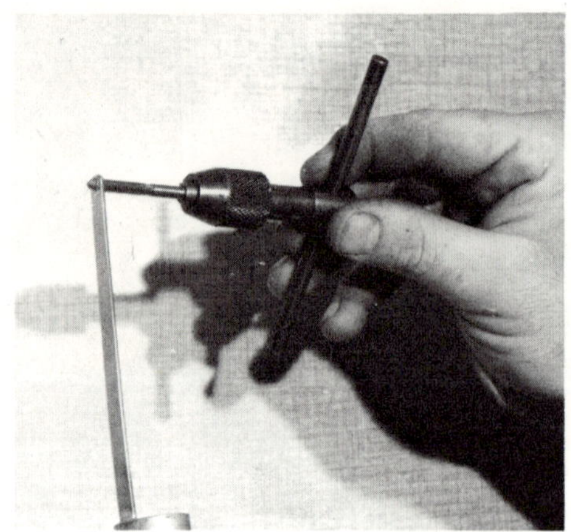

7. Bend the end of the handle extension to a right angle about 1/4 inch from the end.

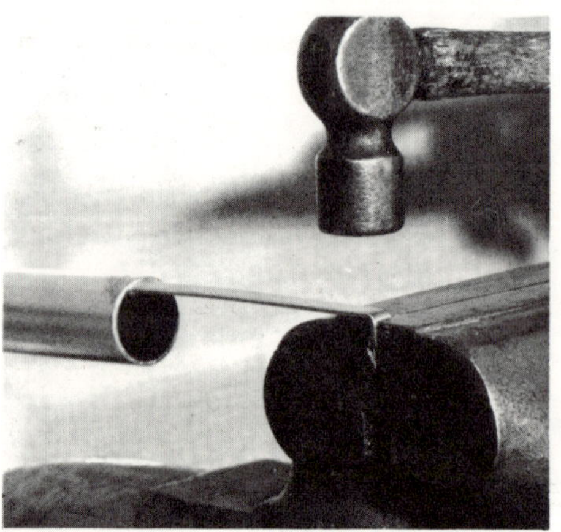

8. Put a 90° twist into each end of the handle extension.

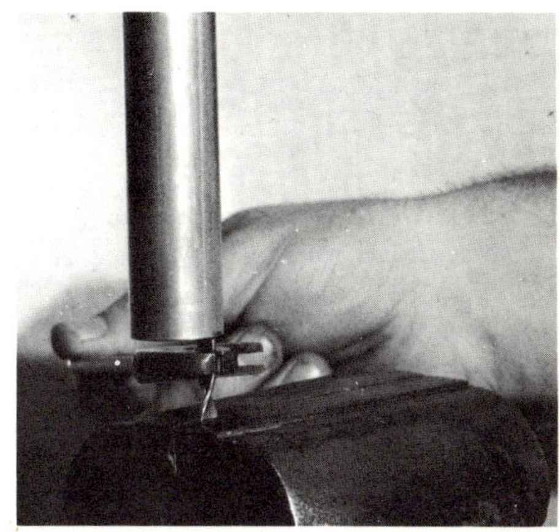

9. Assemble the spoon to the handle with an 8-32 plated machine screw. Cut off the excess length from the screw and peen the end to keep it from coming off. It should be just loose enough so the spoon can swing freely.

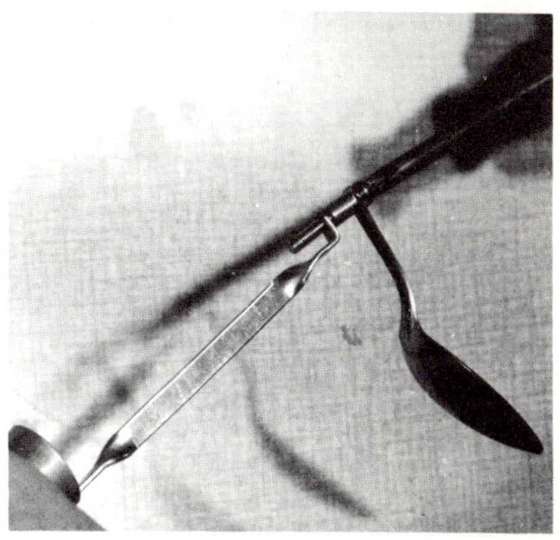

10. Bend the bowl of the spoon to a 90 degree angle to the spoon handle.

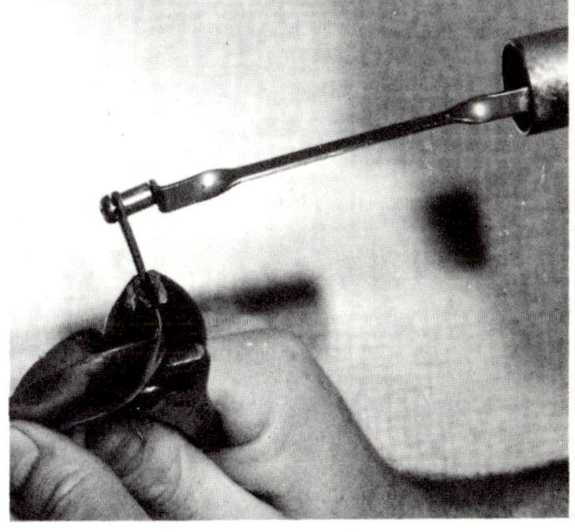

11. It may be necessary to bend the handle extension to meet the needs of the individual patient.

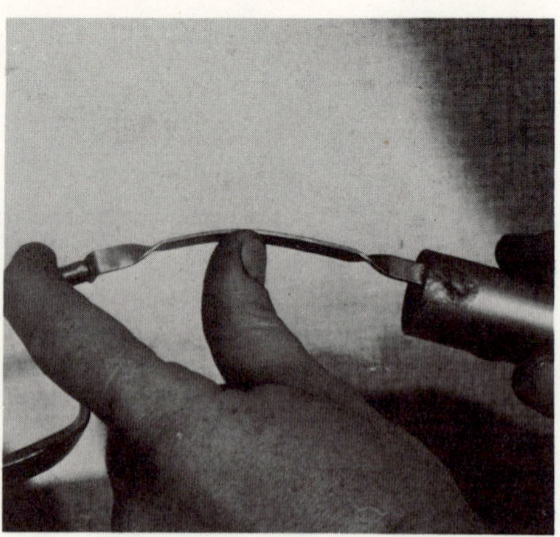

HOW TO MAKE A PENCIL HOLDER

<u>Introduction</u>

The patient who has suffered impairment of the finger muscles is able to write only with great difficulty, if at all. It requires considerable strength to grip a pencil well enough to write with it.

In many such cases, the patient could write if some means could be provided to hold the fingers in proper position and also grip the pencil. The pencil holder is a simple assistive device to accomplish this purpose. It consists of two stainless steel rings which fit the index and middle fingers. These rings are soldered together so the fingers will be held in writing position. To the rings is soldered a clip to hold the pencil.

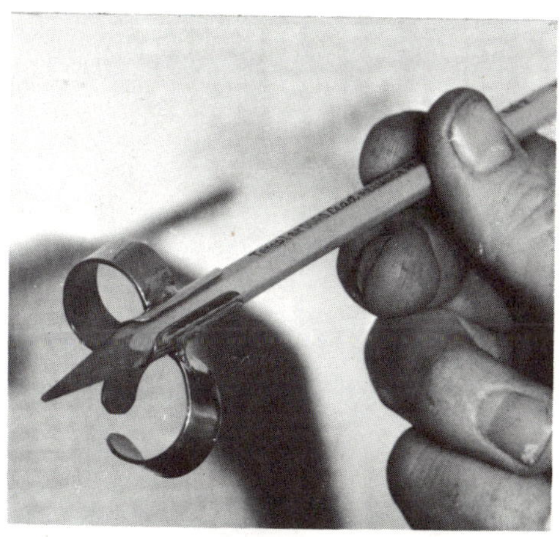
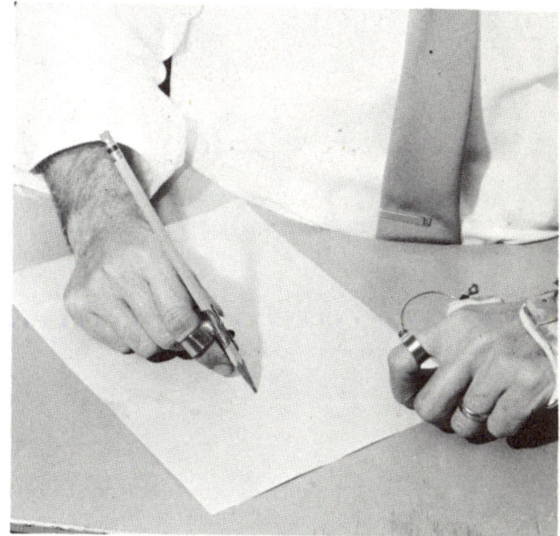

1. On a pice of stainless steel .037 by 1/2 inch, lay out dimensions of the circumferences of the distal phalanx of the thumb, and the middle phalanx of the index finger.

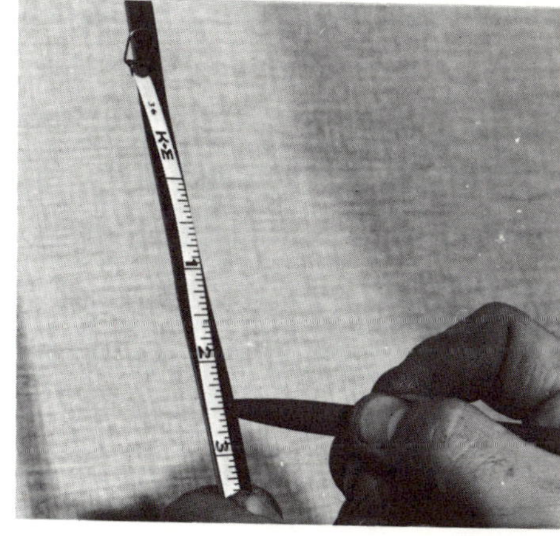

2. Cut off the pieces as laid out in "1".

3. Shape the pieces into rings.

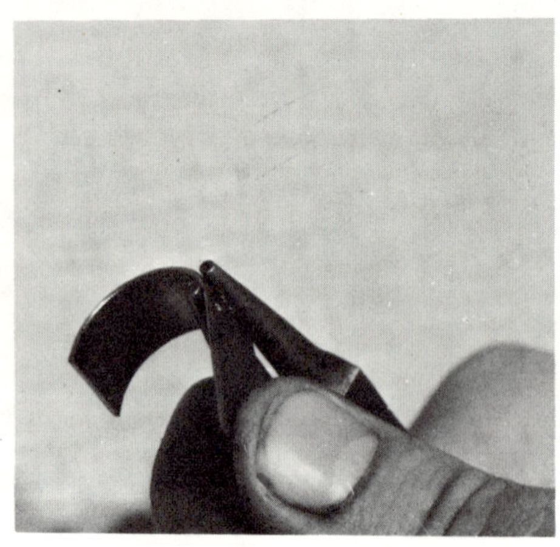

4. The completed rings should look like these.

5. Make the pencil holding sleeve.

    a. Cut a piece of .025 inch stainless steel 1 by 7/8 inches, round the corners and smooth the edges.

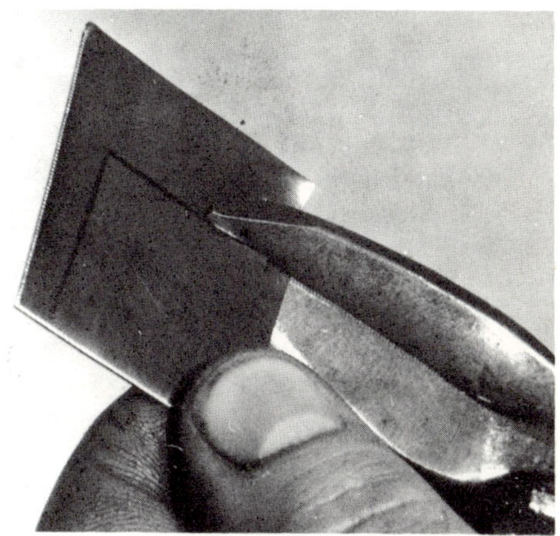

    b. Form the sleeve on a piece of 5/16 inch steel rod.

    c. The sleeve should fit the pencil quite securely, so it will not slip when writing.

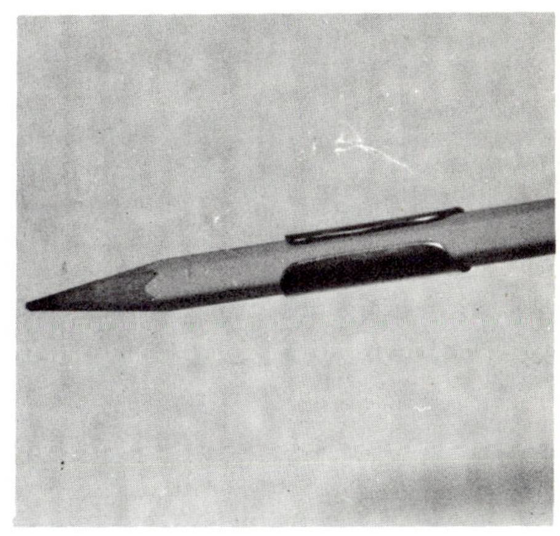

6. Place the rings on the patient's fingers and with the fingers held in the writing position, mark the adjacent surfaces as a guide for soldering. Make a mental picture of the angular relationship between the two rings.

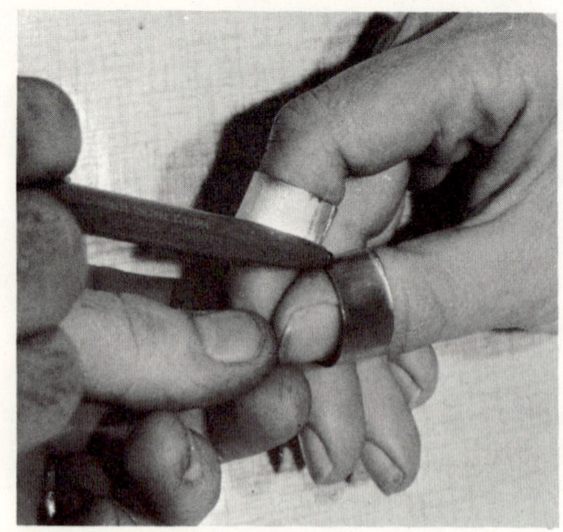

7. Solder the rings together in the same position they were when on the patient's fingers. This usually requires several tries before they fit just right.

8. With the rings on the patient's fingers, place the pencil holding sleeve and pencil in position for writing, and mark to indicate the proper angle and location.

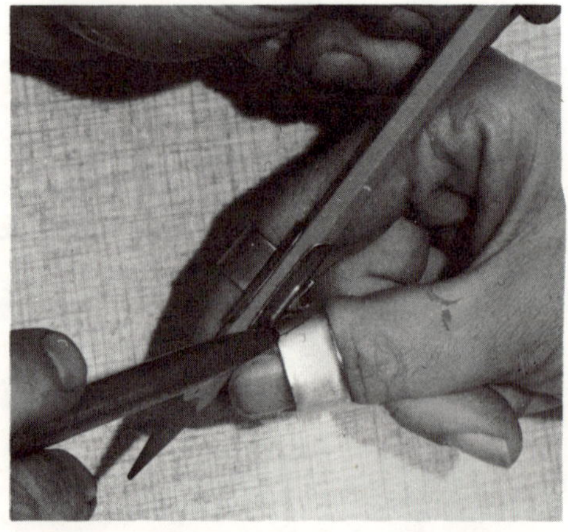

9. Solder the holding sleeve in place.

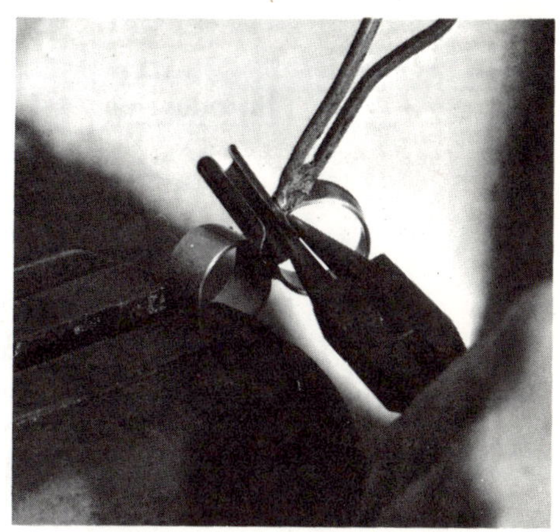

10. Try the pencil holder on the patient. It may be necessary to resolder the rings and sleeve in slightly different positions until they fit perfectly.

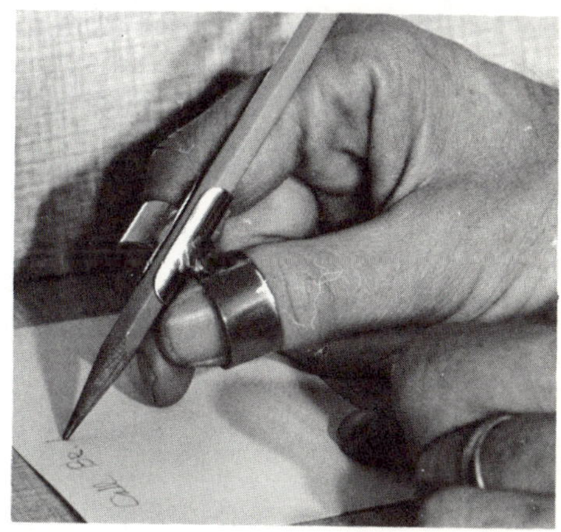

HOW TO MAKE A RAZOR HOLDER

<u>Introduction</u>

The weight of the average electric shaver is such that many patients with weakened hands and arms are unable to shave with this type of razor. The advantages of the electric razor over the safety razor are many; however, with a little ingenuity a device can be made that makes it possible for the patient to shave himself with an electric razor with little or no difficulty.

It should be noted that the razor used for illustration is typical, and it is not implied that this kind or type is better or more suitable. Practically any of the more commonly used electric shavers can be used with razor holders by merely adapting the Royalite clip to the shape of the particular razor to be used.

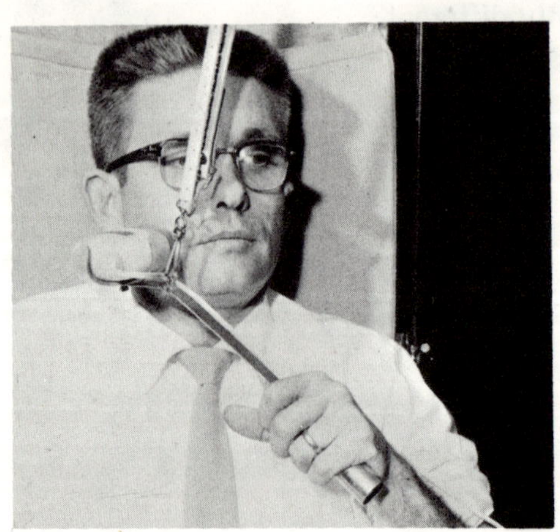

1. Make a standard handle of 1 inch 24ST aluminum tubing 3-1/2 inches long, the same as was made for the swivel spoon.

2. Cut a piece of .064 by 1/2 inch stainless steel for the handle extension, making it long enough for the patient to get the razor to all parts of his face, then hammer one end concave to conform to the inner surface of the handle. Smooth edges, round off one end.

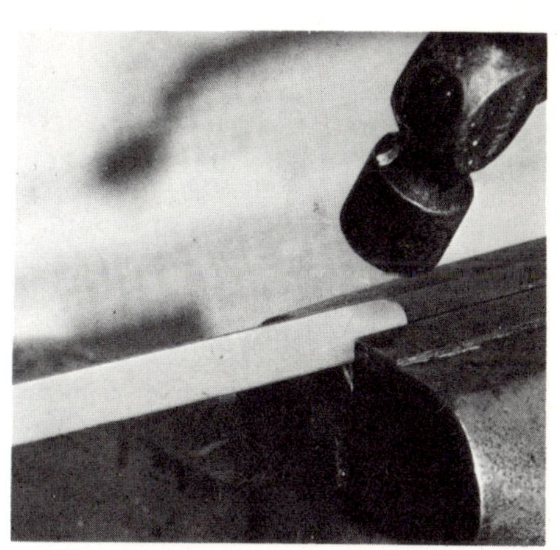

3. Solder the handle extension to the handle using Alumaweld solder and flux, then drill a #29 hole in the end of the handle extension.

4. Form a suspension ring of 1/16 inch
    stainless steel wire and cut it off.

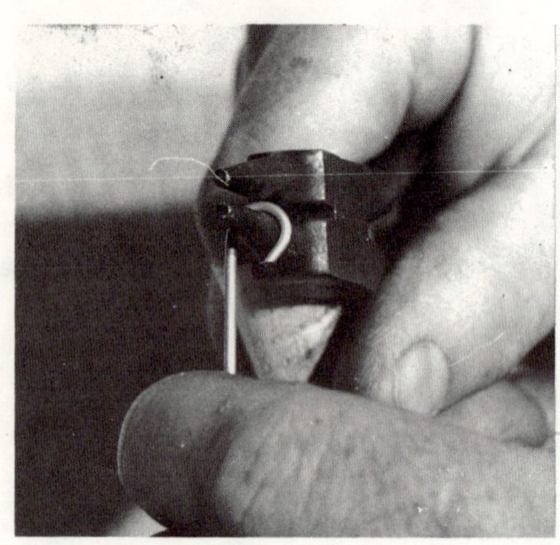

5. Silver solder the suspension ring to
    the handle extension at a point 2
    inches from the #29 hole. It must
    be on the side of the handle exten-
    sion that is soldered to the handle,
    and parallel to it. Smooth and
    polish.

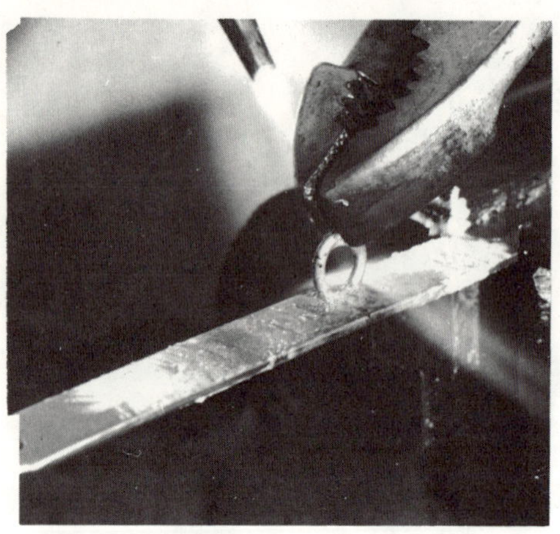

6. Make a paper pattern for the razor
    holding clamp.

    a. Place the razor on a piece of
        paper and trace its outline
        around it.

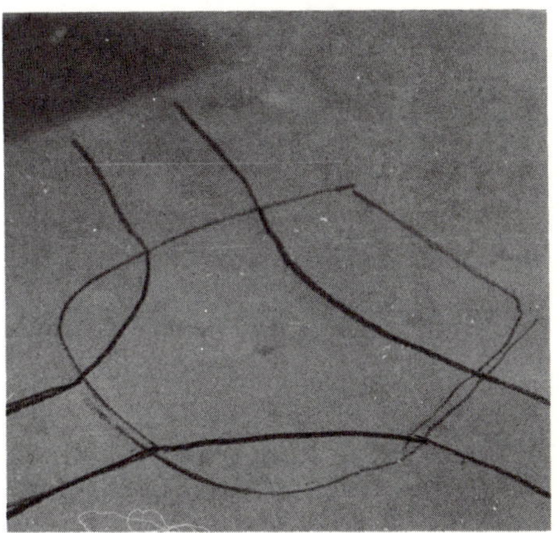

b. Sketch in the three holding
fingers and connect them with
curved lines as shown, then
cut out the pattern and transfer it to a piece of 1/8 inch
Royalite, and cut the latter to
shape, using the bandsaw.

7. Form the Royalite razor holding clamp.

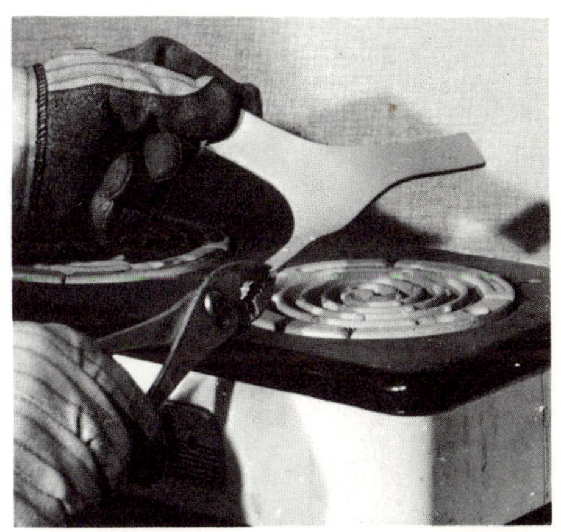

a. Heat the piece of Royalite
in an oven or over a hot
plate until it is pliable. Be
careful not to overheat it,
and wear heavy gloves to
protect the hands.

b. While the Royalite is still hot
and pliable, form it around
the electric razor as shown.
Hold it very snugly to the razor until it cools.

c. Cut off the excess material from the holding fingers and around the corners.

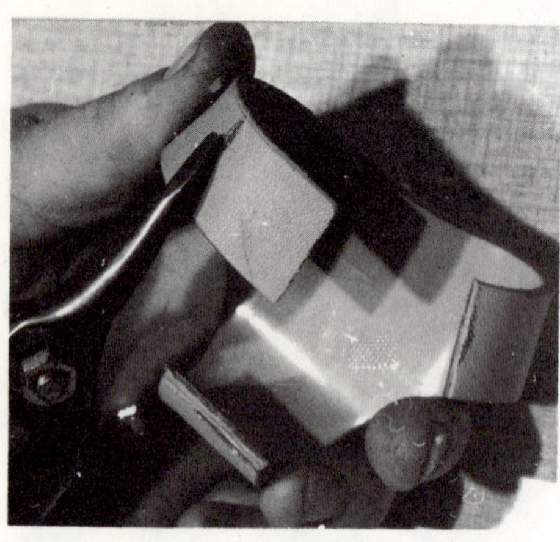

d. Polish the edges by rubbing with a piece of felt saturated with acetone or methylethyl-ketone. Do not use too much of the solvent as it may run on the surface, causing a streak.

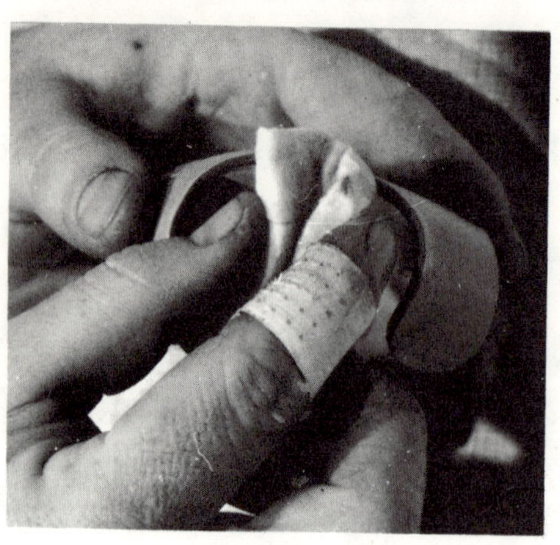

8. Cut a piece of .037 inch stainless steel, 3/4 by 3/4 inch for the swivel plate. Round the corners and smooth the edges.

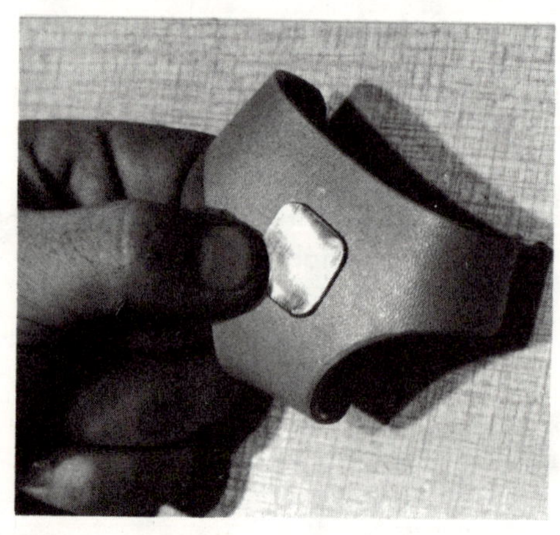

9. Drill and countersink two #40 holes in diagonally opposite corners of the plate; drill a #29 hole in the center. Drill a 17/64 inch hole in the center of the razor holding clamp, center the hole in the swivel plate over this hole, mark the corner holes, and drill #40 holes in the Royalite at the marks.

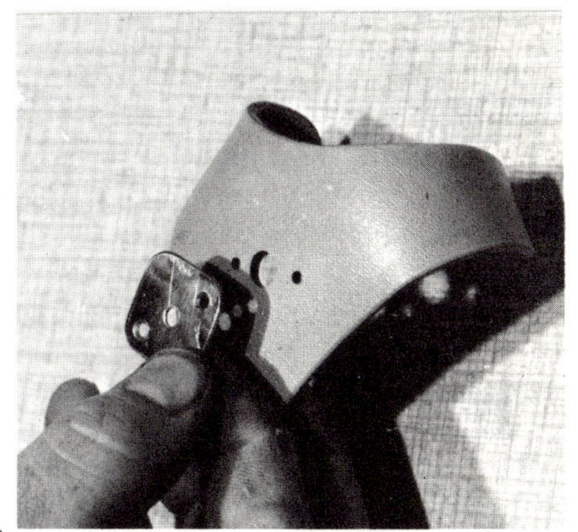

10. Rivet the swivel plate to the holding clip with 3/32 inch soft aluminum rivets.

11. Make a friction spring of two turns of stainless steel wire, inside diameter to fit over a 6-32 screw.

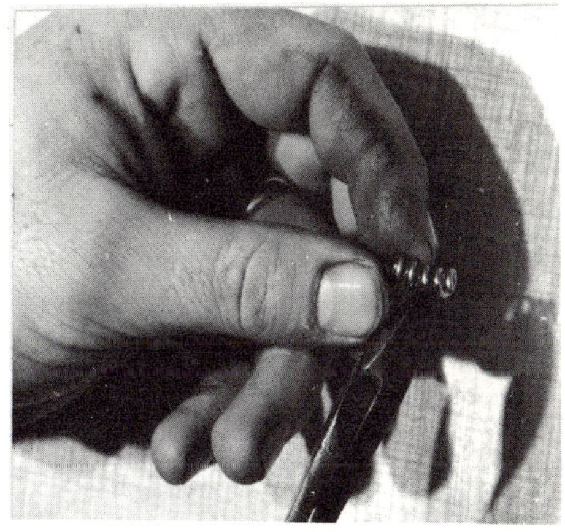

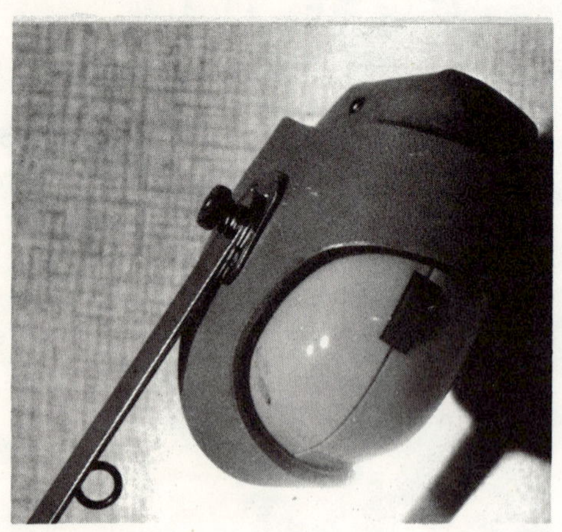

12. Assemble the holder by inserting a 6-32 machine screw into the hole in the swivel plate, putting the screw through the hole in the handle extension, installing the spring and nut, then adjusting the tension so the razor will not swivel loosely but will swivel when the patient wants it to.

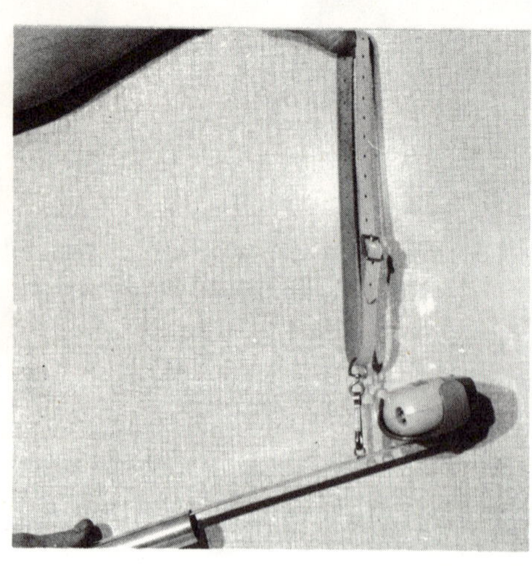

13. Attach a regular suspension feeder strap to the suspension ring with a snap hook. This is used to suspend the razor from an overhead support.

14. The razor can be used by the patient in a wheel chair or any other location where an overhead support is available.

HOW TO MAKE A COMB HOLDER

Introduction

The comb holder makes it possible for the patient with weakened hand muscles to comb his or her hair with a minimum of difficulty. The comb holder may be fitted with a clip to attach to the "T"-bar, or with a standard aluminum handle. In the instructions that follow, the comb holder will be fitted with the latter. The clip described in making the toothbrush holder is substituted for the handle in case it is desired to use the device with the feeder.

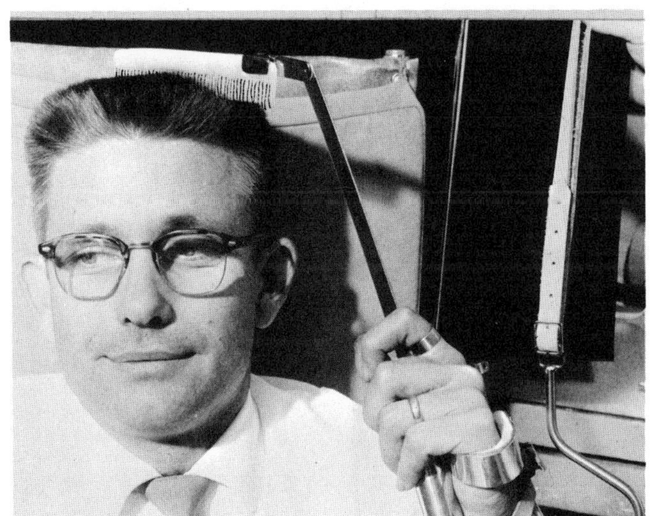

1. Make a handle and handle extension
    the same as was done for the razor
    holder, except make the extension
    long enough so the patient can get
    the comb to all parts of his head.
    Drill a #29 hole in the end of the
    extension, round corners, smooth
    edges, and polish.

2. Make a clip swivel extension from a
    piece of stainless steel .064 by 1/2
    by 1-1/2 inches; drill a #35 hole in
    one end, thread the hole with a 6-32
    tap, round all corners and smooth all
    edges.

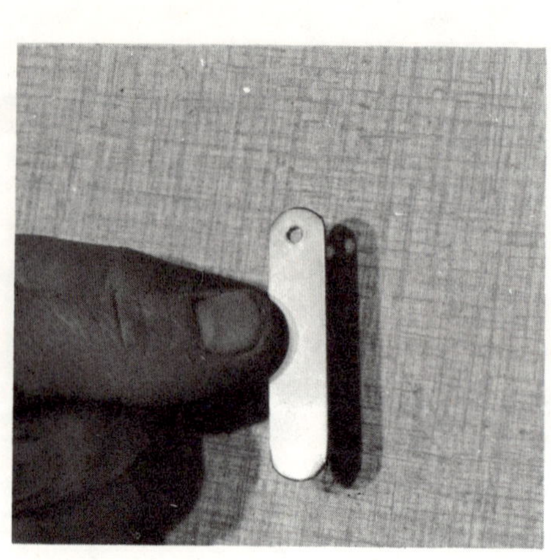

3. Measure the back of the comb from
    the roots of the teeth on one side
    to the same point on the other side.

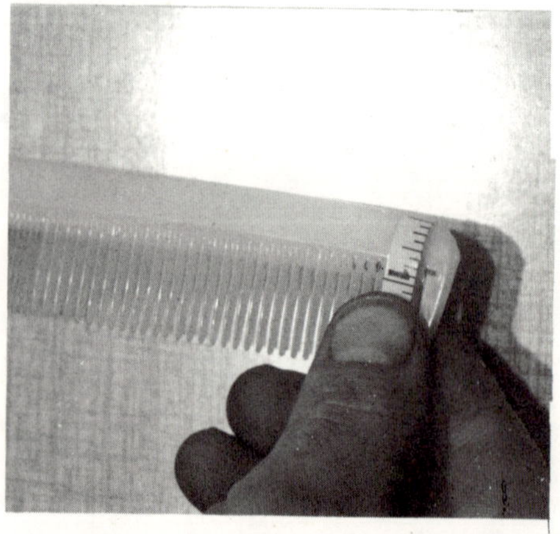

4. Cut a piece of .025 inch stainless steel, 1 inch long, width the same as the dimension obtained in "3". Round all corners and smooth edges. Carefully form the clip over the back of the comb.

5. Shape the clip until it fits the comb as in the illustration. It must be tight enough so it will not come loose accidentally.

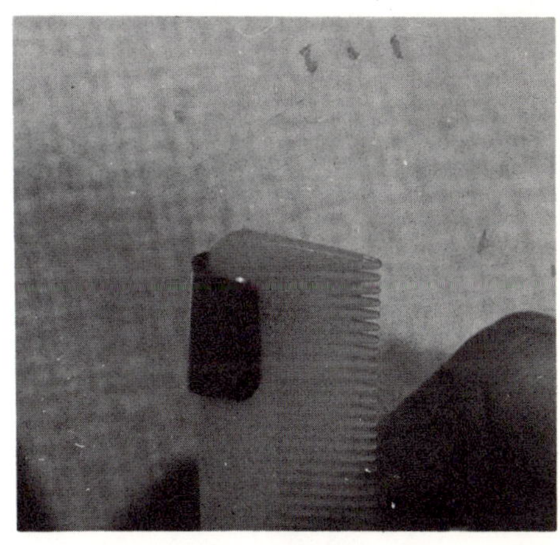

6. Silver solder the clip to the clip swivel extension. Clean and polish.

7. Make a leather friction washer 1/2 inch in diameter with a 1/8 inch hole.

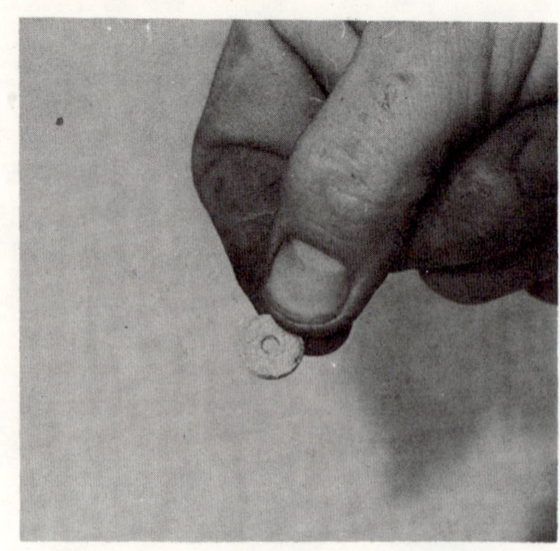

8. Assemble the clip swivel extension to the extension, using a 6-32 screw. Be sure the friction washer is between the two parts.

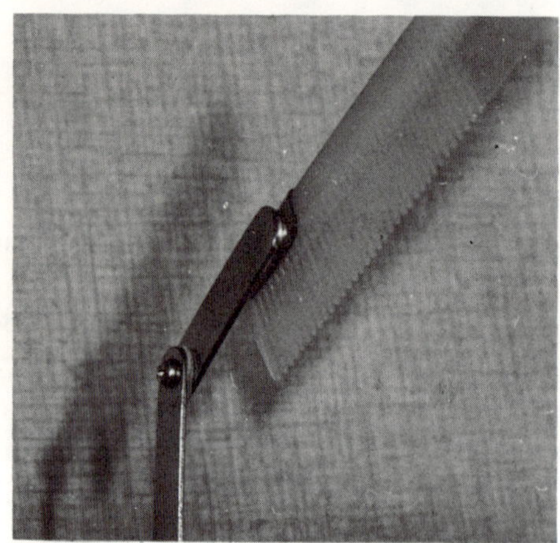

HOW TO MAKE A PLATE GUARD

Introduction

Many patients are barely able to manipulate a swivel spoon or a spoon on a feeder, and it is quite common to find them unable to control the device well enough to avoid pushing food off the plate. In addition, they often have difficulty because the food pushes ahead of the spoon instead of getting into it so they can get it to their mouths. The plate guard is a simple device in the shape of a plastic fence to clip onto the plate.   It prevents food from being pushed off, and gives the patient something to work against in getting food into his spoon.

The plate guard is a simple device that is easily made, and is very useful to the patient who has the difficulties described above.

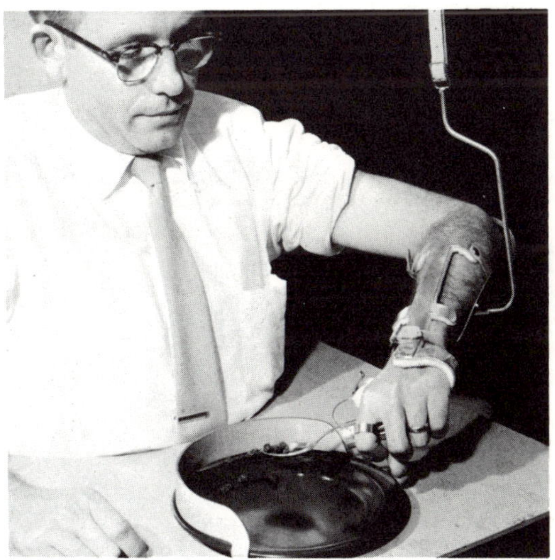

1.   Make a pattern for the plate guard.

   a.   The pattern on the next page is for a plate guard to fit a 9 inch plate. Because of its length it had to be cut in half at the dotted line and printed in two parts.

   b.   To use the pattern, cut the two parts out, then join the dotted lines together with scotch tape.

   c.   Trace the pattern on a piece of 1/8 inch Royalite.

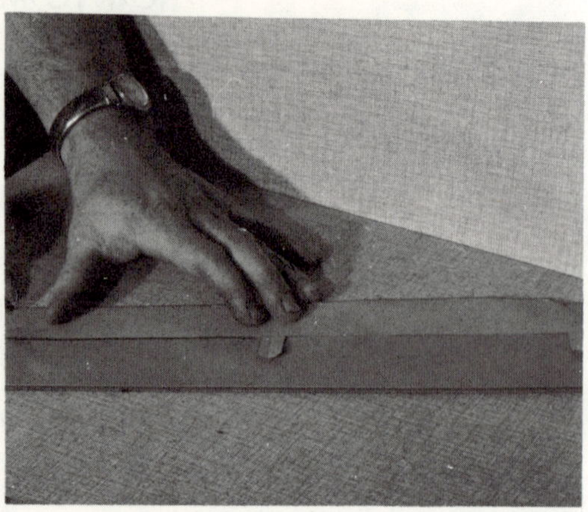

2.   Cut out the plate guard on a bandsaw. The piece of Royalite should look like the illustration.

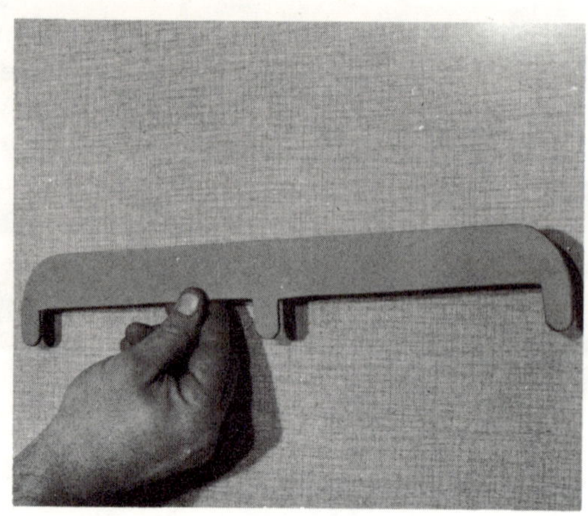

3.   Round the edges of the piece, then polish them with a piece of felt moistened with acetone. Do not use too much acetone, if it runs it will streak the finish.

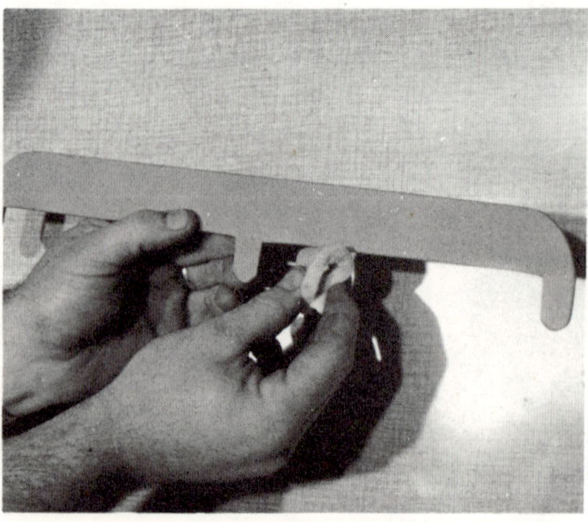

4. Make a form to shape the plate guard.

   a. Cut a piece of 2 inch board into a circle 6-1/2 inches in diameter.

   b. Cut a piece of 1/4 inch plyboard into a circle 7-1/4 inches in diameter.

   c. Fasten the two pieces together with screws or small nails so the plyboard extends evenly about 3/8 inch beyond the edges of the thicker edge.

   d. Make a strap and buckle of 2 inch material and fasten it firmly to the 2 inch thick portion with screws or nails.

5. Heat the Royalite in an oven until it is soft and pliable, then quickly wrap it around the 2 inch face of the form being sure the ends do not lap over the strap. Bring the strap around the piece and buckle it tight. Wear gloves to protect the hands from the heat.

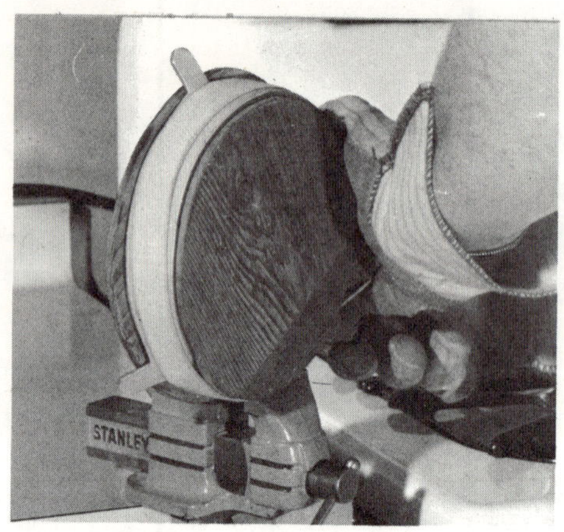

6. Quickly bend the retaining clips over before the material cools too much.

7. Remove the strap and slip the plate guard from the form.

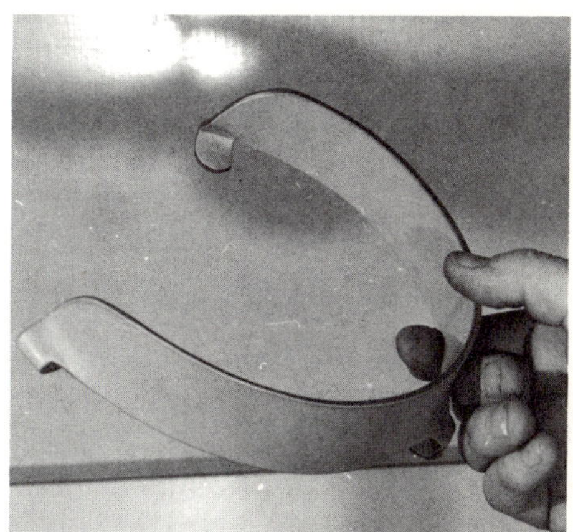

8. Try the plate guard on a plate.

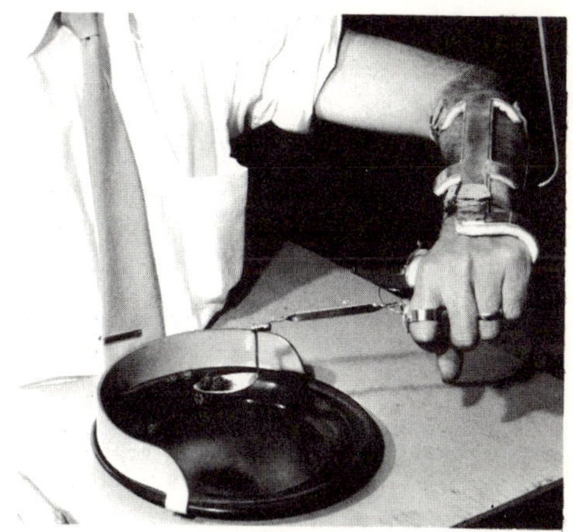

HOW TO MAKE BUILT-UP HANDLES FOR EATING UTENSILS

<u>Introduction</u>

It requires considerably more strength and muscle coordination to grasp a spoon with the small, thin conventional handle than one with a large diameter built-up handle. The larger area of grasping surface helps make it easier to hold and control the spoon, and in addition the larger diameter handle permits the patient to make use of his limited amount of grasp without having to bring the fingers completely together.

Built-up handles can be applied to any eating utensil. In the instructions that follow, a spoon will be used as an example, but the same general procedure would apply to a fork or knife.

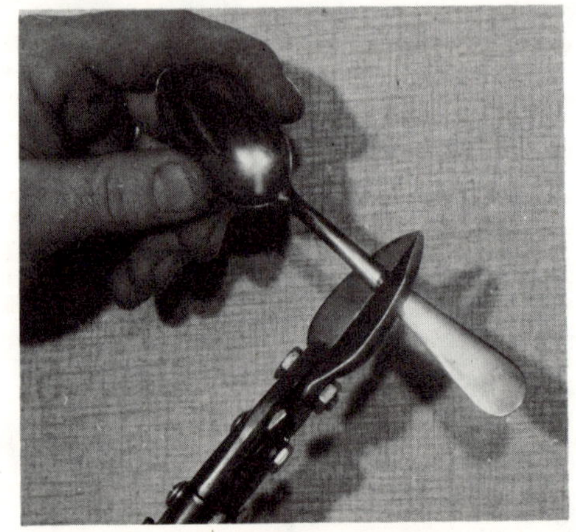

1. Cut off the handle of the spoon, leaving about 1-1/2 inches of the shank. An ordinary stainless steel spoon is best.

2. Cut a 3 inch piece of 1 inch aluminum tubing for a handle, clamp it in the vise, and hold the spoon shank in place with a clip as shown in the illustration.

3. Solder the shank to the handle, using Alumaweld solder and flux.

4. Smooth all edges and polish.

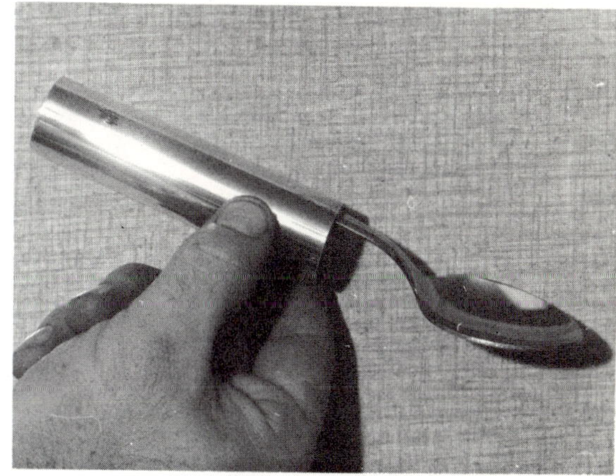

HOW TO MAKE A LIPSTICK HOLDER

Introduction

As with the toothbrush and comb holders, the lipstick holder is an aid to personal care. To the woman patient, being able to enhance her appearance through the use of cosmetics which she can apply herself is of great importance. Its value as a morale-builder should not be under-estimated. To some, spending time and money to make a lipstick holder may seem a trifle friv-olous, but experience has proven that in terms of the progress and happiness of the patient it is time and money well spent.

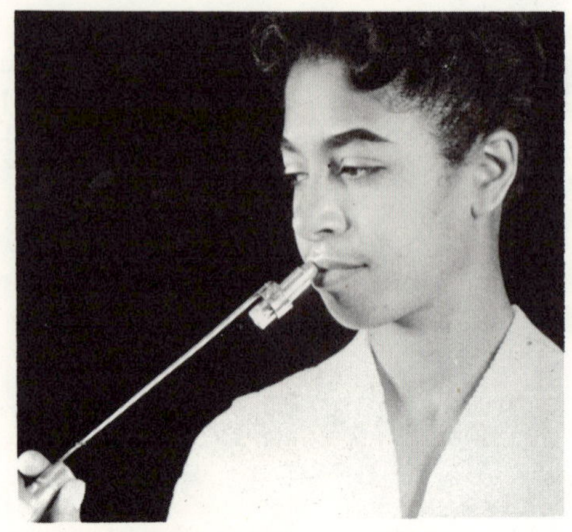

1. Measure the circumference of the lipstick.

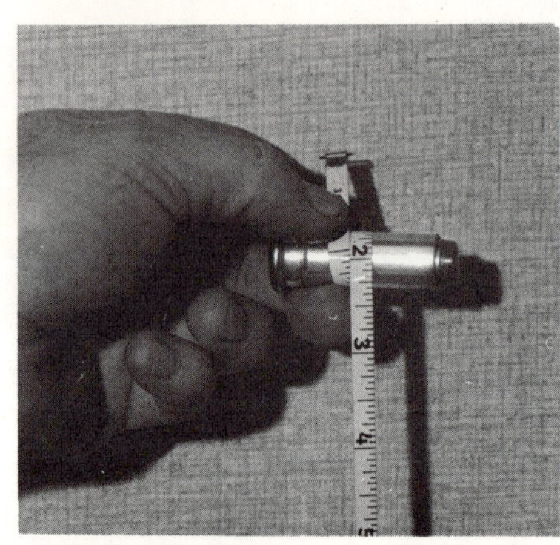

2. On a piece of .064 x 1/4 inch stainless steel, lay out the circum-ference measurement of the lip-stick, plus 1/8 inch.

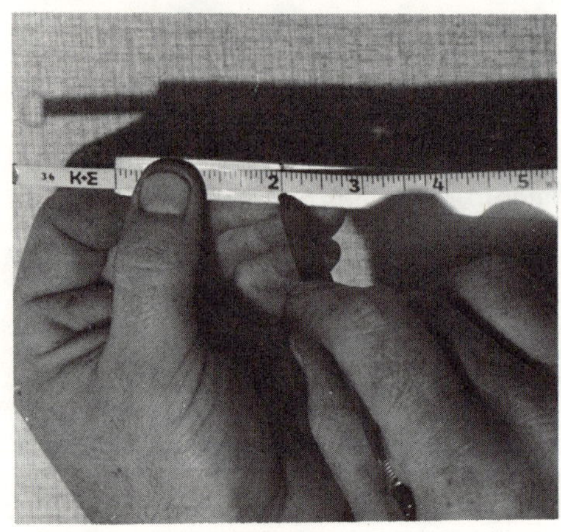

3. Cut off the piece of stainless steel, smooth the edges.

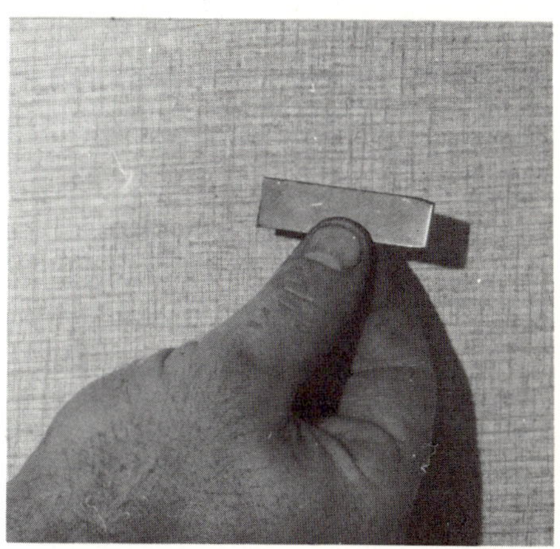

4. Using a piece of round steel approximately the same diameter as the lipstick for a mandrel form the piece of stainless steel into a ring.

5. The completed ring should look like the illustration, and should fit on the lipstick loosely.

6. Make the "T"-bar clip, following
    the same procedure as described
    earlier.

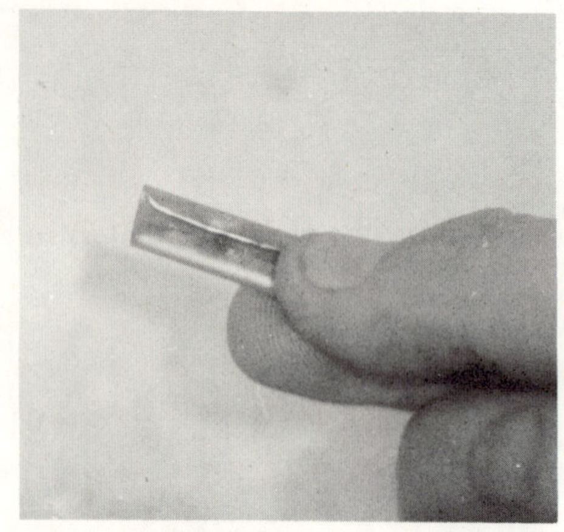

7. Make an extension from .064 x 1/4
    inch stainless steel; it must be long
    enough so the patient can easily reach
    her mouth with the lipstick.

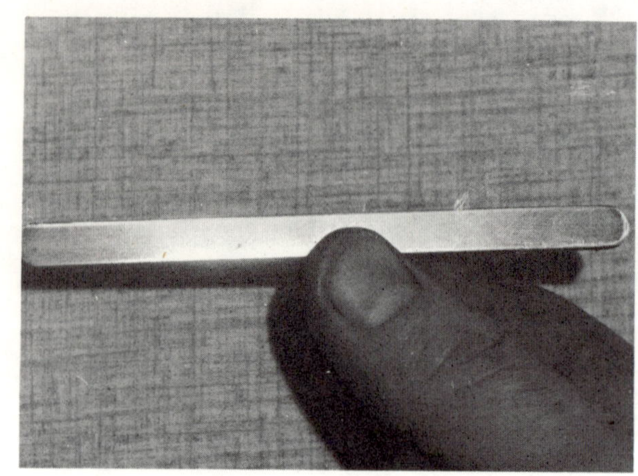

8. Silver solder the clip to one end of
    the extension, then twist the end of
    the extension 90 degrees as illus-
    trated.

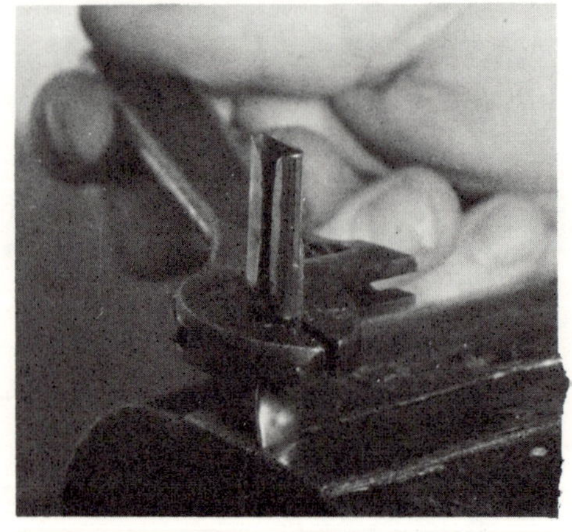

9. Silver solder the ring to the extension and twist it 90 degrees. Be sure to solder the parts so the extension covers the break in the ring.

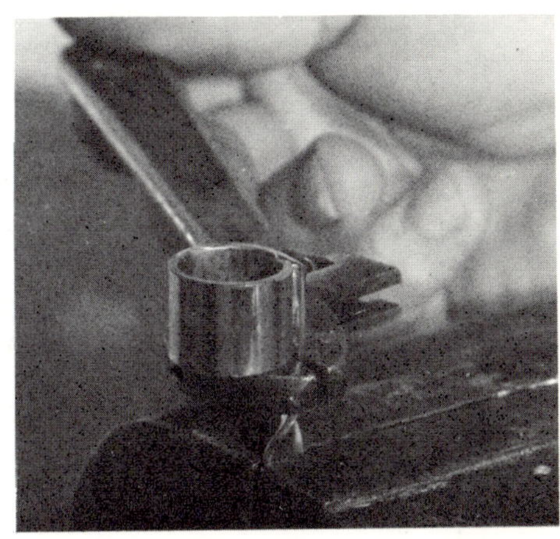

10. Drill a #29 hole through the extension and ring where they are soldered together and thread the hole with an 8-32 tap, then install an 8-32 thumbscrew. Try the lipstick in the holder.

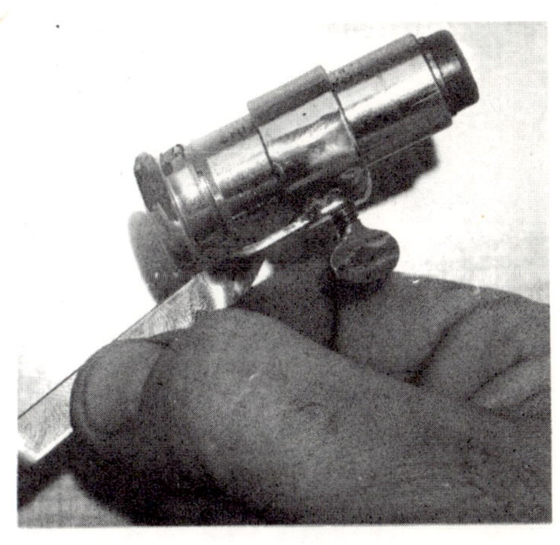

11. Adjust the "T"-bar clip on the "T"-bar and bend the extension as necessary to make it possible for the patient to use the lipstick with minimum difficulty.

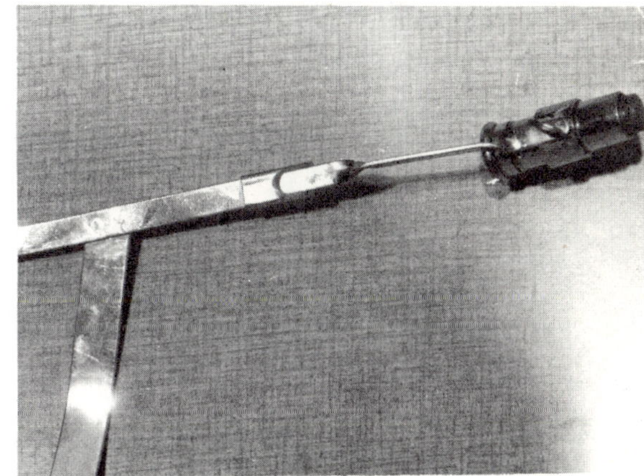

HOW TO MAKE A MOUTH STICK

Introduction

It is not uncommon for a patient to suffer extensive loss of the muscles of the hands and arms while retaining the use of the neck muscles. These patients are able to grip implements of one kind or other in their teeth and manipulate them through movements of the head. With practice, some have become skilled in painting, drawing, and other art work.

It is very desirable to provide the patient with an implement that is shaped so he can grip it firmly with the teeth, is the right length for most convenient use, is sanitary, and light in weight. The mouth stick described here is made of stainless steel with a plastisol coating to permit a good grip. The plastisol resists abrasion by the teeth, and gives good service.

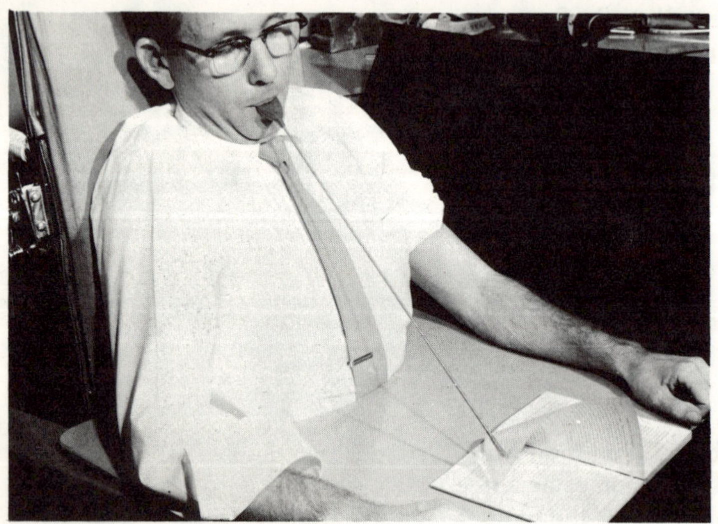

1. Cut a piece of 1/8 inch stainless steel rod about 18 inches long, and cut a 1/4 inch long slot in one end with a hacksaw.

2. Make a pattern for the mouthpiece. Use the drawing at right as a guide.

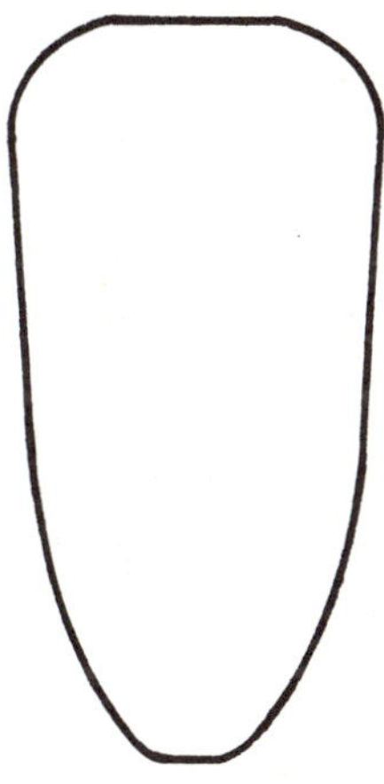

3. Trace the pattern onto a piece of .037 inch stainless steel, then cut it out with shears.

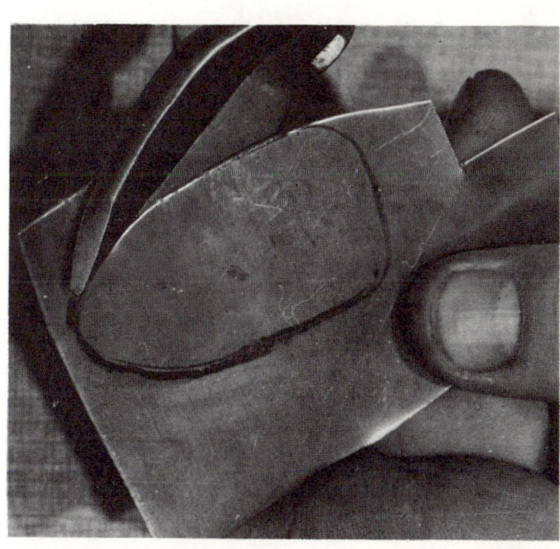

4. Silver solder the mouthpiece to the rod, being sure to get the mouth-piece well down into the slot.

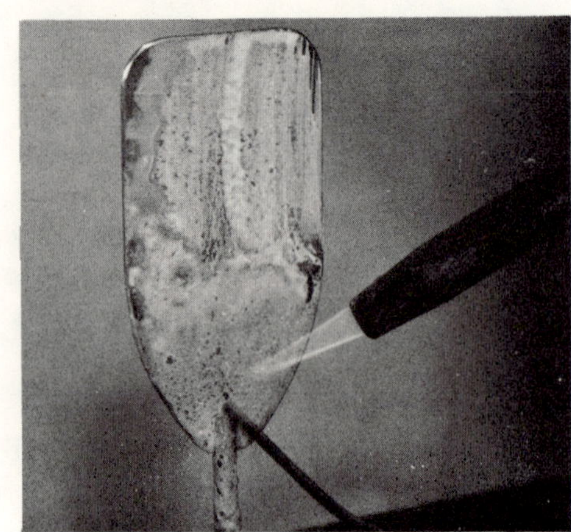

5. Apply plastisol to the mouthpiece.

   a. Heat the mouthpiece with a torch until it just starts to turn a bluish color. The metal must be fairly hot or the plastisol will be too thin and will not cure. If it is too hot, it will burn the material. A little practice will make it possible to get it just right each time.

   b. While the metal is hot, dip the mouthpiece into Plastisol type PK 2802, holding it there for about one-half a minute. This material is available in white or red.

c. Remove the mouthpiece and turn it to prevent the Plastisol from sagging or dripping. It will soon solidify.

6. Apply Plastisol to the tip.

a. Heat the tip with the torch.

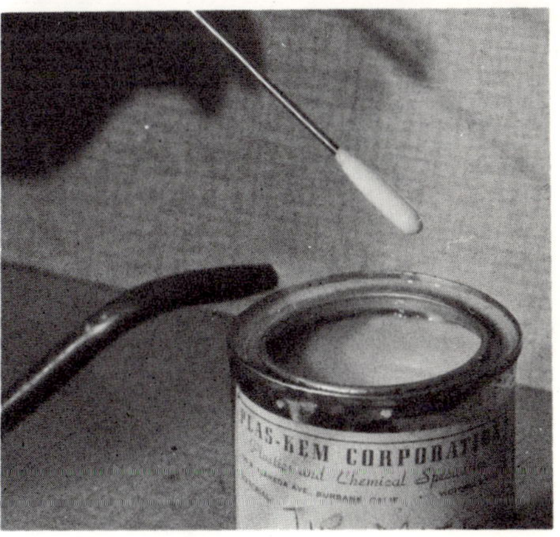

b. While still hot, dip the tip into Plastisol type PK 2686. This material is tackier when cured, which makes it easier to turn the pages of books and magazines.

7. Put the mouthstick in an oven at 350 degrees Fahrenheit for three to five minutes to cure the Plastisol.

HOW TO MAKE A GLASS HOLDER

<u>Introduction</u>

The wheel chair patient who cannot pick up a glass and hold it to his mouth so he can get a drink of water requires a great deal of attention to satisfy this one requirement. If some means can be found to attach a glass of water to his chair so he can get to it without help, he can drink any time he wants to and the only help he needs is to occasionally fill his glass.

A glass holder is a device for holding a glass of water in a convenient position so the patient can get at the drinking tube without having to pick up the glass. It is very useful, and is one assistive device that is very much appreciated by the patient.

1. Obtain an ordinary bathroom type glass holder and cut off the mounting flange and toothbrush holding flanges. The glass holders made of plated brass are easiest to work with. Smooth corners.

2. Leave the gap where the mounting flange was removed, as it permits a coffee cup to fit into the holder, as shown.

3. Make a mounting pivot by cutting off a 1 inch long piece of 3/16 inch inside diameter stainless steel tubing.

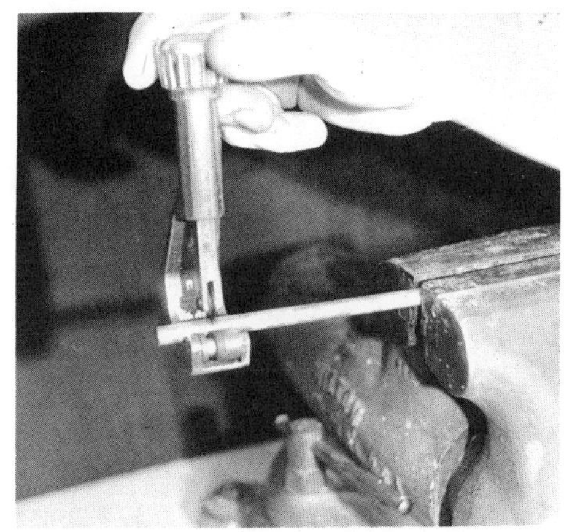

4. Solder the mounting pivot to the bottom of the glass holder.

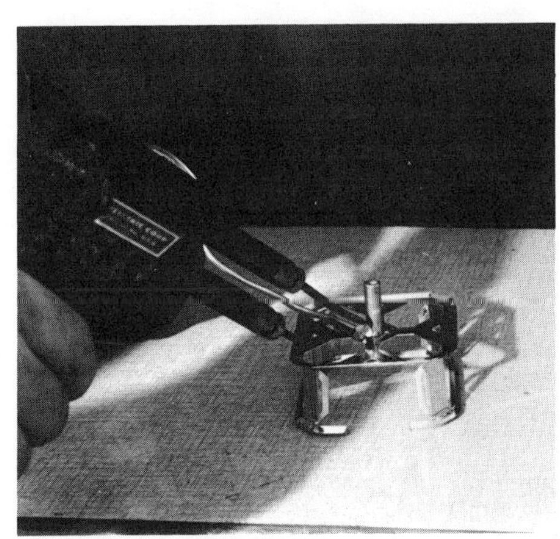

5. Make a swivel rod of a piece of 3/16 inch stainless steel rod 12 to 18 inches long; make a right-angle bend about 3/4 inch from one end. Check to see if it fits into the mounting pivot.

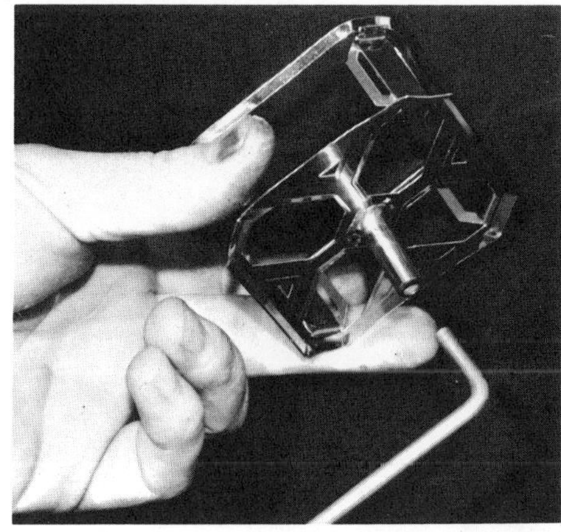

6. Make the swivel rod mounting bracket by cutting off a 2 inch piece of the 3/16 inch inside diameter stainless steel tubing to which a 1 inch square piece of .064 inch stainless steel has been silver soldered.

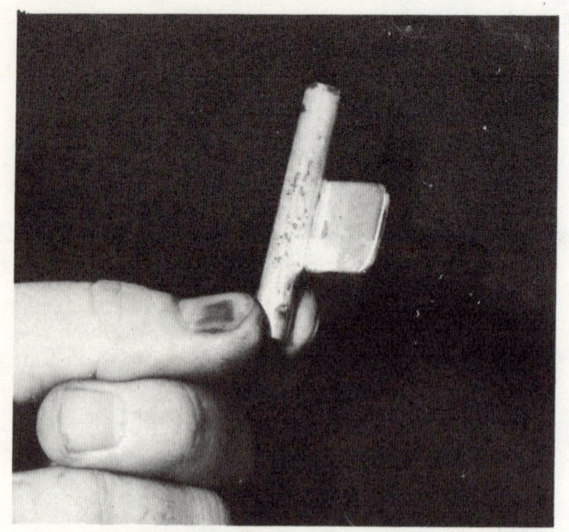

7. Drill a 1/4 inch hole in the mounting bracket, smooth and polish.

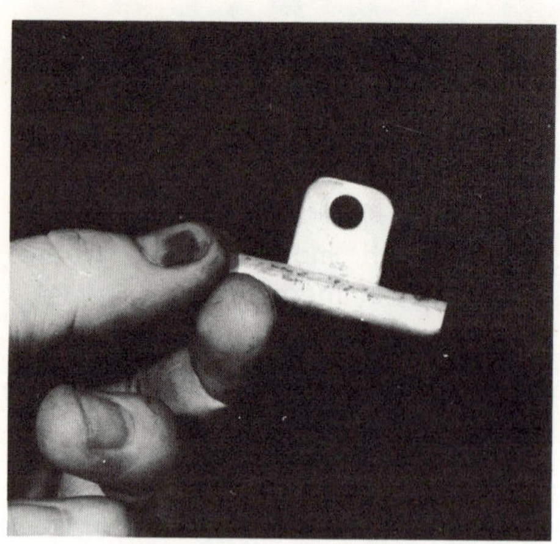

8. Attach the mounting bracket to one of the feeder bracket bolts on the wheel chair.

9. Adjust the swivel rod so the glass is held in a convenient position for the patient.

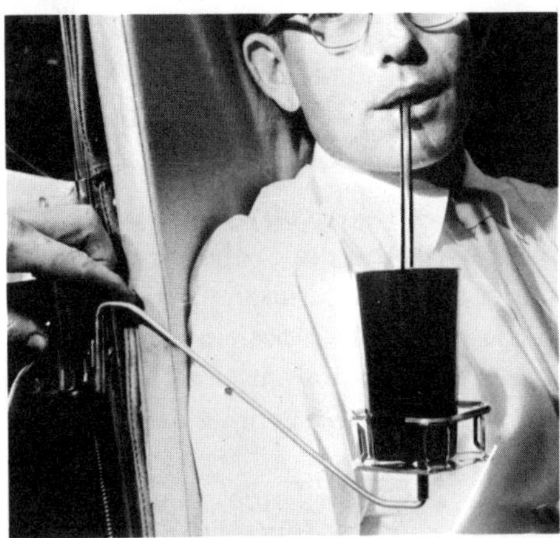

HOW TO MAKE A STRAW HOLDER

<u>Introduction</u>

One difficulty many patients experienced with the use of the glass holder was the tendency of the drinking tube to roll around in the glass into a position in which it pointed away from the patient, making it impossible for him to get his mouth to it.

The straw holder is a simple stainless steel clip that holds the drinking tube in a set position in the glass, thus putting an end to the difficulties described. The clip is simple to make and will last indefinitely.

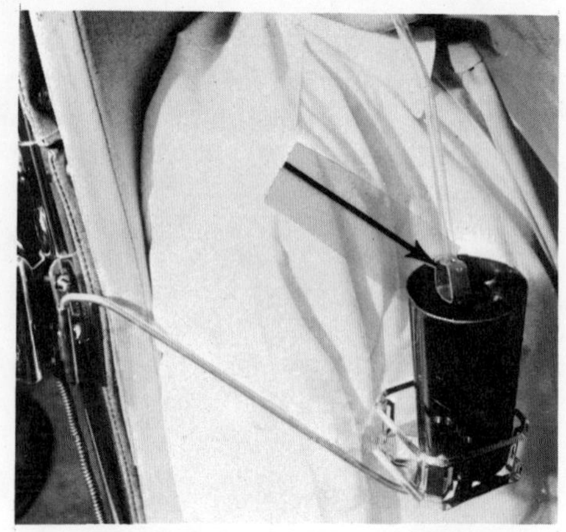

1. Cut a piece of .037 inch stainless steel 1/2 inch wide and 4 inches long, round one end, then bend a 3/4 inch clip in the rounded end as shown.

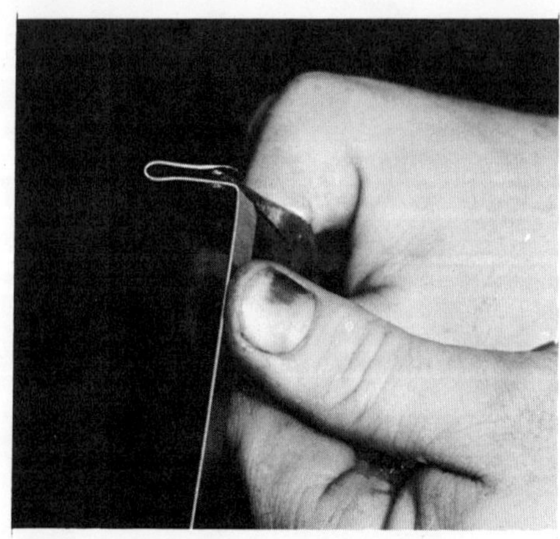

2. Drill two 5/16 inch holes in the piece, the first 5/8 inch from the clip, the second 1-3/8 inch from the first. Most drinking tubes fit into the 5/16 inch holes, but if other diameters are encountered drill holes just large enough to permit the tube to enter.

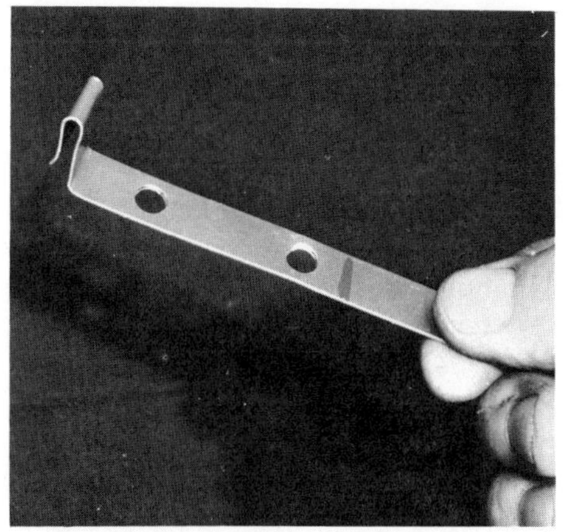

3. Round all corners, smooth edges, polish.

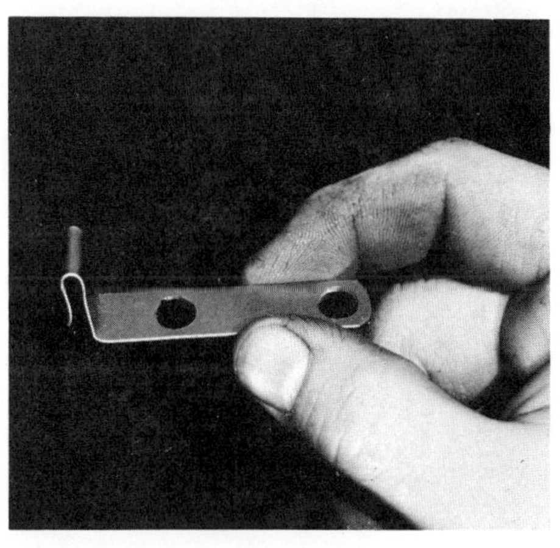

4. Bend the end over at right angles, about 3/16 inch from the edge of the hole.

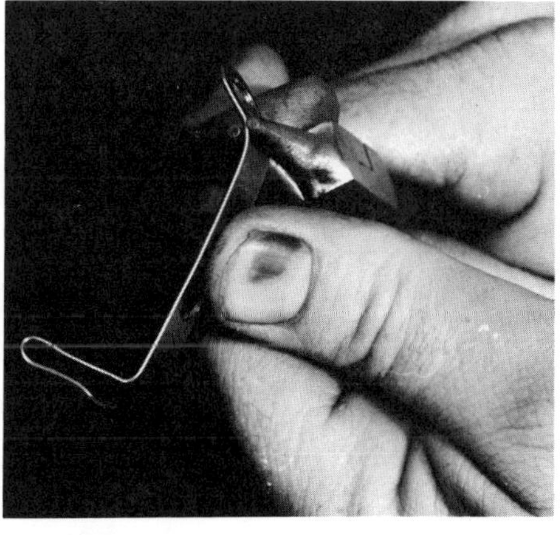

5. Make a second right-angle bend about 3/16 inch from the edge of the second hole, as shown. This should bring the two holes into alignment.

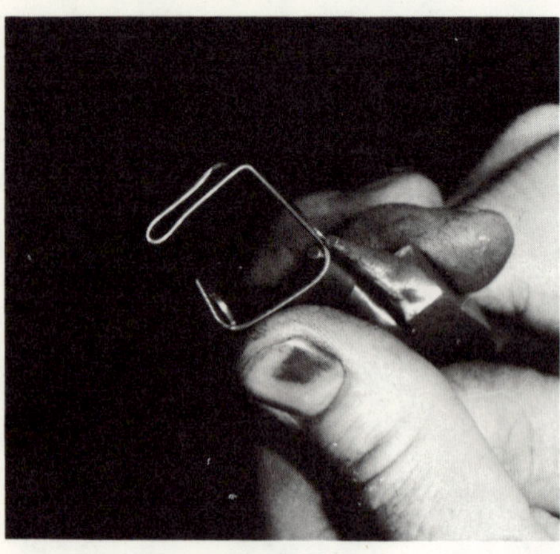

6. Try the holder on a drinking tube and make adjustments so the tube will be held at the proper angle.

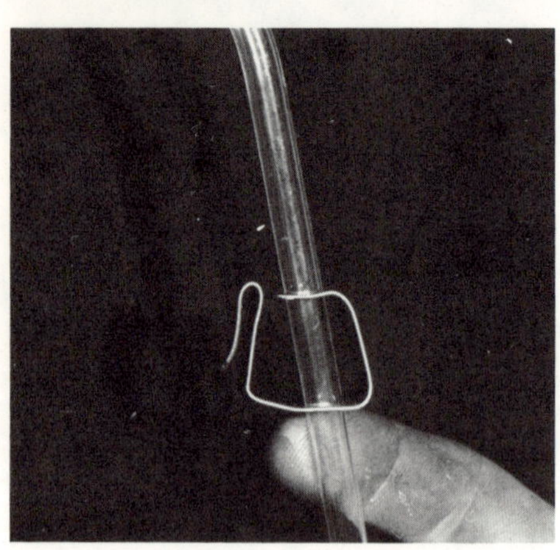

7. Try the straw holder and drinking tube in a glass.

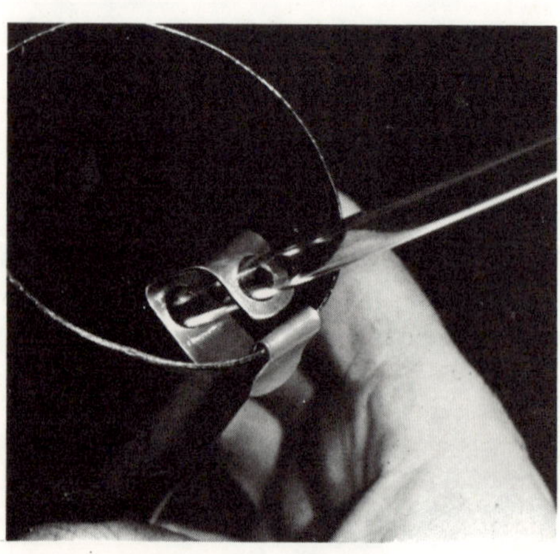

HOW TO MAKE A SHOE HORN

<u>Introduction</u>

It is difficult and impossible for many patients to put their shoes on using the ordinary small shoe horn, because they cannot stoop down to reach their shoes or are unable to grip the shoe horn tightly enough when bent over.

The shoe horn described in this set of instructions is made long enough so the patient does not have to stoop over, and it is large enough so he can get a grip on it even with weakened muscle.

1. Lay out the pattern; draw a 16 inch line on a piece of .051 inch aluminum, scribe a 1/2 inch circle on one end, a 2 inch circle on the other, then draw two lines tangent to these circles and cut the metal to the line.

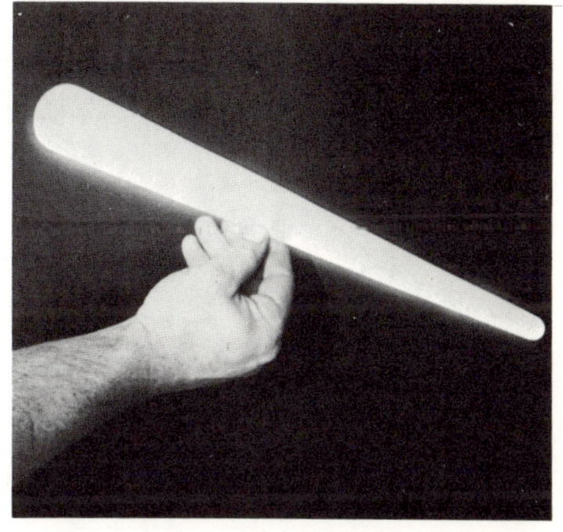

2. Using a concave forming block, hammer the piece along one face to make it hollow its full length.

3. Bend the narrow end over to form a
hook, smooth all edges and polish.

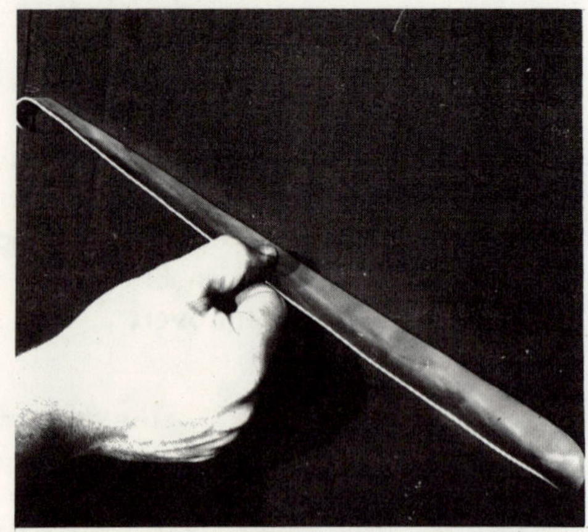

HOW TO MAKE **A** CIGARETTE HOLDER

<u>Introduction</u>

Just as the glass holder enables the patient to get a drink of water without help, the cigarette holder enables him to smoke without having to have someone put the cigarette in his mouth each time. The average cigarette smoker smokes about twenty cigarettes a day. Making it possible for him to do it with a minimum of help gives him additional independence.

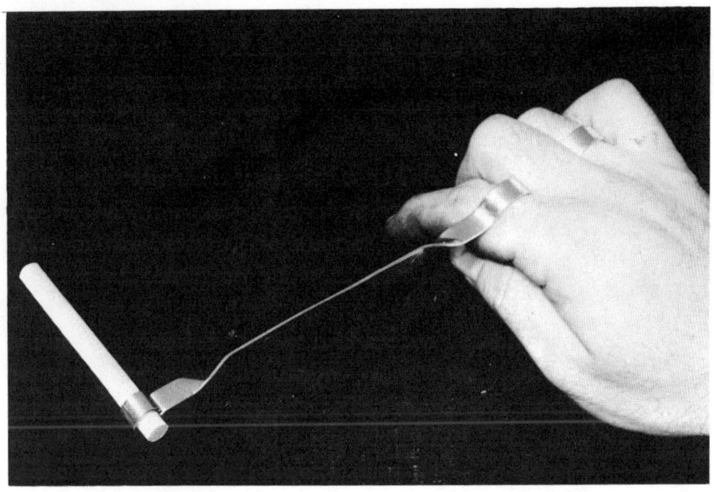

1. Cut a piece of .064 inch aluminum, 3/8 inch wide and 12 to 14 inches long, depending on the distance from the patient's hand to his mouth. Round corners, smooth and polish.

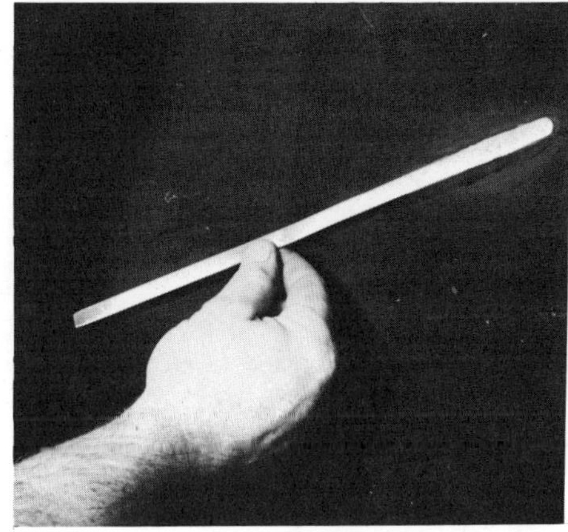

2. Shape a 5/16 inch ring in one end of the piece to hold the cigarette, using a 5/16 inch mandrel in the vise to form it.

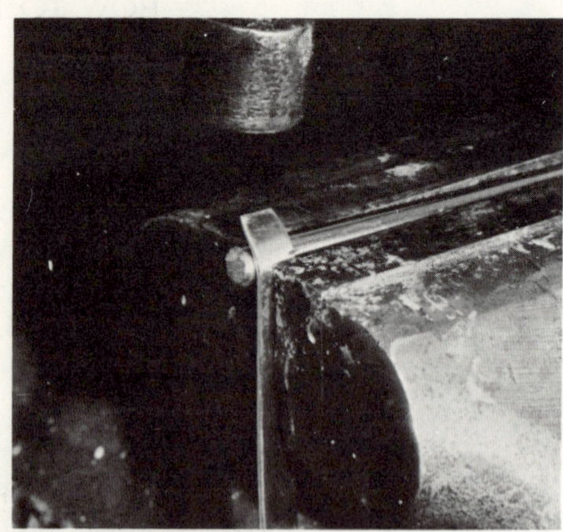

3. Adjust the ring so a cigarette can be inserted easily, but will be held tightly enough so it cannot fall out.

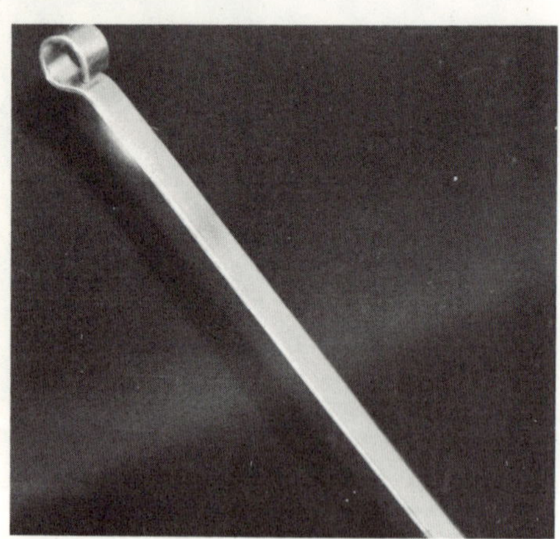

4. Shape the handle so it can intertwine between the fingers.

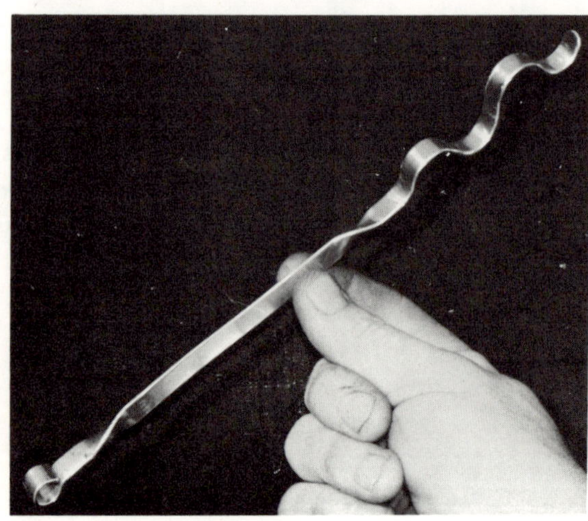

CHAPTER VI. BIOMECHANICS OF FUNCTIONAL HAND SPLINTS AND ARM BRACES

Introduction

Biomechanics is the science that deals with the forces which act on the living body. Mechanics is that branch of the science of physics which deals with the actions of forces on bodies.

We are concerned with biomechanics in our work with functional hand splints and arm braces because there is a complex interplay of forces between these devices and the hands, arms, and shoulders of the patients on whom they are fitted. One of the main functions of a splint or brace is to direct and distribute these forces in such a way that the comfort of the patient is maintained or improved, while at the same time contributing increased function.

Every person responsible for any aspect of the relationship between the patient and the splint or brace with which he is fitted should have an understanding of some of the biomechanical principles involved. The more important of these principles will be discussed in this section with emphasis on their application to the various types of splints and braces.

Force

A force is a push or pull that tends to change the motion of a body, either in velocity or direction, or to change the body's shape. Forces are measured in pounds, ounces, grams or similar units.

Weight

Weight is the force by which a body is attracted toward the earth by gravity. When we refer to the weight of a patient's arm stretching the rubber bands on a brace, we mean that his arm is attracted to the earth by a force sufficient to stretch the bands.

Vectors

Describing and analyzing the interplay of forces between the muscles of the patient's hand and arm and the splint or brace he wears would be very difficult without some way to draw symbols representing forces.

The symbol used to represent a force is called a "vector". A vector is an arrow; the length of the shaft of the arrow determines the _amount_ of force represented, the head of the arrow shows the _direction_ of the force, (Figure 135).

Figure 135.

Before drawing a vector a scale is chosen, such as 1/4 inch = 1 pound; the shaft of the arrow is drawn accordingly. Using the scale mentioned to represent a force of 8 pounds, the shaft of the vector would be drawn 2 inches long.

Force Applied to a Stationary Body

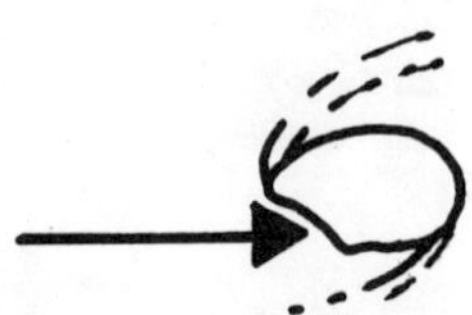

Force applied to a stationary body tends to either move it forward, or to turn it around to a new position, (Figure 136).

Figure 136.

Force Systems

It is common to find two or more forces acting on a body together, forming a "force system". These can be extremely complex and difficult to understand. The elements of the more common combinations are:

1. Two or more forces acting in the same direction and in the same line. The resultant of these forces is found by adding them. (Figure 137 "A".)

2. Two or more forces acting in opposite directions and in the same line. The resultant of these forces is found by subtracting the smaller from the larger. (Figure 137 "B".)

3. Two or more forces acting at an angle to one another. The resultant of such forces may be found graphically, as explained in the next paragraphs. (Figure 137 "C".)

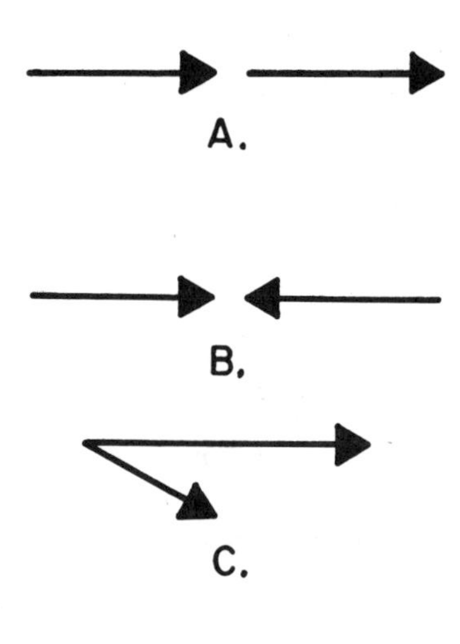

Forces at an Angle: Composition of Forces

When two forces act on a body at an angle, the resultant force may be found by drawing a parallelogram of forces, as illustrated at right. The two forces are drawn to scale and at the correct angles, as in Figure 137 "D". The two vectors are used for two sides of a parallelogram, and the other two sides are drawn as in Figure 137 "E". A diagonal drawn as in Figure 137 "F" forms a new vector that represents the resultant of the two forces with which we started. Note that it is not quite as long as the two put together, and the direction is somewhat of a compromise between the two.

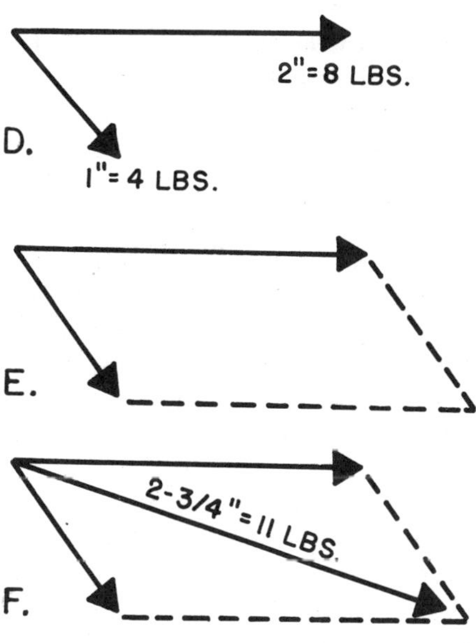

Figure 137.

Moments of Forces; Levers

The lever in one form or another is used very often in both functional hand splints and arm braces. To more easily understand the action of levers it is best to start with the idea of moments of forces, and then see how they act in a lever system.

If a force is exerted at a 90 degree angle against one end of a four-foot bar, and the other end is provided with a bearing mounted on a solid support, the bar will rotate around the bearing. (Figure 138.)

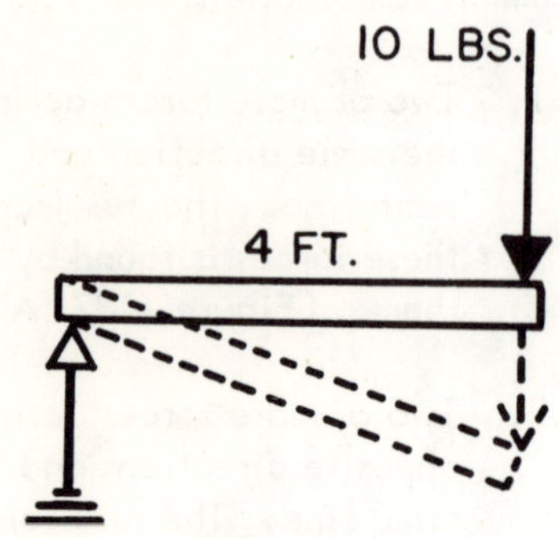

Figure 138.

The measure of the tendency of the force to make the bar rotate on the bearing is the moment, and it is found by multiplying the force by the length of the bar. It is expressed as foot-pounds, or the like. For example, if a ten-pound force is applied to a four-foot bar the moment is forty foot-pounds. This is only true when the force is applied at a 90 degree angle to the bar as illustrated.

If we now extend our bar beyond the bearing, or "fulcrum" as it is called, we have made a simple "First Class" lever. Assume that the extension on our bar is two feet, and we want to exert a force on it great enough to balance the ten-pound force on the four-foot portion. (Figure 139.)

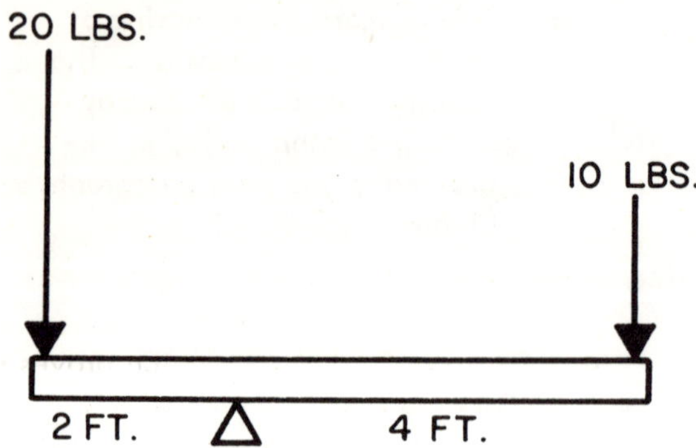

Figure 139.

The moment on the two-foot side must equal that on the four-foot side. To obtain a 40 foot-pound moment on the two-foot side we must exert on it a force of 20 pounds.

The first class lever can be better understood if the idea behind it is stated as a principle. In Figure 140 the lever arm, (the part of the bar between the point of application of the force (F), and the fulcrum) is called "L"; that portion between the fulcrum and the weight to be lifted is "1".

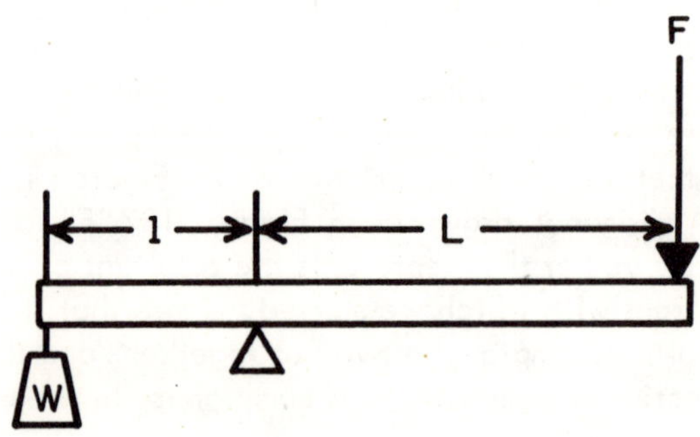

Figure 140.

The weight to be lifted is "W". The relationship between these factors is: F is to W as 1 is to L. Putting it another way, F x L = W x 1. In our first illustration, F = 10 lbs., L = 4 ft., W = 20 lbs., and 1 = 2 ft.; F x L = 40, W x 1 = 40 -- the two moments are equal. An example of the application of the principle of the first class lever is the feeder trough (Figure 141). In the illustration, L is the trough from the pivot to the point where the patient's elbow rests, F is the downward force exerted by the patient, 1 is the distance from the pivot point to the point of grasp, and W is the weight of the patient's hand and whatever object he may be holding in it. It is easily seen that F will never exceed a fixed amount, that is, the maximum effort the patient can make. We apply the principle of the first class lever by lengthening or shortening L and 1 so that he can lift a maximum W through a satisfactory range of motion. A limiting factor, of course, is the distance through which the patient can push his elbow on the end of the trough.

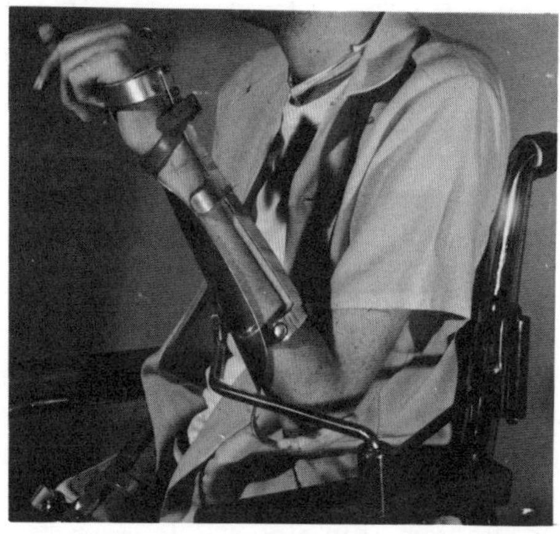

Figure 141.

The second class lever is illustrated in Figure 142. It can exert more force with a given over-all length of bar because it uses the full length of the bar as a lever arm; note in the diagram, L extends from the fulcrum to the point of application of the force. The same relationships still apply, however: $F \times L = W \times 1$. If the weight is 20 lbs., and $F \times 6 = 20 \times 2$; $F = 6\text{-}2/3$, which proves that a six-foot bar used as a second class lever can be made to lift a given weight with less force than a first class lever of the same over-all length.

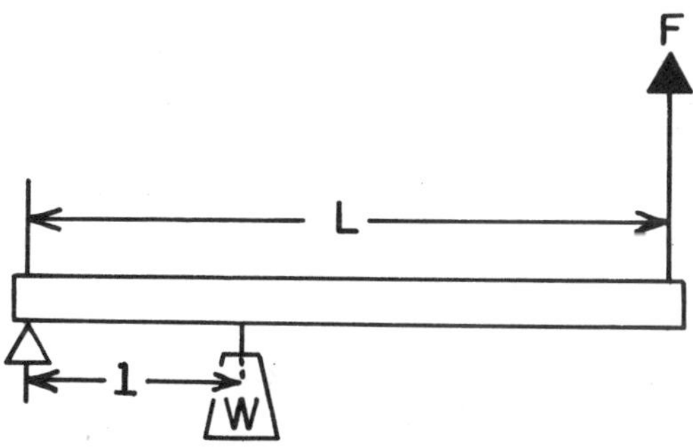

Figure 142.

An example of the application of the second class lever is the interphalangeal extension assist with metacarpophalangeal extension stop, shown in Figure 143.

L is the finger from the point of attachment of the rubber band to the metacarpophalangeal joint; 1 is the distance from the control bar to the metacarpophalangeal joint; F is the tension of the rubber bands; and W is the load exerted against the control bar. It is clear that the latter force is considerably greater than that exerted by the rubber bands. If F exerted

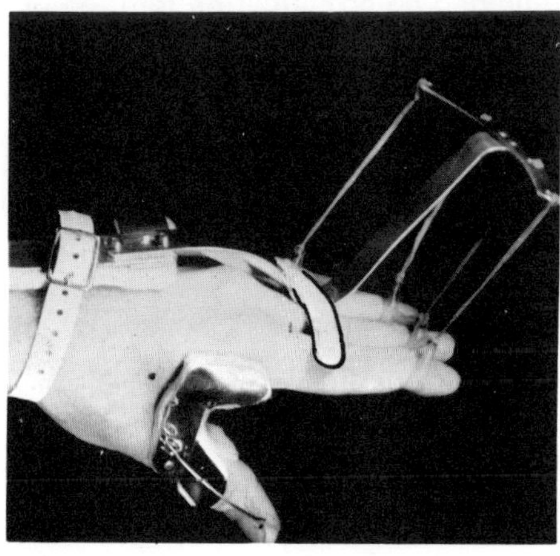

Figure 143.

on the forefinger by the rubber bands of the splint is 5 ounces, L on the finger is 4 inches, and 1 on the finger is 1 inch, how much force is exerted at W?

The third class lever is illustrated in Figure 144. It exerts <u>less</u> force at W than is applied at F. However, levers are not always used to increase the amount of force at W. Sometimes it is desirable to sacrifice force to gain distance traveled at W. In the diagram, the closer F is moved to the fulcrum, the greater will be the excursion of the bar at W for a given amount of excursion at F. However, as F is moved closer to the fulcrum, the amount of force required to lift W increases.

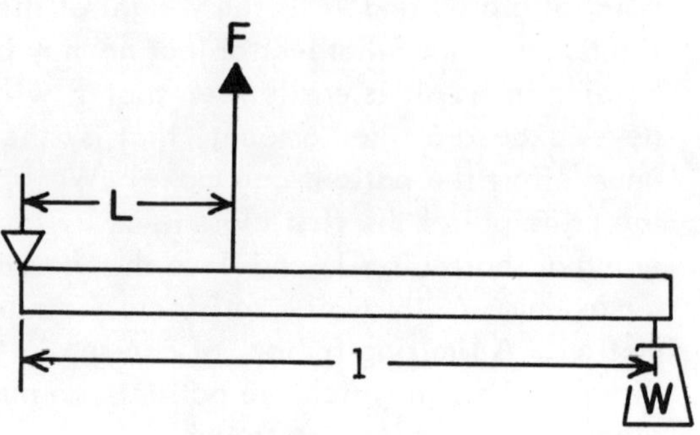

Figure 144.

There are many examples of third class levers in the human body. The brachialis muscle functions in a third class lever system between the humerus and the ulna. In Figure 145, the elbow joint is the fulcrum, F is applied to the ulna slightly distal of the fulcrum, L is the ulna between the fulcrum and the hand. This is a better example than most, because the brachialis pulls directly on the ulna and no complicating rotation of the latter is involved.

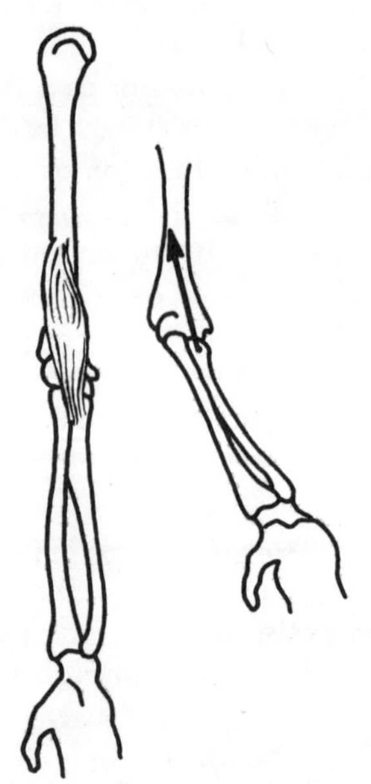

Figure 145.

Forces at an Angle: Resolution of Forces

The moment of force on a lever can be calculated by multiplying the force times the length of the lever only when the force is applied at a 90 degree angle to the lever, as was explained earlier. In many applications in splints and functional arm braces the force is not applied to the lever at right angles. To find the moment in this case it is necessary to break the force into two forces, on at right angles to the lever, the other parallel to it, since the force parallel to the lever cannot create a moment of force, the one at right angles to the lever can be measured and the moment calculated. This procedure is called "resolution of forces".

This principle is illustrated in Figure 146, a functional arm brace with shoulder flexion assist. In the position shown, the rubber bands exert a pull of 7 pounds at an angle of about 15 degrees to the lever arm. Using a scale of 1/4 inch equals 1 pound, a vector is drawn and a parallelogram constructed as shown in Figure 147. The small vector at right angles to the lever is measured and found to be 3/8 inch long, equal to 1-1/2 pounds. If L on this brace is 10 inches and 1 is 12 inches, how much flexion assist is exerted at W?

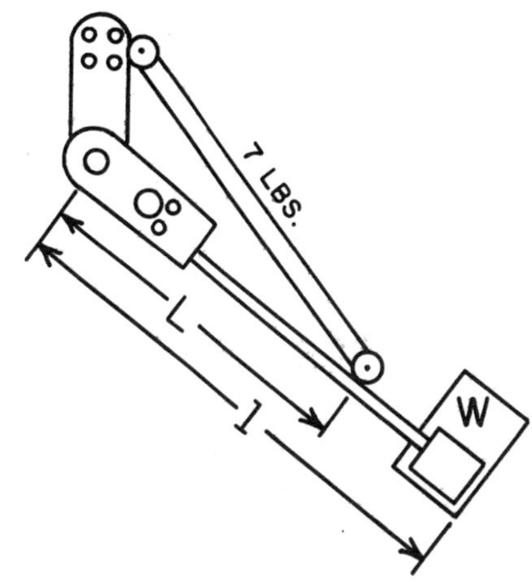

Figure 146.

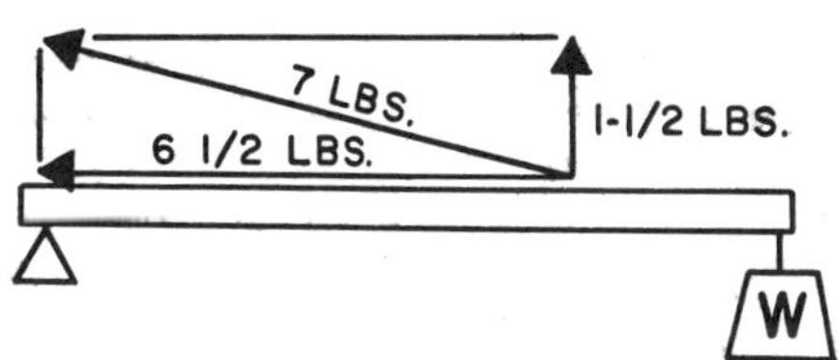

Figure 147.

Pressure

Pressure is the amount of force exerted on a body per unit of area, and is often measured in pounds per square inch. If pressure is concentrated on a small area of a patient's hand or arm, or other part of his body, it will cause pain. The more widely and evenly the pressures exerted by a splint or brace are spread over the surface of the body, the less pain will be caused. For this reason, parts fitted to the patient's body must fit accurately, so excess pressure is not concentrated on high spots. Padding helps distribute the pressures more uniformly because it yields to the high spots and fills in the low ones, and has a certain amount of "give" to accommodate to changes in body contour caused by movement.

Sources of Power

It is important to always remember that successful fitting of functional hand splints and arm braces often requires the utilization of muscular forces so small they are hardly perceptible. In many cases it may seem futile to bother with such small forces, but this attitude must be avoided, as it is often found that enabling and encouraging the patient to perform even a very small movement may, over a period of time, result in increased strength and function.

It is very common to find a patient unable to flex a joint while still retaining some degree of strength to extend it. The reverse may occur, with the patient able to flex, but not extend the joint. Here it becomes necessary to solve the problem of providing power to perform the lost function.

There are two kinds of energy, kinetic and potential. Kinetic energy is possessed by a body in actual motion, as a swinging hammer. Potential energy is possessed by a body that has stored some by being acted upon by the kinetic energy of another body, as when you use your hands to stretch a rubber band. When the rubber band is stretched, it possesses potential energy which is released when the band is allowed to return to its normal length.

The patient who can, for example, flex his finger joints but cannot extend them exerts kinetic energy when he flexes them. If some of this could be stored as potential energy, it could be released for use in extending the finger joints. Functional hand splints and arm braces apply this principle by using springs or rubber bands to store potential energy during the movement the patient is able to accomplish. This stored energy is used later to perform the opposite movement.

This solution to the problem requires that the patient have enough strength in flexion, for example, not only to grasp objects and hold them, but to stretch the rubber bands as well.

Here is where the orthotist and the patient can become very discouraged if the amount of strength available is very slight, yet it is imperative that it be harnessed if it is to be increased through use. Sometimes patients are fitted who have so little strength that they cannot stretch an ordinary rubber band; strips of very thin rubber ("rubber dam") used in dental work must be used instead.

An important point to remember is that most of the useful energy stored in the rubber bands or springs must come from the patient's own muscle contractions. The force exerted by gravity alone cannot be used to get satisfactory function. If a patient has lost both elbow flexion and extension it is useless to fit him with a brace in such a way that the weight of his forearm stretches some rubber bands in the hope that they will then help him to flex the elbow. The forearm will not extend as much as it should, but will stop at a point where the rubber bands are stretched enough to balance its weight. In such cases it may be necessary to use an outside source of power such as the CO_2 powered artificial muscle or electric motors.

Characteristics of Rubber Bands

Ordinary rubber bands are used quite often as a means for obtaining useful function from a partially paralyzed hand or arm. They are inexpensive, can be purchased very easily at any drug or stationery store, and the patient can be taught to replace them himself or they can be replaced by a member of his family.

It is very easy to obtain very small graduations in tension by using the rubber bands in multiples; removing or adding one band is a very small change in the total tension exerted by 32 bands. Rubber bands can be connected to a lever system in parallel so the over-all length is the same but the cross-section increases as bands are added.

As bands are added, the amount of tension-increase per inch of stretch increases, as can be seen in the table below.

LENGTH-TENSION DATA, NO. 32 RUBBER BANDS IN PARALLEL

Length, Inches	Tension, Oz. 1 Band	Tension, Oz. 2 Bands	Tension, Oz. 3 Bands
3	0.0	0.0	0.0
4	4.5	10.0	15.0
5	8.0	16.0	24.0
6	10.0	20.5	31.0
7	11.5	24.5	35.0
8	12.5	28.0	
9	14.0	31.0	
10	15.5	34.0	
11	17.0		
12	18.5		
13	21.0		
14	22.5		
15	23.5		
16	25.0		

In connecting rubber bands for a shoulder flexion assist functional arm brace, a large amount of tension increase for each inch of excursion is needed, because the system is a third class lever with the force applied very near to the fulcrum. However, such a large increase in tension per inch of stretch is not always desired, as with the elbow flexion assist. With this device the rubber bands are attached a considerable distance distally from the elbow joint, which decreases the mechanical disadvantage of the third-class lever and increases the excursion of the joint. This in turn stretches the rubber bands through a greater distance. To avoid such a rapid build-up in tension that the patient would be unable to fully extend his elbow, it is necessary to connect several rubber bands end-to-end or in series, then connect the desired number of these series of bands in parallel.

LENGTH-TENSION DATA, NO. 32 RUBBER BANDS IN
SERIES AND PARALLEL

Length, Inches	Tension, Oz. 3 Bands, Series	Tension, Oz. 2 Series of 3 Bands in Parallel	Tension, Oz. 3 Series of 3 Bands in Parallel
9	0.0	0.0	0.0
10	4.0	7.5	10.0
11	6.0	10.5	15.0
12	7.5	13.0	18.0
13	8.0	15.5	22.0
14	9.0	16.5	24.5
15	9.5	18.5	26.5
16	10.5	20.5	29.0
17	11.0	22.0	31.0
18	11.5	23.0	33.0

By using various combinations of rubber bands in series and parallel it is possible to provide the proper rate of increase in tension per inch of excursion to balance the strength the patient can exert. The adjustment can be made quite accurately by adding or removing one or two bands at a time until the desired amount of tension is obtained.

Unfortunately, rubber bands are somewhat unstable in that they gradually lose their tension when stretched, and they deteriorate with age. Because of these disadvantages, they must be placed every few months, and it is necessary to make frequent adjustments to maintain balance with the patient's musculature. The latter is not too much of a drawback, because it should b ticipated that when the patient uses the device his strength will increase, making adjustments essary anyway.

Springs can be used instead of rubber bands, and they have the advantage of long life without much loss of tension. However, they cannot be adjusted to various combinations of te and length as easily as rubber bands, and they are bulkier and harder on clothes.

Tension Relief

A difficult problem in functional arm braces is that of relieving the patient's extensor muscles of having to maintain the tension on the rubber bands for long periods of time when the arm is extended in normal position, but still have the tension available for use when he wants to flex his shoulder or elbow.

This is accomplished by so arranging the anchor points for the rubber bands and the brace rods and joints that when the joint is fully extended the lower anchor point is in line with the joint axis and the upper anchor point. Under these conditions the force is in a straight line and the third-class lever situation no longer exists (Figure 148). On some braces a stud is placed on the upper portion of the brace so the rubber bands will contact it on extension of the joint, thus preventing it from going into hyperextension and creating a problem in that direction. To bring the rubber band tension into play, the patient swings his trunk slightly so inertia in turn swings his arm forward.

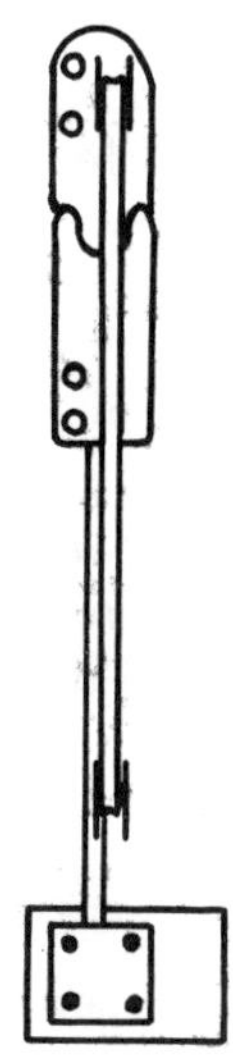

Figure 148.

"Gadget Tolerance"

Patients will not tolerate more than a certain amount of mechanical apparatus fastened to their bodies. This is sometimes referred to as "gadget tolerance". It varies greatly from one individual to another, which makes it impossible to state with certainty the maximum amount of apparatus any given patient will wear and use regularly.

In general, the less the patient's handicap the less tolerant he will be of appliances to offset it. There is a point where the amount of function the patient obtains from a device does not offset the inconvenience of wearing it, and when this point is reached he will refuse to use it.

Another important factor in determining the patient's gadget tolerance is the degree to which he is motivated to do the things the device will enable him to do. The patient who does not care whether or not he can take care of his own needs, and does not at all mind being waited on is a poor prospect for a splint or brace until this attitude is changed.

The severely involved patient will usually tolerate more apparatus than one who is less severely handicapped. The bilateral will usually tolerate more than the unilateral. For this reason the orthotist often finds that many of his most successful cases are the severely involved patients who have little to work with and are extremely difficult to fit, but who will tolerate a maximum of apparatus to obtain some ability to function. The less severe cases are easier to fit but often discard their devices, feeling that the gain in function is not worth the trouble of wearing them.

The physical or occupational therapist is often very helpful in training and motivating a patient to use his brace or splint day in and day out.

An additional complicating factor is the fact that one of the objectives in fitting a patient with a functional splint or brace is to encourage him to use the paralyzed member as much as possible, which in turn often brings about an increase in strength to such an extent that in time the device can be dispensed with. In such cases the greatest success is climaxed by the patient achieving complete independence from splints or braces.

Because of the fact that some patients do not have sufficient "gadget tolerance" and in some cases do not have sufficient motivation, a certain number of failures is inevitable. However, it is almost impossible to predict which patients will be successful and which will fail. For this reason the orthotist, therapist, and physician should be very certain that a patient cannot benefit from an appliance before recommending that he not be fitted.

Mechanical Principles of Ball Bearing Feeders

The lever system used to obtain function with the feeder arm trough has been discussed. The manner in which the swivel arms of the ball bearing feeder enable the patient to reach out, or withdraw after reaching, is another application of the idea of storing potential energy for later use.

When the feeder swivel arm pivot shaft is vertical, the arm will remain in any position in which it is placed. This is because the vertical vector representing the force of gravity is in line with the pivot shaft and no horizontal component is present. The assembly is in equilibrium and no movement occurs, which condition is illustrated in Figure 149 "A". However, any force exerted on the swivel arm in a horizontal plane and at right angles to it will cause it to rotate on the bearing.

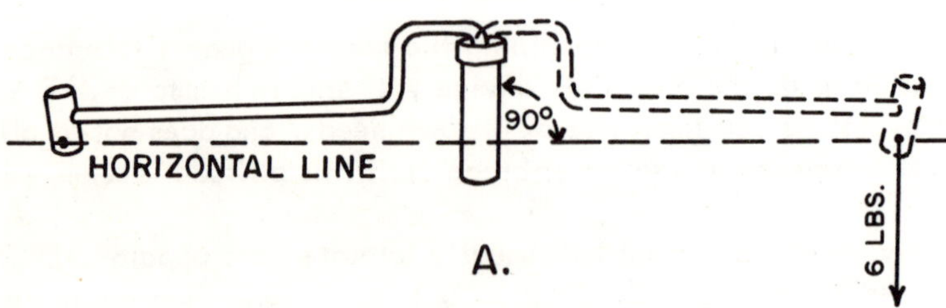

Figure 149.

If we tilt the pivot shaft a few degrees so the end of the swivel arm is raised above the horizontal line, as shown in Figure 149 "B", we create a situation in which the weight at the end of the arm behaves in the same manner as a comparable weight on an inclined plane similar to line "X"-"Y". While we usually think of an inclined plane as something like a ramp down which a barrel will roll because of the force of gravity, careful study of "B" will show that gravity will cause the end of the swivel arm to swing to the lowest point it can reach for exactly the same reasons that gravity makes the barrel roll down the ramp.

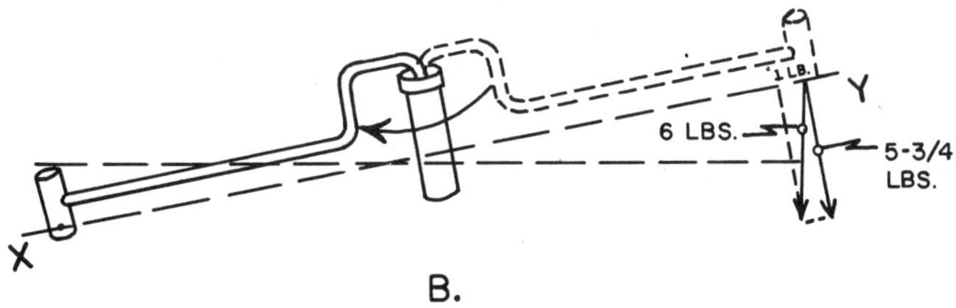

Figure 149.

If we assume that the force of gravity on the end of the swivel arm is 6 pounds, we can find the other components by using vectors to resolve the forces involved. The force at right angles to "X"-"Y" is 5-3/4 pounds, and all of it is exerted in side thrust on the swivel bearing. The 1 pound force on line "X"-"Y", however, will cause the swivel arm to swing on its bearing as indicated by the curved arrow. The end of the arm will follow a curved path, of course, but always on the same plane as line "X"-"Y", until it reaches the lowest point possible. When this happens, the arm will remain motionless at the low point.

If we make our "inclined plane" steeper, by further tilting the pivot shaft as in Figure 149 "C", the force tending to make the swivel arm swing downward is increased. Of the 6 pounds of force, 5-1/2 pounds are dissipated in side thrust against the bearing, but 2-1/2 pounds act on the swivel arm to make it swing downward as shown by the curved arrow.

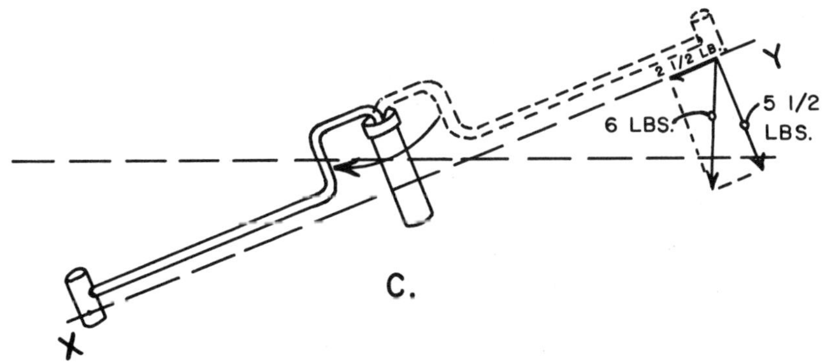

Figure 149.

If we assume that the equipment in the illustration was set up for a patient to use, the force of gravity would help him move his arm from right to left, "down-hill". He would have to exert enough muscular force to return it to its original position, and the amount of force required would be greater than would be the case if the pivot was vertical as in drawing "A". This extra force is used to raise the end of the arm, or "climb the hill", so to speak, thus storing some potential energy for later use. This procedure of over-exerting muscles that retain strength in order to store energy in rubber bands, springs, or at the top of an incline is a very useful one and is the basic principle of a number of functional hand splints and arm braces.

THE USE OF THE ARTIFICIAL MUSCLE IN UPPER EXTREMITIES ORTHOTICS

Introduction

Restoration of voluntary motion to severely involved upper extremity paralytics is achieved by supporting the paralyzed member with braces and transferring motion from external actuators to the physiological member via the supportive structure. Prehension, or closure of one or more fingers against the thumb to provide grasp, and positioning of the hand by the arm are the primary goals of applying power to the upper extremity.

Among activities of primary importance to the patient are mobility, self-feeding, manipulation of small objects, and personal hygiene. These goals may be achieved to a certain degree by supplying motion to one arm and hand with assistive power and control devices. To obtain mobility, the person may drive a commercially available powered wheel chair if he has sufficient control sites for the operation of four transducers.[*]

[*] A device that transmits power from one system to another system.

General Design Considerations

Simplicity of attachment, reliability, low cost, minimum power consumption, safety and appearance are the general factors which most influence the design of any external power system. The attachment and removal of the device from the patient will ultimately be handled by a hospital attendant, relative of the patient, or other person whose technical ability is minimal. Necessity of frequent electronic or mechanical adjustments must be eliminated entirely. The complexity of setting up the control system for operation should be limited to the turning on or off of not more than two switches or valves. Reliability should not be affected by impact loading or shock due to abnormal handling, such as being dropped three or four feet or being stored in a haphazard manner. Materials that exhibit explosive, toxic or other safety hazards should not be used. Cosmesis and comfort are important because of patient acceptance, and as a human factor this varies considerably. The most important factor that will influence the acceptance or rejection of a device will be the degree of usefulness of the device to the patient. Maximum possible function should be restored to the patient without impeding the normal function of an unrelated bodily member that has current usefulness; i.e., a totally flail left arm should not be controlled by the right hand if the right arm and hand retain sufficient strength to perform useful activity.

Due to the extreme variability of involvement and factors of importance such as the psychological adjustment of patients, determination, skills, and basic intelligence, each patient must be custom-fitted. An experienced occupational therapist who is familiar with the patient is usually the best source of information concerning the patient's machine tolerances.

Components

The complete externally powered orthotic device consists of the supportive bracing plus a power source, and at least one transducer, signal shaping network and actuator per channel of motion. A simplified power and control system is summarized in Figure 150, with a selection of available components listed. It should be noted that, except for electrical power sources and possibly some electric motors, the components listed refer to items that are of special design and that commercially available components are generally not applicable.

In selecting an external source of power one must consider the efficiency of the source used, its ease of application, availability and cost of replacement. Carbon dioxide and electricity are, at present, the most promising. Carbon dioxide is a gas which liquefies at 750 psi, 72° F. It is non-toxic, non-combustible, costs approximately ten cents per pound and is readily available almost anywhere in the United States.[1] Electricity can be stored more efficiently

---

[1] R. Snelson, A. Karchak, Jr., and V. L. Nickel: "Application of External Power in Upper Extremity Orthotics", Orthopedic & Prosthetic Appliance Journal, American Orthotics & Prosthetics Association, 15:345-348, December, 1961.

than compressed gases,[1] but actuators for compressed gases at the present state of the art are lighter, cheaper, and more easily adapted to bracing.

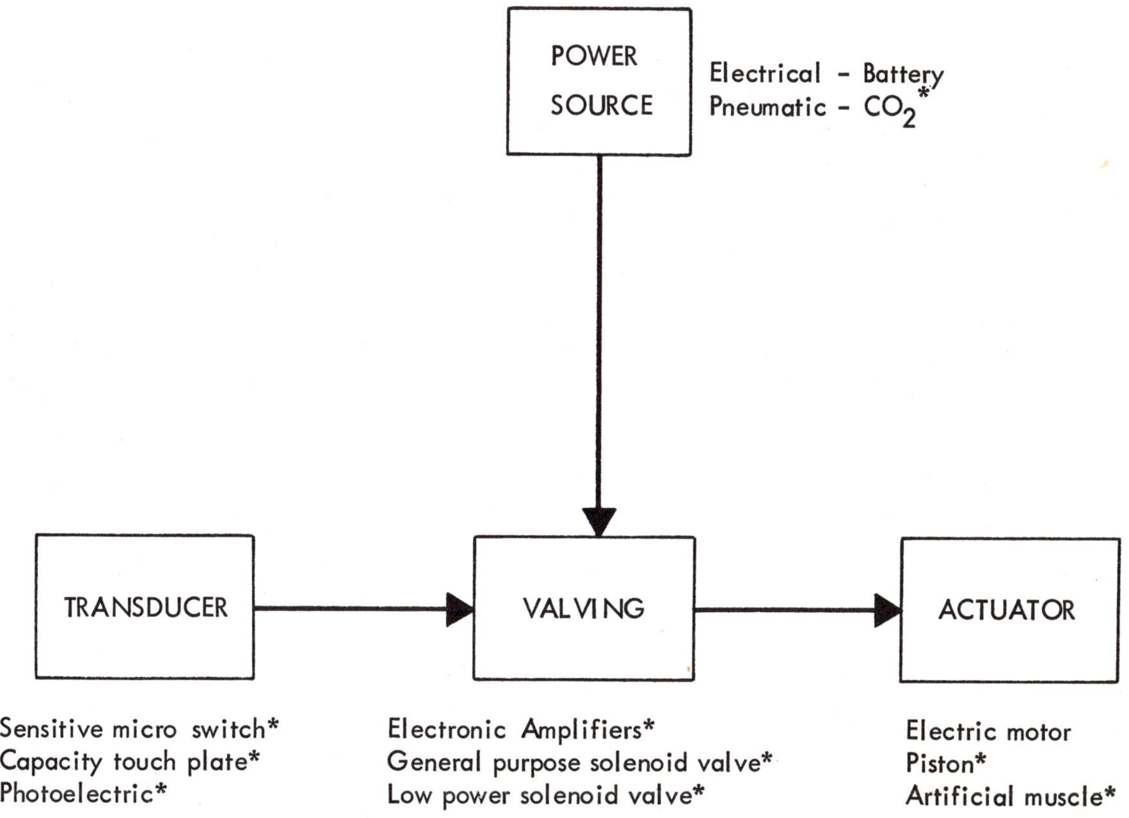

*Indicates specialized components or containers.

Figure 150. Block Diagram of External Power and Control System

[1] W. C. Gorthy, S. W. Alderson, E. A. Kiessling, R. W. Mann, H. A. Mauch, G. M. Motis, J. B. Reswick, H. F. Schulte, and T. A. Smith: "Report of Panel on Power Sources", National Academy of Sciences – National Research Council, Publication 874, pp. 31, September, 1961.

Available actuators include pistons, bellows, artificial muscles and electric motors. The piston and the artificial muscle are the favored choices at present. The artificial muscle (Figure 151) has no static friction, exhibits no alignment problems and is lighter and cheaper than the piston. The piston, however, has greater excursion, and can exert linear expansion as well as contraction force. Electric motors have not, as yet, been used except for experimental purposes. Actuators in the range of 20 to 200 inch pounds with two to six inches of excursion will generally satisfy all the power requirements of externally powered devices (Figure 152-A and 152-B).

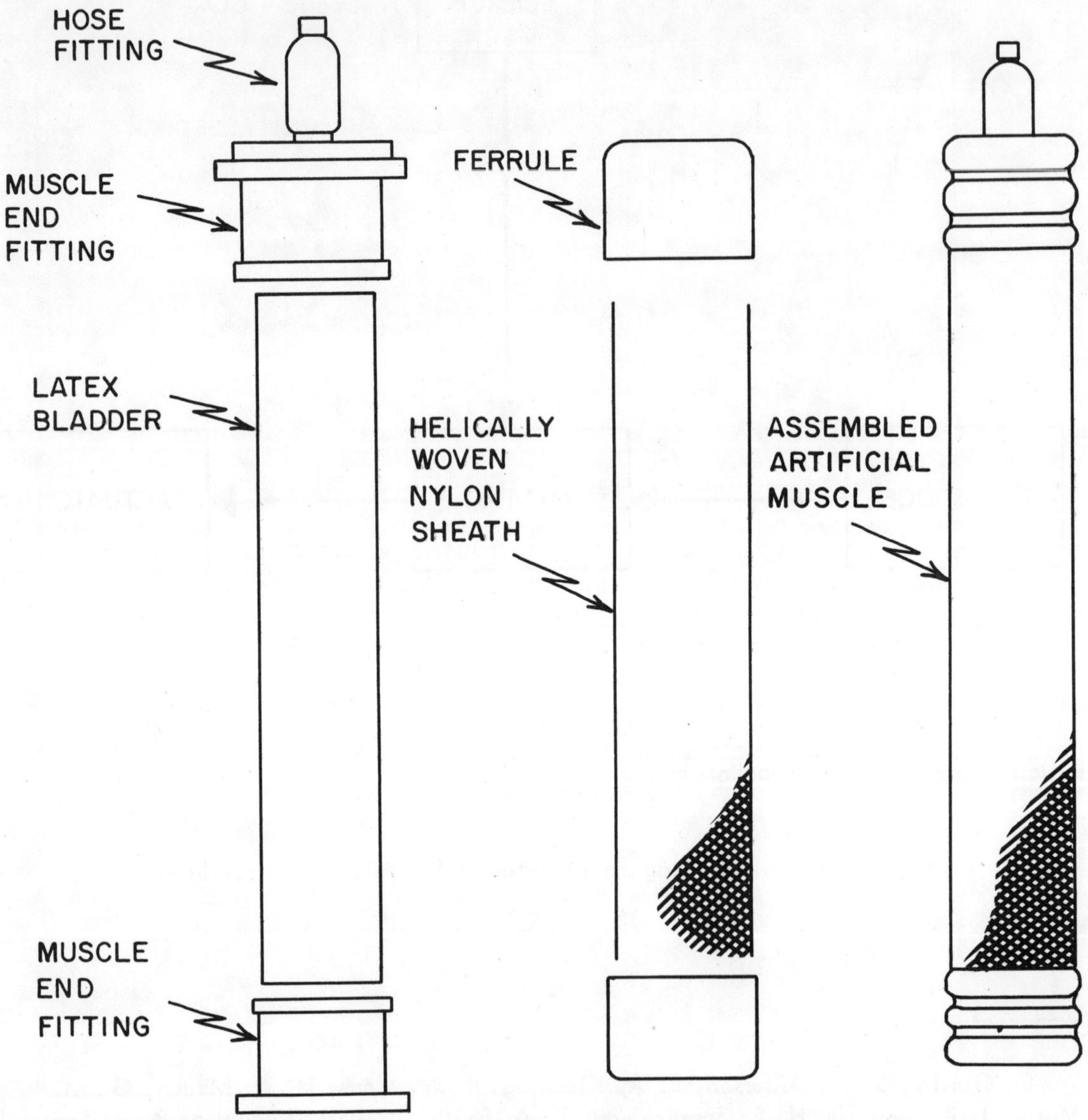

Figure 151. Artificial Muscle Assembly

FORCE (LBS.) VS. EXCURSION (INCHES)
Weave = 8 picks/inch
Muscle length = 8 inches
Rancho Los Amigos Hospital
50 lbs./sq. inch guage
40 "
30 "
FORCE (LBS.)
EXCURSION (INCHES)
80
70
60
50
40
30
20
10
0
0
.5
1.0
1.5
2.0

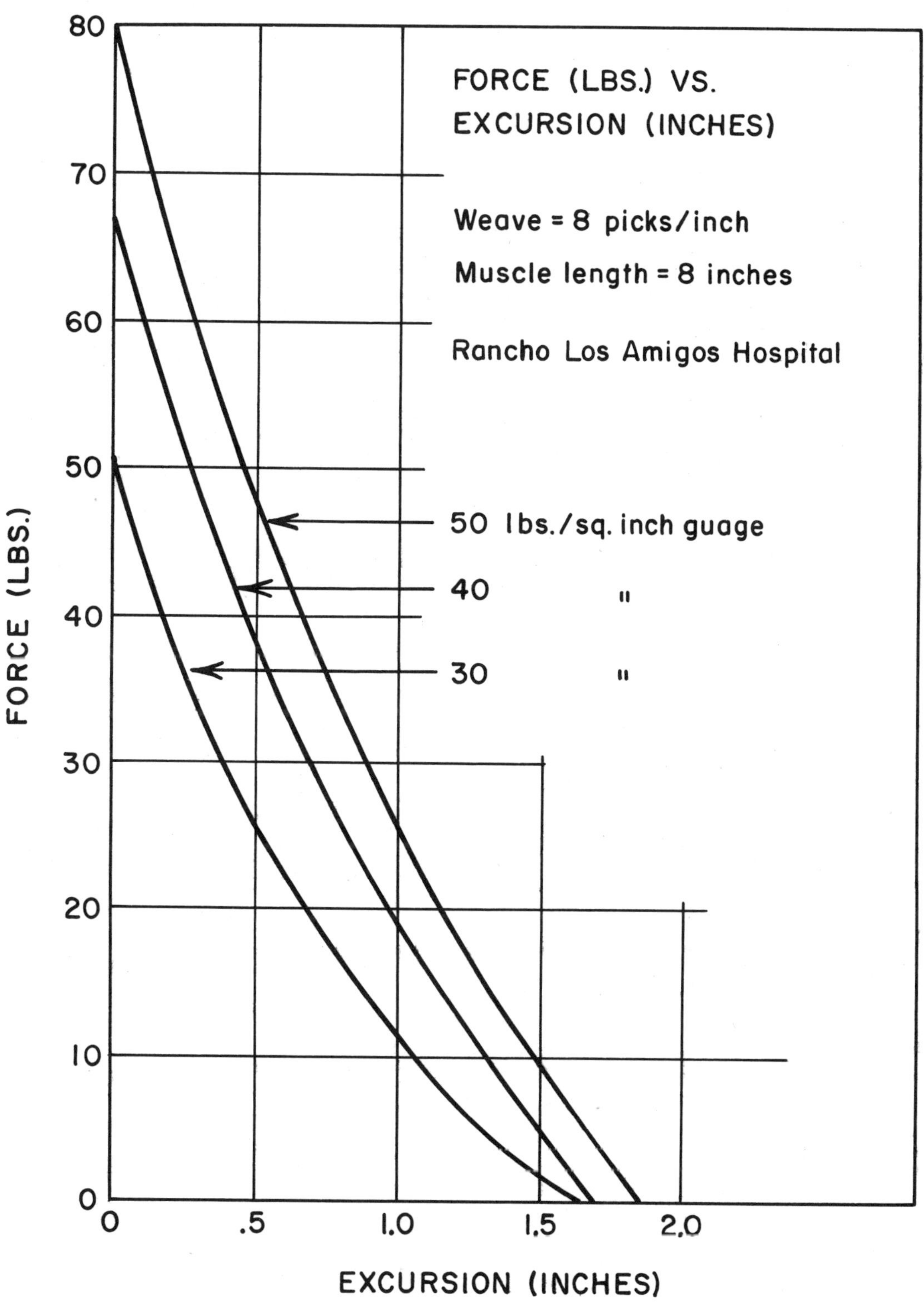

Figure 152-A

FORCE (LBS.) VS.
EXCURSION (INCHES)
Weave = 10 picks/inch
Muscle length = 8 inches
Rancho Los Amigos Hospital
70
60
50
40
30
20
10
0
FORCE (LBS.)
50 lbs./sq. inch guage
40   "
30   "
0   .5   1.0   1.5   2.0
EXCURSION (INCHES)

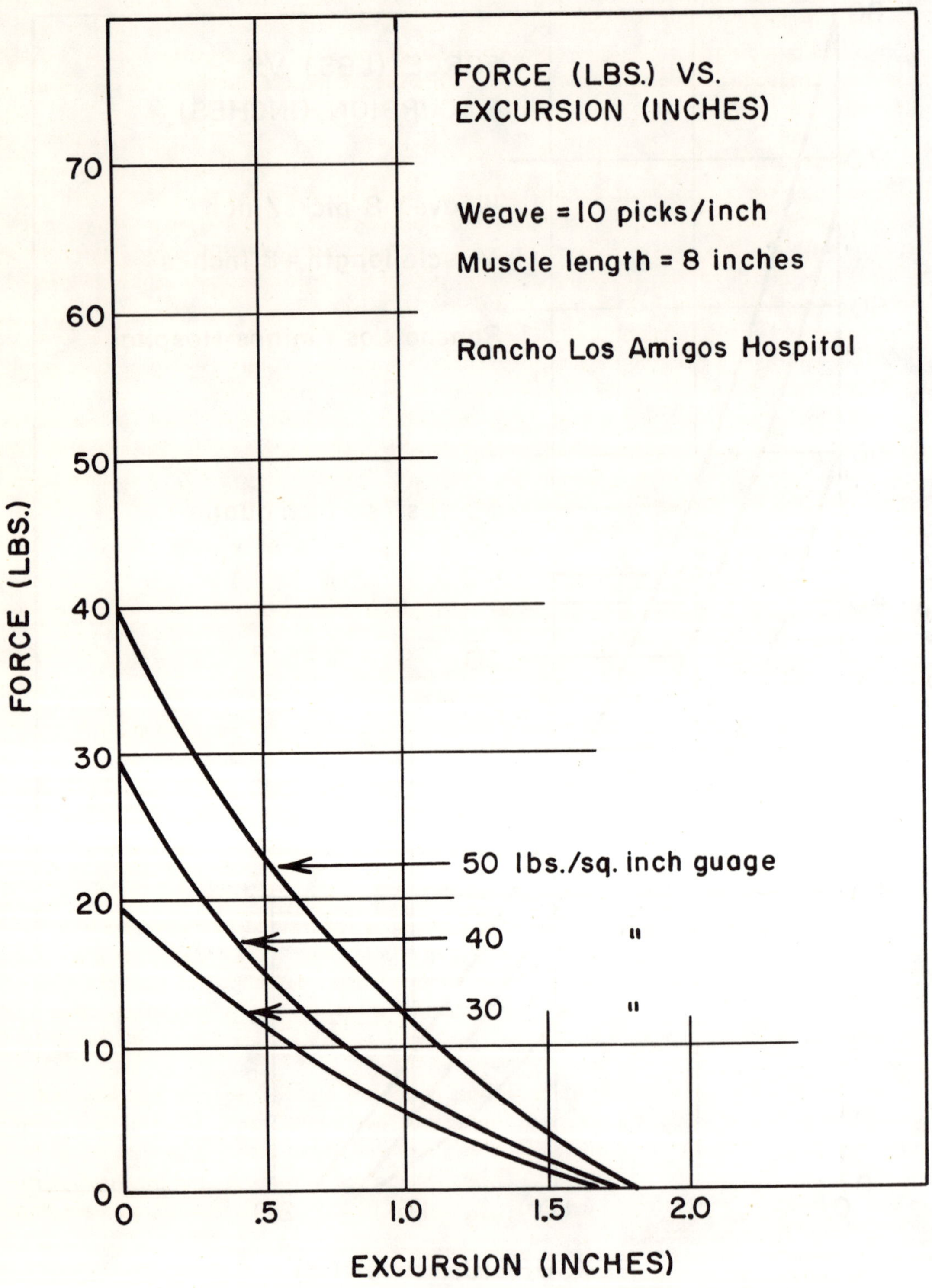

Figure 152-B

There is considerable choice available in the types of transducers, however there are two qualifications that must be met by any transducer that is used. The first is that a provision for a position of "hold"; i.e., the patient must be able to relax his operation of the transducer at any time and by doing so cause the orthosis to become locked in the position that it is in at the time the transducer ceased to be acted upon. The second feature necessary in almost all transducers is that its operation be dependent upon unidirectional motion or pressure.

The type of transducer chosen for a particular fitting is determined primarily by the residual muscle power available at the physiological control site and the skill of the patient. Patients who have control sites that retain four or more ounces of force with two inches of excursion can operate a lever valve (Figure 153 "A") and thus directly control a pneumatic actuator without any additional equipment. The simplicity of this type of control makes it attractive from the standpoint of reliability and cost and is usually employed in those cases where sufficient operating power is available. When the control site has two to three ounces of force with an excursion of 1/8 inch available, the sensitive micro switch can usually be successfully fitted. This switch consists of a pair of micro switches that are arranged so as to have unidirectional operation, a position of hold, and it is packaged in a rectangular housing for convenient fitting (Figure 153 "B"). The lever valve and the sensitive micro switch were developed by the Orthotic Department of the Rancho Los Amigos Hospital, Downey, California, and have been used extensively by that institution. A large percentage of all successfully fitted externally powered upper extremity orthoses employ these two devices.

Many patients are so seriously involved that only traces of residual muscle activity remain. These potential control sites may be only the flicker of a finger or toe or a muscle bulge that is detectable but practically unmeasureable. To harness muscle activity of this low order of magnitude, a somewhat more complicated system must be used.

The capacity touch plate offers some potential application. It has good stability and requires very minute pressures to operate. The patient merely touches an electrode to activate the device. A position of hold is inherent in this type of control, but it is very difficult to obtain unidirectional operation. Use of this device is not recommended except in unusual applications where only one function or a cycling operation is desired.

The most promising time proportional control for those patients with no control sites except head motion is the photoelectric device (Figure 154). The patient operates this device by positioning a light beam from a light source on the head to solar cells that are placed within convenient range. This method of control has the advantage of multiple channel operation from a single activation point and channel selection that is faster than other types of control.

Utilization of electromyographic potentials to control orthotic devices has received considerable attention at various institutions throughout the United States. This method offers the advantage of volitional control and insinuates the possibility of creating graded signal for feeding servo mechanisms. Although the EMG phenomena are attractive, the technical difficulties encountered in attempting to use EMG potentials in a practical application have thus far proved almost insurmountable.

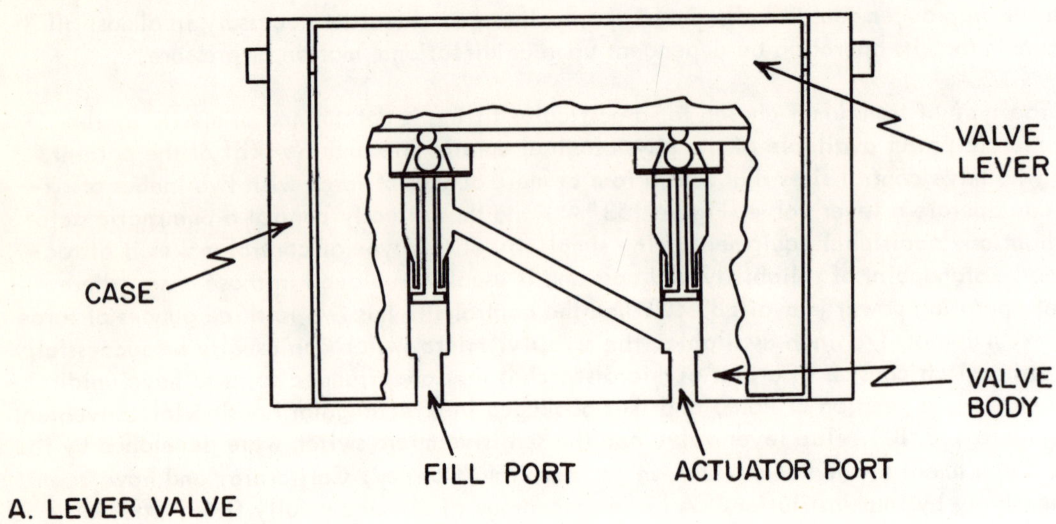

A. LEVER VALVE

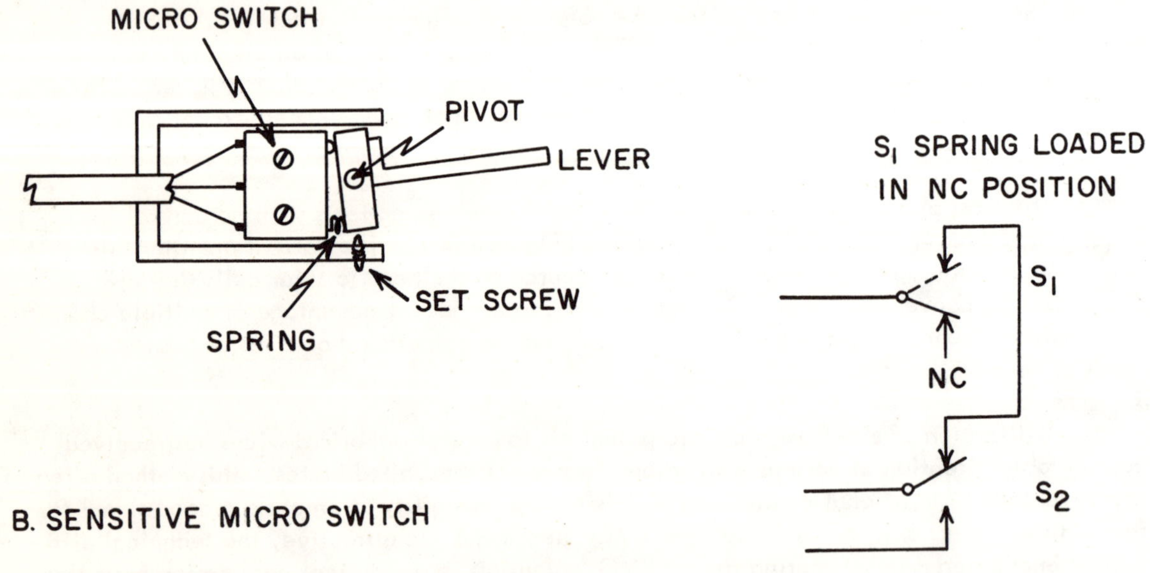

B. SENSITIVE MICRO SWITCH

Figure 153

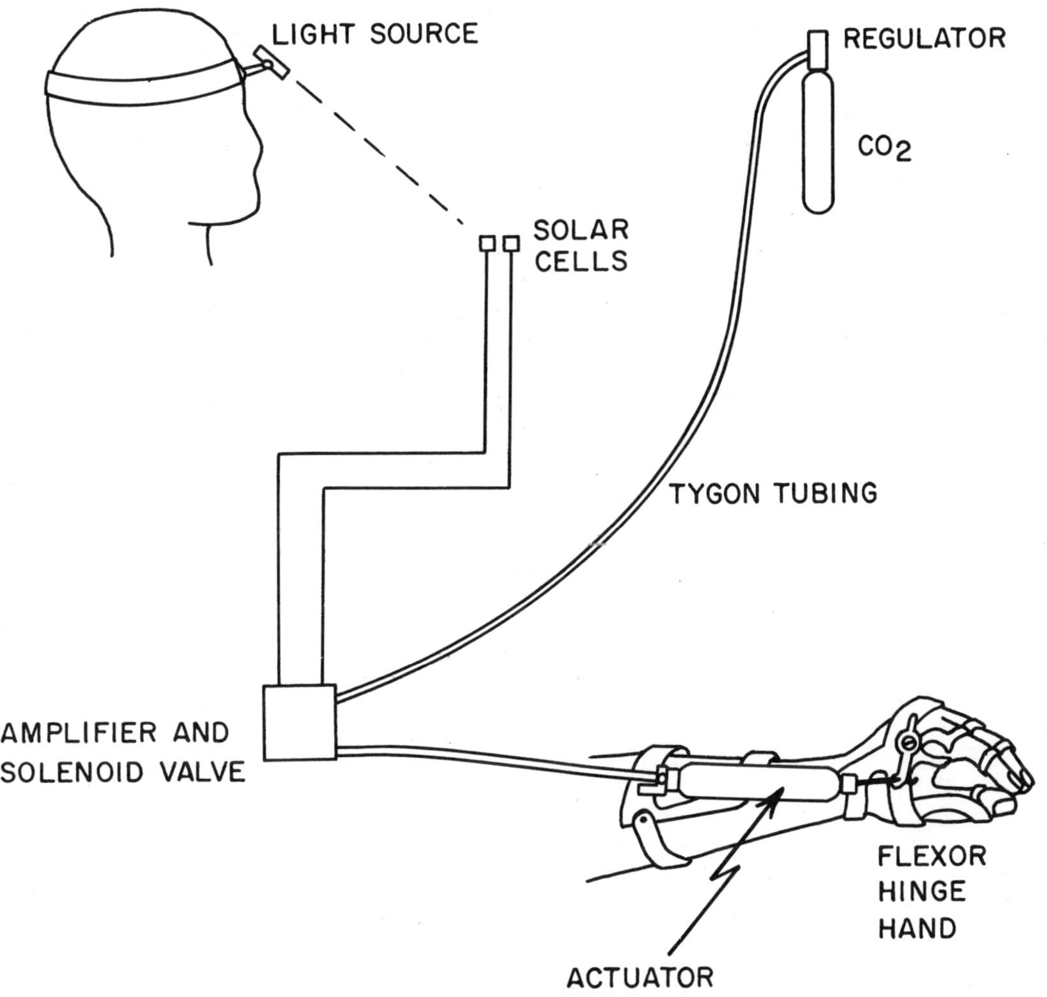

Figure 154. Photoelectric Control Application

Variable resistance, capacitance and inductance transducers for both implanted and external control of servo operated orthoses are in the developmental stage at various research centers. Rancho Los Amigos Hospital has had moderate success in this area but no hardware is available at this time.

The problem of valving dictates the necessity for a range of moderately low power to sensitive solenoid valves for pneumatically powered systems and electronic amplifiers and relays for electrical systems. Miniature industrial electronic components are readily available in wide selection, and specialized components of this nature have not been developed except for special purpose amplifiers that have been designed to meet size, economic and power consumption specifications. Design and fabrication of these amplifiers should present little challenge to the electronic engineer. Persons who do not wish to design their own may acquire special amplifiers for orthotically oriented capacity touch plates, EMG electrode, photo devices, etc., from rehabilitation centers that have upper extremity orthotic research laboratories and in some cases from orthopedic supply houses.

The solenoid valves that are adequate for pneumatically powered orthotic devices were developed by the Orthotic Department of Rancho Los Amigos Hospital. A general purpose valve that fits those applications where it is necessary to operate multiple sets of actuators and where appreciable electric power is available is shown in Figure 155 "A". This valve represents a reduction in size (1" x 1-1/2" x 1-3/4"), weight (3 oz.), and power requirements (8 watts) over its commercially available equivalent. A low power valve (Figure 155 "B") is available that may be used when the situation demands minimum electrical power consumption. This valve requires 1 watt for operation. The general purpose valve and the low power valve operate against 90 lb. of pressure and are packaged in banks of two with interconnected fill and exhaust lines for convenient usage.

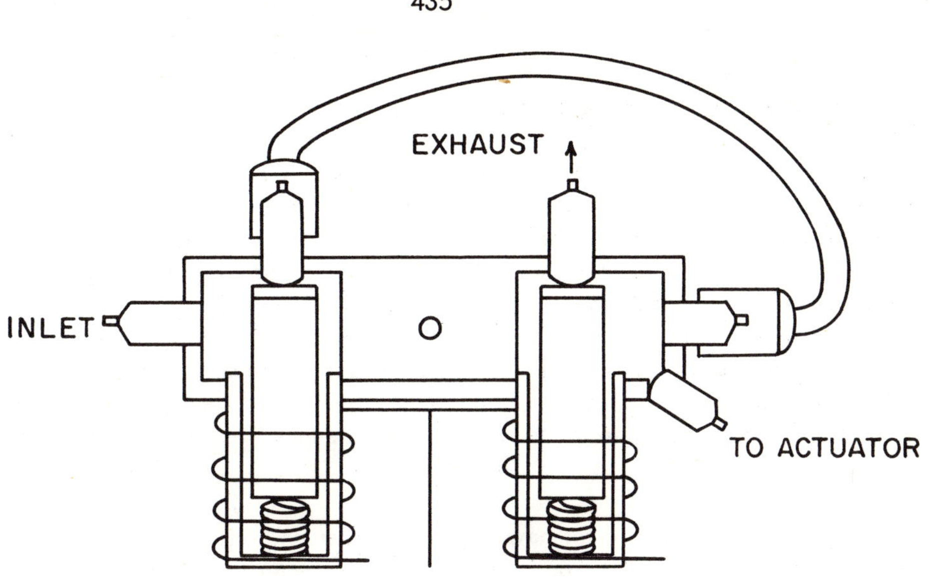

A. GENERAL PURPOSE SOLENOID VALVE

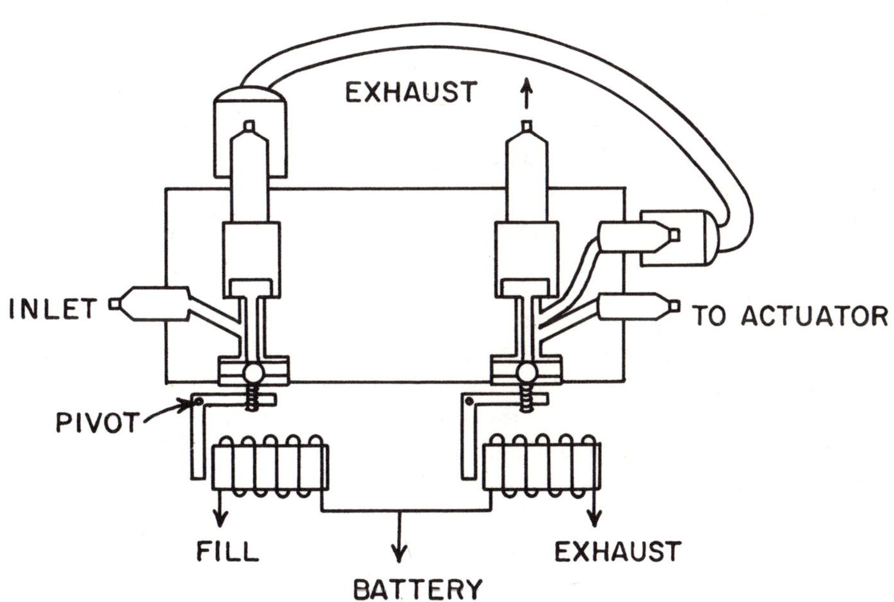

B. LOW POWER SOLENOID VALVE

Figure 155.

<u>Upper Extremity Power and Control System</u>

A total replacement CO_2 powered device is shown in Figure 156. This particular apparatus is representative of the present state of the art of the clinically proven externally powered orthosis and summarizes the upper extremity device to a certain extent. Positioning the arm and hand is accomplished by supplying power to the various joints of the bracing structure with pistons. This arm support is called a CO_2 powered feeder. Pinch is provided with a flexor hinge hand splint (Figure 157) that is powered with an artificial muscle. The flexor hinge splint provides a functional "three-jaw chuck" type of prehension. In this type of prehension, the index and long fingers contact the thumb. While it has limitations, the "flexor hinge hand" grasp has been found to be one of the most frequently used in daily activities.[1]

This fitting has four pistons for powering the feeder and one artificial muscle attached to the flexor hinge hand splint. The patient controls the feeder and hand splint by operating five sensitive micro switches which activate a bank of six pairs of solenoid valves that control the flow of gas to the pistons and artificial muscle. It should be noted that the pivot points lie as close as possible in line with the axes of motion of the arm (Figure 134-C). External power is available for total arm lift, horizontal motion of the proximal arm, and elbow flexion in the vertical and horizontal planes. Elimination of the distal arm eliminates possible horizontal motion produced under the arm trough.[2] The sensitive micro switches are mounted on a plate to which a swivel mounted foot support is attached that permits freedom of rotation in two perpendicular axes. Selection and activation are achieved in the horizontal and vertical planes, respectively. This mounting apparatus is called a Foot Plate (Figure 134-D).

The sensitive micro switches that control this feeder are also used to steer a motor driven wheel chair. The patient alternates the connection of the micro switch control to either the feeder or the wheel chair by activating a spring loaded switch located at the heel of the foot plate which, in turn, operates a latching relay that performs the switching operation. Initial fitting and adjustment of the foot plate, as well as other transducer positioning mounts, are critical and subject to re-adjustments as the physical condition of the patient alters.

Many problems remain unsolved, as is evident by the primitiveness of existing devices. Simultaneous control of multiple sets of actuators to produce controlled motions which more closely approximate normal motions is obviously desirable. Discovery of heretofore untapped physiological control sites, along with the development of artificial muscle and control systems to produce controlled activities above and beyond the simple acts of self-feeding, self-grooming and other personal tasks, are some of the major future goals.

---

[1] M. H. Anderson and L. R. Snelson: "The Flexor Hinge Splint and Artificial Muscle", University of California, School of Medicine, pp. 1, February, 1962.

[2] L. Barber: "Progress Report - Orthotic Department", Attending Staff Association, Rancho Los Amigos Hospital, Downey, California, pp. 6-7, August, 1961.

PROXIMAL ARM
CHAIR ARM
FOREARM TROUGH
TOP VIEW

PISTON
PISTON
PISTON
CHAIR ARM
SIDE VIEW

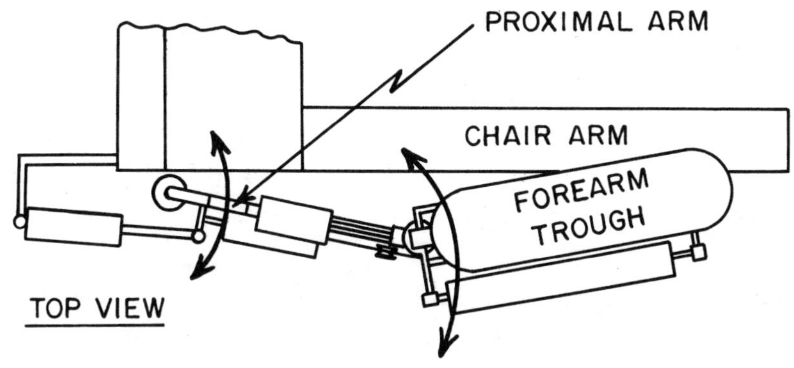

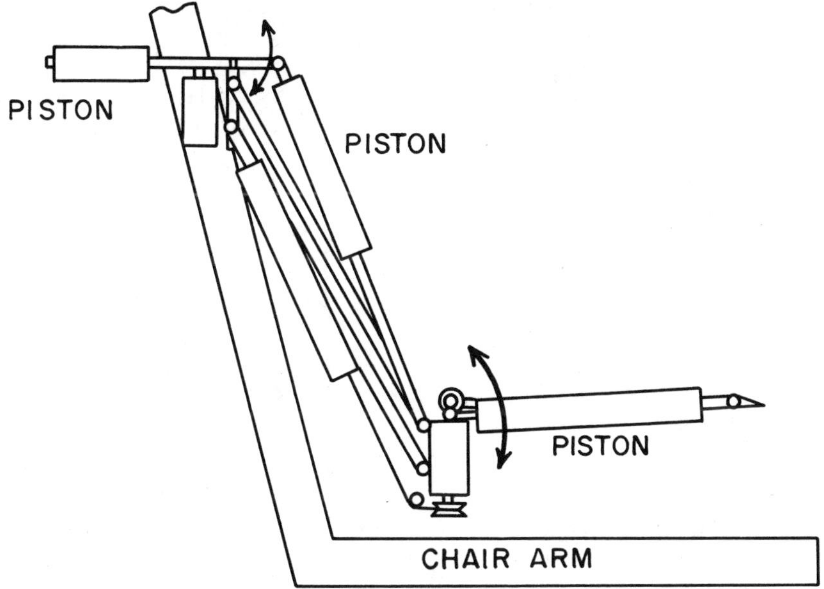

Figure 156.

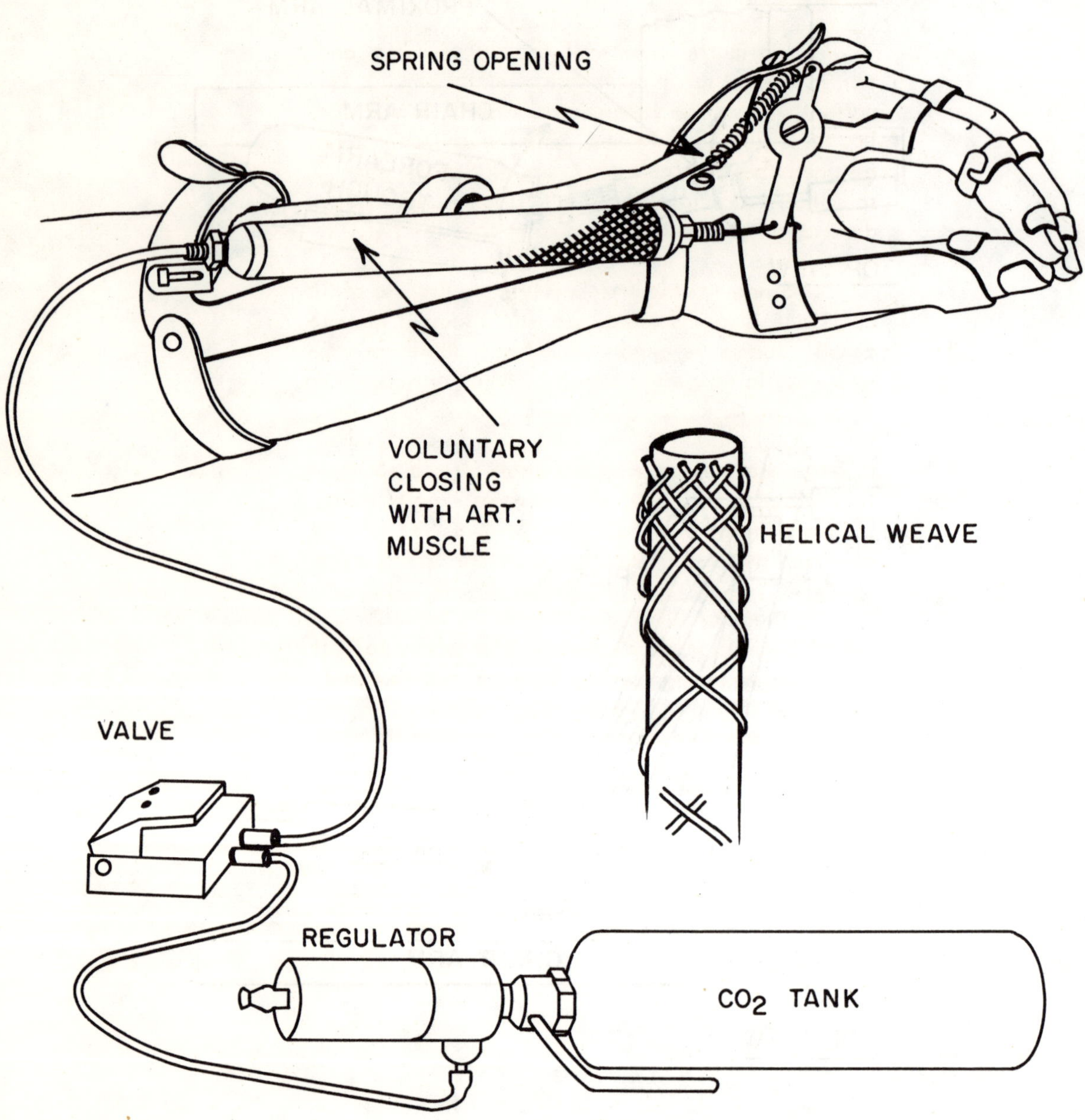

Figure 157. Artificial Muscle Driven Flexor Hinge Splint

CHAPTER VII.   HOW TO TRAIN THE PATIENT TO USE FUNCTIONAL HAND SPLINTS
AND ARM BRACES

<u>Introduction</u>

No matter how well the orthotist fits and adjusts a hand splint or functional arm brace, the patient will seldom obtain much benefit from it unless he is trained in its use. The therapist is usually assigned this responsibility, and it can be safely said that the successful rehabilitation of the patient depends on how well the therapist does the job.

Since training the patient in the use of his appliance is so important, we will discuss it in some detail. Just as with most other tasks, there are various ways in which the training can be done, and no doubt many of them have equal merit. However, since at the start it is less confusing to the student to learn one method that has been proved successful, rather than to talk about several ways of doing the same thing, we shall confine ourselves to one approach that experience has shown to be effective.

There are five major steps in orthotic rehabilitation: functional evaluation of the patient by the clinic team, pre-bracing care as prescribed, functional arm brace use training, re-evaluation of the patient before discharge, and follow-up re-evaluation several months after discharge.   In the following discussion, each of these five steps will be considered in detail.

<u>Functional Evaluation of the Patient</u>

The functional evaluation of the patient should be done by the clinic team: physician, therapist, and orthotist. This evaluation usually consists of eight separate procedures, the results of which, when taken as a whole, constitute a plan for rehabilitating the patient as completely as possible. This plan reflects the best thinking of the members of the clinic team, but is always subject to change as the patient progresses. The evaluation consists of doing and recording the results of the following procedures.

1. Perform a muscle examination to determine manual muscle grades. This task is performed by the physician, the therapist, or both, and is recorded on a form prepared for the purpose. A typical form is shown on the next three pages. This record indicates the status of the patient's muscular strength at the start of treatment, and is useful as a guide in deciding what to do for the patient and as a means for measuring progress as treatment continues.

    Grading methods vary a great deal, but the one used in the Evaluation Check List has proven satisfactory in most instances and is widely used.

2. Make and record measurements of range of motion of joints, both active (A) and passive (P). Full mobility of all joints is very desirable, and precautions must be taken to prevent deterioration of joints in paralyzed members. There is little chance of successfully restoring function by strengthening muscles if the involved joints are allowed to stiffen.

3. Test the patient's ability to use his remaining muscle power to perform useful activities, such as hand functions involved in activities of daily living. These tests are commonly divided into "Eating" (spoon to mouth, drink from cup, etc.), "Means of Communication" (writing, using telephone, etc.), "Hand Activities" (smoking, turning pages, using scissors, etc.).

    Each of these activities is rated, using a scale of this type: N = Normal; A = Adequate with adaptive equipment; X = Slow and laborious; O = No performance. The results of this test give the members of the clinic group an over-all picture of what the patient can and cannot do, and this information is then used to formulate realistic goals for the treatment to follow. A typical list of activities is shown in the "Check List for Splint and Brace Use Training" on pages 444-446.

4. Determine what the patient's basic living needs will be in the environment he will return to when discharged. These can vary according to the social and economic status of the patient and his family, and similar factors.

UNIVERSITY OF CALIFORNIA, LOS ANGELES
School of Medicine, Department of Surgery (Orthopedics)
PROSTHETICS—ORTHOTICS PROGRAM

UPPER EXTREMITIES ORTHOTICS

EVALUATION CHECK LIST

Name: _________________________________ Age _______ Sex _______ Date _________

1. Medical history:

2. Social history:

3. Vocational history:

    A. Education and field of specialization.

    B. Work history.

    C. Patient's present vocational interests.

4. Hobbies and recreation:

5. Patient's report of self care abilities:

6. Upper extremity functional evaluation:

7. Degree of patient's interest in proposed bracing and splinting program, and willingness to devote necessary time and effort:

8. Medical and surgical considerations related to bracing or splinting:

9. Prescription:

FBUE – ECL – 1   Examiner _________________________

FUNCTIONAL EVALUATION OF THE UPPER EXTREMITIES

		Zero L R	Partial L R	Complete L R	Remarks
TRUNK	Bending				
	Twisting				
	Fwd. bending & return				
UPPER EXTREMITIES	Arm overhead				
	Forward reach				
	Hand to mouth				
	Hand behind back				
	Palm up				
	Palm down				
	Open hand				
	Grasp				
	Grasp release				
	Pinch				
	Pinch release				

		RANGE			Remarks
		Active L R	Passive L R	Muscle Strength L R	
SCAPULAR	Protraction				
	Retraction				
	Elevation				
	Depression				
GLENO-HUMERAL	Flexion				
	Extension				
	Abduction				
	Adduction				
	External rotation				
	Internal rotation				
	Horizontal abduction				
	Horizontal adduction				
ELBOW	Flexion				
	Extension				
FOREARM	Pronation				
	Supination				
WRIST	Extension				
	Flexion				
	Ulnar deviation				
	Radial deviation				
THUMB	Opposition				
	Abduction				
	Adduction				
	Flexion M.P.				
	Flexion I.P.				
	Extension M.P.				
	Extension I.P.				

	FINGER		RANGE						STRENGTH						Remarks
			MP A P		IP A P		DIP A P		MP A P		IP A P		DIP A P		
LEFT	Index	Flexion													
		Extension													
	Middle	Flexion													
		Extension													
	Ring	Flexion													
		Extension													
	Little	Flexion													
		Extension													

	FINGER		RANGE						STRENGTH						Remarks
			MP A P		IP A P		DIP A P		MP A P		IP A P		DIP A P		
RIGHT	Index	Flexion													
		Extension													
	Middle	Flexion													
		Extension													
	Ring	Flexion													
		Extension													
	Little	Flexion													
		Extension													

5 - N	Normal	Complete range of motion against gravity with full resistance.
4 - G	Good	Complete range of motion against gravity with some resistance.
3 - F	Fair	Complete range of motion against gravity.
2 - P	Poor	Complete range of motion with gravity eliminated.
1 - T	Trace	Evidence of slight contractility. No joint motion.
0 - O	Zero	No evidence of contractility.

CHECK LIST FOR SPLINT AND BRACE USE TRAINING

Name ____ Date __________

Diagnosis ____

Brace and Splint Equipment ____

Activity	Bilateral	Unilateral	Practical for Patient
Activities of Daily Living			
Dressing:			
Remove brace or splint		x	Most dressing impractical
Replace brace or splint		x	for bilateral patient --
Remove shirt or jacket		x	other bilaterals can learn
Replace shirt or jacket		x	some parts of dressing
Remove shirt or trousers		x	skills
Replace shirt or trousers		x	
Lock and unlock leg braces if used		x	
Remove shoes and socks		x	
Replace shoes and socks		x	
Operate wheel chair if used		x	
Hang clothes on hanger		x	
Fastenings: Zipper		x	
Buttons		x	
Belt buckle		x	
Shoe laces and tie		x	
Tie necktie		x	
Eating:			
Sandwich: hold	x		
to mouth	x		
Spoon: grasp and hold	x		
to mouth	x		
Cup: grasp	x		
lift to mouth	x		
Glass: grasp	x		
lift to mouth	x		
Cut with knife and fork	x	x	
Carry tray	x	x	
Bathroom - Grooming:			
Comb hair	x		
Brush teeth	x	x	
Apply lipstick	x		
Shave	x		
File nails	x	x	
Clip nails	x	x	
Clean glasses	x	x	
On - off water faucets	x		
In - out of tub/shower	x		
On - off toilet	x		

Activity	Bilateral	Unilateral	Practical for Patient
General:			
On – off light switches	x		
Door knob	x		
Screen door latch	x		
Trunk latch – lock	x		
Door chain latch	x		
Night lock	x	x	
Padlock and key	x		
Drawer pulls	x		
Window latch and lift	x		
Window blinds	x		
Wallet from pocket	x	x	
Money from wallet	x		
Coins from table			
Pack of cigarettes	x	x	
Book of matches	x	x	
Paper cup from dispenser	x		
Bottle opener – wall	x		
hand	x	x	
Electric wall plug	x		
Carry suit case (briefcase)	x		
Hammer and nail	x	x	
Communication Skills:			
Write with pencil	x		
Fountain pen	x		
Sharpen pencil – wall	x	x	
table	x	x	
Draw line with ruler	x	x	
Cut with scissors	x	x	
Typewriter	x	x	
Letter – open	x	x	
fold, insert, seal	x	x	
stamp	x	x	
Answer phone	x		
Dial phone	x	x	
Coin in pay phone	x		
Drive car	x	x	
Vocational Skills:			
Type and placement of materials			Check in terms of the specific activity re-
Positions required by person			quired by each
Heights – location of facilities			patient

Activity	Bilateral	Unilateral	Practical for Patient
Homemaking:			
Wash dishes	x	x	
Pour into bowl	x	x	
Hold bowl to stir	x	x	
Lift pot to stove	x		
Kitchen utensils (egg beater)	x	x	
Iron – weight – position clothes	x	x	
Wash clothes	x	x	
Vacuum floors	x	x	
Can opener	x	x	
Other			

Activity	Bilateral	Unilateral	Practical for Patient
Avocational Skills:			
Cards	x	x	
T.V. (Change stations)	x		
Gardening	x	x	
Art activity	x	x	
Craft activity	x	x	
Use of tools	x	x	
Other			

Other activities ____

____

____

Summary and training plan: ____

____

____

____

____

____

5. Determine the patient's vocational goals so that the treatment program can be designed to help accomplish them. The assistance of a vocational rehabilitation counselor is useful in helping patients to arrive at practical and realistic decisions as to future occupation, if any, and he should be brought into the clinic team as an active participant when it is clear that tangible vocational goals are feasible.

6. Prepare prescription. The physician is responsible for the prescription. Pre-bracing care, bracing or splinting, strengthening excercises, joint mobilization activities, functional activity testing and training, and activities of daily living testing and training, as needed, may be specifically required in the prescription. It is quite important that the patient understand as much about these procedures as possible by explaining to him what they are for, how they will feel to him, how he may benefit, what braces are, how they feel, what they do, and so on. Every effort should be made to avoid over-selling the patient on the amount of recovery or function he may expect from the course of treatment recommended. He should be guided toward setting up goals that show promise of being attainable within a reasonable length of time. These efforts are continued during the pre-bracing care period.

Pre-Bracing Care

Pre-bracing care is carried out by the physical and occupational therapists as prescribed by the physician. The purposes to be achieved by this work are, in general, the following:

1. Improve general muscle strength, with special attention to those muscles that will operate the brace.

2. Increase general endurance so the patient can carry the added weight and do the added work required by the brace.

3. Achieve a functional range of motion of the affected joints, except where stability is required to offset weak or flail musculature.

4. Teach the patient what his braces will be like, how they will look, what they are made of, how they work, and how they will feel.

5. Help the patient to set realistic and attainable goals, by pointing out to him the limitations of the braces as well as the possibilities. The patient must not be allowed to become discouraged, but must be assisted in realizing that it takes time to learn.

6. Discover what activities the patient can do without braces, how he does them, what he wants to do in addition, and how these additional things can be done, with braces or by other means.

Training the Patient to Use His Hand Splints and Functional Arm Braces

There are three important points for the therapist to keep in mind in training the patient to use his braces to best advantage.

1. Learning to perform useful functions with the aid of braces can only be accomplished through the patient's own efforts. The will to stick with this task must come from within the patient himself. The job of the therapist is to encourage the patient and guide his efforts to learn. The patient's confidence in his ability to overcome the difficulties in his way must be reinforced from time to time. If this is not done, lack of progress will soon cause discouragement and the patient will give up trying. We all tend to want to please those we meet in our daily lives by doing the things that are expected of us; the therapist should make it clear to the patient that she expects him to make a conscientious effort to learn. The therapist works <u>for</u>, not <u>on</u> the patient.

2. Avoid strict rules and rigidity of procedure in the learning situation. Encourage the patient to help plan his own training program by drawing from him the things he most wants to be able to do, then helping him to choose from these the ones he can accomplish easiest to start with. The activities learned must give the patient a feeling of increased independence. Never ask the patient to learn things that are not practical in terms of usefulness to him as compared to the time and energy he will have to put out to master them. These may not always be the things the therapist thinks he should learn.

3. The patient should not be required to wear the braces too long at any one time, but just for the length of time he can tolerate them comfortably, both physically and psychologically. The goal is to be able to wear them comfortably all day. It is desirable to let the patient have a "vacation" from brace wearing for a day or two occasionally.

Training the patient to do useful things with the help of his braces is called "controls and use training", and for convenience is usually divided into five areas:

1. Controls training
2. Self-care
3. Communicating
4. Vocational skills
5. Avocational skills
6. Using transportation

We will discuss the problems of training the patient in each of these areas in turn.

<u>Controls Training</u>. The patient must become skilled in the operation of his splint or brace as a mechanical device before he can proceed to learn its use in various activities of daily living. The "Controls Training Check List" on the following page is a useful guide in teaching the patient to control his splint or brace. Some devices are so simple that control of them can be taught in five minutes, but the more complicated ones require much more time. As soon as the patient masters the fundamentals of controlling his device he should be started on functional activities, such as eating.

<u>Self-Care</u>. Self-care requires that proficiency be developed in four kinds of activities: putting on and removing the splint or brace; dressing and undressing; bathroom activities; and eating.

Putting on and Removing the Brace

1. The therapist puts the brace on the patient first, explaining which straps must be fastened, how the rubber bands are placed, and where and how each part of the brace fits. It is not wise to spend all of a practice session putting on and removing the brace; after two or three times it is better to leave it on and start to learn other activities.

2. If the patient has a severe involvement and needs complicated braces it may not be practical to try to teach him to put them on and remove them. It is none the less important for him to know exactly how this is done, so he can explain it to someone who may be trying to help him.

3. The therapist teaches the patient to do the easiest steps in putting on and taking off the brace first, doing the more difficult ones herself. The patient is gradually encouraged to master the more difficult operations until he is able to do them all, if he is capable of doing so.

Dressing and Undressing

1. Undressing is easier and less frustrating, so it is best to start with these activities rather than those involved in dressing.

2. Patients with bilateral involvement can rarely learn to dress and undress without help, and attempting to teach this is seldom a practical

UNIVERSITY OF CALIFORNIA, LOS ANGELES
School of Medicine, Department of Surgery (Orthopedics)
Continuing Education in the Medical and Health Sciences
PROSTHETICS EDUCATION PROGRAM

"Functional Bracing of the Upper Extremities"
CONTROLS TRAINING CHECK LIST

Name ____ Date ______________

Brace/Splint Equipment ____

	Able to Perform	Needs More Training	Results
A. Hand Splint Operation			
1. Control of open/close smooth.			
2. Opens and closes completely.			
3. Grasp/pinch (strength adequate).			
4. Prehension pattern with objects skillful.			
B. Forearm Operation			
1. Control of flexion/extension skillful.			
2. Lifts objects from table.			
3. Uses in forward reach.			
4. Cycles elbow lock correctly.			
5. Locking sequence smooth and integrated.			

	Able to Perform	Needs More Training	Results
C. Shoulder Operation			
1. Control of flexion-extension.			
2. Tension relief used skillfully.			
3. Locking sequence smooth and integrated.			
4. Cycles shoulder lock correctly.			
D. Assists or Stabilizers			
1. Pre-positioning skillful.			
E. Work Areas			
1. At table-adequate forward reach.			
2. Adequate pickup, carry, release.			
3. Object to/from high shelf.			
4. Object to/from low shelf. (floor)			

<u>Summary:</u>

Work area limitations:

Endurance for wearing equipment:

Endurance for operating equipment:

Rest position of equipment:

goal. The patient has to put on part of his clothing without the braces on, and usually the flexibility of the braces is not such as to allow the patient to put on outer clothing without help.

3. The therapist should break each undressing and dressing task into steps in the order in which they are done, then teach the patient those he is capable of learning, filling in the more difficult ones herself. Handling buttons, snaps, hooks, zippers, and other fasteners is one of the biggest problems. The special buttonhook is very helpful, and other simple devices, such as large thumb rings on zipper tabs, can often make a difficult task easier.

4. Remedial occupational therapy that requires the same motions the patient needs in dressing may be assigned during other periods of the day. Many crafts can be set up in such a way that these functional movements may be simulated.

Bathroom Activities

1. The unilateral patient, or the one with relatively mild bilateral involvement, can learn to care for many of his needs through the use of special assistive devices, such as holders for comb, toothbrush, electric razor, lipstick, and the like. Most of these patients can learn to wash their faces.

2. The severely involved bilateral patient can rarely do much for himself in this area of activity without considerable help.

3. Toileting is the most serious problem and the most difficult to solve. Most patients with bilateral involvement must have help. In some cases, this matter is best handled by making adaptations without the braces on, and if the patient has already worked out some method that would have to be changed if he is fitted with braces, the method should be carefully analyzed before asking the patient to give it up in order to use the braces.

Eating

1. The severely involved bilateral patient has in many cases learned to eat satisfactorily using

ball-bearing feeders, or similar devices. Changing
these patients over to eating with functional arm
braces should be carefully analyzed before making
the move, as it may be more advantageous to let
them continue with the equipment they are used to.

2. In teaching a patient to eat, it is wise to first select
the applicable special assistive devices, such as a
swivel spoon, combination spoon and fork, plate
guard, and special large handles on knives or forks.

3. Start the training with an easy task, eating a sandwich. A peanut butter sandwich is best at the start,
as it does not fall apart easily. The first step is to
place the sandwich in the patient's grasp, either
his hook or his fingers, according to the type of
brace being worn, then direct him to practice getting the sandwich to his mouth to eat it.

4. When the patient has learned to get the sandwich
to his mouth, he should be taught how to pick up
the sandwich from a plate. To accomplish this,
he must first learn to position his hook if he uses
one, or get his fingers into proper position if he uses
a splint or no finger equipment at all. After instruction in positioning the prehension members,
the patient is taught how to pick up the sandwich from the plate. Sometimes this is made easier if he is instructed to push the sandwich to the
edge of the plate so it overhangs slightly.

5. When the patient has mastered sandwich-eating
he can next learn to eat with fork and spoon.
It is best to practice with pieces of modeling
clay for food. At first the therapist directs the
patient to hold the spoon properly, then places
small pieces of clay in it, after which the patient practices getting it up to his mouth, keeping the spoon level so the pieces of clay do not
spill out. When he can do this well, he is next
directed to push the clay onto the spoon, using
plate and plate guard if necessary, after which
he combines the two steps, pushing the clay onto the spoon, then lifting it to his mouth without spilling it. The final step is to try the new
skill with real food, easy ones like mashed po-
tatoes and apple sauce at first. Keep in mind
the importance of adaptations that will make it
easier for the patient, such as adjustments in

height of table or chair, special assistive devices
such as the swivel spoon and plate guard, and the
like.

6. The chief problem to be overcome by the patient
in learning to eat is the development of smooth,
rhythmical movements, and proper hand and arm
positioning. Here again, other occupational
therapy activities can be used as training exer-
cises to make this easier. Typical of these exer-
cises are leather lacing projects with the therapist helping or with the lace in a vise, and
playing cards using a card holder.

Communicating. This category of activities includes all means whereby the
patient transmits and receives information. There are four areas of communication activities in which many patients can work with profit: writing, typing,
telephoning, and mail.

<u>Writing</u>

1. Even the most seriously handicapped patient
should be encouraged to learn to sign his name,
even though any extensive amount of writing
may never be practical.

2. In the beginning the patient should be given
paper with widely spaced lines on which to
write. If necessary, he should be provided
with a pencil holder with finger rings made
to fit his fingers. A pencil should be used
instead of a pen at the start. The patient
should first practice large circular motions,
followed by large writing. Always remember that much of the motion used in writing
comes from the shoulder, so if shoulder rotation is weak the patient will have difficulty in progressing across the paper. A
regular nib type ink pen is better than a
ball-point pen when the patient is ready to
start writing with a pen, because it does not
require as much pressure.

3. The patient may improve his ability to write
by doing other occupational activities, such
as using wax crayons, painting, and finger
painting with braces off.

Typing

1. Typing with a standard typewriter is difficult if not impossible for the patient who is severely involved; so for the majority of patients an electric typewriter is the only kind that can be used successfully.

2. Patients who have such serious involvements that they must use complex braces and splints may be provided with a rubber-tipped stick to hold for use in striking the keys.

3. The patient's handicap should be studied so that adaptations can be devised to make it easier for him. Handling materials such as paper, carbon paper, and the like is usually difficult, and can be made easier by providing a "U"-shaped desk, for example, which minimizes the amount of across-the-desk reaching required. Table and desk height are often very important considerations.

4. When starting, put the paper in the machine for the patient, letting him remove it when he is finished, as it is much easier for him to remove it than insert it, depending of course on the amount of shoulder flexion strength he has. Corrections are hard to make because of the difficulty in manipulating the roller, and because of the pressure and accurate placement necessary to erase an error properly. The use of "Snopake" or "Ko-Rec-Type" which cover the error with white opaque material may be better than erasing. It is a good idea to encourage the patient to type more slowly but with great accuracy so mistakes can be avoided.

5. The patient must learn the keyboard before he can type letters or other material. As his skill increases he can master inserting the paper into the carriage, and handling corrections.

Telephoning

1. Use of the standard hand-set or "French" phone, the most widely used type, requires that the patient learn to lift the receiver and hold it to the ear, dial the number, and replace the receiver when the call is completed.

2. The first step is to see how much the patient
   can accomplish, using his braces and attempting the task in the normal manner. Some patients can get the hand-set to their ears but
   cannot dial the number. Others cannot do
   the first operation, but can do the second.
   Others have troubles with both.

3. For the patient who cannot hold the hand-
   set to his ear adaptive devices are available
   that hold the transmitter-receiver assembly
   at head-height so that all the patient has to
   do is lean against it. With the device comes
   a cam and lever system that enables the patient to open and close the telephone circuit in the same way the hand-set ordinarily
   does when it is replaced in the cradle. For
   the patient who cannot operate the dial,
   sometimes a round stick that he can grasp
   will enable him to manipulate the mechanism satisfactorily. The telephone company
   has a number of adaptive devices that are
   helpful in many cases. To obtain their help
   contact the Special Equipment Section of
   the Sales Department, Bell Telephone System.

4. The use of pay telephones of the usual type
   is very difficult for the patient who wears
   braces. The greatest problem is getting
   the coin in the round opening, and the placement of the phone makes it next to impossible for the average patient to accomplish this operation.

Mail

1. Handling letters received and those to be
   mailed is a useful activity that goes well
   with typing, and patients should be encouraged to learn how to do these things
   as well as how to type.

2. Start by teaching the patient how to remove
   a letter from an envelope that has already
   been slit with the letter opener, then how
   to unfold the letter and place it in the "IN"
   basket. Next, the patient can learn how
   to fold a letter, put it in the envelope,

stamp, and seal it. Last and most difficult is
slitting the envelope of a letter with the opener.
Placing the stamp on the envelope is also difficult, particularly for bilaterals using hooks.
However, a stamp dispenser using stamps in rolls
and a drum moistener is a helpful aid.

<u>Vocational Skills</u>. The problem of preparing the patient for gainful employ-
ment can often be a very difficult one to solve. This work is usually called
"vocational rehabilitation", and is carried on by workers specially trained
to know the requirements of a great many occupations and to analyze the
various potentialities of the handicapped person to see in which occupation
he might have the best chance for success. These special workers are called
vocational rehabilitation counselors, and it has been found very advantageous to use their services in helping the patient make a wise selection of
an occupation and to obtain training in preparation for it. The state vocational rehabilitation services are able to provide for the costs of counseling
and guidance, orthopedic appliances where needed, and vocational training in business or trade school or other institutions as may be necessary to
prepare the patient for employment. Since there are many cases where vocational goals influence the choice of orthotic equipment, the physician in
charge of the clinic team is wise to include the vocational rehabilitation
counselor as a team member as early in a case as possible where vocational
goals are a factor.

<u>Avocational Skills</u>. Recreational activities are very important to the hand-
icapped patient, and the therapist should give some thought to making an
inventory of those the patient prefers and seeing what adaptations can be
made to the bracing to enable him to continue them. The controls on radio
and television sets can be adapted easily to enable the patient to operate
them. The handles on fishing poles and landing nets can be modified to be
more easily manipulated. Reading stands, page-turners, and special lights
are available for those who like to read. Devices for holding playing cards
are also available. The possibilities are limited only by the amount of ingenuity of the therapist and the patient.

<u>Transportation</u>. One of the most difficult problems of the severely involved
patient is handling money when getting onto a bus or streetcar. If the am-
bulant patient can arrange to pay his fare, public transportation offers
little difficulty. Special coin purses are helpful in solving these problems
in some cases. Many patients are able to drive cars equipped with power
steering, but for those who cannot, adaptive equipment is now available
that makes it possible to steer the car with the feet.

Re-Evaluation of the Patient Before Discharge

Before the patient is discharged for treatment, his bracing should be re-evaluated by the therapist, then later by the entire clinic team. The most important points to consider in the re-evaluation are:

1. Is the bracing comfortable, and can it be worn with comfort for long
periods of time?

2. Is the bracing providing as much assistance as devices of this particular type can be expected to provide?

3. Is the bracing functioning mechanically in a smooth and efficient
manner?

4. Does the patient feel that he gains enough function from wearing the
bracing to make it worth while to continue to use it?

Sometimes it is helpful to have the patient write a list of things he can do without his bracing and then make a list of those things he can do only by wearing his equipment. Many times the patient does not realize how much the bracing helps him until he makes this analysis.

Follow-up Re-Evaluation

When several months have passed since the patient was discharged, he should be brought back for another evaluation in which the same points are covered as in the re-evaluation just described. At this time it is often found that the patient's strength has increased or his needs are now different, requiring that the bracing be changed. It is well to analyze the patient care-fully, to determine if new bracing or revisions of the old ones will not give him added function and independence.

To summarize the functions performed by the clinic team in rehabilitating any upper extremities orthotics patient are:

1. Patient Evaluation Conference (Use "Evaluation Check List")

   a. History

   b. Determination of level of patient function

      (1) Muscle examination (standard procedure)

      (2) Motion examination

         (a) Grasp: finger flexion and extension, thumb
             flexion, extension, and opposition

             (b) Wrist motion: flexion, extension; stability

             (c) Forearm rotation: pronation, supination

             (d) Elbow motion: flexion, extension; stability

             (e) Scapulo-humeral motion: internal, external rotation; forward flexion; abduction, extension

             (f) Scapulo-thoracic motion: scapular elevation depression, abduction, adduction

c. Functional needs of patient: self-care, communication, vocational, avocational

d. Surgical considerations

e. Psychological and physical adaptiveness

f. Other possible limiting factors:

    (1) Fatigability

    (2) Respiratory (vital capacity) limitation

    (3) Ambulation problems

    (4) Trunk balance

    (5) Skin condition

    (6) Use of bracing for exercise purpose only

2. Prescription Conference (Use "Evaluation Check List")

a. Discuss tentative goals to be attained by bracing

b. Discuss problems of fitting, weight of device

c. Cosmesis problems

d. Selection of components to best meet needs and solve problems

e. Psychological preparation for bracing: explanation of time required, face fact that functional replacement is always incomplete, indoctrination on goals to be achieved by bracing.

f. Treatment after bracing: improve general muscle function and endurance, strengthen motor power to be harnessed, acquire and maintain functional joint range of motion.

3. Fabrication and fitting of brace or braces by orthotist

4. Evaluate mechanics of the brace, referring back to the goals and problems discussed in the prescription conference.

5. Therapist does controls training. (Use "Controls Training Check List")

6. Therapist does use training. (Use "Check List for Splint and Brace Use Training")

7. Re-Evaluation Conference
   a. Re-evaluation of muscles and motions, compare with earlier evaluation
   b. Re-evaluation of mechanics of splint or brace
   c. Re-evaluate control and use of appliance
   d. Re-evaluation of goals of bracing--self-care, communication, vocational, avocational
   e. Discussion of need for brace changes, different components, replacement of worn and broken parts
   f. Discussion of need for further training

LIBERTY MUTUAL
RESEARCH CENTER
LIBRARY